CW00641228

The Universe

A Travel Guide

The Universe Contents

A star being distorted by its close passage to a supermassive black hole at the centre of a galaxy.

Welcome to the Universe

Bill Nye

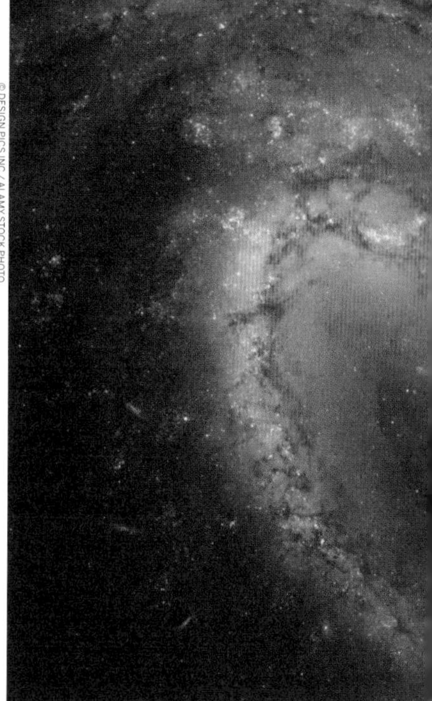
© DESIGN PICS INC / ALAMY STOCK PHOTO

Lonely Planet's *The Universe* gives us more perspective, often breathtaking, more insight, often deep – and more unusual facts, often ones you can't find anywhere else, regarding the profound happenstance of our existence. Simply put, the remarkable sequence of cosmic accidents required to enable us to be here on this planet and publish books like this one is astonishing. Unique to these pages are wonderful comparisons of Earth with the other worlds of our solar system and even those exoplanets orbiting other stars. They drive home the jaw-dropping idea that you and I, and everything we can observe around us, are made of the dust and gas blasted spaceward by exploding ancient Suns. And from the stardust and drifting gas, the extraordinary diversity of living things, including animals like you and me, emerged. You and I are at least one way that the cosmos knows itself. An utterly amazing idea that fills me with reverence every time I think on it.

While you are going about your business every day, thinking about what's happening on Earth right now, this book will help you think about a much grander timeline as well. From the comfortable surface of Earth, our deep-thinking ancestors observed our planet and its relationship, their relationship, to the night sky and the Sun. They learned where to live and how to survive. From the icy blackness of space, our spacecraft, built by our best scientists and engineers, make further observations that relentlessly show us Earth is like no other place in the solar system, and remains the only place we can live and thrive. By understanding the changes here over recent millennia, we can see that, if we're going to continue to thrive, we must preserve our environment. Otherwise, we'll go extinct, like 90% of the species that gave it a go on Earth before we showed up.

This cosmic perspective induces all of us to compare Earth to our neighbouring worlds out there. It's one thing to consider Earth as a pretty big place, especially if you tried to walk around it. It's another thing to think that 1300 Earths would fit inside a sphere the size of Jupiter, and over a million Earths would fit inside the volume of the Sun. While we're appreciating the visible differences of the traditional planets, what you might call their qualitative differences, this book helps us take it all in by the numbers, the planets' (and exoplanets') quantitative differences, and beyond that, the differences between our own Sun and the uncountable stars above, visible and invisible. In here, these essential distinctions are spelled out – or counted up.

The rocky and metallic compositions of Mars, Venus, and Mercury are very much

Barred spiral galaxy NGC 1300

like Earth's, but the environments of these other worlds are completely different. The text and pictures here will help you understand why. The unique chemical composition of the rocks, craters, and sands of the other worlds in the solar system has caused these extraterrestrial environments to have chemistries that are literally other worldly. These processes have conspired to produce radically different surface temperatures on Mars and Venus. Our discoveries in planetary science offer us a planet-sized lesson in the importance of the greenhouse effect, how our planet became habitable, and how the biochemistry of life changed the chemistry of the atmosphere and sea.

The story carries out away from the Sun, where we find the gas-giant planets: Jupiter and Saturn. They don't seem to even have surfaces as such. There's nowhere to stand, but they're so massive that, if you got too close, their gravity would crush you quick. On out further from the Sun we find Uranus and Neptune. They're very large and very cold, with enormous icy storm systems and winds moving at fantastic speeds. All of these other worlds in our Solar system, the ones that are not Earth are very different, very interesting – and utterly hostile.

As you turn these pages, learning the facts of everything from our solar system to the far reaches of intergalactic space, consider that there's no other planet that we know of anywhere, upon which you could even catch a breath to be taken away, or seek a deciliter of water to be sipped – let alone be afforded an opportunity to live long and prosper. The Earth is unique, amazing, and our home.

From a cosmic perspective, we are a pretty big deal. We've changed the climate of a whole planet. Run the numbers for yourself. Climate change is our doing. If we're going to make it much farther on this world, we're going to have to engage in some un-doing. Right now, it's our chance to change things. We are but a speck in the cosmic scheme. But it's our speck, and the more we know and appreciate it, the better chance we have keeping it hospitable for species like us.

Introduction to the Universe

With 2 trillion estimated galaxies and uncountable stars, our Universe is filled with wild examples of exoplanets, stars, black holes, nebulae, galaxy clusters and more, which scientists are still probing.

Our Universe began in a tremendous explosion known as the Big Bang about 13.7 billion years ago. We know this by observing light in our Universe which has travelled a great distance through space and time to reach us today. Observations by NASA's Wilkinson Anisotropy Microwave Probe (WMAP) revealed microwave light from this very early epoch, about 400,000 years after the Big Bang.

A period of darkness ensued, until about a few hundred million years later, when the first objects flooded the Universe with light. The first stars were much bigger and brighter than any nearby today, with masses about 1000 times that of our Sun. These stars first grouped together into mini-galaxies; the Hubble Space Telescope has captured stunning pictures of earlier galaxies, as far back in time as ten billion light years away.

By about a few billion years after the Big Bang, the mini-galaxies had merged to form mature galaxies, including spiral galaxies like our own Milky Way. It had also expanded, racing under the force of the so-called Hubble constant. Now, 13.7 billion years from the Big Bang, our planet orbits a middle-aged Sun in one arm of a mature galaxy with a supermassive black hole in the middle. Our own solar system orbits the Milky Way's centre, while our galaxy itself speeds through space.

Under the Milky Way in San Pedro de Atacama, Chile.

Scale of the Universe

Throughout history, humans have used a variety of techniques and methods to help them answer the questions 'How far?' and 'How big?'. Generations of explorers have looked deeper and deeper into the vast expanse of the Universe. And the journey continues today, as new methods are used, and new discoveries are made.

In the third century BC, Aristarchus of Samos asked the question 'How far away is the moon?' He was able to measure the distance by looking at the shadow of the Earth on the moon during a lunar eclipse.

It was Edmund Halley, famous for predicting the return of the comet that bears his name (p. 344), who three centuries ago found a way to measure the distance to the Sun and to the planet Venus. He knew that the planet Venus would very rarely, every 121 years, pass directly between the Earth and the Sun. The apparent position of the planet, relative to the disc of the Sun behind it, is shifted depending on where you are on Earth. And how different that shift is depends on the distance from both Venus and the Sun to the Earth. This rare event, the transit of Venus, occurred again most recently on June 8, 2004. It was knowing this fundamental distance from the Earth to the Sun that helped us find the true scale of the entire solar system for the first time.

When we leave the solar system, we find our star and its planets are just one small part of the Milky Way Galaxy. The Milky Way is a huge city of stars, so big that even at the speed of light, it would take 100,000 years to travel across it. All the stars in the night sky,

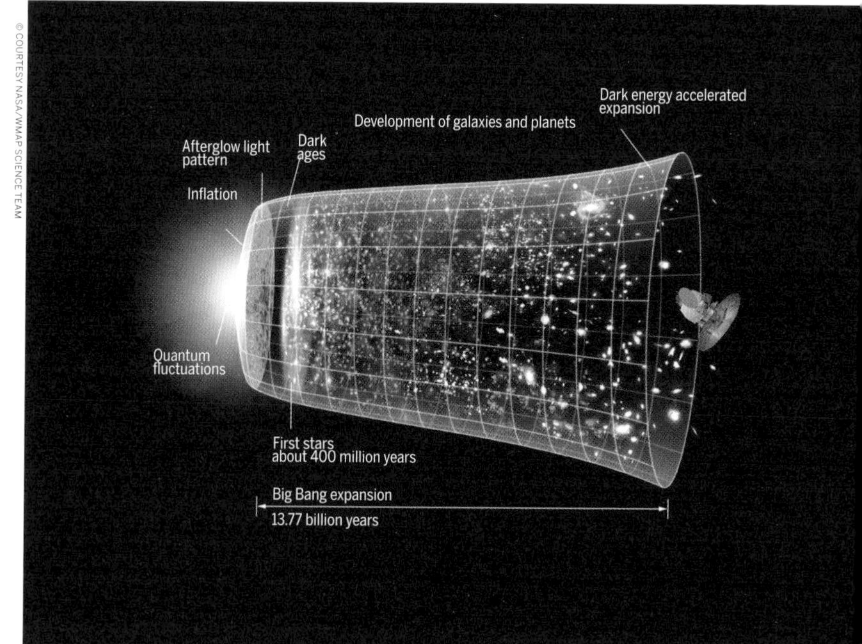

A timeline of the Universe since the Big Bang.

This Hubble Space Telescope image captures the effect of gravitational lensing by dark matter in a galaxy cluster.

including our Sun, are just some of the residents of this galaxy, along with millions of other stars too faint to be seen.

The further away a star is, the fainter it looks. Astronomers use this as a clue to figure out the distance to stars that are very far away. But how do you know if the star really is far away, or just not very bright to begin with? This problem was solved in 1908 when Henrietta Leavitt discovered a way to tell the 'wattage' of certain stars that changed their pulse rate linked to their wattage. This allowed their distances to be measured all the way across the Milky Way.

Beyond our own galaxy lies a vast expanse of galaxies. The deeper we see into space, the more galaxies we discover. There are billions of galaxies, the most distant of which are so far away that the light arriving from them on Earth today set out from the galaxies billions of years ago. So we see them not as they are today, but as they looked long before there was any life on Earth.

Finding the distance to these very distant galaxies is challenging, but astronomers can do so by watching for incredibly bright exploding stars called supernovae. Some types of exploding stars have a known brightness – wattage – so we can figure out how far they are by measuring how bright they appear to us, and therefore the distance to their home galaxy. These are called 'standard candles'.

So how big is the Universe? No one knows if the Universe is infinitely large, or even if ours is the only Universe that exists. And other parts of the Universe, very far away, might be quite different from the Universe closer to home. At the time of publication using our most advanced technology and given the current size of the ever-expanding Universe, scientists estimate it is roughly 46 billion light years, or 440 sextillion km (274 sextillion mi). If it's hard to wrap your head around that number, welcome to the club. The Universe is almost inconceivably big, and we have only observed a small portion of it (astronomers estimate we have observed roughly 4% of the known Universe).

Modern Observational Methods

In 1609 an Italian physicist and astronomer named Galileo became the first person to point a telescope skyward. Although that telescope was small and the images fuzzy, Galileo was able to make out mountains and craters on the moon, as well as a ribbon of diffuse light arching across the sky – which would later be identified as our Milky Way Galaxy. After Galileo's and, later, Sir Isaac Newton's time, astronomy flourished as a result of larger and more complex telescopes. With advancing technology, astronomers discovered many faint stars and the cal-

Today's observatories have significantly larger apertures than the basic telescopes of Galileo's day, but the principle is the same.

© JURGEN FALCHLE/ALAMY STOCK PHOTO

Hubble Space Telescope in orbit.

culation of stellar distances. In the 19th century, using a new instrument called a spectroscope, astronomers gathered information about the chemical composition and motions of celestial objects.

Twentieth century astronomers developed bigger and bigger telescopes and, later, specialised instruments that could peer into the distant reaches of space and time. Eventually, enlarging telescopes no longer improved our view, because the atmosphere which helps sustain life on earth causes substantial distortion and reduction in our ability to view distant celestial objects with clarity.

That's why astronomers around the world dreamed of having an observatory in space – a concept first proposed by astronomer Lyman Spitzer in the 1940s. From a position above Earth's atmosphere, a telescope would be able to detect light from stars, galaxies, and other objects in space before that light is absorbed or distorted. Therefore, the view would be a lot sharper than that from even the largest telescope on the ground.

In the 1970s the European Space Agency and the National Aeronautics and Space Administration began working together to design and build what would become the Hubble Space Telescope. On 25 April 1990, five astronauts aboard the space shuttle Discovery deployed the eagerly anticipated telescope in an orbit roughly 600 km (380 mi) above the Earth's surface. That deployment and, later, the unprecedented images that Hubble delivered represented the fulfillment of a 50-year dream and more than two decades of dedicated collaboration between scientists, engineers, contractors, and institutions from all over the world.

Since Hubble was launched, a number of other space telescopes have been successfully deployed to advance our knowledge of the Universe. These include the Spitzer Space Telescope, named for the man whose idea sparked a new era in telescopes and observation.

Today's Telescopes

Around the world, astronomers, space scientists and astrophysicists plying the depths of the Universe work in a variety of scientific fields, combining physics, chemistry, biology and other sciences to advance human knowledge of space. Much of their work relies on data from telescopes devoted to the observation of celestial objects. These can be either ground-based (located here on our planet) or space-based, rotating in orbit around Earth.

Ground-based telescopes are typically located in places around the world that meet a certain set of observing conditions. Broadly speaking, this includes locations with good air quality, low light

For non-professional astronomers, the Zeiss Telescope at Griffith Observatory, CA, offers a glimpse at the heavens.

Lowell Observatory in Arizona.

pollution, and often high altitude to reduce the impact of the atmosphere on observations. Generally, you'll find the world's top observatories on mountains, in deserts, and/or on islands – sometimes a combination of all three. Well-known locations with multiple ground-based telescopes include Mauna Kea in Hawaii, the Atacama Desert in Chile, and the Canary Islands.

Space-based telescopes are, as their name suggests, located outside the Earth's atmosphere in orbit. As such, they often have much greater ability to capture high-resolution images of celestial objects, unaffected by the interference of our atmosphere. The most popular space telescopes include the Hubble and Spitzer Space Telescopes, both operated by NASA's Jet Propulsion Lab (JPL) in California. Other space telescopes include the Transiting Exoplanet Survey Satellite (TESS) and forthcoming James Webb Space Telescope (which will replace the Hubble).

Types of Telescopes

Astronomers gain knowledge by looking across the spectrum of light frequency. Typically, the tools they use fall within two broad categories: optical telescopes and radio telescopes. The instruments used to gather this data comb across the entire electromagnetic spectrum. Visible light rays (what we see when we view the stars with the naked eye) are actually only a small part of this spectrum; radio waves, infrared, ultraviolet, X-rays and gamma-rays are all also examined for the information they contain about far-off objects.

Ground-based observatories often focus on radio waves, which can be captured by antennas, and visible and infrared light, which are gathered at large optical telescopes. The technique of spectroscopy can help parse the information encoded in these rays. Other electromagnetic waves such as X-rays are best received in space, and these are monitored by telescopes in orbit where Earth's atmosphere doesn't get in the way.

How to Use This Book

Like its namesake, the book you hold is big – and like our understanding of the Universe, it is also, by necessity, incomplete. Astronomers continue to explore the Universe with ever-improving technology, unlocking previously unknown secrets and mysteries. In these pages, you'll discover some you likely don't already know, and undoubtedly have questions and hypotheses about what we'll discover next.

As you work through this text, the general organisation of the book will lead you from home on our Earth out into the far reaches of the solar system, then into our neighbouring stars and planetary systems and finally into the rest of our galaxy and the Universe as a whole, via carefully selected examples of known exoplanets, stars, nebulae and galaxies, as well as

This artist's illustration gives an impression of how common planets are around the stars in the Milky Way.

An artist's concept of our Milky Way Galaxy.

even more exotic deep-sky objects. You'll discover as much as we know about our celestial neighbourhood, and our place in it. In addition to planets and moons, get to know our Sun, explore the asteroid belt and the Kuiper Belt, and learn what lays beyond, in interstellar space.

Outside our solar system, the book guides you to some of the notable neighbouring stars, stellar systems, and exoplanets we've discovered. You'll understand how we search for planets where life might exist and the stars they orbit. Some of these are located within the Milky Way; others we've observed from our particular perspective in the Universe though they live far beyond the boundaries of what we consider our galaxy.

Finally, the book steps out to the edge of the observable Universe – at least what we've observed with the technology available today. You'll get to know the structure of the Milky Way as well as an orientation to neighbouring galaxies like the Andromeda Galaxy which is visible from Earth. You'll explore other galactic formations and zoom even further out to learn about galactic clusters and superclusters. By the end of the book, you'll have a sense for the structure of the entire Universe as well as some of the big questions we still have as we ponder our place in it. You may not be able to plan your next vacation on the basis of the planetary moons, exoplanets and stunning nebulae featured, but you'll find lots to amaze and awe.

Naming Conventions

As you work through this book, you'll discover objects that go by a variety of names. Some you will recognise, but others may seem encoded. Celestial nomenclature has long been a controversial topic. At its inaugural meeting in 1922 in Rome, the International Astronomical Union (IAU) standardised the eighty-eight constellation names and their abbreviations. Over half of these date back to Ptolemy, with the rest being more recent additions. Since standardising the constellations, the IAU has gone on to certify the names of other astronomical objects, though common usage doesn't always immediately follow suit

Additionally, there's the added layer of complication between an object's name

A portrait of the astronomer Charles Messier.

1661 'map' of Andromeda constellation.

© GIULIO ERCOLANI/ALAMY STOCK PHOTO

Pictured is the Helix Nebula, otherwise known as NGC 7293.

and its official designation. Generally, a celestial object's name refers to the (usually colloquial) term used in everyday speech, while designation is solely alphanumerical and used almost exclusively in official catalogues and for professional astronomy.

The cataloguing of stars has a long history. Since prehistory, cultures and civilisations around the world have given their own unique names to the brightest and most prominent stars in the sky. As astronomy developed over the centuries, a need arose for a universal cataloguing system, whereby stars were known by the same labels, regardless of the country or culture from which the astronomers came.

To solve this problem, astronomers during the Renaissance attempted to produce catalogues of stars using a set of rules. The earliest example that is still popular today was introduced by Johann Bayer in his *Uranometria Atlas* of 1603. Bayer labelled the stars in each constellation with lowercase Greek letters, in the approximate order of their (apparent) brightness (for example, Alpha Tauri).

Nearly 200 years after the introduction of Bayer's Greek letter system another popular scheme arose, known as Flamsteed numbers, named after the first English Astronomer Royal, John Flamsteed. In this scheme, stars are numbered in their order of right ascension within each constellation (for example, 61 Cygni). Other designation schemes for bright stars have been used, but have not seen the same degree of acceptance. One such scheme was introduced by the American astronomer Benjamin Gould in 1879. Only a handful of stars are occasionally referenced with the Gould scheme today (for example, 38G Puppis).

Stars discovered more recently are fainter than those catalogued under the Bayer or Flamsteed schemes, as they are found by higher-powered telescopes and means of detection. As astronomers discover these new stars, it is standard practice to identify them with an alphanumeric designation. These designations are practical, since star catalogues contain thousands, millions, or even billions of objects. There are also special rules regarding binary and multiple stars, variable stars, novae and supernovae. All of these approaches build on the very first such systems. Messier's 1771 catalogue is one of the best known compilations of non-stellar deep-sky objects, created to list known 'nebula'. At the time this meant objects that were neither comets nor stars. The later 1888 New General Catalogue (NGC) compiled by John Louis Emil Dreyer added to the number of classified objects previously found by the Herschel family, from star clusters and nebulae to galaxies. The prefixes 'M' and 'NGC' indicate objects from these respective catalogues, and many deep-sky objects carry both numbers as well as a more common name.

Within this book, you will find both the name and the more official catalogue designation; if you see only a designation, that means the object has not been given a formal name. Equipped with this foundation of knowledge, you're ready to begin your exploration. Whether you stay close to home or jump straight to the distant edges of our knowledge, the Universe is an endlessly interesting space to let your mind journey.

Top Highlights

Get to Know Our Sun

1 In the Universe, our Sun may not be special or unique – but it is very important to us on Earth. Learn about the Sun, what's happening deep inside the fiery corona, and the research we're currently doing to better understand our nearest, life-giving star.

Learn About Mars, Our Next Home?

2 Zoom in to Mars, the neighbouring planet that continues to captivate our hopes for an off-Earth colony. Learn more about why Mars is such a compelling destination for the human race and discover what may have changed this planet from an Earth-like place to the one it is today.

Meet Other Objects in the Solar System

3 Get to know beloved Pluto and other overlooked dwarf planets in greater detail, as well as the asteroids and comets that punctuate the solar system. You'll discover there are a variety of other objects in our area than just the

The Mars Curiosity rover at Namib Dune.

A colourised image of barred spiral galaxy NGC 1672; its arms are home to stellar nurseries.

Sun, eight planets, and 193 known moons.

Discover Earth-Like Exoplanets

4 Peer deep into other stellar systems like TRAPPIST-1 and Kepler-22, where exoplanets sit within the 'habitable zone'. Scientists believe some of these exoplanets may have conditions similar to those on Earth – including those which led to life on our planet.

Journey to the Nearest Star

5 Visit Proxima Centauri, the closest star to our Sun, a mere 4.243 light years away. With neighbouring Alpha Centauri and Beta Centauri as well as a potentially Earth-like exoplanet of its own, Proxima Centauri gives us a good example of the variety of solar systems in the galaxy.

Wonder at Supernovas and Black Hole Quasars

6 Stars come in more types and stages than the yellow dwarf we are most familiar with. Explosions like Kepler's Supernova leave fantastical nebulae in their wake. ULAS J1120+0641 is a distant, supermassive black hole-powered quasar emitting jets of intense radiation.

Orient Yourself in the Milky Way

7 Learn about the arms of our spiral Milky Way Galaxy and where we sit within ours, the Orion Arm. You'll gain context on which parts of the galaxy we can see – and the mysteries our galaxy still holds for researchers.

Understand How Galaxies Interact

8 Taking a wider view, gain an understanding of the variety of different galactic formations in the Universe and how, on a larger scale, vast clusters which gather thousands of galaxies bound together by gravity.

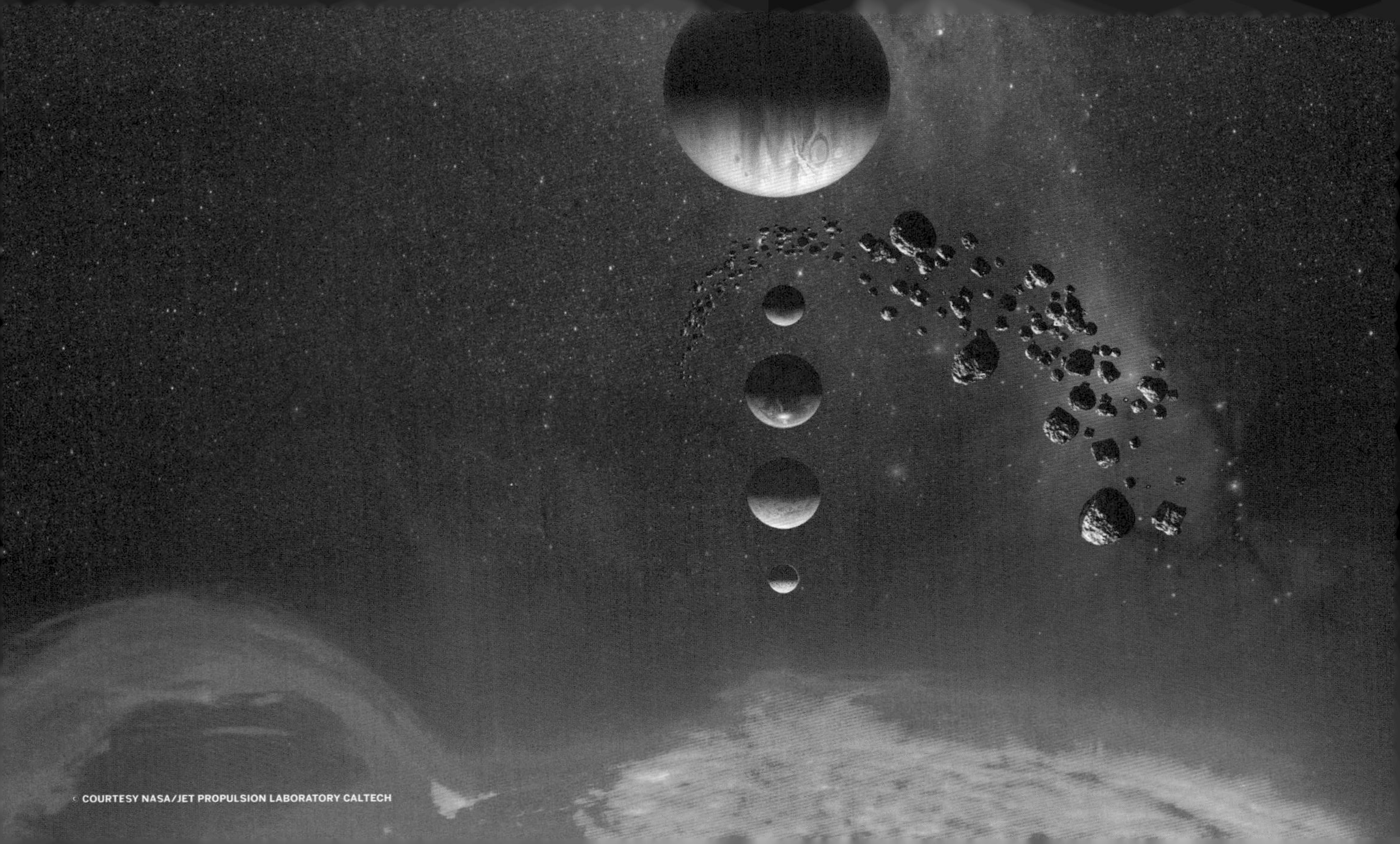

THE SOLAR SYSTEM

Solar System Highlights

Living on Planet Earth

1 Our Universe has many wonders, but the greatest wonder of all is the one that's closest to home. As far as we know for now, it's the only place amongst all the stars, moons, asteroids and planets where life has developed in all its diverse, strange and dazzling forms. So if you only take one thing from your trip around the Universe, it has to be this: Planet Earth is a very special place indeed.

Watching 'Earthrise' from the Surface of the Moon

2 There are plenty of iconic photographs out there, but few can have had the epoch-changing impact of Earthrise – the image of our own planet appearing above the moon's horizon, a brilliant blue globe suspended in the infinite blackness of space, revealing for the first time the wonder – and fragility – of the planet we call home.

Following in Neil Armstrong's Footsteps

3 As every schoolkid knows, the Mare Tranquillitatis, or Sea of Tranquillity, is the landing site of the Apollo 11 mission, where Neil Armstrong and Buzz Aldrin took 'one small step for man, one giant leap for mankind.' Thanks to the moon's lack of atmosphere, the landing site will be preserved for as long as the moon is there – meaning that humanity's first footsteps outside Earth will be visible long into the future. That's quite a thought, isn't it.

Climbing the Crater of Olympus Mons on Mars

4 Three times as high as Mount Everest, and covering an area roughly the size of Arizona, Olympus Mons is the biggest volcano on Mars and – as far as we currently know – the biggest anywhere in the solar system. But with a slope of around 5%, it would be relatively easy to climb – and one day, perhaps, the volcano might be a highlight on the itinerary of every Mars space tourist.

Hiking through the Valles Marineris on Mars

5 The Grand Canyon is big, but it's not nearly as big as this mighty Martian valley – an epic five times the length of the Grand Canyon and four times its depth, running for nearly a fifth of the planet's equator. It's the largest canyon in the solar system, and hiking beneath its towering rust-red cliffs would undoubtedly be a memorable experience.

Storm-chasing the Great Red Spot on Jupiter

6 Even the worst storms on Earth pale in comparison to Jupiter's gargantuan Great Red Spot – an enormous tempest that's been raging for centuries and still going. In fact, it's so big, it could fit the entire Earth inside its diameter with room to spare.

Sailing through Saturn's Rings

7 The unmistakeable sight of Saturn's seven mighty rings is one of the solar system's great spectacles, swirling out around the planet at a distance of up to 282,000 km (175,000 mi). Formed of ice, rock and dust, they're visible from Earth with a half-decent pair of binoculars – but just imagine how incredible they'd look up close if you were to sail past in a spacecraft.

Getting a Tan on Mercury

8 Frazzled and dazzled by its proximity to the Sun, the solar system's littlest planet probably isn't a place you'll want to linger for long. Sunlight here is 11 times brighter than on Earth, and daytime temperatures can hit scorching pizza oven levels – so if you really want to get that all-over, lasting tan, Mercury is most definitely the planet to head for.

Watching the Backwards Sunrise on Venus

9 Despite its beatific name, Venus is an altogether inhospitable place: a hothouse world ravaged by a rampant greenhouse effect that has created a hellish

Earthrise from the moon.

atmosphere hot enough to melt lead – so not really ideal for sunbathing. But there is at least one reason for taking a trip to Venus: due to its unusual backwards rotation, the Sun here appears to rise in the west and set in the east.

Being Dazzled by the Aurorae of Uranus

10 Just like the Northern Lights on Earth, Uranus has its own aurorae – spectacular, shimmering light displays caused by charged particles interacting with gases in the planet's atmosphere. But many scientists think they could be way more spectacular than any you'd see on Earth, due to Uranus' unique sideways rotation and lopsided magnetic field.

Getting Swept Up in Neptune's Vortices

11 Like Jupiter, Neptune is a 'gas giant', with a swirling atmosphere composed mainly of hydrogen, helium and methane. The tempestuous atmosphere creates powerful storm systems known as 'vortices', where winds are believed to reach speeds many times faster than any storms on Earth (up to 2400 km/h, or 1500mph). One of these, dubbed the Great Dark Spot, was seen by Voyager 2 in 1989 – and although it had disappeared by 1994, other similar vortices have since been spotted.

Diving into Europa's Hidden Ocean

12 Of all the moons in our solar system, Europa may be the likeliest place to find life outside Earth. Beneath its icy crust, the moon is thought to conceal a huge, salty ocean where volcanic or hydrothermal vents are believed to exist on the seafloor – creating very similar conditions to the ones in which many scientists believe life on Earth began.

Marvelling at Gigantic Ice Geysers on Enceladus

13 Enceladus, the sixth-largest moon of Saturn, is in many ways the most interesting one. Encased in a thick shell of ice, it's the brightest object in our solar system – and one of the coldest, with surface temperatures of about -201°C (-330°F). It also has a unique feature: giant ice geysers that erupt at its South Pole, intensifying its already dazzling albedo effect (that's a measure of how it reflects solar energy).

Watching the Lava Flow on Io

14 Well over 150 volcanoes have been observed on the little moon of Io, although scientists think that's just a fraction of the total. Squeezed and contorted by the gravitational pull of Jupiter, it's the most geologically active body in the solar system – making it a must-see moon for avid vulcanologists.

If You Like...

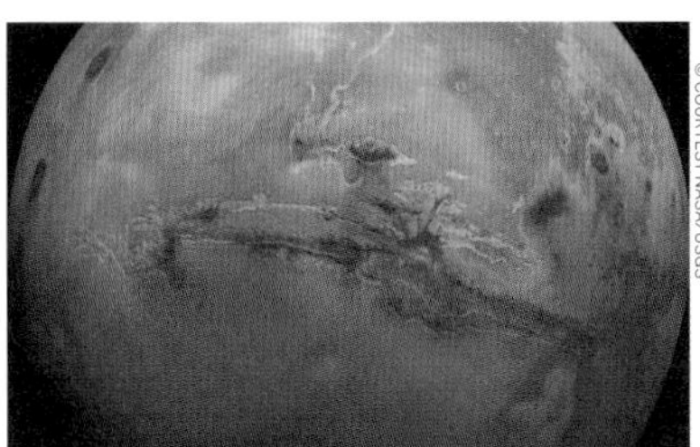
© COURTESY NASA/USGS

This mosaic of Mars shows the entire Valles Marineris canyon.

Geological Wonders

Mt Everest At 8848 m (almost 5.5 mi), Everest is the highest mountain on Earth.
Valles Marineris, Mars Deeper than six Grand Canyons, this mighty 400 km-long (250 mi) valley is a Martian wonder.
Verona Rupes On the Uranian moon Miranda, this is the tallest cliff in the entire solar system, rising over 10 km (6 mi) straight up.
Caloris Basin, Mercury 1545 km (960 mi) across, this vast impact basin was caused by a meteorite strike.
Enceladus This Saturnian moon has spectacular geysers of ice spouting at its South Pole.
The Methane Lakes of Titan Liquid lakes and rivers of methane are thought to flow on Saturn's moon Titan.

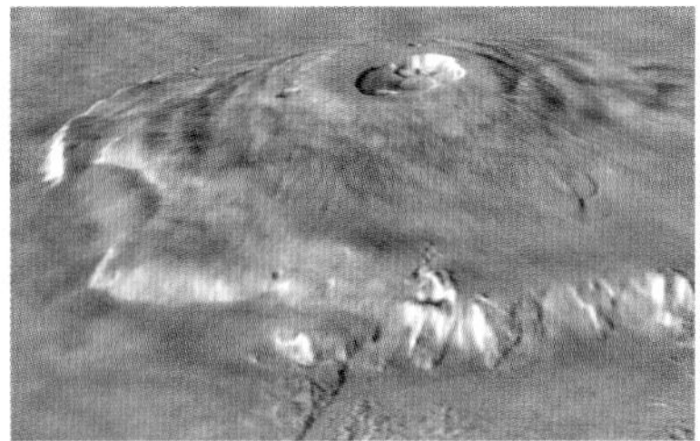
© COURTESY NASA/MOLA SCIENCE TEAM

Olympus Mons on Mars rises up in this Viking image mosaic.

Volcanoes

Kilauea One of Earth's most active volcanes can be found on Hawaii's Big Island.
Olympus Mons, Mars Though dormant, this is the largest volcano (and mountain) in the solar system, 22 km (14 mi) high and 600 km (375 mi) across.
Anywhere on Io This Jovian moon has hundreds of active volcanoes to choose from: don't get too close.
Maat Mons, Venus A massive shield volcano, and the second-highest mountain on Venus.
Triton Along with Venus, Io and Earth, Neptune's moon Triton is one of the few places in our solar system known to be volcanically active.
Titan This Saturnian moon is home to 'cryovolcanoes' which spew not fire, but ice.

© PETE SEAWARD/LONELY PLANET

An ice floe in Wilhemina Bay.

Ice

Antarctica Earth's continent of ice holds approximately 90% of the planet's fresh water.
Uranus One of two 'ice giants' in the outer solar system, Uranus' surface is a swirling, dense fluid made of icy materials.
Neptune The second of our ice giants, more than 80% of Neptune's mass is a dense fluid of icy water, methane and ammonia.
Ganymede The largest moon in our solar system, Jupiter's moon Ganymede has a coat of ice lined by many ridges and grooves (known as 'sulcus').
Europa The surface of this Jovian moon is completely encased in ice – but beneath it, a huge watery ocean exists.
Triton Ice volcanoes on Triton spout liquid nitrogen, methane and a dust that falls back to the surface as snow.

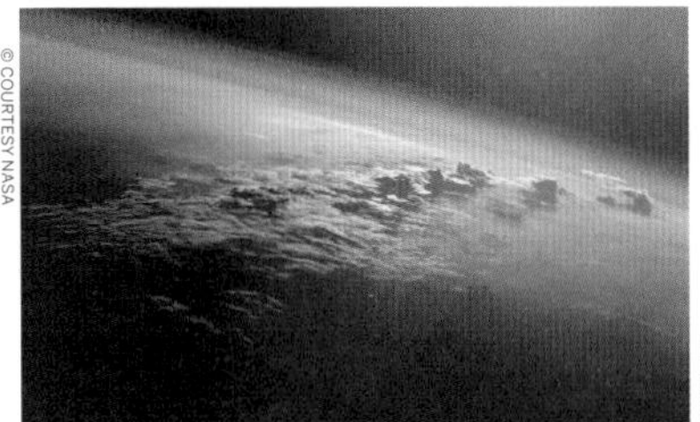

Earth's very own Indian Ocean as viewed from space..

Seas & Oceans

Earth An astonishing 70% of the Earth's surface is ocean.
Europa Though covered by ice, Europa's ocean may be 60 to 150 km (40 to 100 mi) deep.
Titan Giant seas of methane churn on the surface of Titan – and may contain forms of life.
Callisto Like its sister moon Europa, Callisto's icy crust may conceal a liquid salty ocean.

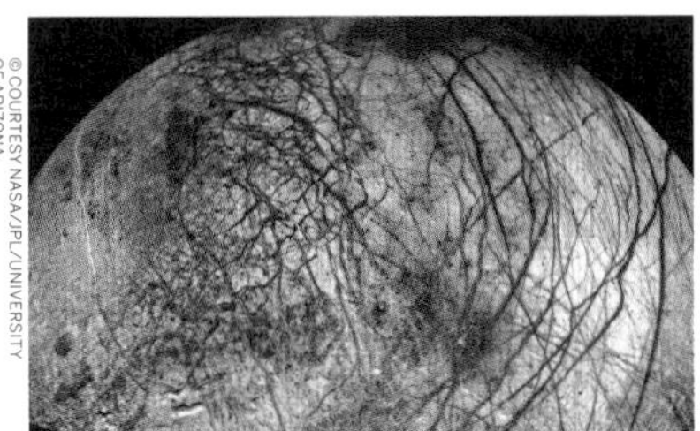

Europa's frozen surface presents dramatic cracks.

Weird Landscapes

Utopia Planitia, Mars 3300 km (2050 mi) across, this enormous impact crater is the largest anywhere in the solar system.
The Cracks of Europa The surface of this frosty Jovian moon is covered in huge striations, ridges and cracks, some measuring hundreds of miles long.
The Yin-yang world of Iapetus This Saturnian moon is deeply peculiar: it's half-black, half-white.
The Canyons of Miranda Miranda's deeply fissured surface is split by giant canyons, including some 12 times deeper than the Grand Canyon.

The 'Death Star' Moon A huge impact crater gives the Saturnian moon of Mimas a disturbing resemblance to Darth Vader's planet-killing space station.
The Ice Mountain of Ceres A lone dome of ice exists on the dwarf planet Ceres.

Saturn's C-ring, pictured here, is only one of several.

Rings

Saturn Seven in number, Saturn's giant rings are many thousands of kilometres long, but only around 10 m (30 ft) thick.
Uranus After Saturn, Uranus has the most impressive rings: 13 in all, split into three distinct zones.
Neptune Neptune's rings are so faint, they're practically invisible.

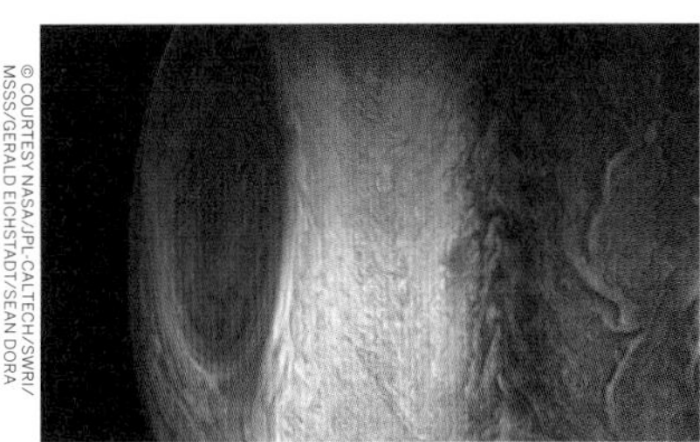

The Great Red Spot would be the trophy of any storm chaser.

Storms

Tornado Alley, Earth Cutting across the southern United States, this belt of land is home to some of the most powerful winds on Earth.
The Great Red Spot, Jupiter Storms on Jupiter really are on a different scale: some are big enough to swallow up entire planets.
The Vortices of Neptune Like Jupiter, Neptune is ravaged by truly apocalyptic winds and storms.

Transits & Eclipses

Viewed from our position on Earth, the solar system can be a pretty tricky place to get your head around. From where we are, it always seems like we're standing still, in a fixed position in space – when in fact, we're rotating at thousands of miles an hour in a complex dance of interplanetary orbits, circling in space around a common central point, the Sun. Because the distances and scales involved are so large, it's hard to notice the fact that the positions of the stars and planets are in fact changing all the time literally before our eyes, and the night sky we see is subtly different every single time we view it.

A solar eclipse at totality.

There is one phenomenon that brings the constant motion of our solar system into sharp focus, however: during an eclipse, we actually get to see the effect of another planetary body in motion right before our eyes.

These spectacular events come in two flavours, lunar and solar. A lunar eclipse occurs when Earth moves between the Sun and the moon, blocking out the Sun's light and casting the Earth's shadow onto the lunar surface. A lunar eclipse can either be partial, when only some of the Earth's shadow appears on the moon, or total, when the moon and the Sun are on exact opposite sides of Earth and the Earth's shadow completely covers the moon. During a total lunar eclipse, the Earth's atmosphere blocks out most of the blue light coming from the Sun, which makes the moon appear to turn a vivid shade of blood-red. Total lunar eclipses are quite rare, but at least two partial lunar eclipses happen every year, each lasting several hours at a time.

Even more spectacular is a solar eclipse, when the moon moves into a position between the Sun and the Earth. A solar eclipse also comes in two main flavours, partial and total. A partial solar eclipse happens when the Sun, moon and Earth are not exactly lined up. During a partial solar eclipse, the moon's crescent moves over the Sun's surface, and only a part of the Sun is covered up by the moon.

During a total solar eclipse, the moon moves directly in line with the Sun relative to our position on Earth. The moon appears first as a crescent, then continues to move across the Sun's face until it becomes a solid black disk, leaving only a thin halo of sunlight around its outer edge, known as the corona. The sky becomes very dark and night-like as the moon's shadow falls onto the Earth. A total solar eclipse is only visible from a small area on Earth: the people who are standing in the centre of the moon's shadow when it hits Earth (known as the umbra) will witness a total eclipse, while those standing outside this zone (in the shadow zone known as the penumbra) will only see a partial eclipse. Unlike lunar eclipses, solar eclipses only last for a few minutes.

There is actually a third type of solar eclipse: the annular solar eclipse, which happens when the moon is furthest from

Earth. Because the moon is further away from Earth, it seems smaller, and doesn't completely block the Sun, making the moon appear as a dark disk on top of a larger solar disc. This creates what looks like a ring around the moon.

Eclipses are not unique to Earth, the moon and the Sun, however. The same phenomena, known as transits, occur between different planets and moons. Like an eclipse, a transit occurs when one object appears to pass in front of another object. But in a transit, the apparent size of the first object is not large enough to cast the second into complete shadow. Instead, a much smaller dark shadow makes its way across the face of the further planet or star. On occasion, these events can be seen from Earth (with the help of telescopes and other instruments) or by satellites.

The most recent transit was that of Venus across the face of the Sun in 2012. There have been many such events recorded by astronomers throughout history, although they occur very rarely - transits of Venus, for example, come in pairs, eight years apart, separated by a gap of approximately 120 years. From Earth, the only transits we can see directly are those of Venus and Mercury. The next transits of Mercury are scheduled for 11 November 2019, then again on 13 November 2032. The next transits of Venus won't be until 11 December 2117 and 8 December 8 2125 - so if you missed the one in 2012, unfortunately you may be out of luck.

Perhaps the most famous example of transit-hunting was Captain Cook's voyage across the Pacific in 1769. The official purpose of the expedition was to observe the transit of Venus from the newly discovered island of Tahiti; the hope was that it would enable astronomers to measure the size of the solar system, one of the great mysteries of 18th-century science. But after observing the transit, Cook was also charged with another task: searching for the fabled Terra Australis Incognita, the unknown 'south land' that was believed to exist on the southern edge of the Pacific Ocean. Unfortunately, after months of searching, Cook didn't find it, as it didn't exist! But he did reach Australia and New Zealand instead, and nearly wrecked his ship on the Great Barrier Reef in the process.

The search today is much easier. Those interested can see the next solar eclipses on 4 December 2021, over Antarctica; 20 April 2023 in Western Australia and West Papua; 8 April 2024, across North America; and 12 August 2026 over parts of Europe. While many of these eclipses will be narrow paths over land, planning to see a total solar eclipse is a venture well worth undertaking. Lunar eclipses, by contrast, are much more widely visible.

Why Does NASA Study Eclipses?

Hundreds of years ago, when people observed the moon during an eclipse, they discovered that the shape of Earth is round. Even after all these years, scientists are still learning from lunar eclipses. In December 2011, NASA's Lunar Reconnaissance Orbiter gathered data about how quickly the moon's day side (the side that always faces Earth) cools during a lunar eclipse. NASA can determine what the moon's surface is made of from this data. If an area of the moon's surface is flat, it will cool quickly. Scientists use this data to know which areas of the moon are rough with boulders and which are flat.

NASA also studies solar eclipses. Scientists use solar eclipses as an opportunity to study the Sun's corona (the Sun's top layer). During an annular eclipse, NASA uses ground and space instruments to view the corona when the moon blocks the Sun's glare.

Introduction to the Planets

Of the 4.5 billion years since the formation of the solar system, humans have been around for a mere sliver of time, and our studies of our solar system neighbours occupy an even shorter period.

An artist's illustration of the planets (and Pluto) as they might be imagined to reach into space from Earth.

Ancient civilisations charted the course of the Sun and moon as far back as megaliths from 10,000 to 11,000 years ago, while ancient Sumeria had its own lunar calendar. In the subsequent centuries, astronomers around the world made astonishing discoveries with modest technology. All were equipped with enquiring minds and in some cases, such as the 6th century Indian scholar Aryhabata, a dazzling grasp of mathematics. The advent of the telescope, in the 17th century, allowed the likes of Galileo to observe planetary features with his own eyes. In 1963, Yuri Gagarin was the first human in space; in 1969, Neil Armstrong became the first person on the moon, the first to walk on the surface of another celestial body. But humans have yet to visit another planet. If any should take your fancy, here's a snapshot, in order of their distance from Sun, of what you need to know.

Mercury

Mascot
The winged messenger.

Size isn't everything
The smallest planet in our solar system is also, at 0.4 astronomical units away, the nearest to the Sun. So close, in fact, that sunlight on Mercury is seven to eleven times brighter than it is on Earth depending on distance from the Sun, yet without an atmosphere that's still not sufficient to make it the hottest body in the solar system – that's Venus. Mercury is, however, the fastest planet and speeds around the Sun at a blistering average clip of 170,505 km/h (105,947mph). Following an elliptical orbit, in which its proximity to the Sun can vary by as much as 24 million km (15 million mi), it makes a complete orbit of the Sun in just 88 days. That means each Earth year is more than four Mercurian years.

When was it discovered?
As a bright satellite visible to the naked eye, Mercury was known to antiquity as a non-fixed object in the sky. It was one of the Greek planets (though they weren't defined in exactly the same way) as well as being known to the Babylonians. The first sighting of Mercury in the scientific age came in 1631, thanks to the invention of the telescope. This led to simultaneous observations, by the English astronomer Thomas Harriott and the Italian scientist

Galileo Galilei. That same year it was observed transiting before the Sun.

Fascinating fact

Despite its reputation as a hot, speeding mass, more recent observations have shown Mercury to be a planet of contrasts. For example, the interiors of the deep craters found at Mercury's north and south poles lie in permanent shadow; while the rest of the planet is exposed to temperatures as high as 427°C (800°F), the craters may contain water-ice according to 2012 observations by the MESSENGER mission.

Any features of particular interest?

The intriguing spines of Pantheon Fossae on Mercury's surface catch the eye. The 40 km (24 mi) crater was discovered during MESSENGER's first flyby of the planet, in 2008, and is situated upon Caloris Planitia, the 'fire plains' of northern Mercury. The feature takes its name from the ancient Roman Pantheon, the celebrated roof of which was built by the equally celebrated Apollodorus of Damascus. It consists of a central core, from which sections radiate outwards. In the case of Pantheon Fossae, the main crater sends out long, thin troughs - tectonic fault lines that earned Pantheon Fossae the nickname of the 'spider crater'.

Are there moons or rings?

Small Mercury makes its orbit of the Sun all on its own, without the company of any moons or rings.

Could it support life?

Mercury's environment is not conducive to life as we know it. Battered by the Sun's heat and lacking an atmosphere to protect it from the solar wind or from meteor impacts, life on sweltering Mercury isn't considered to be a possibility.

Mercury in popular culture

With little hard data to go on, Mercury's appearances in popular culture relied on a blend of guesswork and artistic license. Isaac Asimov was drawn to Mercury's, well, mercurial nature, famously harnessing it for 'I, Robot', a story whose anti-hero is a robot built to withstand extreme solar radiation. (Not to be confused with 'Sonny', the robot with human aspirations, who appears alongside Will Smith in the 2004 movie that borrows Asimov's title.) Elsewhere, the Zanzibar-born Farrokh Bulsara took the planet's name as his own, and thus gave the world Freddie Mercury. The show *Invader Zim* included the planet being turned into a spaceship.

Has there been a key mission?

More recent flybys of Mercury by MESSENGER have provided scientists with a wealth of data about the composition of the planet: we now know, for example, that Mercury's large, metallic core represents about 85% of its total radius. Images from MESSENGER have also shown 'crater rays', vast streaks on the planet's surface formed by ejecta material following meteor strikes. In 2015, MESSENGER was crashed into the surface of Mercury, bringing its mission to an end.

What are the scientists saying?

The next mission to Mercury, which scientists hope will yield yet more of the planet's secrets, is already on its way. Launched by the European Space Agency (ESA) in 2018, the BepiColombo mission will make numerous flybys of Mercury when it eventually catches up with the speeding fireball - which, if all goes to plan, should be in December 2025. The mission, which also plans to test Einstein's theory of relativity, is quite literally hot stuff for the ESA - the first time it has dispatched a mission to such a hot part of the solar system. It is sure to serve up some testing challenges, not least counteracting the Sun's massive gravitational pull. The mission includes two orbiters, one created by the Japan Aerospace Aerospace Exploration Agency (JAXA) to analyse the planet's magnetic field, or magnetosphere. ESA's orbiter intends to study the surface and interior, learning more about its metallic core and impacts on the surface.

Venus

Mascot

The goddess of love.

The hothead

For many years, scientists thought of Venus as a kind of sister planet to Earth. The two bodies are similar in size and structure and, by celestial standards, are similar distances from the Sun: Earth is 1 astronomical unit (AU) away from our star and Venus, the second closest planet to the Sun, is .72 AU. The fact that it is closer than Earth led to the belief it might offer conditions similar to the tropical regions of our home planet. Newer findings disprove the theory of a mildly tropical environment. The extremely high levels of carbon dioxide in the Venusian atmosphere, combined with a lack of surface water, leave the planet unable to regulate its temperature; put simply, it keeps getting hotter. It's a prime instance of the greenhouse effect. We now know Venus is the hottest planet in the solar system, where surface temperatures can reach 470°C (878°F).

When was it discovered?

Venus looms large in the history of astronomy due, in part, to its 'transits': every hundred years, the planet dramatically passes across the face of the Sun. Mayan astronomers as far back as 650 BC were charting Venus' movement in order to create a planetary calendar, which turned out to be very accurate.

Fascinating fact

Venus' reputation for extreme temperatures can be misleading once you reach the level at which clouds form, roughly 48 km (30 mi) into the atmosphere: the temperature drops markedly, to about the same as the surface temperature of Earth. The high concentrations of carbon dioxide in the Venusian atmosphere mean clouds here are composed of sulphuric acid. These, in turn, race across the surface of the planet, whipping around the Venusian globe at speeds of up to 354 km/h (220mph). A complete circumnavigation takes four days – the equivalent 'cloud journey' on Earth takes four times as long.

Any features of particular interest?

Heat has played a major part in shaping the surface features of Venus. Take the long and winding monster that is Baltis Vallis, a channel likely cut by lava – the longest of its kind in the solar system. It was discovered by the Soviet Venera orbiters in 1983; the twin spacecraft plotted Baltis Vallis' course for almost 1000 km (620 mi). Current measurements put it at around 7000 km (4350 mi) long, varying in width between one and three kilometres. With both ends now obscured by overlying rock formations, Baltis Vallis may be even longer. Either way, it is longer than Earth's Nile River, the longest on our home planet, which comes in at a paltry 6650 km (4132 mi) by comparison.

Are there moons or rings?

Like Mercury, Venus has no moons or rings.

Could it support life?

Venus is far too hot to support life now on its surface, though some scientists theorise that its atmosphere might be temperate enough to support airborne microbial life. This might be similar to the organisms that feed in Earth's ocean vents. Venus might have supported life of a more robust type earlier in its evolution, before the planet heated up and lost its hypothesised oceans.

Venus in popular culture

'Men are from Mars,' according to the American relationship guru John Gray, but 'women are from Venus'. Historically, the planet has had feminine status conferred upon it; only three Venusian features are not named for women. (One, the volcano Maxwell Montes, is named for a man, the other two for letters of the Greek alphabet.) In music, Venus has inspired Lou Reed, who dressed her in furs. The pulp-comic-book artists of the 1950s had the same idea, populating Venus with semi-clad 'Amazonian' women, in what proved a highly prolific – if highly dubious – genre.

Has there been a key mission?

Following its launch in May 1989, the Magellan spacecraft took 15 months to reach its orbit around Venus. It was worth the wait. The high-quality radar images of Venusian terrain it began sending back in September 1990 offered detailed views of this mysterious planet. The name of the mission was apposite: Ferdinand Magellan, the 16th century Portuguese explorer, is synonymous with mapmaking, providing as he did the first charts of much of the world's oceans. By the time his interplanetary counterpart was finished, NASA's Magellan mission had recorded images of close to 85% of the Venusian surface.

What are the scientists saying?

While scientists have debunked the theory of a sister planet, more recent research hints at habitable conditions sometime in Venus' ancient history. Models created by NASA's Goddard Institute for Space Studies (GISS) suggest a shallow ocean and Earth-like temperatures may have existed on the Venusian surface for up to 2 billion years. The findings of the GISS research has, potentially, significant implications for future NASA missions tasked with locating possible habitable planets and studying their evolving atmospheres; these include the Transiting Exoplanet Survey Satellite, and the James Webb Space Telescope.

Earth

Planet life

Earth might be the only planet in our solar system that's home to oceans, but its name comes from early European words for 'ground'. Earth is the planet against which we measure all others, being precisely one astronomical unit away from the Sun, an average distance of 150 million km (93 million mi). The fifth-largest planet in the solar system may be the only planet that can sustain life, but it is utterly dependent on the Sun – which keeps us warm, enables plants to photosynthesise, and lights our way. However, it's not just sunlight that makes the eight-minute journey to Earth. The Sun also releases a 'solar wind' packed with harmful radiation. Luckily, help is at hand, in the shape of Earth's atmosphere, a shield against radiation and incoming meteors. It's a relationship we'll ultimately live to regret. In 5 billion years' time, the Sun will have increased by as much as 100 times, at which point the Earth, the densest planet in our solar system, will be vaporised.

What have we discovered and when?

Discoveries about Earth made by our ancestors have led to some interesting ideas about the planet and its place in the universe. One common misconception is that, until the Middle Ages, many people thought the Earth to be flat. In fact, humans have known the Earth is round for more than two millennia. During the summer solstice, the ancient Greeks calculated Earth's circumference by measuring shadows. Aristotle observed that the stars appeared different in Egypt than his native Greece, suggesting a curvature in the planet's surface. The same conclusion was drawn by Roman sailors, who noted only higher ground was visible from a distance. As early as the 6th century, the Indian scholar Aryabhata calculated the circumference of the Earth. He got it wrong - but only by 172 km (107 mi).

Fascinating facts

The 70% of Earth covered by the planet's oceans supports an abundance of life. With an average depth of 4 km (2.5 mi), the oceans act as a vast reservoir for all but 3% of the planet's water. Our seas are also home to some of Earth's most imposing land formations. For example, Earth's longest mountain range, the Mid-Ocean Ridge, is located deep beneath the waves, shared by the Arctic and Atlantic oceans. Longer than the Andes, Rockies and Himalayas placed end to end, it runs for a mind-boggling 65,000 km (40,389 mi).

Are there moons or rings?

Earth's one moon is the only other body in the solar system that humans have set foot upon. It also controls Earth's tides.

Any features of particular interest?

Nowhere are Earth's natural features better showcased than at its extremes. The planet's highest point is Mount Everest, at 8848m above sea level. Around 800 people a year make a bid for its summit, many taking life-threatening risks. The lowest point is Challenger Deep, in the Mariana Trench in the Pacific Ocean. Its full depth has yet to be ascertained - to date, we know it is at least 10,994m. The water pressure at that depth is 1000 times greater than at sea level. In fact, it's as much as 8 tons per square inch, the equivalent of trying to lift 50 jumbo jets off the ground. Then there's the Great Barrier Reef, one of the few natural features on the planet visible from space. It is located in the Coral Sea, off the northwest coast of Australia. Coral, a scaly, rock-like structure made up of living organisms, has been growing on this part of Earth's surface for 25 million years. Spanning a vast area of around 344,400 sq km (133,000 sq mi), it is a staggering 2300 km (1400 mi) long.

Are there key missions for Earth?

NASA may be best known for its work in space, but in the immediate future some of its most important missions will be those focused on our home planet. NASA Earth Science is a division of the agency dedicated to monitoring our air, land and oceans. Key upcoming projects include the Surface Water and Ocean Topography (SWOT) mission. Due to launch in September 2021, SWOT will undertake the first global survey of the Earth's surface water, to better understand how bodies of water, including polar oceans, change. Another vital mission is TEMPO (Tropospheric Emissions: Monitoring Pollution), wherein an instrument designed to monitor air pollutants across North America will orbit 35,400 km (22,000 mi) above the Earth.

What are the scientists saying?

In many ways, life on Earth faces an uncertain future. The planet's climate has warmed up faster in the past few decades than it had in the two previous millennia. The Earth's 'paleoclimate', evidence stored in tree rings, ocean sediments and polar ice caps, suggests the planet is warming 10 times faster than previously thought, mainly due to the impact of human activities. As a result, sea ice is vanishing, the oceans are getting warmer and countries around the world are experiencing more extreme weather events in both frequency and intensity. A warmer planet isn't the only threat. Accelerating deforestation is also having dire consequences, not least in Brazil's Amazon basin, one of Earth's most biodiverse regions. A recent two-year study found that, although one new plant or animal species is being discovered in the Amazon every other day, the rate of deforestation means that many new species may become extinct before they are found. The new geologic age has been named the Anthropocene to indicate that humans are now the dominant influence.

Mars

Mascot
The god of war.

The Red Planet
The fourth-closest planet to the Sun, at 1.52 astronomical units, Mars is known as the 'Red Planet'. The nickname is hardly new. The ancient Egyptians dubbed it Har Decher, the 'red one', more than four millennia ago. It's all thanks to iron minerals in the Martian soil – when these react with oxygen, they rust, giving the surface a distinctly red appearance. Mars is currently playing host to NASA's InSight lander, which arrived on the surface in November 2018. InSight is the latest in a series of missions trying to determine whether Mars might one day sustain a human population. Current conditions on Mars are far from ideal. Average temperatures are well below freezing; violent storms fill the air with dust, and a thin atmosphere offers little protection against incoming debris. Oh yes, and there's no liquid water. Recent missions to Mars, however, suggest a very different history, including ancient floods and evidence of saline groundwater.

When was it discovered
Though 1877 was an important year for observations of Mars, it was a mixed bag. In August of that year, the American astronomer Asaph Hall made an astonishing discovery – the existence of not one Martian moon but two. He named them Deimos and Phobos, after the sons of Ares, the Greek god of war. Meanwhile, in Italy,

Hall's fellow astronomer Giovanni Schiaparelli noticed a network of channels clearly visible on the Red Planet's surface. When he reported his findings to the wider scientific community though, something got lost in translation: in English, features Schiaparelli referred to as 'canali' became 'canals', leading to suggestions they had been created by alien engineers.

Fascinating fact

The Martian was right: you can grow potatoes on Mars, at least in theory. In fact, the science necessary to cultivate food in space is already up and running. The International Space Station is currently using a light-based 'plant farm', known as 'Veggie', to grow lettuce, along with a series of other crops. Similar technology has proved to be highly effective on Earth. In countries where farmland is in relatively short supply - for example in the Netherlands - crops are grown with LED lighting, producing less heat and CO_2 emissions than conventional methods.

Any features of particular interest?

Once believed to be oceans, the largest of Mars' famous 'dark spots' is Syrtis Major Planum. The name is derived from 'Syrtis maior', the ancient Roman name for the Gulf of Sirte, now Sidra, off the coast of modern-day Libya. In 1659, Syrtis became the first Martian feature to be recorded from Earth, when it was sketched by the Dutch astronomer Christiaan Huygens. It extends north from the planet's equator for 1500 km (930 mi, then west to east for 1000 km (620 mi). The darkness of the spot is due to the presence of basalt. It is made more visible by the fact that, by Martian standards, the air is relatively free of dust. Syrtis Major Planum appeared on an early map of the planet, created when Mars made its close approach to Earth in 1877; more recently, due to its clear conditions, it has been identified as a possible landing site for future Mars missions.

Mars in popular culture

For aliens in popular culture, 'Martians' is the usual shorthand. Somehow, 'Venusians' or 'Neptunians' don't pack quite the same punch. For a time, Martians seemed to occupy a very real place in the human psyche. The radio play of *The War of the Worlds*, broadcast in 1938, reportedly had some Americans believe they were listening to news. It's hard to see them responding in panic to marauding 'Plutonians'. Later, in the febrile political climate of the 1950s, Martians were the perfect big-screen baddies: ruthless, relentless, hell-bent on taking over the world.

Has there been a key mission?

The InSight mission to Mars aims to do something never before attempted: a complete health-check on another planet. Since landing in 2018, InSight has begun, through a suite of instruments, to conduct experiments on Mars' interior. These will effectively take the planet's temperature, as well as check its pulse and reflexes. The research should yield vital information about the formation of the Red Planet, in turn providing clues to the makeup of similar terrestrial planets in the inner solar system, namely Mercury, Venus and Earth. The data returned by Insight may also tell us something of the origins of exoplanets, those planets orbiting stars outside of our solar system.

What are the scientists saying?

NASA's Orion Multi-Purpose Crew Vehicle (MPCV), currently being tested, is meant to bring astronauts to Mars for the first ever time. It was initially designed to bring humans to the International Space Station. In 2018, NASA announced plans for the Lunar Orbital Platform-Gateway, a new space station that will orbit the Moon. Known as 'Gateway', it is intended as a base from which to launch Orion into deep space, and on missions to Mars in particular. Orion will look broadly similar to the Apollo spacecraft, but will benefit from significantly updated technology. The habitable part of the craft has room for six astronauts and, if all goes to plan, components for Gateway will hopefully be launched in the early 2020s. Crewed deep-space missions aboard Orion would then commence in the latter part of that decade.

Jupiter

Mascot

King of the gods.

The big beast

The 'gas giant' Jupiter is the biggest planet in our solar system. Eleven times the size of Earth, its surface is covered by distinctive striped clouds, first noted by Galileo in the 17th century. Scientists have subsequently discovered that these markings, comprised of water and ammonia, extends into deep layers. Jupiter's atmosphere is rich in hydrogen – a factor which, in other galaxies, combined with the presence of helium, might have led to the birth of a star. Conditions on Jupiter's surface, however, are more likely to feel like the end of a world rather than the beginning. Of the vast storms that cover much of the planet, the most well-known is the Great Red Spot, believed to have started 300 years ago.

When was it discovered?

Sightings of Jupiter can be traced back to the 8th century BC, when the planet was first recorded, by Babylonian astronomers. Named for the ancient Roman king of the gods, and father of Mars, the first detailed observations of Jupiter by telescope were made in 1610 by Galileo. He was the first to identify Jupiter's four largest moons, two of which outstrip Mercury in size.

Fascinating fact

A consequence of Jupiter's vast size is its extreme gravity, which is so powerful that it's causing the planet to contract. As it does so, Jupiter's internal matter is compressed, generating both enormous friction and vast amounts of heat. In fact, Jupiter gives out more heat than it absorbs from the Sun.

Any features of particular interest?

Io, the third largest of Jupiter's known 79 moons, is the most volcanically active celestial body in the universe. Io is home to hundreds of volcanoes (and those are just the ones we know of), each releasing lava tens of kilometres into the planet's atmosphere. Scientist believe the lava is primarily molten sulphur and silicates, and that Io's thin atmosphere is mainly sulphur dioxide. Io may also have an iron core large enough to give the moon its own magnetic field. In fact, the orbit of Io bisects Jupiter's magnetic field (which is 20 times stronger than Earth's), generating electrical currents of 400,000 volts.

Are there moons or rings?

Not only does Jupiter boast the four immense Galilean moons, it has (at least) 75 others in orbit. In 1979 it surprised scientists when Voyager 1 discovered that Jupiter also had a ring system, the third to be discovered in the solar system. As dust rings they're quite faint and hard to see, however.

Could it support life?

Gas giant Jupiter, with its magnetosphere and lack of a proper crust, is itself not a candidate for life. Its moons are another story. With possible oceans and a stable orbit, these are considered prime hunting grounds for life elsewhere in the solar system. The moons Ganymede, Europa and Callisto have a faint atmosphere as well.

Jupiter in popular culture

The planet's sheer size, not to mention Jupiter's numerous moons, have inspired countless writers and artists. Io, in 2019, spawned its own eponymous Netflix feature film, in which it features as the destination for humans abandoning a hideously polluted Earth. The planet's mainland was the location for *Jupiter Ascending*, a 2015 Wachowski epic and an example of the genre sci-fi buffs call 'space opera'. Author David Mitchell sent characters to other moons in *Cloud Atlas*, his 2004 novel, which made the Booker Prize shortlist and was later adapted into a movie, also by the Wachowskis.

Has there been a key mission?

Jupiter may have been observed for centuries but it is only recently that we have begun to develop a more detailed understanding of conditions there. In 2000, the Saturn probe Cassini was able to create true-colour images of the planet, although, from almost 10 million km (6.2 million mi) away, it was only able to produce composite mosaics. Bear in mind Jupiter is 587 million km (365 million mi) from Earth at its closest point of approach. In 2016, the NASA spacecraft Juno got more up-close-and-personal. Launched in 2011, Juno's ongoing orbit of the gas giant continues to send back detailed data on all aspects of the planet's composition and features.

What are the scientists saying?

We all love a good Jovian moon; after all, there are so many to choose from: seventy-nine and counting, more than any other planet in the solar system. Twelve new moons were discovered in 2018, by scientists looking for very distant objects. Specifically, a planet some think lies far beyond Pluto, the so-called 'Planet X'. Having further researched Jupiter's intriguing (and amazingly numerous) moons, scientists believe some may offer the conditions required to support life. Top of the list is Europa, which may be home to a vast ocean hidden beneath the moon's surface. NASA's upcoming Europa Clipper mission-in-planning intends to search for subsurface lakes on the moon once it has launched, sometime in the 2020s, to settle the question of possible oceans once and for all.

Saturn

Mascot
God of the harvest.

Lord of the rings
Saturn's most prominent features are its seven rings. Composed of ice and dust, they can be seen from Earth with the aid of a telescope. Less obvious is the fact that, while Saturn's rings are thousands of kilometres long, they are only around 10m (30 ft) thick on average. Like Jupiter, Saturn is a 'gas giant'. So, while it has a rocky core made up of metals such as iron and nickel, it has no true surface. Instead, Saturn is covered with swirling gases, primarily hydrogen and helium. To human eyes, Saturn's 'surface' appears to be striped, an effect caused by its extreme winds and storms.

When was it discovered?
The oldest records documenting observations of Saturn are those of Assyrian astronomers in the ancient Middle East, dating from around 700 BC. They described a ringed planet, which they named the 'Star of Ninib', in honour of an Assyrian solar deity. Ptolemy also included the visible planet in his writings on astronomy circa about 150 AD. Galileo Galilei observed the planet's rings in 1610, the first to see them, but was unable to theorise their true origin. In 1659 the astronomer Christiaan Huygens was finally

able to determine that the mysterious objects were in fact a ring system.

Fascinating fact

Saturn is home to a weather feature unique in the solar system. Located over Saturn's northern pole is a six-sided, hexagonal jetstream. Observations made first by the Voyager I spacecraft and later by Cassini have helped scientists to record its size; measuring 32,000 km (20,000 mi) across, this hexagonal jetstream has a powerful storm at its centre, and can reach speeds of 320 km/h (200mph).

Any features of particular interest?

Saturn's moons are themselves fascinating. Mimas, one of the planet's lesser moons, has a diameter of just 396 km (240 mi); until the arrival of the Voyager probes, in the 1980s, it appeared as no more than a small dot on the moon's surface. Mimas is best known for its uncanny resemblance to the Death Star, scourge of the Rebel Alliance in the *Star Wars* movies. The surface of Mimas shows evidence of numerous impact strikes. The largest, Herschel Crater, is roughly one-third of the diameter of Mimas.

Are there moons or rings?

Saturn is the gold standard of ringed planets, with four main rings and three ring groups that are fainter and narrower. Some of the rings contain particles that are as large as a house, but all of them are relatively narrow. What they lack in height they make up for in width; from the planet to the edge of the rings is 80,000 km (50,000 mi). Saturn also has 62 known moons, from bitsy moonlets to giant Titan, which is larger than the planet Mercury. The rings and moons interact with each other, as the moons draw from the dust and particles of the rings to accrete surface material and grow in size. Additionally, the E ring is formed by icy debris ejected by the moon Enceladus.

Could it support life?

Saturn is too windy and high-pressure to host life itself, but its moons may be another story. If Enceladus has a liquid subsurface ocean it might be a candidate.

Saturn in popular culture

Saturn's appearances in popular culture are extremely varied. In *The Rings of Saturn*, by the late German author W.G. Sebald, the planet keeps watch as the author contemplates the nature of time and memory, walking around the English county of Suffolk. Writers such as Arthur C. Clarke and Isaac Asimov have set stories on Saturn's moons. In classical music, Gustav Holst's monumental composition *The Planets* features a movement for every planet besides Earth. Of the seven, the Saturn movement is regarded to have been the composer's favourite.

Has there been a key mission?

It was the arrival of the NASA Cassini mission, in 2004, that began to reveal the secrets of Saturn's moons. The mission was named in honour of the 17th century Italian astronomer Giovanni Cassini; from its arrival in orbit until the mission's end in 2017, the Cassini spacecraft orbited the great gas giant close to 300 times, locating a number of moons where the presence of water might be capable of sustaining life.

What are the scientists saying?

The particularly dark moon Phoebe could open a window on the solar system's past. It is an example of a captured object – a smaller celestial body trapped by a larger planet's gravity, much like the Martian moon Phobos. The dark material Phoebe appears to be made of is common in the outer reaches of the solar system. This suggests Phoebe dates back to the formation of the system itself and that, due to its peripheral position, Phoebe was not pulled into the gravity that created the various planets. In fact, if Phoebe was not subjected to the heating process that took place during planetary formations, its chemical composition might have remained unchanged for billions of years. This means Phoebe could contain evidence about the creation of the Milky Way. Titan is also of great interest, as the only known moon with a robust atmosphere. It is the second-largest satellite in the solar system, after Jupiter's Ganymede.

Uranus

Mascot

The father of the Titans.

What's in a name?

An 'ice giant' four times the size of Earth, Uranus is the third-largest planet in our solar system. Named for the Roman god of the sky and heavens, scientists speculate Uranus originally formed closer to the Sun before moving to the outer solar system. Now it is the seventh-furthest planet from the Sun. The planet's distinctive blue-green colour is caused by methane clouds moving across Uranus' surface, where temperatures bottom out at a perishing -200°C (-328°F). Uranus' atmosphere is composed mainly of helium and hydrogen, similarly to Saturn and Jupiter but with a thinner layer. The planet makes an orbit of the Sun once every 84 Earth years. Its orbit ranges from 2.5 billion km (1.7 billion mi) to 3 billion km (1.89 billion mi) away from the Sun.

When was it discovered?

The discovery of Uranus is a relatively late entry in the annals of astronomy as far as planets go. The first discovered planet not to be recorded by ancient civilisations, Uranus was first observed in 1781, by British astronomer William Herschel. Six years later, Herschel also discovered the two largest of Uranus' moons, Oberon and Titania. Under conditions of very good visibility, it can be observed with the naked eye if one knows just where to look.

Fascinating fact

Despite the many decades scientists have been studying Uranus, the composition of the planet's striking blue cloud cover only came to light as recently as 2017. In large planets that orbit closer to the Sun, clouds have a very high concentration of ammonia. Uranus, in contrast, has clouds that contain hydrogen sulphide, a highly toxic gas.

Any features of particular interest?

Like Venus, Uranus rotates from east to west – that's the opposite direction to most other planets. Where Uranus is especially unique is that it turns on a 90-degree angle to its orbit, effectively rotating on its side like a ball. One theory is that, billions of years ago, Uranus was struck by a large object – possibly as big as the Earth – that pushed the planet into an extreme tilt, one that today means Uranus exhibits some interesting seasonal variations. At Uranus' northern pole, winter brings 21 years of darkness, while summer means 21 years of daylight. Spring and fall are even more dramatic, with 42 years of light and dark respectively.

Are there moons or rings?

Uranus has five major moons, all named after characters in Shakespeare. All told it is known to have 27 moons, most of which were only discovered in the space age. Some of them may be remnants from Uranus' possible collision, knocked off on impact and captured in orbit around the planet. It also has its own, relatively young, ring system, 13 in total. They are all very faint and dark, composed of a mix of fine dust and large particles. Only discovered in 1977, they are thought to arise from collisions. The dust bands have brief lifespans of 100 to 1000 years unless they are renewed with new material. Some of the moons may act as 'shepherds' for these rings.

Could it support life?

Ice giant Uranus, which receives only a faint amount of sunlight and has radically long seasons thanks to its sideways rotation. Wind speeds would also be an issue; blowing in the reverse direction of the planet's rotation, they go up to 900 km/h (560mph). All told Uranus is not on anyone's list for hosting possible life forms. Its large moons may have some potential, though it will be a long while yet before they can be properly evaluated.

Uranus in popular culture

Uranus features prominently in the *Captain Underpants* series of children's books, making appearances in such classics as *Captain Underpants and the Attack of the Talking Toilets*, and *Captain Underpants and the Perilous Plot of Professor Poopypants*. Uranus also appears in various Marvel comics series, and is a regular destination for Earth's favourite Time Lord, Doctor Who. In music, the 1967 Pink Floyd track 'Astronomy Domine' contains a reference to the Uranian moon of Titania. While in Gustav Holst's *The Planets*, the composer subtitled Uranus' movement 'The Magician'.

Has there been a key mission?

At 2.6 billion km (1.6 billion mi) from Earth, Uranus' remote location means no spacecraft has conducted extensive orbital operations there. NASA's Voyager 2, launched on 20 August 1977, is the only spacecraft to have made a flyby of Uranus. After a total journey time of over nine years, on 24 January 1986, Voyager 2 came within 80,000 km (50,000 mi) of the great ice giant. In a window of just six hours, the spacecraft was able to gather the first data on the planet, including close-up images of its rings and moons.

What are the scientists saying?

Given Uranus' distance from Earth, studying its 'mini moons' – which in some cases are no more than 12 km (8 mi) across – is a serious challenge. Nonetheless, scientists at the University of Idaho think they may have discovered two more tiny moons, orbiting somewhere near the planet's outer rings.

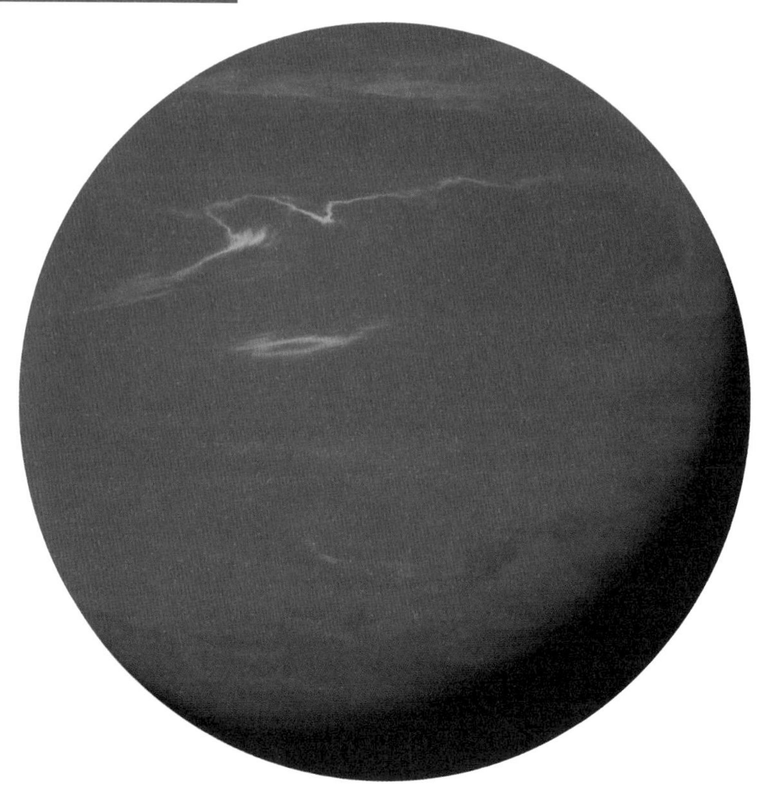

Neptune

Mascot

The god of the sea.

The Bluest Planet

At 4.3 billion km (2.7 billion mi) from Earth at its closest point, Neptune is the only planet in the solar system not visible with the naked eye. And, as the furthest planet from the Sun, it's hardly surprising the great ice giant is a bleak wilderness. In fact, Neptune is 30 times further from the Sun than Earth is; by the time solar energy reaches Neptune, it is 900 times weaker than the strength of solar energy on Earth. On Neptune, it never gets truly light. Or, for that matter, warm, with average temperatures of around -200°C (-392°F). Only its pole can get warmer, high enough that methane gas here isn't frozen and can leak out of the deep atmosphere. With a year on Neptune lasting 165 Earth years, the 40-year long summers and winters allow for a moderate ten degrees warming at the poles, as the north or south pole receive greater sunlight over these stretches. On the plus side, Neptune is extremely pretty, its surface a dazzling blue that would put the bluest of Earth's seas to shame.

When was it discovered?

Neptune was discovered by the German astronomer Johann Gottfried Galle, in 1846. Galle was working on the theory of a French counterpart, a mathematician named Urbain Joseph Le Verrier. The two men believed an as-yet-unseen planet was influencing the orbit of nearby Uranus, and were able to accurately predict Neptune's existence and location based on their scientific calculations alone.

Fascinating fact

The cold and dark are likely to prove challenging for any visitor to Neptune, but new arrivals may well get blown off the face of the planet before they get the chance to worry about either. Wind speeds on Neptune are as much as four times stronger than the strongest experienced on Earth, with Neptune's icy blasts reaching astonishing speeds of 1930 km/h (1200mph).

Any features of particular interest?

The Neptunian moon of Triton is the coolest of operators. Containing 99.5% of the mass orbiting Neptune, it is named after the son of Poseidon. During a flyby in 1989, Voyager 2 recorded surface temperatures on this frozen satellite that were even cooler than those on Neptune. That makes Triton one of the coldest objects in the solar system. So cold, in fact, that almost the entire moon is covered in ice, formed from condensed nitrogen. Triton's ice reflects three-quarters of incoming sunlight, yet its extreme distance from the Sun means it appears relatively dull.

Are there moons or rings?

Neptune has 14 moons, of which only two were discovered by ground telescope. Triton was found just 17 days after Galle sighted the planet for the first time, but the next to be found, Nereid, wasn't seen until 1949, and the rest were discovered by Voyager 2 or Hubble. Neptune is also the fourth planet found to have rings, discovered in 1984. There are five dark rings, mostly faint and dusty, more similar in make-up to the rings of Jupiter than those of Saturn.

Could it support life?

While having no solid surface and frigid, windy 'weather' make Neptune a likely no-go, its moon Triton would possibly be an interesting site for a space colony.

Neptune in popular culture

The planet serves as the backdrop for the 1997 sci-fi horror film *Event Horizon*, in which a pre-*Matrix* Laurence Fishburne is dispatched to investigate a spacecraft that has disappeared in orbit around Neptune. In the H.G. Wells short story 'The Star' (1897), Neptune is destroyed following an interplanetary collision. Meanwhile, in the 2001 *Star Trek* TV reboot, there are spacecraft capable of flying from Earth to Neptune and back in six minutes.

Has there been a key mission?

Nearly 150 years after Neptune was first observed (from an observatory in Berlin), NASA's Voyager 2 got the opportunity, in 1989, to study Neptune more closely. To date, it is the only probe to undertake a near-reconnaissance of Neptune, and Voyager 2 was able to shed new light on Neptune's composition and moons. Since Voyager 2's visit, scientists have had to rely on the Hubble Space Telescope to monitor the gas giant. It's simply too far to target for orbiters at this time.

What are the scientists saying?

Neptune remains distant and mysterious. Its volatile weather suggests it is likely to experience violent and long-lasting storms, much like the approximately 300-year-old Great Red Spot on Jupiter. A dark area in the southern part of Neptune, located by Voyager 2 and roughly the size of Earth, has been christened the Great Dark Spot. Subsequent attempts to locate it once more have proved inconclusive. But Hubble has since discovered another spot, this time in the north. Towards the edge of the spot, methane pushed high into the atmosphere appears to have cooled to form ice-crystal clouds. A zone of clear gas towards the centre of the spot could be a window to different cloud formations closer to Neptune's surface. There have now been six great storms observed on Neptune, some in the very process of formation. They seem to roam widely before dissipating.

Manned Space Flight

The Quest for Orbit

Exploring outer space has been a dream of humankind for almost as long as we have gazed up at the stars – but the act of actually putting astronauts into space has taken centuries of research, planning, development and scientific innovation. The race to put a man into space wasn't a strictly scientific endeavour during the 1950s and 1960; as nuclear tensions developed down here on Earth between the USA and the USSR, the struggle to control space became a crucial part of the escalating arms and information race.

Following the end of the Second World War, both the USSR and the USA poured huge amounts of money and expertise into their respective space programmes, building on the ballistic missile technology that had been developed in the wake of the war to deliver intercontinental nuclear weapons. Rather ironically, these apocalyptic weapons of war also provided the means for what must surely be man's greatest peacetime adventure: putting humans into space.

Initially, American efforts focused on developing jet aeroplanes that were capable of flying into orbit, but ultimately the technological (and physical) challenges involved proved too great. Subsequently, both the USA and USSR settled on a

Soviet Union postcard issued for the anniversary of the space flight of Yuri Gagarin.

Sputnik I on display.

A scale reproduction of the Vostok 1 rocket used for Yuri Gagarin's first space flight.

much simpler design: placing a small capsule on top of an extremely big rocket and launching it into orbit. The huge rockets would provide the necessary thrust to escape the Earth's gravitational pull and would separate from the capsule after launch; once in orbit, the capsule could circle the Earth, before retro-rockets slowed it down enough to re-enter the atmosphere with its pilot still alive and well. At least, that was the theory, later to be tested.

It wasn't until 1957 that the first nation finally achieved orbit. Though the US had a head start thanks to their advanced atomic programme, it was the Soviets who got there first. On 4 October 1957, the USSR sent the first man-made satellite, Sputnik 1, into space, where it orbited for nearly three months before eventually plunging back to Earth. A second satellite, Sputnik 2, was launched in November of the same year, carrying scientific instruments and a rather unfortunate dog by the name of Laika – the first living creature sent into space (and, sadly, also the first to perish there). The US wasn't too far behind, however; their first satellite, Explorer 1, went into orbit on 31 January 1958. It was followed soon afterwards by a third Soviet Sputnik, launched on 15 May 1958.

But the race was still on to be the first nation to put a man into space (and, just as importantly, bring him back alive). Three years later, on 12 April 1961, the Soviets again claimed the prize: with the launch of Vostok 1, Lieutenant Yuri Gagarin officially became the first human sent into space, orbiting for a total of 89 minutes and reaching an altitude of 327 km (203 mi). The psychological impact of these two Soviet space firsts was huge, especially with the Cold War still raging back on Earth.

In fact, it was a pretty close race: the Soviets only beat the Americans by a matter of weeks. On 5 May 1961, Alan Shepard – aboard Freedom 7 – finally became the first American in space.

Race to the Moon

Perturbed by their inability to stay ahead of the Soviet space programme, the Americans decided to take a quantum leap. On 25 May 1961, a mere 20 days after the first US manned spaceflight, President John F. Kennedy proclaimed to a stunned nation his intention to land a man on the moon by the end of the decade. It was a bold, almost impossibly ambitious aim, intended to give hope and courage to a worried nation. Whether it could be achieved was another question entirely.

To accomplish this goal, NASA established its landmark Apollo programme. The plan was to use small teams of three astronauts, who would be sent into orbit in small command modules installed on top of massive Saturn rockets (based on designs initially developed by the German rocket scientist Wernher von Braun). The Apollo missions would also be supported by a parallel programme, Project Gemini, whose two-man teams helped develop spaceflight capabilities that were ultimately translated over into the Moon mission.

US hopes were boosted by another ground-breaking flight that followed on 20 February 1962, when John Glenn – in the Friendship 7 – became the first American to orbit Earth. But the Soviets weren't done yet. In March 1965, the USSR achieved yet another astonishing first when Soviet cosmonaut Alexei Leonov made the first spacewalk

From left, astronauts Neil Armstrong, Michael Collins,and Edwin (Buzz) Aldrin.

in history. Leonov spent a total of 10 minutes outside the Voskhod 2 capsule. The USSR beat the USA to this milestone by almost three months, when at last NASA astronaut Ed White took a spacewalk from Gemini 4. Almost two decades later, cosmonaut Svetlana Savitskaya became the first woman to perform a spacewalk in 1984.

In response, the Apollo teams redoubled their efforts, and the programme moved forward astonishingly fast – but not without costs. The first low-orbit test of the Apollo command and service module was ready and set to launch on 21 February 1967, nearly six years after Kennedy's announcement. But the mission never flew; on 27 January, a catastrophic fire on the Apollo 1 command module during a launch rehearsal killed all three astronauts: Gus Grissom, Ed White and Roger Chaffee. After the tragic setback, the programme pressed on to several successful launches over the next 18 months. On Christmas Eve 1968, Frank Borman, Bill Anders and Jim Lovell became the first men to orbit the moon, aboard Apollo 8; and after two further exploratory Apollo missions, the stage was set for a moon landing sometime in 1969.

On 20 July, the world watched in awe as the three-man crew of Apollo 11 achieved the seemingly impossible. While astronaut Michael Collins remained behind on the command module, Neil Armstrong and Buzz Aldrin piloted the Eagle lander down onto the moon's surface, where they performed a textbook touchdown in an area known as the Sea of Tranquillity. Then, a few minutes after landing, Neil Armstrong made his unforgettable exit from the module, uttering his immortal words: 'That's one small step for man; one giant leap for mankind.'

It was a landmark moment in human history, never to be forgotten by anyone who witnessed it. Humankind had taken our very first steps onto another world.

The Shuttle Programme

The Apollo programme continued until 1972. In total, six moon landings were made, and 12 astronauts walked on the Moon. The total cost of the Apollo initiative was $25.4 billion USD, roughly equivalent to around $150 billion today: a truly enormous sum, more than the entire cost of the Marshall Plan to rebuild Europe in the wake of WWII.

To offset the immense cost of manned space exploration, NASA shifted its focus to developing a reusable space shuttle that could be deployed over many different missions.

The space shuttle, officially called the Space Transportation System (STS), began its flight career with Columbia roaring off NASA's launch pad at Kennedy Space Center in Florida on 12 April 1981.

The orbiter, most commonly referred to as the space shuttle, was the only part of the shuttle 'stack' that made the trek into orbit. Its boosters were jettisoned into the Atlantic Ocean, where they could be retrieved and reused. The external tank was the only part of the stack not used again, as it was designed to burn up on re-entry. Crucially, when the shuttle returned to Earth, it didn't need parachutes as the Apollo capsules had; instead, it was able to glide back to Earth on its own wings, just like an ordinary aeroplane.

Between the first launch on 12 April 1981, and the final landing on 21 July 2011, NASA's space shuttle fleet – Columbia, Challenger, Discovery, Atlantis and Endeavour – flew 135 missions, carrying out ground-breaking space walks; launching, recovering and repairing satellites; conducting cutting-edge scientific research; and building the largest structure in space, the International Space Station. But the programme wasn't without its tragedies: on 28 January 1986, Challenger disintegrated 73 seconds after launch, killing all seven astronauts on board, while on 1 February 2003, Columbia disintegrated during re-entry, killing its entire seven-person crew.

The Shuttle Astronauts

355 people have flown on NASA's space shuttles, representing 16 different countries. A total of 306 men and 49 women have travelled on the shuttles; Story Musgrave is the only astronaut to have flown on all five shuttles, while astronauts Jerry Ross and Franklin Chang-Diaz have flown the most shuttle missions (seven each). The oldest person to travel in space was John Glenn. At the age of 77 in 1998, 36 years after he became the first American in orbit, Glenn flew on shuttle Discovery's STS-95 mission

The space shuttle Atlantis launches from NASA's Kennedy Space Center.

The shuttle programme represented another huge leap forward in space flight, but again, the costs involved were truly eye-watering: each 2010 shuttle mission is estimated to have cost around $775 million. Ultimately, despite its many achievements, it was deemed too expensive to continue. The final space shuttle mission, STS-135, flown by Atlantis, landed on 21 July 2011. In total, NASA's space shuttles travelled 872,906,379 km (542,398,878 mi), making 21,152 Earth orbits. The entire programme is estimated to have cost around $113.7 billion – similar to, or perhaps even more than, the Apollo moon missions. But the scientific advances have been enormous, and many everyday technologies on which we now rely – including space blankets, enriched baby food, artificial limbs, laser eye surgery, digital cameras, solar panels and even handheld vacuum cleaners – can trace their origins back to our quest to explore outer space.

The Present Day

Today, Russia and the US aren't the only countries taking steps into the stars: the European Union, India, Japan and China all have their own advanced space programmes. In 2019, China achieved its own first, by landing its Chang'e-4 spacecraft on the dark side of the moon, a feat never before achieved.

Concurrently, a number of private companies have moved into the space-exploration business, including Richard Branson's Virgin Galactic, which aims to open up space flight to ordinary people, and Elon Musk's SpaceX, which is working on the next big step: a mission to Mars.

A Mars landing has been talked about since the Apollo programme, and while the technological challenges involved are immense, Mars missions are currently in development by several national space agencies and private companies. NASA has stated its intention to land humans on Mars sometime in the 2030s, but no date has yet been fixed in stone.

Elon Musk at a SpaceX event.

Richard Branson of Virgin Galactic launches the SpaceShip Two VSS Unity spaceship.

The International Space Station

The first proposal for a manned space station occurred in 1869, when an American novelist told the story of how a 'Brick Moon' came to orbit Earth to help ships navigate at sea. In 1923, Hermann Oberth was the first to use the term 'space station' to describe a wheel-like facility that would serve as the jumping-off place for human journeys to the moon and Mars. In 1952, Wernher von Braun published his concept of a space station in *Colliers* magazine. He envisioned one with a diameter of 76 m (250 ft), an orbit more than 1600 km (1000 mi) above the Earth, and which would spin to provide artificial gravity through centrifugal force.

The Soviet Union launched the world's first space station, Salyut 1, in 1971 – a decade after launching the first human into space. The United States sent its first space station, the larger Skylab, into orbit in 1973; it hosted three crews before it was abandoned in 1974. Russia continued to focus on long-duration space missions and in 1986 launched the first modules of the Mir space station.

In 1998, the first two modules of the International Space Station (ISS) were launched and joined together in orbit. Other modules soon followed and the first crew arrived in 2000.

That station has been continuously occupied since November 2000. An international crew of three to six people live and work while travelling at a speed of 8 km/s (5 mps), orbiting Earth about every 90 minutes. In 24 hours, the space station thus makes 16 orbits of Earth, travelling through 16 sunrises and sunsets. It can be seen transiting above with the naked eye. Astronauts and cosmonauts have conducted more than 216 spacewalks (and counting!) for space station construction, maintenance and repair since December 1998, performing needed upgrades.

The space station is 109 m (357 ft) end to end – one yard shy of the full length of an American football field including the end zones. The living and working space in the station is larger than a six-bedroom house (and has six sleeping quarters, two bathrooms, a gym and a 360-degree-view bay window). To mitigate the loss of muscle and bone mass in the human body due to microgravity, the astronauts work out at least two hours a day. Astronaut Peggy Whitson set the record for spending the most total time living and working in space, an incredible 665 days, on 2 September 2017.

Astronauts Anne McClain and Serena Auñón-Chancellor work aboard the ISS.

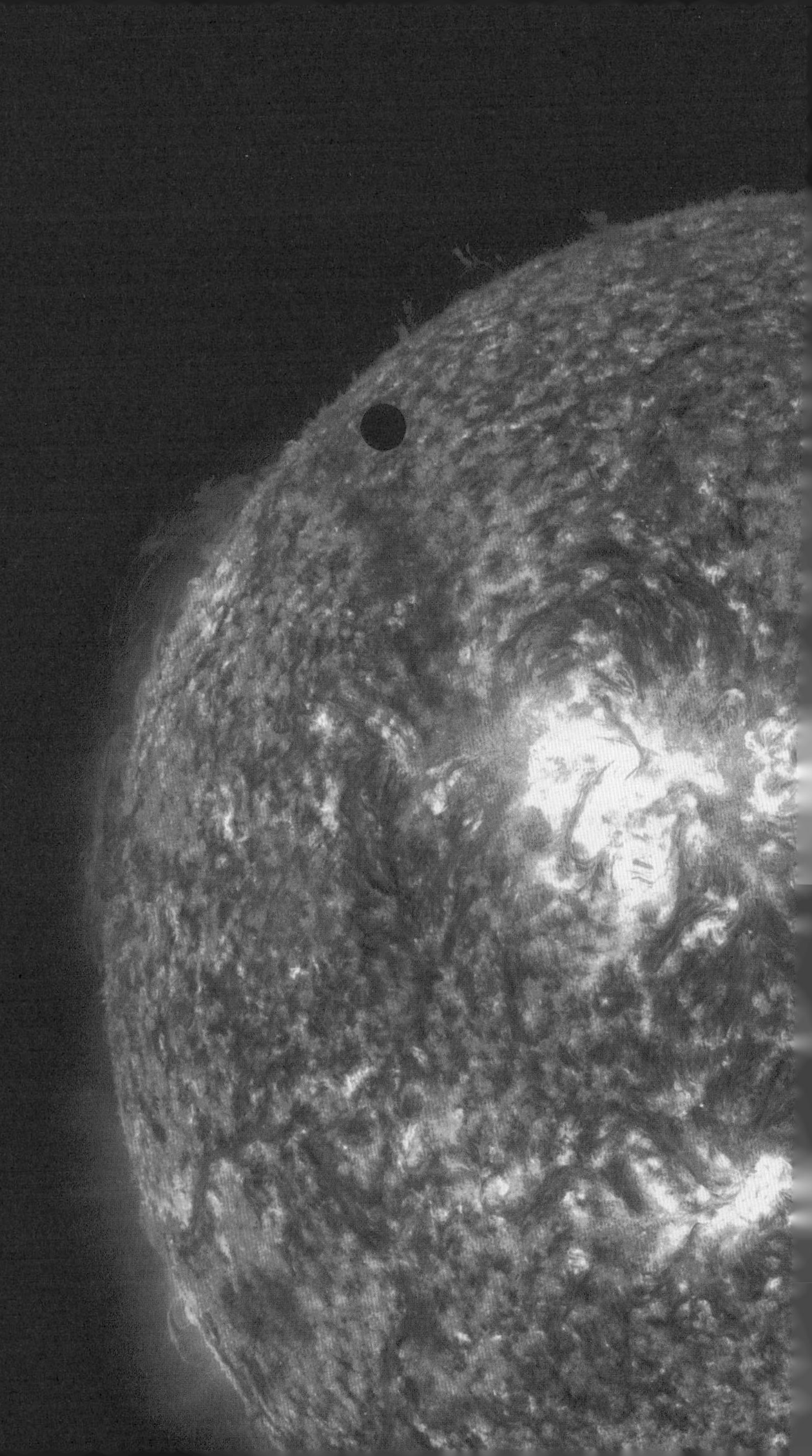

SUN

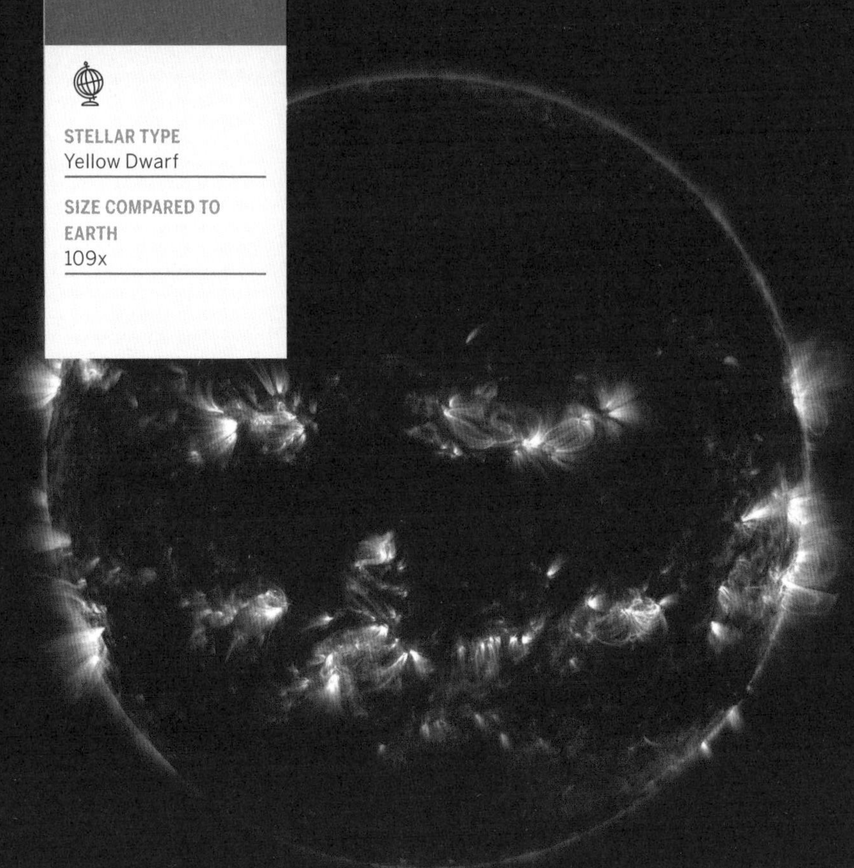

An image of the Sun's corona.

The Sun at a Glance

The Sun is a 'yellow dwarf' star, a hot ball of glowing gases at the heart of our solar system, and our own life source.

Its gravity holds the solar system together, keeping everything, from the biggest planets to the smallest debris particles, in its orbit. Electric currents in the Sun generate a magnetic field that is carried through the solar system by the solar wind, a stream of electrically charged gas ejected outward from the Sun in all directions. The connection and interactions between the Sun and Earth drive the seasons, ocean currents, weather, climate, radiation belts and aurorae. Without the Sun's intense energy, there would be no life on Earth.

The Sun is the centre of our solar system and makes up 99.8% of the mass of the entire solar system. While it has no moons or rings, it is orbited by eight planets, at least five dwarf planets and tens of thousands of asteroids, not forgetting three trillion comets, all in thrall to our Sun's gravitational pull. Meanwhile, the Sun might seem special to us here on Earth, but there are billions of stars just like it scattered across the Milky Way.

In November 2018, NASA's Parker Solar Probe made its first close pass by the Sun – coming within a record-breaking 24 million km (15 million mi) of the surface of the star. As it accelerates, Parker will repeatedly break the speed record for an unmanned spacecraft,finally hitting a top speed, relative to the Sun, of 690,000 km/h (430,000mph). At the closest point of approach thus far, the spacecraft has encountered temperatures as hot as 2,000,000°C (3,600,000°F). Parker was built to be protected from the intense solar radiation by its heat shield, along with various autonomous systems designed to keep the spacecraft safe without guidance from Earth. These include the automatic retraction of its solar panels to regulate temperature.

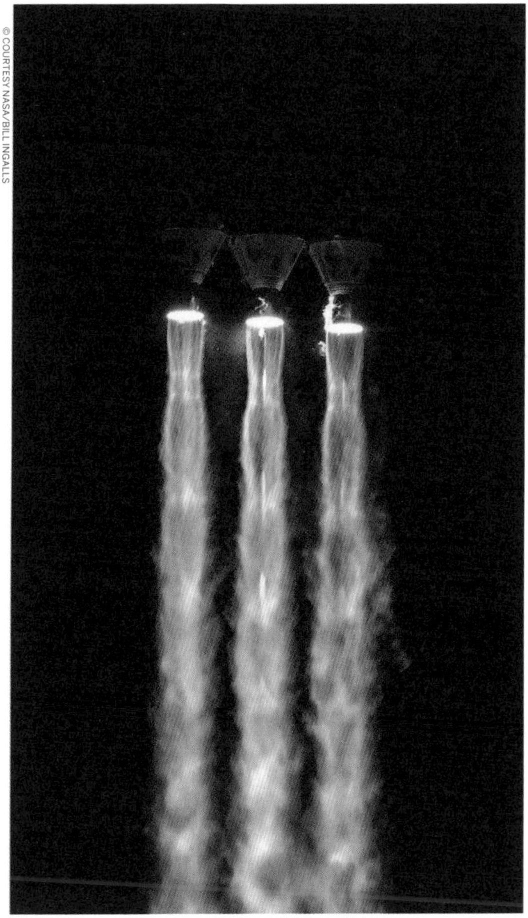

The launch of the Parker solar probe.

AVERAGE DISTANCE FROM EARTH
1 AU

DISTANCE FROM GALACTIC CENTRE
26,000 light years

NEIGHBOURHOOD
Orion Spur

LENGTH OF ONE ROTATION
25 Earth days at equator; 36 Earth days at poles

ATMOSPHERE
Hydrogen, helium

Top Tip

Leave the jewellery at home. The surface of the Sun may be nearly 15 million degrees Celsius (27 million degrees Fahrenheit) cooler than its core, but at about 9941°F (5500°C) that's still hot enough not just to melt diamonds but boil the resulting liquid.

Getting There & Away

Travelling to the Sun by conventional methods would lend new meaning to the term 'long haul'. Flying at its average cruising speed of about 885 km/h (550mph), a modern airliner would take 19 years to reach the Sun's outer reaches. And the cabin might become quite uncomfortably hot, to make a rather significant understatement.

© YULIA MARKOVA/SHUTTERSTOCK

A scale comparison of the Sun and the planets.

Orientation

Our Sun is not an especially large star. Many stars are significantly greater in size, while others positively dwarf it – VY Canis Majoris, the largest known star in our universe, is 2000 times bigger. Yet the Sun packs quite a punch. It is still massive when compared to our home planet; in terms of mass alone, it would require 332,946 Earths to match that of the Sun. The Sun's volume would need 1.3 million Earths to fill it.

The Sun is 150 million km (93 million mi) from Earth, the measurement that determines the distance of one astronomical unit (AU). Its nearest stellar neighbour is Alpha Centauri, the triple-star system. Of those neighbours, Proxima Centauri is 4.24 light years away, and Alpha Centauri A and B (two stars orbiting each other) are 4.37 light years away. A light year is the distance light travels in one year, which is the equivalent of 9.3 trillion km (5.8 trillion mi).

The Sun, and everything that orbits it, is located in the Milky Way galaxy. More specifically, our Sun is in a 'spiral arm', called the Orion Spur, that extends outward from Sagittarius A* (pronounced 'A-star'), the location acknowledged as the Milky Way's galactic centre. The Sun orbits this centre, bringing the planets, asteroids, comets and other objects along with it. Our solar system is moving at an average velocity of 724,000 km/h (450,000mph), but even at this speed it takes about 230 million years to make one complete orbit around the Milky Way.

The Sun rotates as it orbits the centre of the galaxy. With respect to the plane of the planets' orbits, the spin of the Sun has an axial tilt of 7.25 degrees. Since the Sun is not a solid body, different parts of the Sun rotate at different rates. At its equator, the Sun spins around once about every 25 Earth days, but at its poles the Sun rotates once on its axis every 36 Earth days.

The Sun *vs* Earth

-- Radius --
109x
EARTH

-- Mass --
333,000x
EARTH

-- Volume --
1.3 million x
EARTH

-- Surface gravity --
28x
EARTH

-- Mean temperature --
171x
EARTH

-- Surface area --
11,917x
EARTH

-- Surface pressure --
1/1000x
EARTH

-- Density --
25%
EARTH

-- Orbit velocity --
55x
EARTH

-- Orbit distance --
n/a

The Sun, like other stars, is a ball of gas. In terms of atoms, it is made of 91% hydrogen and 8.9% helium. By mass, the Sun is about 70.6% hydrogen and 27.4% helium.

The Sun's enormous mass is held together by gravitational attraction, producing immense pressure and temperature at its core. The Sun has six regions, three of which are located in its interior; starting from the centre, these are the core, the radiative zone, and the convective zone. The three outer regions, which make up the visible surface, are called the photosphere, the chromosphere and the outermost most of all, the corona.

At the core, the temperature is about 15 million degrees Celsius (27 million degrees Fahrenheit), which is sufficient to sustain thermonuclear fusion. This is a process in which atoms combine to form larger atoms, releasing staggering amounts of energy in the process. Specifically, in the Sun's core, hydrogen atoms fuse to make helium.

The energy produced in the core powers the Sun and produces all the heat and light the Sun emits. Energy from the core is carried outward by radiation, which bounces around the radiative zone, taking about 170,000 years to get from the core to the top of the convective zone. The temperature drops below 2 million degrees Celsius (3.5 million degrees Fahrenheit) in the convective zone, where large bubbles of hot plasma (a soup of ionized atoms) move upwards. The surface of the Sun – the part we can see – is about 5500°C (10,000°F). The temperature varies, just as the Sun's visible surface sometimes has dark sunspots, which are areas of intense magnetic activity that can lead to dramatic solar explosions.

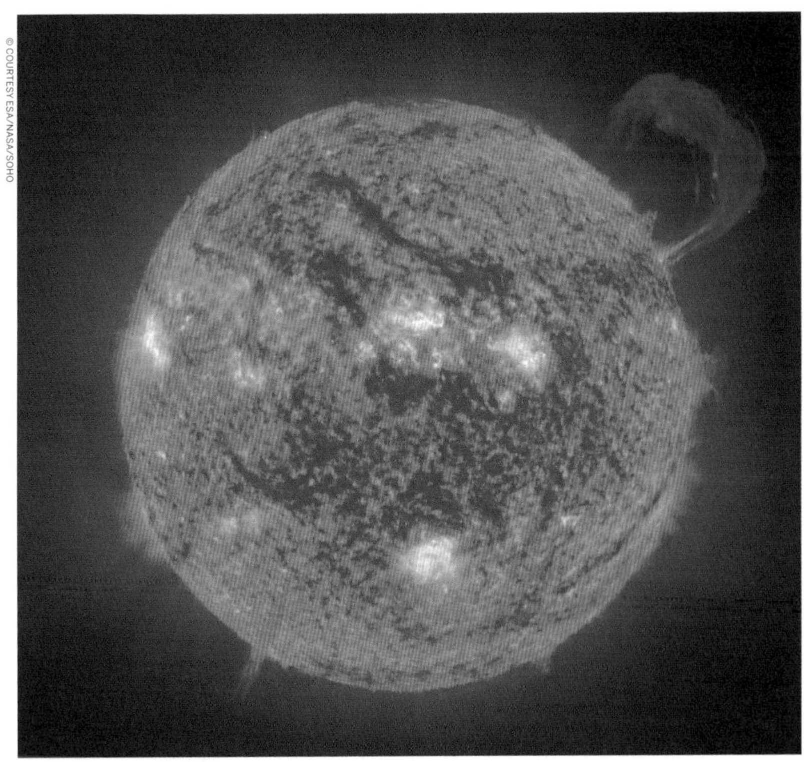

A prominence of relatively cool dense plasma is seen here suspended in the Sun's corona.

Atmosphere

Unlike the inner terrestrial planets, the Sun does not have a solid surface from which an atmosphere might begin. The surface of the Sun, such as it is, makes up the photosphere – a region 483 km (300 mi) deep from which the Sun's radiation escapes outward, becoming hotter as the altitude rises. We see radiation from the photosphere as sunlight when it reaches Earth about eight minutes after it leaves the Sun. Above the photosphere lie the chromosphere and the corona (crown), and together these three regions make up the Sun's relatively thin atmosphere. This is where we see features such as sunspots and solar flares.

During total solar eclipses, when the moon covers the photosphere, the chromosphere looks like a red rim around the Sun, while the corona forms a beautiful white crown with plasma streamers narrowing outward, forming shapes that look like flower petals.

Interestingly, the temperature in the Sun's atmosphere increases with altitude, despite the increased distance from the core. Temperatures in the corona can get as high as a few million K but the source of this coronal heating is a scientific mystery.

The Stuff of Life

While the potential for life on the Sun is impossible, life anywhere else would be impossible without the energy the Sun produces. Sunlight is one of the building blocks of life for countless organisms on Earth, which in turn form the start of many food chains.

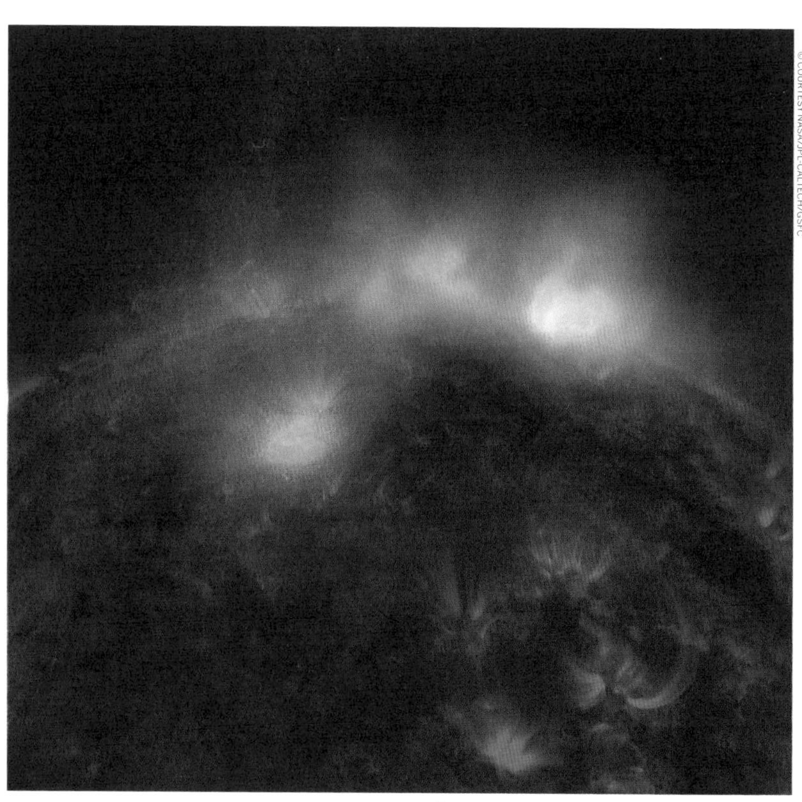

X-rays are seen streaming off the sun in this NuSTAR image.

Heliosphere

The electric currents in the Sun generate a complex magnetic field that extends into space to form the interplanetary magnetic field. The volume of space controlled by the Sun's magnetic field is called the heliosphere.

The Sun's magnetic field is carried out through the solar system by the solar wind, a stream of electrically charged gas blowing outward from the Sun in all directions. Since the Sun rotates, the magnetic field spins out into a large rotating spiral, known as the Parker spiral (it's named after astrophysicist Eugene Parker, who discovered the solar wind).

The Sun doesn't behave the same way all the time. In fact, it goes through phases of its own solar cycle. About every 11 years, the Sun's geographic poles change their magnetic polarity. When this happens, the Sun's photosphere, chromosphere and corona undergo changes, from being relatively calm to violently active. The height of the Sun's activity, known as 'solar maximum', is a time of solar storms; these can be classified, broadly, as sunspots, solar flares and coronal 'mass ejections'. These are caused by irregularities in the Sun's magnetic field and can release huge amounts of energy and particles, some of which reach us here on Earth. This 'space weather' can damage satellites, corrode pipelines and affect power grids.

Take a Pause

The heliosphere is a vast region that extends as far as 17.7 billion km (11 billion mi) from the Sun. It is roughly spherical in shape but can extend outwards in the shape of a comet's tail. The outermost limit of the heliosphere is known as the heliopause.

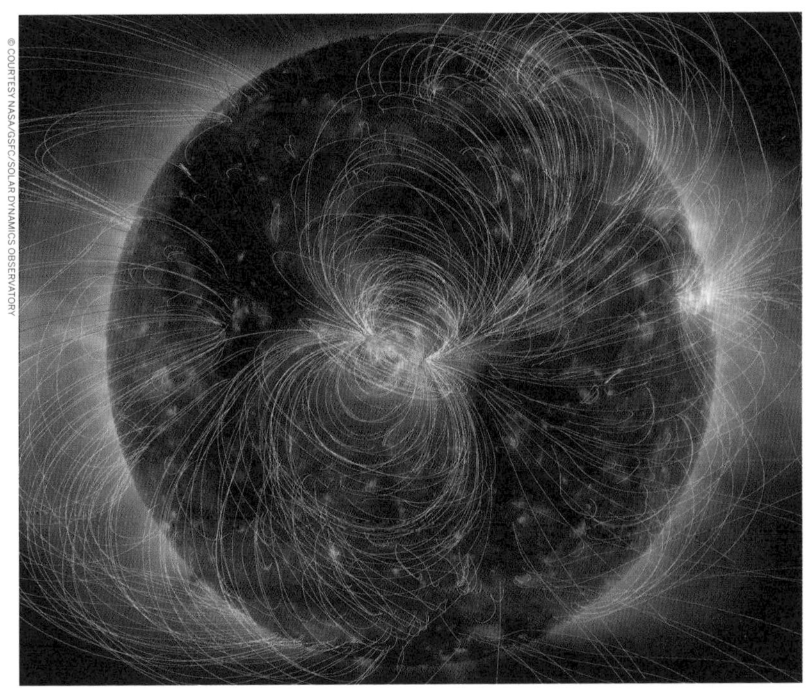

The Sun's heliosphere, as generated by NASA's Solar Dynamics Observatory models.

Solar Flares: A Spotter's Guide

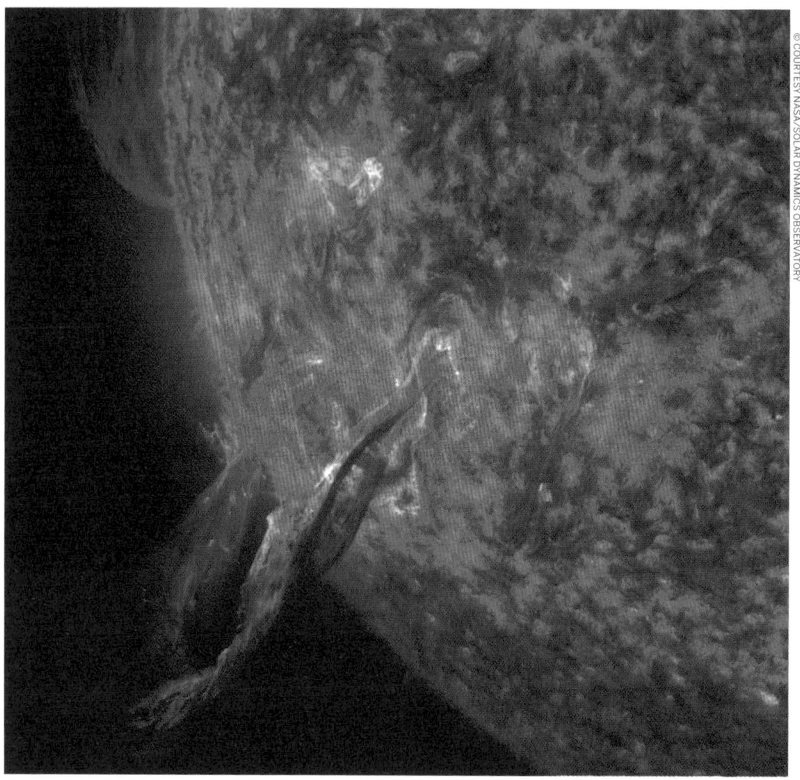

The hyder flare event seen here lasted for three hours on November 1, 2014.

Solar flares are giant explosions on the sun that send energy, light and high-speed particles into space. These flares are often associated with solar-magnetic storms, known as coronal mass ejections (CMEs). While these are the most common solar events, the sun can also emit streams of very fast protons, known as solar energetic particle (SEP) events. Other types of disturbances in the solar wind are known as corotating interaction regions (CIRs). Such solar activity can lead to 'storms' on Earth that can, if strong enough, interfere with shortwave radio communications, GPS signals and power grids.

The National Oceanic and Atmospheric Administration has devised categories for solar flares, an alphabetical classification system similar to the Richter scale for earthquakes. This divides solar flares according to their strength, and each letter – in ascending order, A, B, C, M and X – represents a tenfold increase in energy output. So, an X-class flare is ten times more powerful than an M-class flare and 100 times more powerful than a C. Within each letter classification, there is an additional scale, from 1 to 9.

Classes of Solar Flare

A- and B-class

The smallest flares, creating no problems on Earth and, understandably, the hardest to see.

C-class

While C-class flares are significantly stronger than A- or B-class flares, they are still too weak to greatly affect events on Earth.

M-class

Now we're getting somewhere. M-class flares created by the Sun have been known to cause brief radio blackouts at the poles, as well as minor radiation storms. They might also endanger astronauts.

X-class flares

These are the big guns. X is the most powerful type of flare, though some flares are more than 10 times the power of an X1, so X-class flares can be rated higher than 9.

The most powerful flare measured with modern methods was in 2003, during a solar maximum. It was so powerful that it overloaded the sensors measuring it; the sensors cut out at X15, but the flare was estimated to be as high as an X45 by some. The biggest X-class flares are by far the largest explosions in the solar system and are awesome to watch. As the Sun's magnetic fields cross over each other and reconnect, loops many tens of times the size of Earth leap from the Sun's surface. In the biggest events, this reconnection process can produce as much energy as one billion hydrogen bombs.

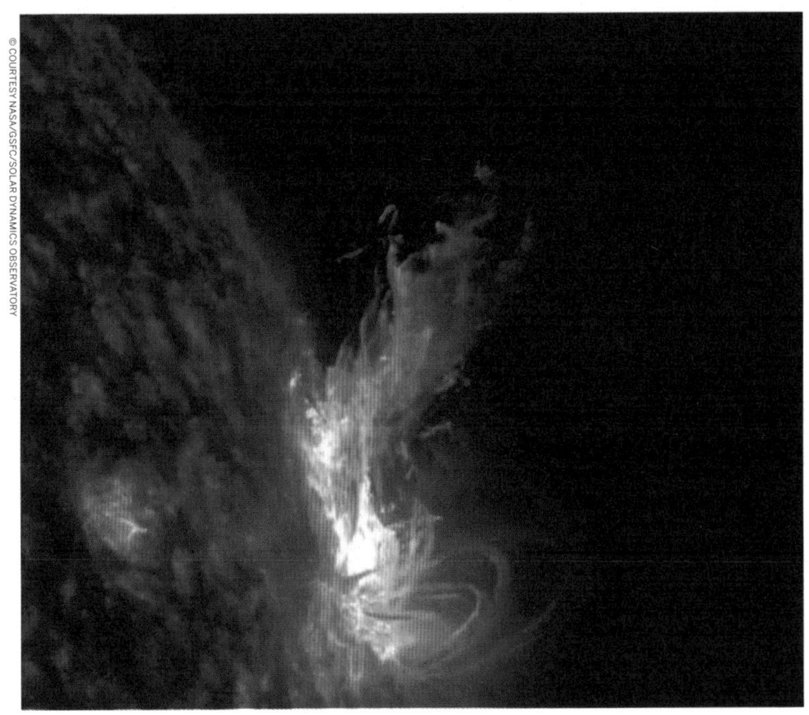

M5-class medium-sized flares in an active region of the Sun.

History

The Sun and the rest of the solar system formed from a giant, rotating cloud of gas and dust called a solar nebula, about 4.5 billion years ago. As this nebula collapsed, due to its overwhelming gravity, it spun faster until it eventually flattened into a disk shape. Most of the material was pulled towards the centre to form our Sun, which accounts for all but 0.2% of the mass of the entire solar system. Scientists predict the Sun is a little less than halfway through its lifetime, and estimates suggest its state has not altered significantly in 4 billion years. Since its formation, the Sun has

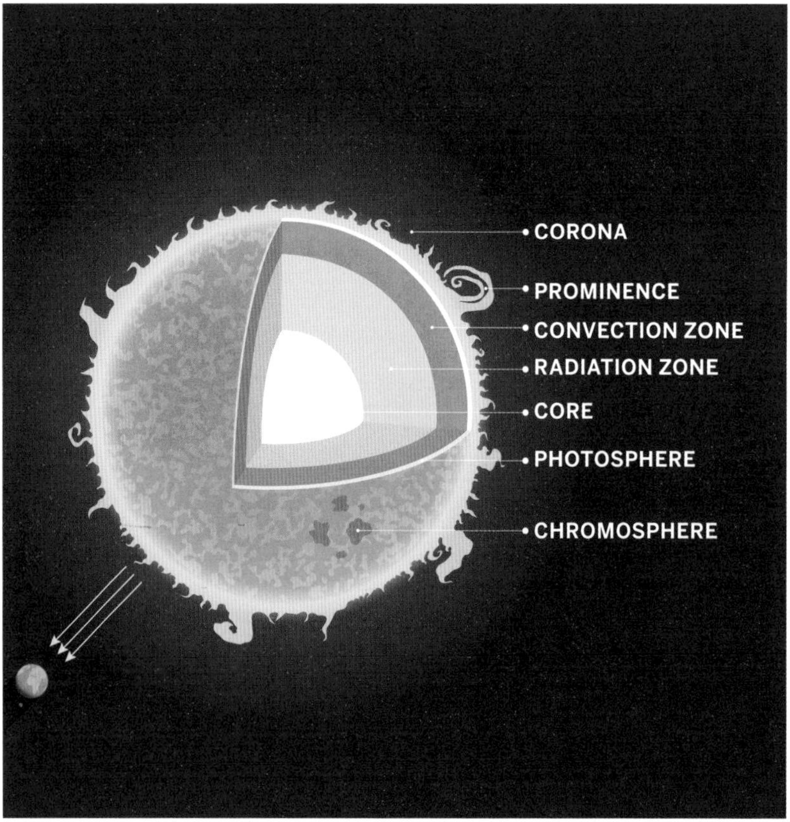

CORONA
PROMINENCE
CONVECTION ZONE
RADIATION ZONE
CORE
PHOTOSPHERE
CHROMOSPHERE

The Sun's layers move from the active corona and chromosphere to its high-pressure core.

acted as a power plant for the solar system, and the energy it continues to produce, though common for a star, is still mind-boggling. Essentially a nuclear reactor, the Sun fuses approximately 600 million tons (544 million metric tons) of hydrogen into helium every second. In that time, 4 million tons (3.6 million metric tons) of matter is converted into the energy that is the source of the Sun's light and heat. The hydrogen fusion taking place at the Sun's core is diminishing, albeit slowly. Scientific models suggest that, over the next 5 billion years, the core of the Sun will see dramatic increases in both temperature and density. The outer layers will continue to expand, turning the Sun into a type of star known as a red giant. This could spell bad news for some of its planetary neighbours, as the Sun engulfs both Mercury and Venus and, potentially, Earth. The Sun's ultimate fate will be to become a white dwarf, a dense, cooling star that while no longer producing energy still provides heat and light from its previous solar activity.

Spotting the Sun

467 BCE

Observations of the Sun, made by the Greek philosopher Anaxagoras, are consistent with the solar features we know today as sunspots.

28 BCE

Sunspots are recorded by astronomers in ancient China.

150 CE

Greek scholar Claudius Ptolemy writes *The Algamest*, a treatise formalising a model of the solar system with Earth at its centre. This view was accepted until the 16th century.

1543

Nicolaus Copernicus publishes *On the Revolutions of the Celestial Spheres*, describing a model of the solar system that is 'heliocentric', with the Sun at its centre.

Polish astronomer Copernicus.

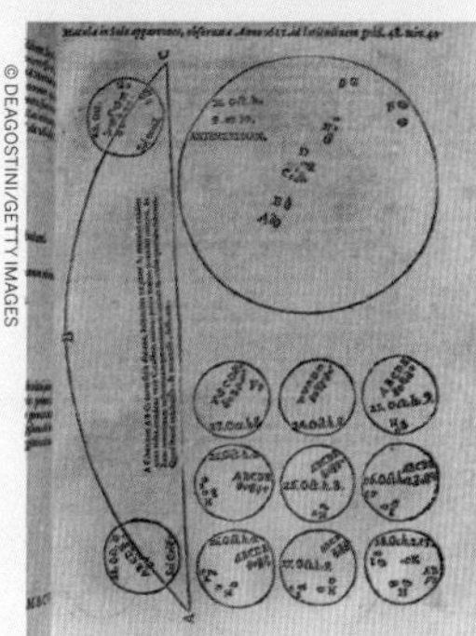

Galileo recorded sunspot observations.

1610

First observations of sunspots, made separately by Galileo Galilei and the English mathematician Thomas Harriott, both through telescopes. (And, in both cases, a terrible idea.) According to the Italian astronomer's 'sunspot letters', held by the British Library, Gallileo had heard about telescopes just the previous year, in 1609. It was the observations of Christoph Scheiner, a Jesuit mathematician, in March 1611 that prompted Galileo to write his famous letters. Scheiner argued, in 1612, that sunspots were in fact satellites of the Sun. Galileo disagreed, and following further observations correctly concluded, in 1613, that sunspots were markings at the surface of the Sun itself.

Galilleo also confirmed that, as the position of these sunspots moved, the Sun must necessarily rotate. Modern astronomers, using more advanced technology, have been able to determine that the cyclical appearance of sunspots over long and short periods is connected with increased levels of magnetic activity.

The Sun in Popular Culture

Humanity's need to understand the Sun lay at the heart of many ancient civilisations. The Mayans, for example, built the Kukulkan pyramid, in El Castillo in Mexico, between 1000 and 1200AD. Seventy-five feet tall, it performs a central role in a dramatic shadow-and-light display during the spring and fall equinoxes. The pyramid's northwest and southwest corners are, respectively, oriented toward the rising point of the Sun for the summer solstice, and its setting point during the winter solstice. The rock formation of Fajada Butte, in New Mexico, projects a dazzling 'sun dagger' when struck by sunlight. Using a spiral petroglyph, a clever piece of rock art, the indigenous Anasazi people were able to track the changing position of the Sun throughout the year. The summer solstice is also the key to Stongehenge, while twice yearly New Yorkers now celebrate Manhattanhenge as the Sun aligns to the city grid.

For the ancient Egyptians, the Sun god, Ra, was the most worshipped of deities and the father of creation. The magnificent temple at Karnak, which is dedicated to Ra, is home to enormous pillars designed to work in harmony with sunlight. More recently, in art and culture, the Sun – or solar activity – has been used as a metaphor for just about every aspect of life. In literature, the novels and poems featuring the Sun as a subject, or in the title, are too numerous to mention; in music it has inspired artists across all genres, from Jim Morrison to composer Edvard Grieg.

© IRAKLI SHAVGULIDZE/SHUTTERSTOCK

Karnak's temple to the sun-god Ra was built for the winter solstice.

In cinema, the Sun's yellow glow has been bolstering the powers of Superman since 1941 (three years after his debut in the DC comic series), while the Sun offers a warming cameo in most films featuring outer space. Less so in Danny Boyle's film *Sunshine* (2007), in which a dying Sun threatens to send the Earth into a state of permanent deep-freeze. To save humanity, a crew of astronauts is sent to reignite the Sun – with a bomb, of course. The Sun is also a staple of advertising agencies the world over, adorning everything from boxes of Californian raisins to the logos of Australian banks (and just about everything else in between.)

© MIHAI ANDRITOIU/GETTY IMAGES

Crowds gather to watch the sunset on Manhattanhenge.

Seasons in the Sun

While references to the Sun abound in popular culture, some works play fast and loose with the Sun's image as the supreme bringer of life; and with scant regard to the complex science of studying the great yellow dwarf star. Here we showcase a few examples, and reflect – oh, yes – on their solar credentials.

Hemingway's evocative title is one of many to utilise the Sun.

The Sun Also Rises – Ernest Hemingway

Technically this is the subtitle of this 1926 novel, at least in Britain, where it was published as *Fiesta* (a nod to the central event, the running of the bulls in Pamplona). The Sun features often, notably during a sweaty bus ride. Other warming elements are provided by the local wines, consumed with pretty much every meal.

Waiting for the Sun – The Doors

The title of the band's third album – best known for tracks such as 'Five to One' and 'The Unknown Soldier' – and also a song of the same name, which featured on the subsequent *Morrison Hotel*. The latter displayed strong solar credentials, its lyrics echoing the wisdom of ancient civilisations in acknowledging the Sun's life-giving powers: 'Can you feel it, now that spring has come / It's time to live in the scattered sun'.

'The Sun Always Shines on TV' – A-ha

A worldwide smash for 1980s Norwegian pop legends, this catchy single followed hot on the heels of their US-Billboard-topping debut, 'Take on Me'. The song's principal theme is the relationship between a young woman and a young man, the slight hiccup being she's human and he's a cartoon. 'Believe me', trills lead singer Morten Harket, 'The Sun always shines on TV'. About as solar as it gets.

Half of a Yellow Sun – Chimamanda Ngozi Adichie

With Nigeria in turmoil during the 1960s Biafran war, *Half of a Yellow Sun* offers a powerful and wide-ranging critique of colonial Africa. Its blend of acute political analysis and painful human drama led, rightly, to much praise. It won the 2007 Orange Prize for Fiction, among a host of other plaudits. The title refers to the Biafran flag.

East of the Sun and West of the Moon – Kay Nielsen

This classic 1914 collection of Norse folk tales with art by Kay Nielsen is considered a highlight of the golden age of illustration. Its title tale, about a prince imprisoned in a castle 'east of the sun and west of the moon' and the young peasant girl on a mission to free him from his wicked stepmother, is a fine example of how folk understandings of celestial objects have filtered into diverse world mythologies.

The half-sun of Biafra's flag.

MERCURY

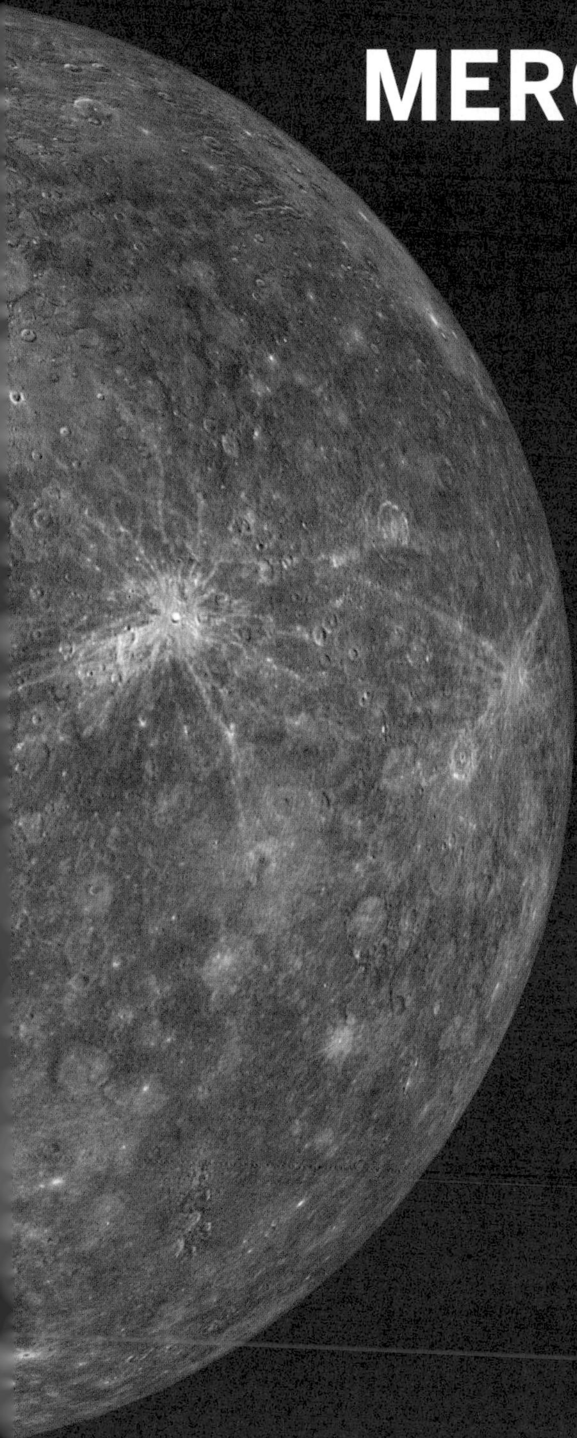

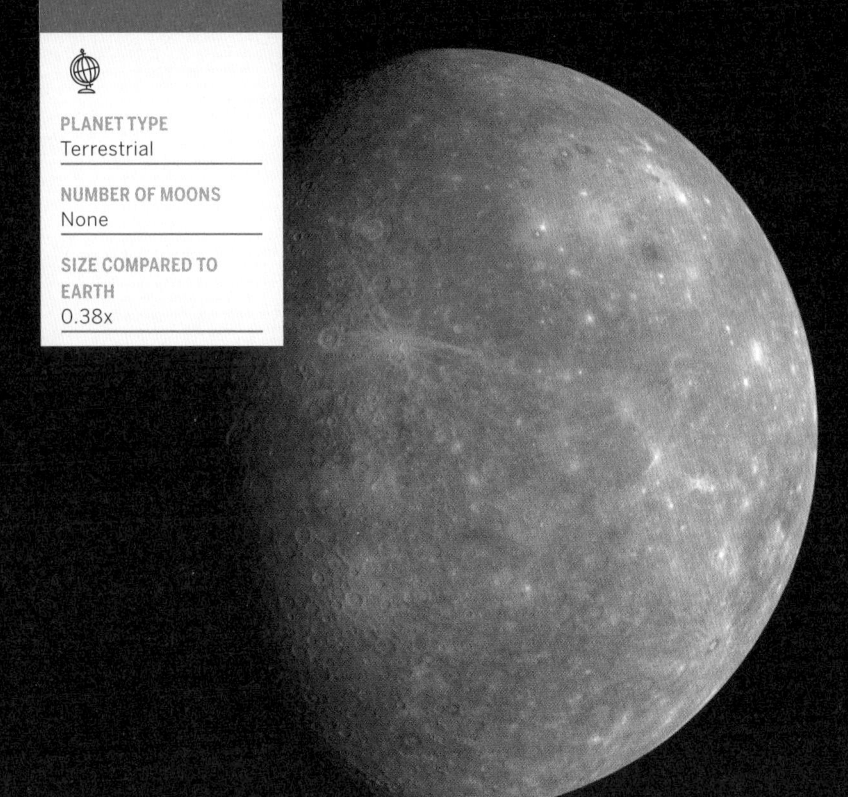

Mercury's surface is visibly pockmarked with craters from meteorite impacts.

Mercury at a Glance

The smallest planet in our solar system and the nearest to the Sun, Mercury is only slightly larger than Earth's moon. The fastest planet as well, Mercury zips around the Sun once every 88 Earth days, travelling through space at a rate of nearly 47 km (29 mi) per second.

By far the smallest of the planets, when it comes to speed, Mercury can take its place on the top tier of the podium. It is thus appropriately named for the swiftest of the ancient Roman gods. With a radius of 2440 km (1516 mi), Mercury is a little more than one-third the width of Earth. At its average distance from the Sun of 58 million km (36 million mi), Mercury is 0.4 astronomical units away from our star (one astronomical unit, abbreviated as AU, is the distance from the Sun to Earth, 150 million km or abut 93 million miles). At this distance, it takes sunlight 3.2 minutes to travel from the Sun to Mercury. From the surface of Mercury, the Sun would appear more than three times as large as it does when viewed from Earth, and the sunlight would be as much as 11 times brighter. Mercury also has a highly eccentric, elliptical orbit. This takes the planet as close as 47 million km (29 million mi) to the Sun, and as distant as 70 million km (43 million mi). That may sound far, but it's less than half the average distance of Earth's orbit. Yet despite its close proximity to the Sun, its thin atmosphere means Mercury is not the hottest planet in our solar system; that would be neighbouring Venus.

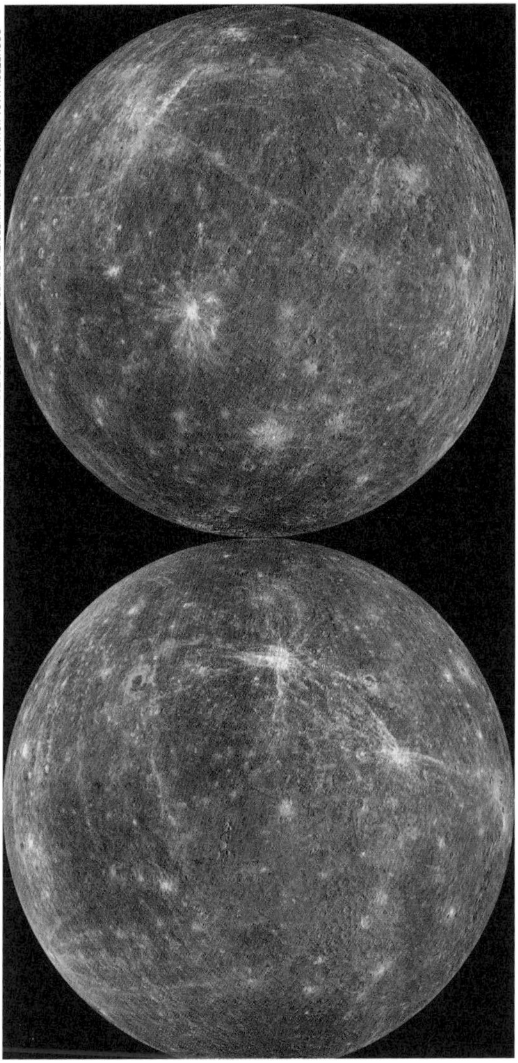

MESSENGER captured this image of Mercury using its GRNS instrument.

DISTANCE FROM SUN
0.4 AU

LIGHT-TIME TO THE SUN
3.2 minutes

LENGTH OF DAY
59 Earth days

LENGTH OF ORBITAL YEAR
88 Earth days

ATMOSPHERE
Trace fluctuating amounts of sodium, hydrogen, helium, potassium and oxygen

Top Tip

Visitors to Mercury needn't worry about packing for the right seasons – it doesn't have them. In terms of the plane of its orbit around the Sun, Mercury's axis of rotation features a tilt of just 2 degrees. That means it spins nearly perfectly upright and so does not experience seasonal changes.

Getting There & Away

A range of options are available for the discerning Mercury-phile. First-class travel aboard New Horizons, the fastest craft ever to leave Earth, albeit bound in the opposite direction, would get you there in 40 days. The budget slingshot option, aboard MESSENGER, including orbital flybys, might take a little longer. As in, 1220 days longer. You'll want to get off-board before the probe runs out of fuel and crashes, too!

© COURTESY NASA/JOHNS HOPKINS UNIVERSITY APPLIED PHYSICS LABORATORY/CARNEGIE INSTITUTION OF WASHINGTON

An artistic rendering of sunrise from the surface of Mercury.

Orientation

After Earth, Mercury is the solar system's second densest planet. It has a large metallic core that's about 85% of the planet's entire radius. There is evidence that this large core is partly molten, or liquid. Mercury's outer shell, comparable to Earth's mantle and crust, is only about 400 km (250 mi) thick. With its lack of a protective atmosphere, Mercury's surface resembles that of Earth's moon, scarred by many impact craters as the result of collisions with meteoroids and asteroids.

Devoid of water and covered by a brittle, high-in-iron crust, most of Mercury's surface appears greyish-brown to the human eye. Bright, visible streaks are called 'crater rays' and are formed by space objects striking the surface. The tremendous amount of energy released in such an impact digs a big hole in the ground, and also crushes a huge amount of rock beneath the point of impact. Some of this crushed material is thrown far from the crater, falling to the surface to form these rays. Fine particles of crushed rock are more reflective than large pieces, so the rays look brighter. The space environment – dust impacts and solar-wind particles – causes the rays to darken with time.

While the surface features on Mercury are not strictly mercurial, they're certainly quirky in their naming conventions. Unlike the grand Latinate names that abound throughout the solar system, Mercury's craters are named for artists, writers and musicians, including such diverse personalities as children's author Dr Seuss and dance pioneer Alvin Ailey.

Mercury *vs* Earth

-- Radius --
38% EARTH

-- Mass --
5.5% EARTH

-- Volume --
5.6% EARTH

-- Surface gravity --
38% EARTH

-- Mean temperature --
134.2°C (273.7°F)
HIGHER than **EARTH**

-- Surface area --
14.5% EARTH

-- Surface pressure --
1 quadrillionth EARTH

-- Density --
98% EARTH

-- Orbit velocity --
1.6x EARTH

-- Orbit distance --
39% EARTH

Mercury is different in other ways as well. When moving at its fastest in its elliptical orbit around the Sun, each rotation is not accompanied by a sunrise and sunset like it is on most other planets. The morning Sun appears to rise briefly, set, then rise again from some parts of the planet's surface. For other parts of the surface, the same thing happens, in reverse, at sunset. One Mercury solar day (one full, day-night rotational cycle) equals 176 Earth days. Mercury's 88-day orbit of the Sun effectively means one day on Mercury is equal to two of the planet's years.

Magnetosphere

Relative to its equator, Mercury's magnetic field is offset, that is, not centred on the planet's inner core. The cause of this anomaly is not yet known. While Mercury's magnetic field at the surface has just 1% the strength of Earth's, it will occasionally interact with the charged particles of the solar wind to create intense magnetic tornadoes that funnel the fast, solar-wind plasma down to the surface of the planet. When the ions strike the surface, they knock off neutrally charged atoms and send them on a loop, high into the sky.

A close-up view of Mercury's surface craters in the southern hemisphere.

Exosphere

Instead of an atmosphere, Mercury has a thin 'exosphere', a transient region made up of atoms blasted off the surface by the solar wind and striking meteoroids. Mercury's exosphere is composed mostly of oxygen, sodium, hydrogen, helium and potassium.

Going to Extremes

Temperatures on the surface of Mercury are extreme, both hot and cold. During the day, temperatures on Mercury's surface can reach 430°C (800°F). And, because the planet has only a transient exosphere, heat is rarely retained after sundown. This means night-time temperatures on the surface can plummet to a brisk -180°C (-290°F).

Land of Ice and Fire

Mercury may have water-ice at its north and south poles inside deep craters, but only in regions of permanent shadow. Given sufficient shade, it might be cold enough to preserve this ice despite the extremely high temperatures on the sunlit parts of the planet. Evidence of water-ice has been detected in Mercury's polar craters by Earth-based radar observations from Arecibo Observatory in Puerto Rico.

History

Along with all of the planets, Mercury formed about 4.5 billion years ago, when gravity pulled gas and dust together to make this small planet. Like its fellow terrestrial planets, Mercury has a central core, a rocky mantle and a solid crust. Very large impact basins, including Caloris and Rachmaninoff, were created by asteroid strikes on the planet's surface early in the solar system's history. While there are large areas of smooth terrain, there are also long cliffs called lobate scarps, formed during a period when Mercury's crust was contracting as the planet cooled and soaring up to a 1.6 km (1 mi) high. Over the billions of years since Mercury formed, these features have risen as the planet's interior has cooled and contracted over time.

Mercury is the second densest planet in the solar system, with the top spot belonging to Earth. However, if you were to remove the effects of gravity, then the composition

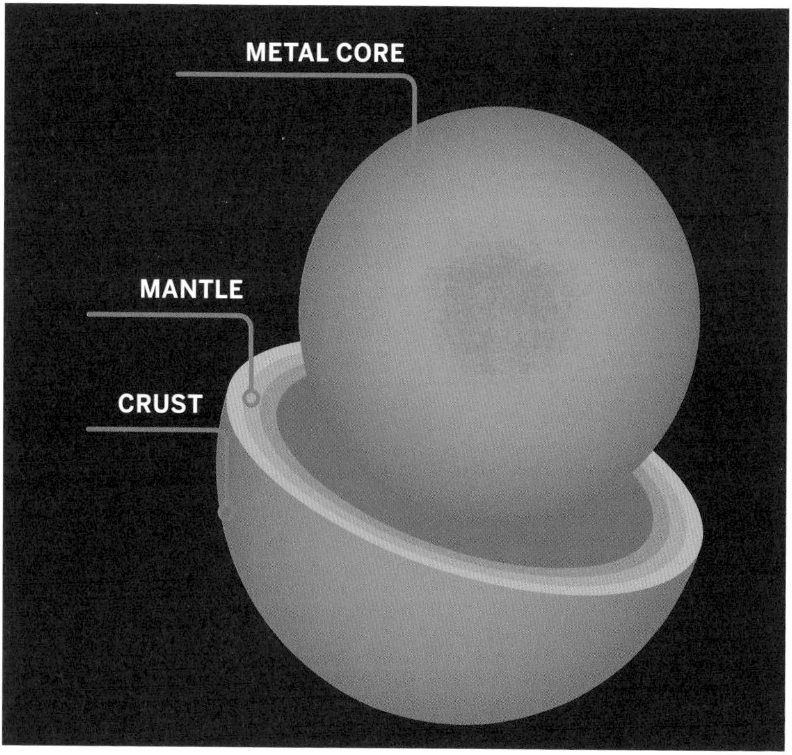

Mercury's core takes up a huge percentage of the planet's volume.

of Mercury would likely be denser than that of our home planet. This gives scientists vital information as to what materials came together to form Mercury's core, estimated to account for more than half the planet's volume. That's around 55%, far greater than that of Earth, whose core makes up only 17% of its volume. Allowing for the lack of gravitational compression, this means Mercury's core was formed from dense materials, including large amounts of iron. Mercury's core is surrounded by a mantle and an additional crust. Earth's crust is about the same thickness as Mercury's, despite the disparity of the other core elements.

Exploring Mercury

Mercury's close proximity to the Sun makes it hard to directly observe the planet from Earth. The best times are dawn and twilight, when little Mercury is not outshone by the Sun's brightness. However, about 13 times each century, interested observers on Earth can watch Mercury pass across the face of the Sun, an event called a 'transit' and a highlight for many amateur astronomers.

The first spacecraft to visit Mercury was Mariner 10, which imaged about 45% of the surface. The MESSENGER spacecraft flew by Mercury three times, beginning in 2008. It orbited the planet for four years, imaging the remainder, before crashing on Mercury's surface on running out of fuel.

Discovery Timeline

1610
England's Thomas Harriott and, separately, the Italian scientist Galileo Galilei, observe Mercury by telescope. But Mercury was already known to the ancients as a wandering star. Long before the Greeks, Assyrian astronomers recorded observations of Mercury as early as the 14th century BC.

1631
The French philosopher Pierre Gassendi uses a telescope to watch as Mercury crosses the face of the Sun.

1965
Astronomers use Earth-based radar to discover that Mercury rotates, three times for every two orbits. For centuries, it was believed Mercury presented only one side to the Sun.

1974–1975
During three flybys, Mariner 10 photographs roughly half of Mercury's surface.

1991
Scientists using radar find signs of water-ice, locked in permanently shadowed craters in Mercury's polar regions.

2008–2009
MESSENGER photographs Mercury during its own three flybys.

2011
MESSENGER begins its orbital mission around Mercury, yielding a treasure trove of images, compositional data and scientific discoveries.

2015
Having expended all its propellant, MESSENGER is deliberately crashed into Mercury, thus ending its mission.

An artist's rendering of MESSENGER in orbit around Mercury.

© ESA/ATG MEDIALAB

An illustration depicts the moment of BepiColombo separating from its Ariane 5 launcher.

Focus on BepiColombo

Launched in October 2018, BepiColombo is a European Space Agency (ESA) mission to Mercury, in cooperation with the Japanese Aerospace Exploration Agency (JAXA). The mission is named after Professor Giuseppe (Bepi) Colombo (1920-1984) from the University of Padua, Italy, the first to see the unsuspected resonance responsible for Mercury's habit of rotating on its axis three times for every two revolutions it makes around the Sun. Space scientists have identified the mission as one of the most challenging planetary projects, since Mercury's proximity to the Sun makes it difficult for a spacecraft to reach; probes must contend with the Sun's much greater gravitational field on their course. It must then survive the harsh environment of solar radiation. When BepiColombo arrives at Mercury in late 2025, it will endure temperatures in excess of 350°C (662°F) during its one year mission, with a possible one-year extension. The data gathered will provide valuable clues for understanding the formation of our solar system in a way impossible with distant observations from Earth.

The information gleaned when BepiColombo arrives will not only throw more light on the composition and history of Mercury, but also on the history and formation of the inner planets in general. The mission consists of two separate spacecraft that will orbit the planet. ESA has built one, the Mercury Planetary Orbiter (MPO), and JAXA is contributing the other, the Mercury Magnetospheric Orbiter (MMO). The MPO will study the surface and internal composition of the planet, while the MMO will study Mercury's magnetosphere, the region of space around a planet dominated by its magnetic field.

On its (very) long journey, the $2 billion BepiColombo mission will make a series of flybys, including two of Venus and one of Earth. The spacecraft's voyage will include an extended 'cruise' phase, to help the craft counter the Sun's gravity and achieve a steady orbit around Mercury. By making precise measurements of the spacecraft's orbit and position, BepiColombo also plans to test Einstein's theory of relativity. It was partly based on Mercury's orbit, posited to change due to space-time curvature.

Mercury in Popular Culture

The smallest planet in our solar system has loomed large in our collective imagination. Historically the Roman god Mercury was depicted as a wing-footed messenger (the Greek god Hermes was a forerunner). Mercury is also a widespread metal; on the periodic table it is the element Hg. For many years it was, somewhat fittingly given the planet's proximity to the Sun, a principal component in thermometers. Scores of science-fiction writers have been inspired by Mercury. The planet features repeatedly in the work of Isaac Asimov, most notably in an *I, Robot* story about a robot able to withstand extreme solar radiation; Mercury also crops up in the writing of Ray Bradbury, CS Lewis, Arthur C Clarke and HP Lovecraft.

Television and film writers, too, have used the planet as a location for their storytelling, tempted by its closeness to our Sun. In the animated Nickelodeon show *Invader Zim* (2001), Mercury is turned into a prototype giant spaceship by Martians. And in the 2007 film *Sunshine*, the *Icarus II* spacecraft goes into orbit around Mercury in order to rendezvous with its predecessor, *Icarus I*.

The British composer Gustav Holst's orchestral suite *The Planets* featured Mercury in a short segment, at a little over four minutes, that reflects its quick orbit around the Sun. In Bill Watterson's beloved comic strip *Calvin and Hobbes*, Calvin and his classmate Susie give a presentation about Mercury, in which Calvin's contribution contains some questionable information. 'The planet Mercury was named after a Roman god with winged feet,' explains Calvin. 'Mercury was the god of flowers and bouquets, which is why today he is a registered trademark of FTD florists', he adds, citing the US flower company. 'Why they named a planet after this guy, I can't imagine.'

Farrokh Bulsara might understand; he incorporated the planet into his stage name to become Freddie Mercury. Meanwhile, the Mercury Prize is awarded annually to the top album released in a given year in Britain and Ireland; the prize owes its name to a former sponsor, rather than the Queen frontman.

Freddie Mercury of Queen in concert.

Mercury Highlights

Caloris Planitia

1 Nearly 4 million years old, these plains occupy the floor of an impact crater that is one of the largest in the solar system. Its full extent, including dramatic lava flows from shallow magmatic action, has only recently been understood. The feature is believed to have been created by a rocky body estimated at least 100 km (62 mi) in diameter.

Pantheon Fossae

2 Named after the Pantheon in ancient Rome, this unique feature was originally nicknamed the 'spider crater', on account of a central depression from which a series of thin trenches radiate.

Raditladi Basin

3 One of the younger features on Mercury, at 263 km (164 mi) in diameter it may not be a giant by planetary standards, but it is home to unique rock formations.

Rachmaninoff Crater

4 The lowest point on Mercury takes its name from Russian composer Sergei Rachmaninoff. An area of extreme scientific interest, the region offers clues to Mercury's volcanic past.

Caloris Montes

5 This series of rough-hewn massifs, the so-called Heat Mountains, owes its existence to tectonic activity but its gnarly texture to incoming space debris. It sits at the edge of the Caloris Planitia impact crater.

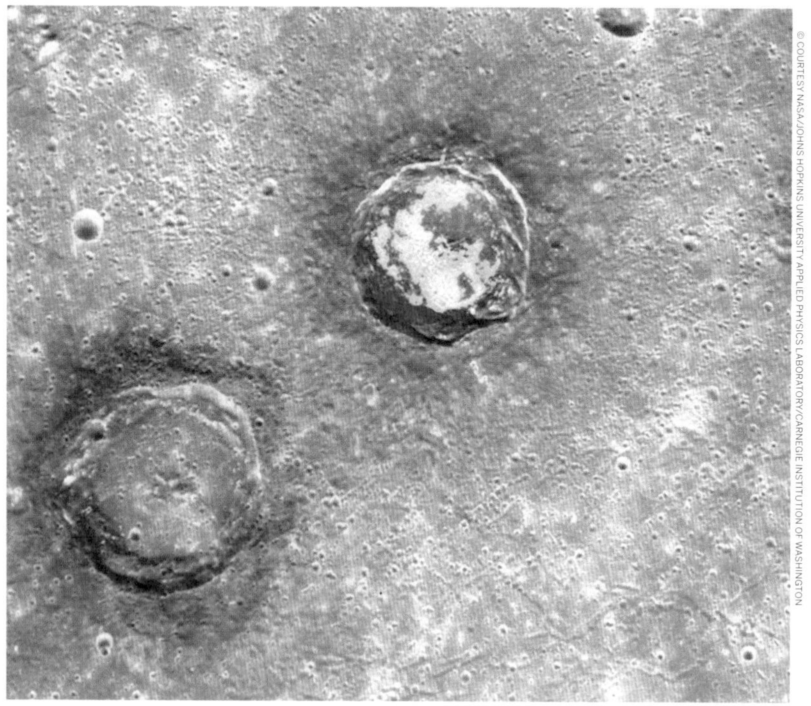

This enhanced colour mosaic shows (from left to right) Munch, Sander, and Poe craters, all in the northwest portion of the Caloris.

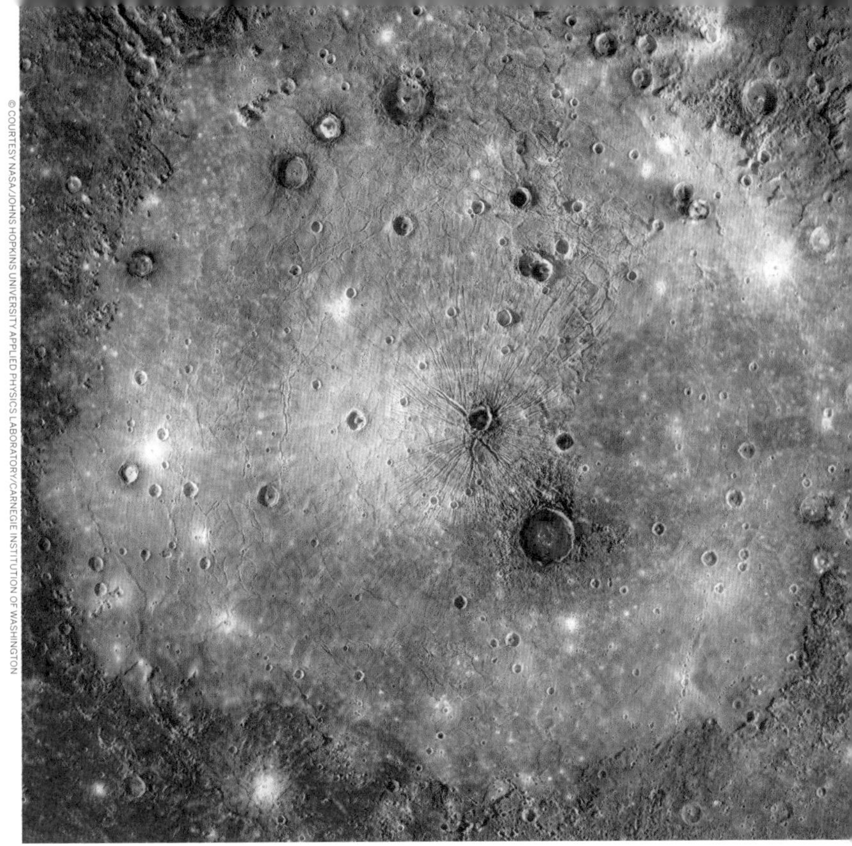

Caloris Planitia seen by MESSENGER.

Caloris Planitia

Created by an interplanetary pummelling, these ancient plains are routinely baked by the passing Sun. Mountains and lava formations offer much for the geologically curious.

Measuring about 1545 km (960 mi) across, the vast plains of Caloris Planitia were first observed by the Mariner 10 probe in 1974. At the closest point of Mercury's orbit, the Sun passes directly overhead, hence the plains take their name from the Latin word for heat, *calor*. The floor of the basin features dramatic lava flows, seen on Earth as dark shapes, and similar in appearance to the 'seas' we observe on the Moon. Walled in by the ridges of Caloris Montes, Caloris Planitia was created by an impact strike. Given the relatively few other craters in the area, this suggests it was formed around 3.8 billion years ago, perhaps after the theorised period known as the 'Late Heavy Bombardment'.

The first observations of the Caloris basin were made as the region passed between day and night, leading astronomers to significantly underestimate its size. It is, in fact, one of the larger impact craters in the solar system that we are aware of.

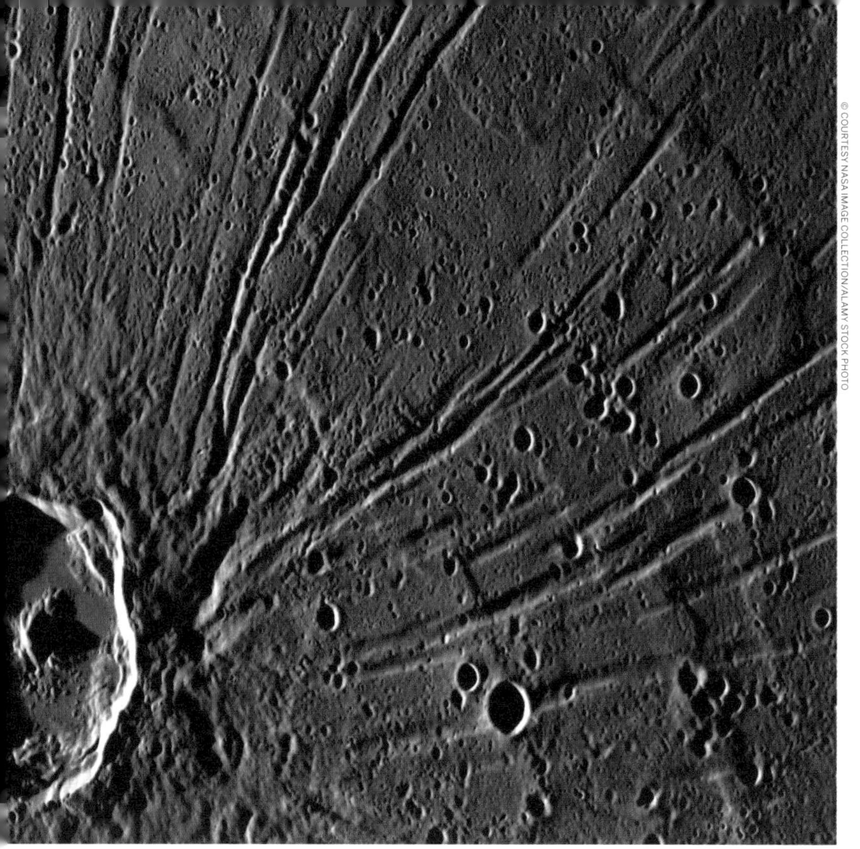

The Crater Apollodor and Pantheon Fossae.

Pantheon Fossae

Taking its official name from an architectural wonder of the ancient world, this eerie, intergalactic 'arachnid' is a unique feature on Mercury's surface.

Located at the centre of Caloris Planitia, this intriguing structure consists of a crater 40 km (24 mi) across, from which radiate long, spiny troughs, thought to be fault lines in the planet's crust. This led to the feature originally being dubbed the 'spider crater', and it was not until MESSENGER's 2008 mission to Mercury allowed a more detailed look that the far grander Pantheon Fossae came into view.

It is named after the Pantheon of ancient Rome, whose domed roof featured sections radiating outward from a central core. Mercury's Pantheon Fossae is reminiscent of this pattern, while *fossea* is the Latin word for 'trenches'. The Pantheon is believed to have been the work of Apollodorus of Damascus, the famous Greek engineer, hence the central crater connecting Pantheon Fossae is known as the Apollodorus crater.

It was the first *fossae* found on Mercury; now that the entire planet's surface has been photographed, Pantheon Fossae remains unique on the planet. The discovery of Pantheon Fossae was another of the highlights of MESSENGER's first flyby of the planet.

The Raditladi Basin is on the small side as far as Mercury's craters go.

Raditladi Basin

Featuring blues and hues that wouldn't disappoint Picasso, Mercury's own 'art gallery' is named for a celebrated African playwright and is one of the youngest of the planet's features.

At a trifling one billion years of age, the Raditladi Basin is one the younger features on Mercury. West of Caloris Planitia, it is named after Leetile Disang Raditladi, a celebrated Botswanan playwright (recall that the International Astronomical Union's convention is to name Mercury's craters after prominent artists). Raditladi is a peak ring crater; despite the name it has no central peak, but rather a ring-shaped plateau that rises from the centre of the crater floor.

A diameter of 263 km (164 mi) makes it relatively modest, at least when compared with some of the behemoth-like craters found elsewhere in the solar system. Yet Raditladi is home to some fascinating features. From an aesthetic point of view, you could say the basin represents Mercury's 'blue period'; the massifs that rise around the basin rim reveal a strikingly blue, rocky material, as do the numerous mounds that litter the crater floor.

MESSENGER is shown against the Rachmaninoff Crater on Mercury's surface.

Rachmaninoff Crater

Something of a low point, but in a good way. The lowest elevation on Mercury provides evidence of geological processes and, nearby, the chance to make a snowman of fire.

Lying at 5.3 km (3.3 mi) below the planet's average elevation, the Rachmaninoff Crater is the lowest point on Mercury. Named after the Russian composer Sergei Rachmaninoff, the crater has proved of extreme interest to space scientists since its discovery on MESSENGER's third flyby. The centre of the crater is home to a peak ring some 130 km (80 mi) in diameter, with bright, red plains that indicate lava flows at some point in their history; their fresh appearance indicates they are relatively young, an exception on this low-volcanism planet.

A complex of volcanic vents, just to the northeast of Rachmaninoff, suggests further evidence of the region's explosive history. Portions of the vents are blanketed in fine-grained material thought to be composed of pyroclastic particles of the kind that are discharged by volcanoes, almost like snow. Created by volcanic activity, that would make for some very fiery, and very hot, snow indeed.

Mariner data provided the images to create this mosaic featuring the Caloris Montes.

Caloris Montes

The Caloris Montes or 'Heat Mountains' sound like something from James Cameron's Avatar. *In fact, this is a series of tall massifs, sculpted over time by incoming projectiles.*

Located on the rim of the Caloris Basin, the wonderfully named 'Heat Mountains' rise up to 2 km (1.2 mi) in height. A series of massifs, rather than a single range, they are each about 10 to 50 km (6 to 31 mi) in length. The cause of their formation is not certain, although it is believed to include a fracturing of Mercury's crust, possibly the result of volcanic activity.

On the surface, the massifs are extremely rough, a texture geologists refer to as 'hackly'. This has been attributed to the impact of secondary projectiles, which struck Mercury's surface long after the impact that created the Caloris Basin. A similar theory has been applied to the Imbrium Sculpture, which is found on Earth's moon. One mystery scientists have yet to solve is the cause of a pronounced gap in the southern part of Caloris Montes, not attributable, thus far, to the processes that have created similar gaps in other planetary mountains.

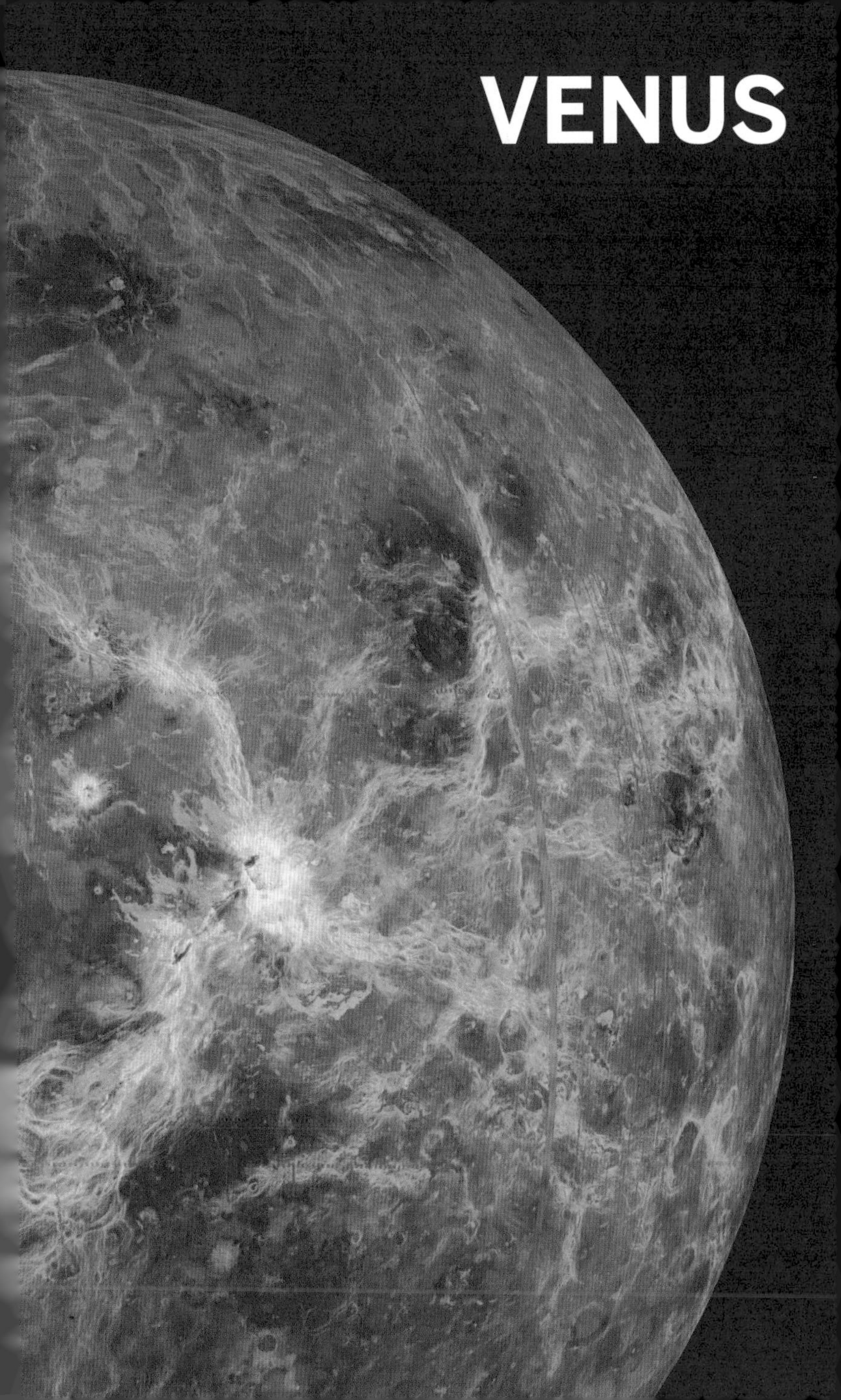
VENUS

A view of Venus captured by the Magellan mission.

Venus at a Glance

The second planet from the Sun, Earth-sized Venus is a pressure cooker, a dim world of intense heat and volcanic activity. It has a toxic, hot atmosphere full of carbon dioxide and clouds made of sulphuric acid.

Named for the Roman goddess of love and beauty, Venus has historically been considered female. Today, Venus is arguably the only planet keeping up with the times, as almost all its features are named after noteworthy women. One volcanic crater is named for Sacajawea, the Native American who helped Lewis and Clark explore the American west, and a deep canyon is named for the Roman goddess Diana.

Venus is similar in structure and size to Earth, facts which once led scientists to think of it as a 'sister' planet. Since it orbited closer to the Sun, most scientists assumed Venus was warmer than Earth, and some believed its climate was comparable to Earth's tropics, though others began predicting a greenhouse effect. Radio observations taken in 1958 showed temperatures closer to 300°C (572°F).

When Mariner 2 flew by Venus soon after, in 1962, it measured the temperature at the planet's surface at between 150°C and 200°C (302°F to 392°F); it also recorded an atmospheric pressure 20 times greater than that on Earth. Subsequent spacecraft revealed temperatures and pressures far exceeding those initial measurements. Mariner 2 also found that Venus had no appreciable magnetic field, unlike Earth; unable to trap belts of radiation to protect its surface, Venus is constantly bombarded by cosmic rays. Venus is different in other ways, too. It spins in the opposite direction from most planets, and extremely slowly.

In fact, its sidereal day is longer than its orbital year. It takes about 243 Earth days to spin around just once and reach the same position; this so-called sidereal day measures the length of time before distant stars appear again at a fixed point in the planet's sky, completing a full rotation. Because it's so close to the sun, a year goes by fast: it takes only 225 Earth days for Venus to go all the way around the sun. When we think of a day, we normally think of one cycle of daytime to nighttime, aka a solar day. On Venus, one day-night cycle takes 117 Earth days, so lengthy because Venus rotates in the opposite direction of its orbital revolution around the Sun.

Almost always covered in clouds, glimpses through the clouds reveal a surface that is home to volcanoes and mountains. From our vantage point on Earth, Venus is one of the brightest objects in the sky as it rises and sets, earning the monikers Morning Star and Evening Star.

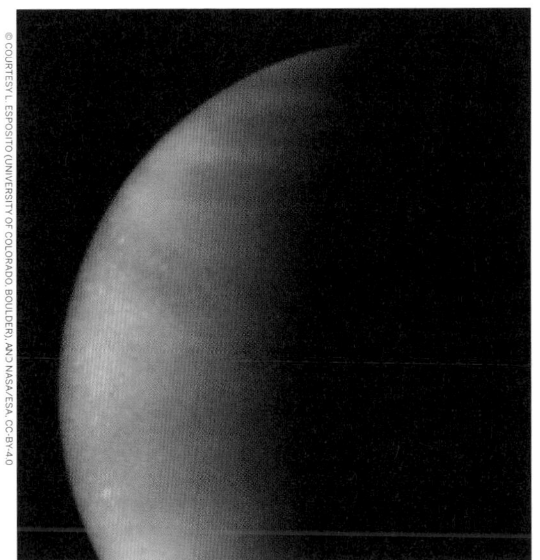

Venus cloud tops as seen by Hubble.

DISTANCE FROM SUN
0.7 AU

LIGHT-TIME TO THE SUN
6 minutes

LENGTH OF DAY
243 Earth days (sidereal); 116.75 Earth days (solar)

LENGTH OF ORBITAL YEAR
224.7 Earth days

ATMOSPHERE
Carbon dioxide, nitrogen

Top Tip

If you're looking for a party planet, Venus is the destination for you. It rotates from east to west, in the opposite direction of its orbital revolution around the Sun. This means the Sun neither rises nor sets – so there's no reason to go home.

Getting There & Away

On average, Venus is 40 million km (25 million mi) away from Earth, approaching its closest distance to our planet every 584 days. You can reach the hottest planet in the solar system in three months. At least, that's how long it took NASA's Mariner 2, way back in the 1960s. A more pressing issue is likely to be the price of a ticket. NASA estimates put the cost of a fully fledged Venus Flagship mission at a cool $3 billion USD.

Orientation

The solid surface of Venus is a volcanic landscape covered with extensive plains featuring high mountains, lava flows and vast ridged plateaus. With a radius of 6052 km (3760 mi), Venus is only slightly smaller than Earth; there is only a 5% difference in their radii. At an average distance of 108.2 million km (67.24 million mi), Venus is 0.7 AU away from the Sun, and it only takes sunlight a quick six minutes to travel from the Sun to Venus.

Both the orbit and rotation of Venus are unusual. Venus is one of just two planets – Uranus is the other – that rotate from east to west. This means a 'backwards' rotation, compared with other planets, and Venus completes one rotation (its sidereal day) in 243 Earth days. That's the longest day of any planet in our solar system; and because Venus makes one complete orbit around the Sun in just 225 Earth days, one sidereal day on Venus is, technically, longer than a Venusian year.

To confuse visiting Earthlings even further, on Venus the Sun doesn't rise and set every day like it does on most other planets, because Venus rotates in the direction opposite of its orbital revolution around the Sun; that's why on Venus, one day-night cycle takes 117 Earth days. With an axial tilt of just 3 degrees, Venus spins nearly upright, and so does not experience any noticeable seasons either, staying above tropical without variation. In keeping with this constancy, the orbit of Venus around the Sun is the most circular of any planet, nearly a perfect circle. The orbits of the other planets are elliptical, or oval-shaped.

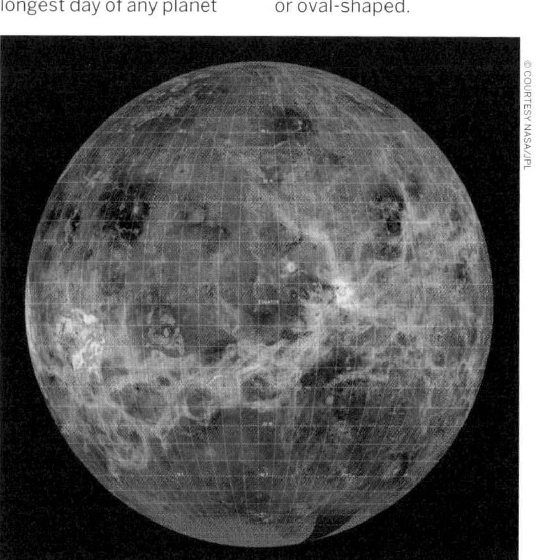

© COURTESY NASA/JPL

Another view of Venus from Magellan's mapping instruments.

Venus *vs* Earth

-- Radius --
95%
EARTH

-- Mass --
81%
EARTH

-- Volume --
85.7%
EARTH

-- Surface gravity --
90.5%
EARTH

-- Mean temperature --
455°C
(851°F)
HIGHER than **EARTH**

-- Surface area --
90%
EARTH

-- Surface pressure --
92x
EARTH

-- Density --
95%
EARTH

-- Orbit velocity --
117%
EARTH

-- Orbit distance --
72%
EARTH

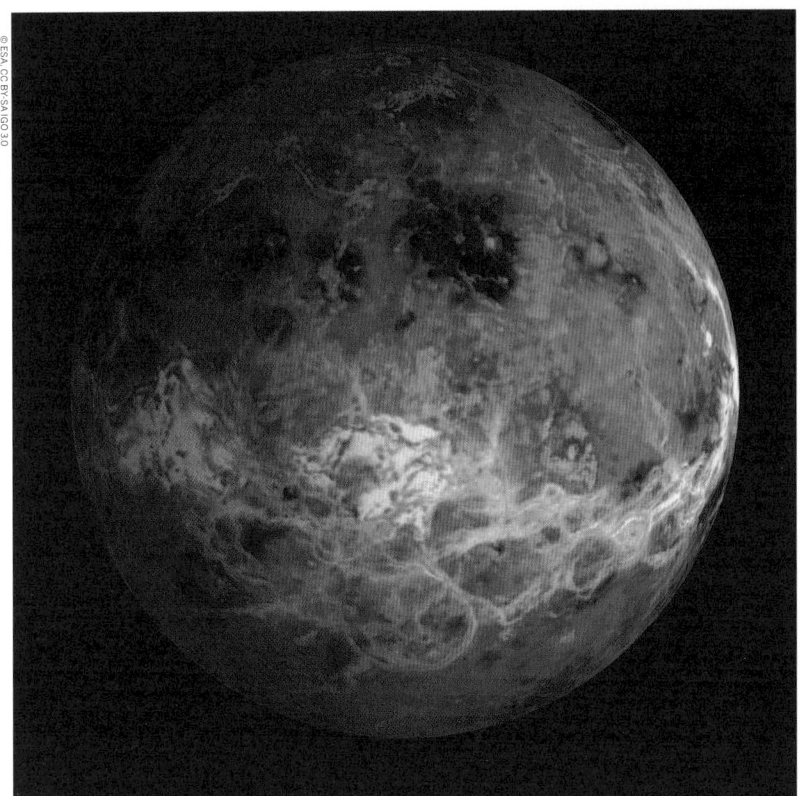

An artist's impression of the Venusian surface.

While the surface conditions of Venus are markedly different from those on Earth (only one planet has a surface hot enough to melt lead, after all), the structures of the two planets are very similar. Venus has an iron core that is approximately 3200 km (2000 mi) in radius; Earth's equivalent is around 1220 km (760 mi). Above Venus' core is a mantle, made of hot rock, which churns slowly, due to the planet's interior heat. Estimates put this at a similar temperature to that of Earth's core, around 4000°C (7230°F) at the core-mantle boundary.

The surface of the planet is a thin crust of rock that bulges and moves as Venus' mantle shifts. This often creates volcanoes, of which Venus has tens of thousands. The highest mountain on Venus, Maxwell Montes, is 11 km (6.8 mi) higher than the mean planetary radius, taller than Earth's titan Mount Everest. Quite unlike our own planet, the landscape on Venus is dusty and dry, and the mean surface temperature, around 471°C (880°F), can get hotter still; although if you were sunbathing at this temperature, would you notice a ten degree rise?

From space, things look positively dreamy. Venus appears a bright white, due to its thick clouds that reflect sunlight. On the surface, the rocks are various shades of grey, like those on Earth. Venus' heavy, thick atmosphere filters sunlight so that, on the planet's surface, its rocks would look orange in colour.

Venus is also covered in craters, the smallest of which are around 1.6 km (1 mi) across. Small meteoroids burn up in the dense atmosphere, so only large meteoroids can reach beyond it to create craters.

Atmosphere

An artist's concept of lightning on Venus, which has a tumultuous atmosphere.

The atmosphere on Venus consists mainly of carbon dioxide, while its clouds are composed of sulphuric acid. The atmosphere traps the Sun's heat, resulting in those shocking surface temperatures of 471°C (880°F) or more. Venus' atmosphere actually has a number of layers, each with differing temperatures and conditions. At cloud level, for example, about 48 km (30 mi) from the surface, the temperature is about the same as that on the surface of the Earth.

Driven by hurricane-force winds of roughly 360 km/h (220mph), the highest clouds zip around Venus in the span of what would be four Earth days. (On Earth, an equivalent mass of air takes about 2 weeks to circumnavigate our globe.) The slow rotation of Venus means that winds circulate at around 60 times the speed of the planet's rotation; by comparison, the fastest winds on Earth are no more than 20% of our home planet's rotation speed.

Atmospheric lightning adds to the drama on Venus, illuminating the quick-moving clouds. As you descend, however, the excitement diminishes. Windspeed decreases with altitude; by the time you reach the surface, it is estimated to drop to just a walking pace. On the ground, conditions would look like a hazy, overcast day on Earth. The atmosphere is so heavy that the pressure would make you feel as if you were 1609 m (5280 ft) under water. Even though Venus is similar in size to the Earth, and has a roughly equivalent iron core, the slow rotation of Venus means its magnetic field is much weaker.

Another oddity are the occasionally observed holes in Venus' ionosphere, which may be the result of the Sun's magnetic field lines penetrating through the planet, It's yet another reason to be grateful for Earth and its robust magnetic field, which protects our planet from experiencing a similar effect.

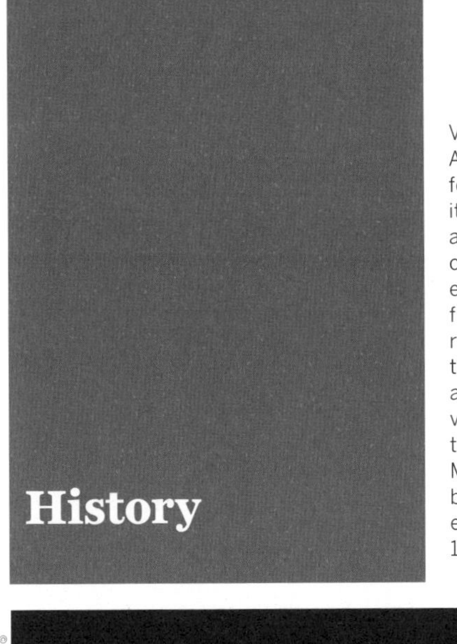

History

Venus is the hothead of the solar system. As with its fellow terrestrial planets, it formed when the solar system settled into its current layout, around 4.5 billion years ago. Gravity was the key, pulling together clouds of gas and dust to form solid planetary bodies. Venus, the second planet from the Sun, consists of a central core, a rocky mantle and a solid crust. It is thought that, between 300 and 500 million years ago, Venus was completely resurfaced by volcanic activity. In fact, Venus is home to more volcanoes than any other planet. More than 1600 major volcanoes have been mapped on its surface to date, with estimates of the total number ranging from 10,000 to many millions.

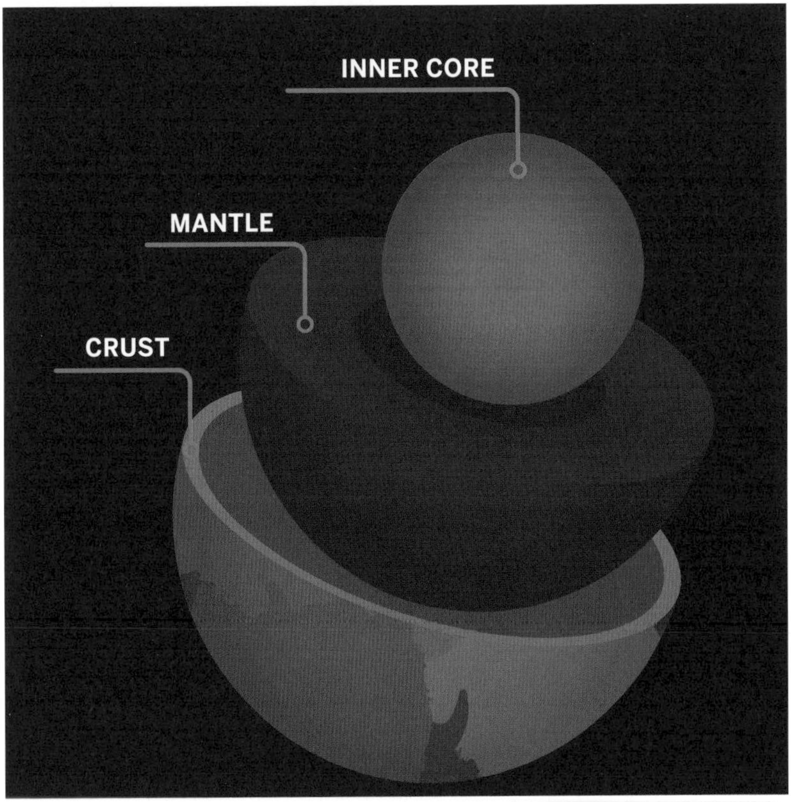

INNER CORE
MANTLE
CRUST

Venus is especially volcanic and without plate tectonics; to our knowledge, its crust is one single surface.

Venus has two large highland areas, Ishtar and Aphrodite Terra. Ishtar Terra, located in the northern polar region, is about the size of Australia. Aphrodite Terra, which straddles the planet's equator and extends for almost 1000 km (6000 mi), is about the size of South America.

The extremely hostile surface environment means no human has yet visited Venus. And the spacecraft that have been sent to the surface of Venus haven't lasted very long. Venus' extreme surface temperatures – as high as 471°C (880°F) – would overheat spacecraft electronics in a short time, so it seems unlikely that a person would survive for long on the Venusian surface. Recent orbital observations and simulations from a JAXA research team indicate that Earth and Venus do share some of the same mechanism at work in their cloud layers, suggesting that jet streams may be formed at high latitudes of Venus, influencing cloud layers.

The current conditions on Venus may be in stark contrast to those that existed millions of years ago, likely to have been much more temperate than today's hothouse. The possibility of life having once existed on Venus (albeit in the distant past) was boosted by climate modelling in 2016 that suggested Venus may once have been home to oceans, as well as far cooler surface temperatures, before a runaway greenhouse effect took over.

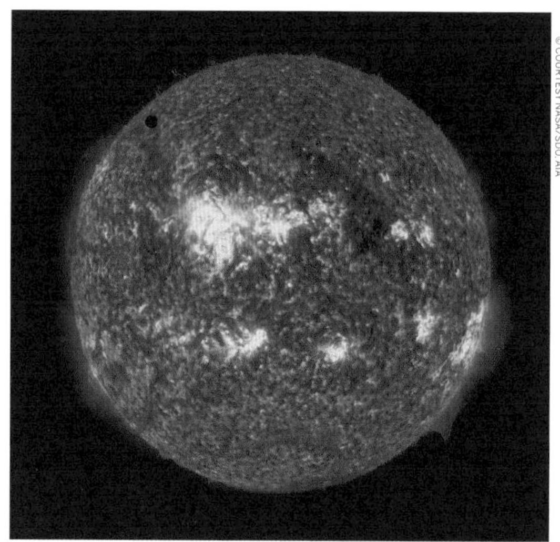

An image of the 2012 Venus transit across the Sun.

Exploring Venus

Venus has been watched from Earth for millennia. After the Moon, it is the brightest object in the night sky. (Mercury, Venus and the Moon are the only bodies between Earth and the Sun.) Every hundred years or so, Venus passes across the face of the Sun, a rare phenomenon known as a 'transit'. Observing these transits of Venus has helped astronomers study the planet and better understand both the solar system and Earth's place in it. Transits of Venus occur in pairs; individual transits are eight years apart, with roughly a century separating each pair. Recorded transits have occurred in 1631 and 1639; 1761 and 1769; 1874 and 1882; and 2004 and 2012.

In ancient times, the transits of Venus were interpreted as having mystical, or spiritual, significance. Modern space science takes a more empirical view and can even pinpoint specific dates. The next transits of Venus will take place on 11 December 2117, and 8 December 2125. Long gaps between transits occur because the respective orbits of Earth and Venus around the Sun are inclined differently. This means that Venus passes between Earth and the Sun without crossing the Sun's face more often than it crosses directly in front of the Sun.

Spacecraft from several nations have visited Venus, including the Soviet Union's successful Venera and Vega series, 16 missions in total. Venera 3, in 1965, was the first man-man object to land on another planet. NASA's Magellan mission, which studied Venus from 1990 to 1994, used radar to map 98% of the surface. More recently, JAXA's Akatsuki mission has, since 2015, been studying the atmosphere of Venus from its orbit.

Discovery Timeline

650 BC

Mayan astronomers make detailed observations of Venus' movements, leading to the creation of a highly accurate planetary calendar.

1610

Galileo documents the phases of Venus in *Sidereus Nuncius* (Sidereal Messenger). Sidereal time is the system used by astronomers to locate celestial objects, based on the Earth's rotation and relative to the fixed stars. Galileo's book is better known by an approximate translation, *The Starry Messenger*.

1639

The first predicted transit of Venus is observed in England by astronomers Jeremiah Horrocks and William Crabtree.

1761–1769

Two European expeditions, one British (led by Captain James Cook) and the other Swedish, travel to Tahiti to watch the 1769 transit of Venus. Observations lead to the first realistic estimate of the Sun's distance from Earth.

1961

Radar returns from Venus are used to determine the Sun's exact distance from Earth. The findings, more accurate than any previous ones, are published a year later.

1962

NASA's Mariner 2 reaches Venus and during its first flyby, on 14 December, reveals the planet's extreme surface temperatures. It is the first spacecraft to send back information from another planet.

1962–1983

Venera, a series of robotic spacecraft sent to Venus by the Soviet Union, has several successful missions over the program's duration. The information that is sent back is the first data transmitted from the atmosphere of another planet.

1990–1994

NASA's Magellan spacecraft, in orbit around Venus, uses radar to map 98% of the planet's surface.

2005–2014

The European Space Agency (ESA) launches Venus Express, a mission to study the planet's atmosphere and surface. The orbiter reaches Venus in April 2006 and continues to study Venus until the end of 2014.

2015

After launching in 2010, the Akatsuki orbiter – Akatsuki means 'Dawn' – achieves orbit around Venus. It is the first mission to Venus undertaken by the Japanese Aerospace Exploration Agency (JAXA).

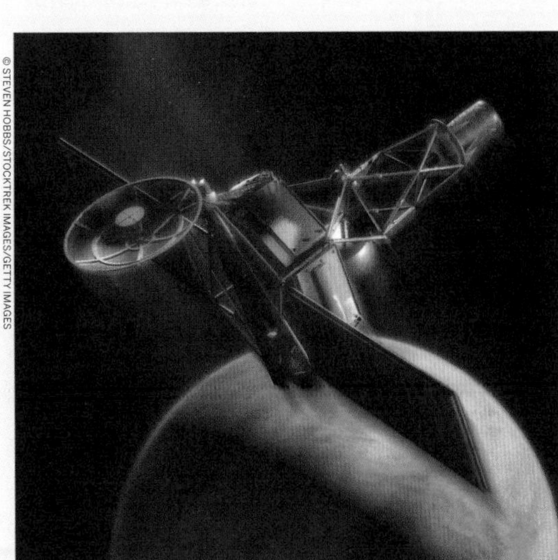

Mariner 2, pictured here, was the first human-made object to fly by another planet.

© PETER HORREE/ALAMY STOCK PHOTO

Botticelli's *The Birth of Venus*.

Venus in Popular Culture

The idea of a female Venus owes much to Botticelli's 15th-century masterpiece *The Birth of Venus*. Named after the goddess of love and beauty, Venus was, for a long time, shorthand for all things feminine. 'Men are from Mars' as the famous relationship guide explained, but 'Women are from Venus'.

The planet was certainly a popular destination for 20th century science-fiction aficionados. Before we knew what lay beneath Venus' mysterious cloud cover, the planet sustained an entire industry of pulp writers and artists, all speculating about what lay in store for potential visitors. For some it was giant, man-eating ants; for many comic-book artists, Venus was home to roving tribes of lightly clad 'Amazons'. Once space exploration had shown Venus to be uninhabitable, the grubbier end of the genre fell away, and it became the territory of more high-minded proponents of the art form, including sci-fi writers Ray Bradbury and Isaac Asimov.

Venus' supposedly racy nature has been the subject of at least one raunchy novel, 1870's *Venus in Furs* by author Leopold von Sacher-Masoch. The author's following, though small, stretched to one Lou Reed, who was inspired in 1967 to write a song of the same name for his band The Velvet Underground.

Matthew Broderick even starred in *The Starry Messenger*, a Venus-inspired play, penned by Kenneth Lonergan, about a frustrated astronomer. Venus has also been a backdrop for video games, such as *Transhuman Space* and *Battlezone*. And in the Disney animated film *The Princess and the Frog*, Ray the firefly falls in love with Venus, the evening star, when he mistakes the planet for a female firefly named Evangeline.

Focus on Mariner

In 1962, Mariner 2's flyby of Venus made scientific history: the first close-up observations of another planet. The Mariner mission included two spacecraft, and Mariner 1 never even left Earth's atmosphere. Almost immediately things went wrong with the first launch; the Atlas-Agena rocket carrying the probe veered dangerously off course, and controllers were forced to destroy the craft just a few minutes into its flight. The fault, when it was traced, was painfully simple – legend has it the blame lies with a hyphen missing from the coding of the spacecraft's communications software.

Mariner 2 followed just over a month later, on 27 August, and this time the launch was successful. NASA's fledgling Jet Propulsion Laboratory (JPL), in Pasadena, California, had planned and developed the first mission to Venus in little more than a year. The 202 kg (447lb) Mariner 2 carried six scientific instruments, weighing an additional 18 kg (40lbs). These were designed specially in order to study Venus' atmosphere and temperature, to search for a possible magnetic field and to study cosmic rays.

During its 108-day journey, Mariner 2 sent back valuable information about its interplanetary environment, including confirmation of the existence of the solar wind, the stream of charged particles that emanates from the Sun. The road to Venus was anything but easy. Of the many hardware anomalies the spacecraft encountered, several inexplicably fixed themselves. But, by the time Mariner 2 was approaching Venus, one of its solar arrays had failed and the vehicle was dangerously close to overheating. On 14 December, Mariner 2 passed within 34,704 km (21,564 mi) of Venus, sending back the data it had collected at a then-blistering rate of 8 bits per second; compare that to NASA's current communications network, which relays data at around 300 megabits per second, twice the rate of a typical high-speed internet connection in modern homes. The spacecraft sent its last transmission on 3 January 1963, shortly before contact was lost. As for the Jet Propulsion Laboratory, Mariner 2 would be the first of dozens of JPL missions exploring planets from Mercury to Neptune. It provided the template for everything to come.

An artist's conception of the trailblazing Mariner 2.

Focus on Magellan

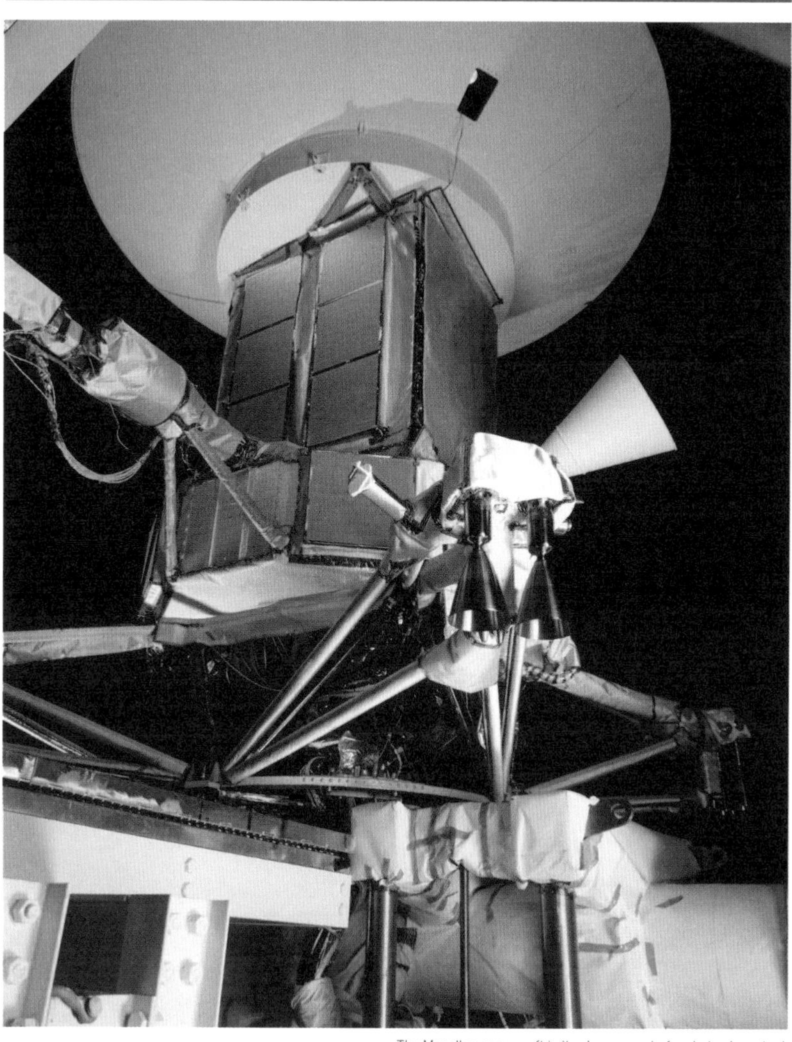

The Magellan spacecraft in its clean room before being launched.

Launch date:
4 May 1989
Orbit Insertion:
10 August 1990

NASA's Magellan mission was named for Ferdinand Magellan, the 16th century Portuguese explorer whose name is synonymous with pioneering voyages. His forays to the East Indies eventually led to the first circumnavigation of the Earth; such were his skills as a navigator he is also well represented in space, lending his name to everything from dwarf galaxies to Martian craters. For NASA, Magellan was the first deep-space probe launched by the United States in almost eleven years. It was also the first launched by a space shuttle – in Magellan's case, Atlantis. Magellan was developed for up-close-and-personal planetary reconnaissance, designed to use a 'synthetic aperture radar' – radar that recreates 3D shapes – to map the Venusian surface down to a resolution of 120 to 300 m (394 to 984 ft).

Magellan was deployed from the payload bay of Atlantis a day after launch, on 5 May 1989. Thereafter, thrusters set it on course for its rendezvous with Venus. Magellan arrived in its Venusian orbit on 10 August 1990. As Mariner had shown three decades earlier, however, deep-space operations can prove extremely challenging. Six days after entering orbit, Magellan suffered a communications outage lasting 15 hours. After a second, 17-hour interruption, NASA sent up new preventative software to reset Magellan's systems.

Restored to health, Magellan began its surveillance on 15 September 1990, and was soon returning high-quality radar images of Venusian terrain. These showed evidence of vulcanism and tectonic movement. Magellan completed its first 243-day cycle (the time it took for Venus to rotate once under Magellan's orbit) of radar mapping on 15 May 1991. These images, providing the first clear views of the surface, covered 83.7% of it. In all, the spacecraft returned an astonishing 1200 gigabits of data, 1.8 gigabits with every orbit. At the time, the data from all previous NASA planetary missions combined amounted to 900 gigabits!

The spacecraft's second mapping cycle, already beyond the original goals of the mission, ended on 15 January 1992, increasing coverage of the surface to 96%. When a third cycle ended, on 13 September that year, the total coverage of the Venusian surface was an incredible 98%.

Magellan found that at least 85% of the Venusian surface is covered with volcanic flows. Despite the very high surface temperatures and high atmospheric pressures (92 bar), the complete lack of water makes erosion an extremely slow process. As a result, surface features can persist for hundreds of millions of years. Contact with Magellan was finally lost on 13 October 1994, after the spacecraft was directed to plunge into the atmosphere to gather data. The spacecraft burned up in the Venusian atmosphere several hours later, ending one of NASA's most successful deep-space missions.

The Supreme Navigator:

Magellan's Successes

» Magellan mapped 98% of Venus' surface at a resolution of 75 m/pixel.

» Using synthetic aperture radar, a technique that simulates the results of a much larger radar antenna, Magellan found that 85% of Venus' surface is covered with volcanic flows.

» The spacecraft found evidence of tectonic movement, turbulent surface winds, lava channels and so-called pancake domes (flat-topped volcanic domes).

» Magellan produced high-resolution gravity data for 95% of Venus, and tested a manoeuvering technique known as 'aerobraking', which uses atmospheric drag to adjust the orbit of a spacecraft.

Magellan data created this 3D-perspective view of Lavinia Planitia on Venus.

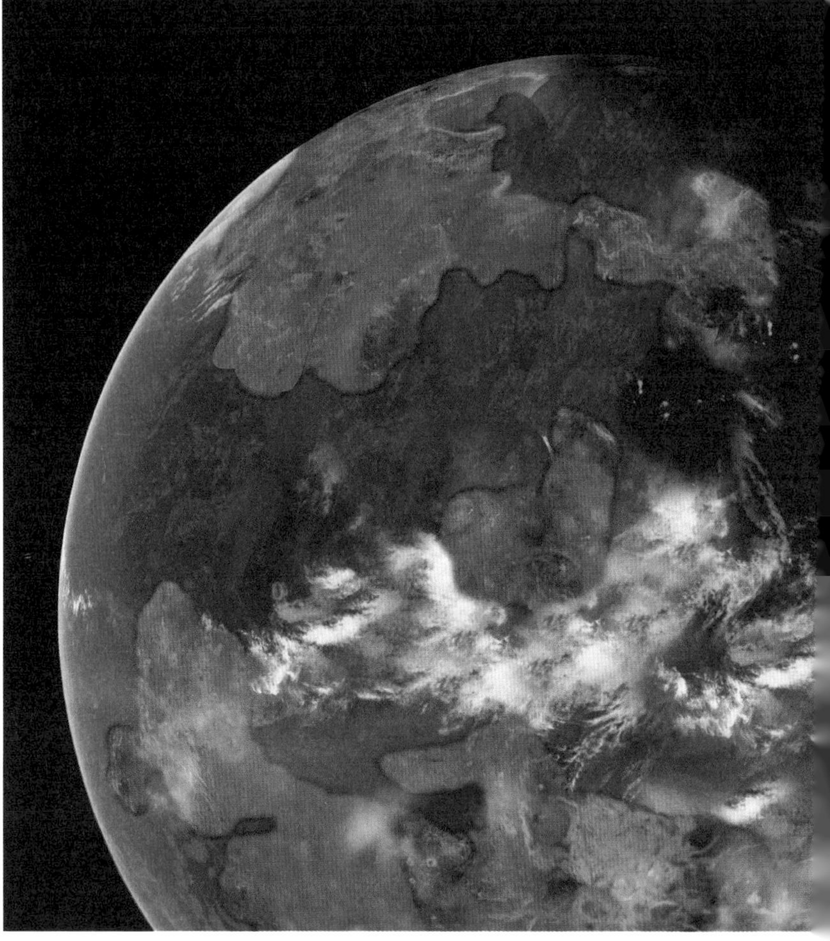

This land-ocean pattern is one possible past configuration of the Venusian surface.

Signs of Life:

A Habitable Venus?

It is Venus' similarities to Earth, both in size and structure, as well as its close proximity, that led to the long-standing idea of a tropical, sister planet. And while various Venus missions have debunked that theory, other research suggests that, in its ancient history, habitable conditions may once have existed there.

Applying modelling used to predict climate change on Earth, NASA's Goddard Institute for Space Studies (GISS) found that Venus may have had a shallow, liquid-water ocean and even habitable surface temperatures for up to 2 billion years of its early history. 'Many of the same tools we use to model climate change on Earth can be adapted to study climates on other planets, both past and present', according to Michael Way, a researcher at GISS. 'These results show ancient Venus may have been a very different place'. At present, modern Venus is a hellish world – with a crushing carbon dioxide atmosphere over 90 times as thick as Earth's,

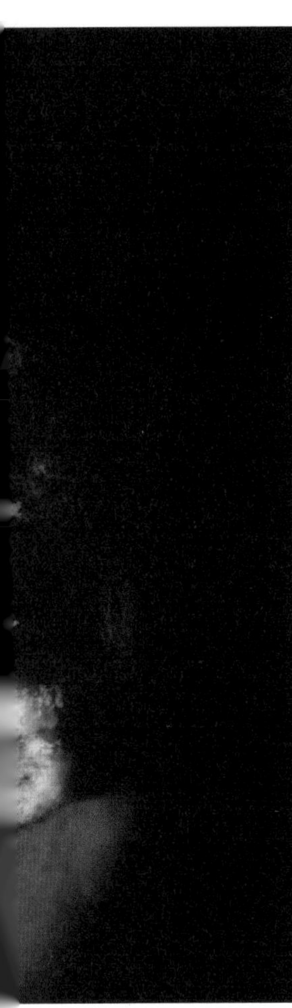

almost no water vapour and average temperatures of 471°C (880°F).

Previous studies have shown that the speed at which a planet spins on its axis is a determining factor in whether or not it has a habitable climate; Venus certainly has a long solar day at 117 Earth days. Until recently, it was assumed that a thick atmosphere, like that of modern Venus, was a causal factor in the planet's slow rotation rate. Newer research has shown that, even with a thin atmosphere like Earth's, Venus' rotation would be the same. That means ancient Venus, with an Earth-like atmosphere, could have had the same rotation rate it has today. Another factor that impacts a planet's climate is topography. The GISS team assumed ancient Venus had more dry land overall than Earth, especially in the tropics. That would have limited the amount of water evaporated from the oceans and, as a result, the greenhouse effect by water vapour.

This type of surface appears ideal for making a planet habitable; there would have been enough water to support abundant life, with sufficient land to reduce the planet's sensitivity to changes from incoming sunlight.

GISS scientists simulated conditions of a hypothetical early Venus – with an atmosphere similar to Earth's; a day as long as Venus' current day; and a shallow ocean, consistent with early data from NASA's Pioneer missions (see sidebar). The researchers added information about Venus' topography from radar measurements taken by NASA's Magellan mission in the 1990s. They filled the lowlands with water, and left the highlands exposed, effectively as Venusian continents. The study also factored in an ancient Sun, up to 30% less bright than today. That said, ancient Venus still received about 40% more sunlight than Earth does even today.

How to boil an ocean

While Venus has a similar formation to Earth, it followed a different evolutionary path. It was measurements by NASA's Pioneer space-weather missions in the 1980s that first suggested Venus may once have had an ocean. However, Venus is closer to the Sun than Earth and so receives far more sunlight. As a result, the planet's early ocean evaporated, water-vapour molecules were broken apart by ultraviolet radiation, and hydrogen escaped to space. With no water left on the surface, carbon dioxide built up in the atmosphere, leading to the so-called runaway greenhouse effect that created Venus' present environmental conditions.

'In the GISS model's simulation, Venus' slow spin exposes its "dayside" to the Sun for almost two months at a time', explained Anthony Del Genio, a GISS scientist. 'This warms the surface and produces rain that creates a thick layer of clouds, which acts like an umbrella to shield the surface from much of the solar heating. The result is mean climate temperatures that are actually a few degrees cooler than Earth's today'. The findings have direct implications for future NASA missions, such as the Transiting Exoplanet Survey Satellite and James Webb Space Telescope, which will try to detect possible habitable planets and characterise their atmospheres.

Venus Highlights

Baltis Vallis

1 This long, winding channel, at 6800 km (4225 mi) long, has no equal anywhere in the solar system. The boilling river that once ran through it would have melted even the toughest of tour boats.

Maat Mons

2 Venus' highest volcano is named for the ancient Egyptian goddess of justice and truth – who, in her spare time, also made sure the stars were aligned and the planets were in their rightful places.

Alpha Regio

3 The finest example of Venus' tiled tessera, the curious rock formations found here bring a touch of celestial style to any planet's surface.

Maxwell Montes

4 The highest point on Venus offers relief from the punishing surface temperature and pressure. It might be the coolest place to spend time, but you still wouldn't last two minutes.

Aphrodite Terra

5 The largest Venusian 'continent' highland area has jumbled volcanic marvels over its land mass, which is divided into two regions: Ovda Regio and Thetis Regio.

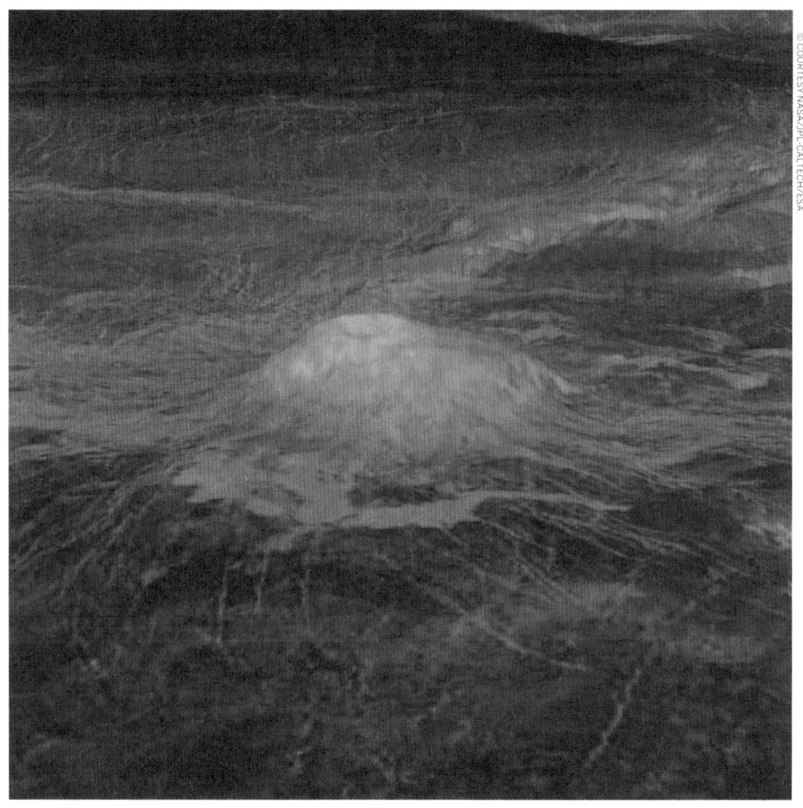

One of the many Venusian volcanoes.

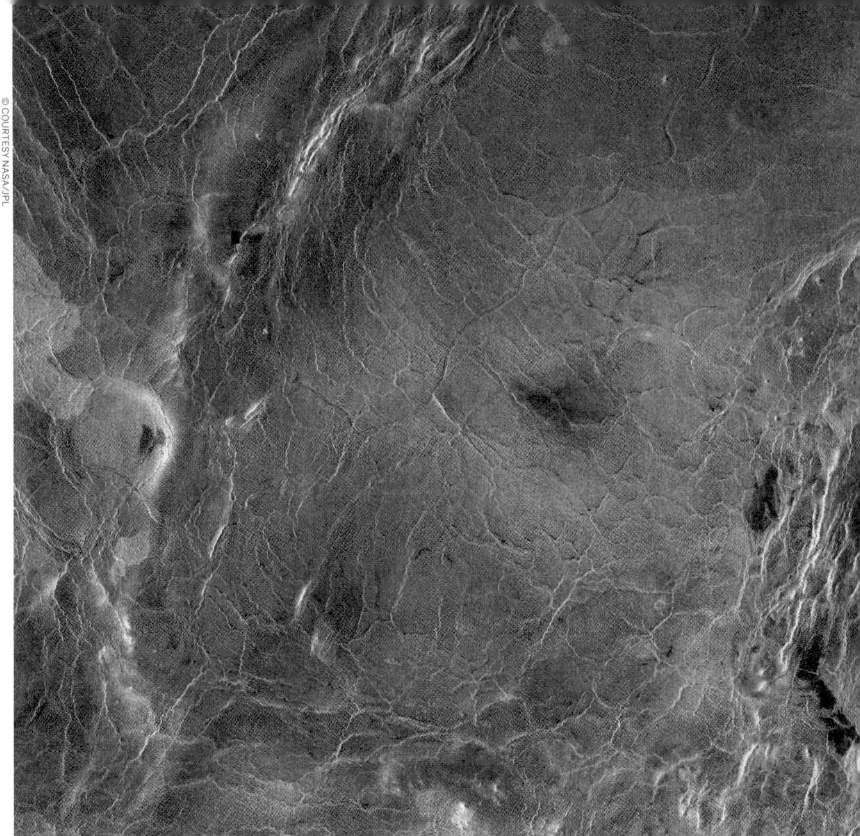

An imaged sector of Baltis Vallis.

Baltis Vallis

The solar system's longest channel may once have been home to a roiling, roaring river of lava. Longer than Earth's Nile River, scientists are only beginning to understand its formation.

This great, snaking channel is the longest of its kind in the solar system. It was discovered in 1983, by the Soviet orbiters Venera 15 and Venera 16. Despite the relatively crude technology of the time – imaging was only available down to a resolution of one km – those early orbiters still succeeded in plotting 1000 km (620 mi) of the immense channel. At 6800 km (4225 mi) long, and ranging in width from between 1 km and 3 km (over half a mile to almost 2 miles wide), Baltis Vallis is longer than Egypt's Nile River – and it's possible it once used to be even longer. With both ends of the channel now obscured under subsequent rock formations, scientists believe a river of lava once flowed through it, carving the channel to an average depth of 100 m (328 ft). The fact that some sections appear to show lava flows travelling uphill suggests the channel has, at various points in its history, undergone periods of significant tectonic uplift.

A 3D perspective of looming Maat Mons.

Maat Mons

Venus' highest volcano may not be erupting currently, but don't let that fool you – evidence suggests it may have let off a whole lot more than steam, and not all that long ago.

This massive shield volcano, the highest on Venus, soars 8 km (5 mi) above the mean planetary radius. The summit is home to a large caldera, some 31 km (19 mi) across, as well at least five smaller calderas created by structural collapse. The volcano is named in honour of Ma'at, the Egyptian goddess of justice, who was said to be responsible for arranging the stars and maintaining order in the universe. There is evidence to suggest that much of Venus is likely to be currently volcanically active. Reconnaissance by the Magellan mission (1990–1994) found ash flows close to the summit, which might indicate that, while Maat Mons is not erupting at the moment, it has been active relatively recently. The Pioneer Venus probes that collected data in the late 1970s also found extremely high concentrations of both sulphur dioxide and methane in Venus' upper atmosphere. These may well have been caused by Maat Mons erupting and spewing gases high into the air.

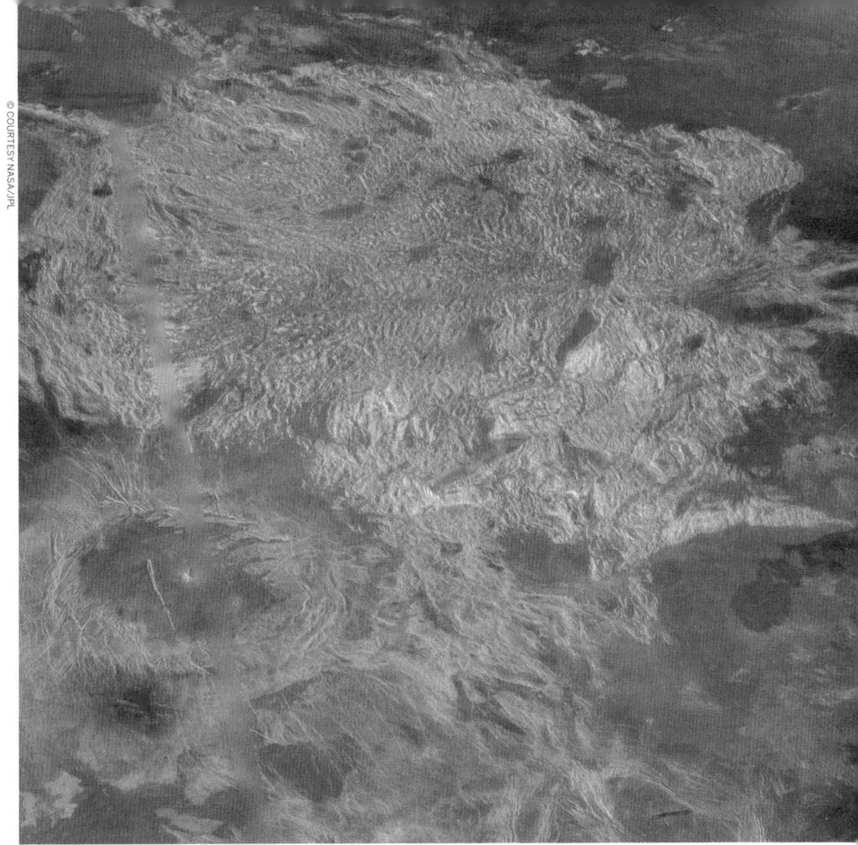

A 3D view of Alpha Regio.

Alpha Regio

At an elevation of 1000 to 2000 m (3280 to 6561 ft) above the surrounding volcanic plains, the mysterious rock formations of Alpha Regio have even been compared to celestial parquet flooring due to their tiled appearance.

First discovered in 1963, Alpha Regio is perhaps Venus' finest example of features known as tessera. The word is derived from the Greek for 'tiled', and is most easily explained by reactions to the sightings of Alpha Regio in maps sent back from the Russian orbiters Venera 15 and Venera 16: the pitted, concertina-like rock formations of this plateau resemble a parquet floor. During its mission to the planet, in 2006, the orbiter Venus Express helped create the first infrared maps of the area, which showed the rocks of Alpha Regio to be lighter in colour than most of the rest of the planet. Based on weathering, this indicates they might be significantly older than rocks found elsewhere on Venus. Alpha Regio is one of only three features on Venus not named after a woman or goddess, the others being the neighbouring tessera of Beta Regio and Maxwell Montes.

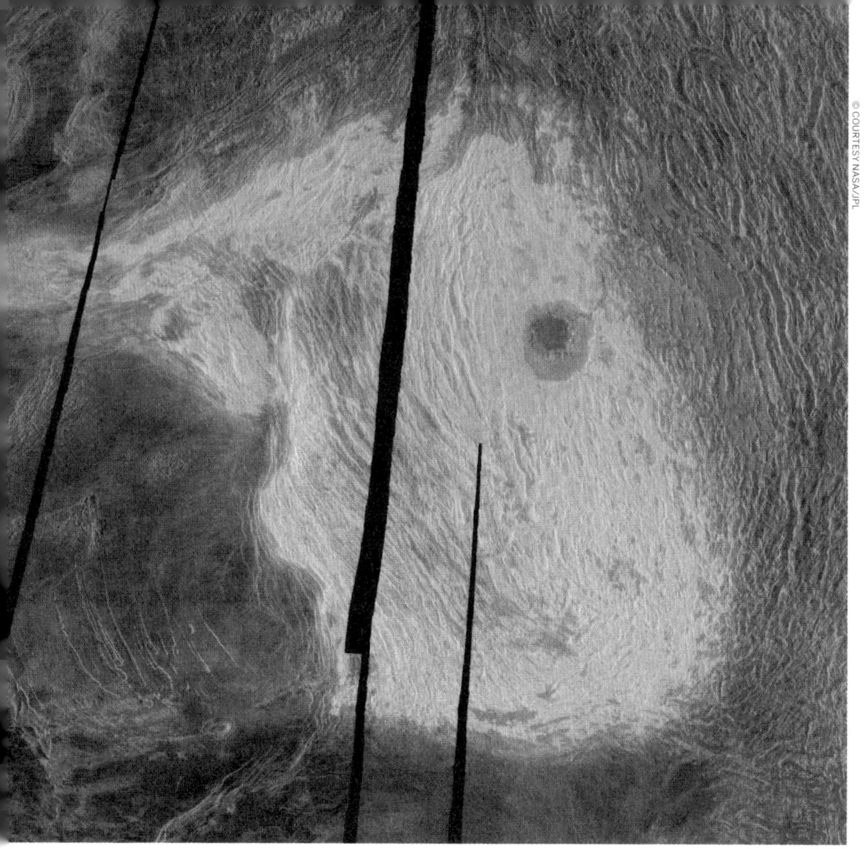

© COURTESY NASA/JPL

An overhead view of the broad Maxwell Montes and nearby Cleopatra Crater.

Maxwell Montes

This vast mountain massif is, quite literally, the coolest place to hang out on Venus. If you're lucky, you might even arrive in time to get a dusting of metal 'snow' on its heights.

The highest point on the surface of Venus, the massif of Maxwell Montes is located on the highlands of Ishtar Terra. Rising, at its highest point, 11 km (6.8 mi) above the planet's mean radius, the extreme elevation makes it the coolest place on the planet – although the coolest place on the solar system's hottest planet still pushes the thermometer to around 380°C (716°F); similarly, the atmospheric pressure, the lowest on the planet, is still a crushing 45 bar. Maxwell Montes is named after James Clerk Maxwell, the mathematician and physicist whose work led to the discovery of radio waves. The mountain was first discovered, by the Arecibo Radio Telescope, in 1967. More than a decade later, in 1978, the orbiter Pioneer Venus 1 confirmed Maxwell Montes as Venus' highest point. Radar surveys of the mountain return extremely bright results, probably due to a high mineral content, which scientists believe has been deposited in a kind of metallic snow.

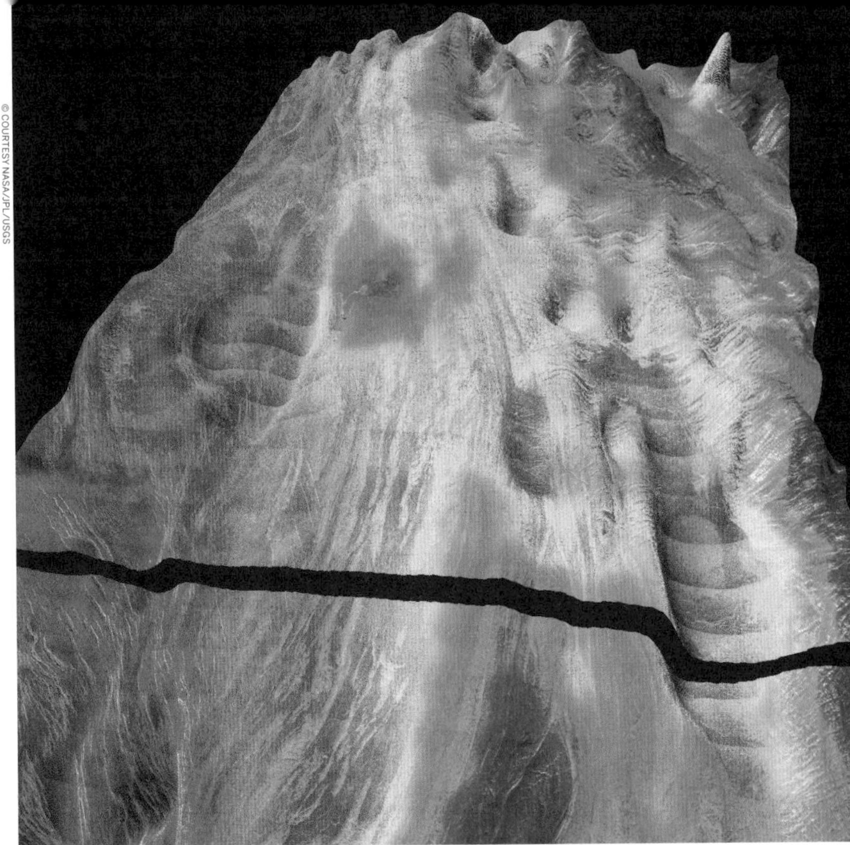

A view from Magellan of Ovda Regio, the western hemisphere for Aphrodite Terra.

Aphrodite Terra

The second and largest Venusian continent, highland area Aphrodite Terra holds unique and intriguing volcanic treasures along the Venusian equator.

The second of the two Venusian 'continents' (the other is the highland region Ishtar Terra), Aphrodite Terra is the region named, of course, for Aphrodite, Greek goddess of love. Having merely a continent named after you, rather than an entire planet, would seem, on the face of it, enough to make any goddess feel insecure. However, the continent in question is quite respectable in area, between the sizes of South America and Africa.

While it is undoubtedly less dramatic than the soaring mountains of Ishtar Terra – where looming Maxwell Montes reigns supreme – Aphrodite Terra has volcanic treasures of its own. Specifically, the rough, jumbled terrain whose numerous lava flows point to its volatile creation. Aphrodite Terra is further divided into two hemispheres: Ovda Regio to the west, and eastern counterpart Thetis Regio. Of the two, Ovda Regio is the more distinctive, characterised by numerous scars in the land, deep rift-type valleys that intersect across the continent.

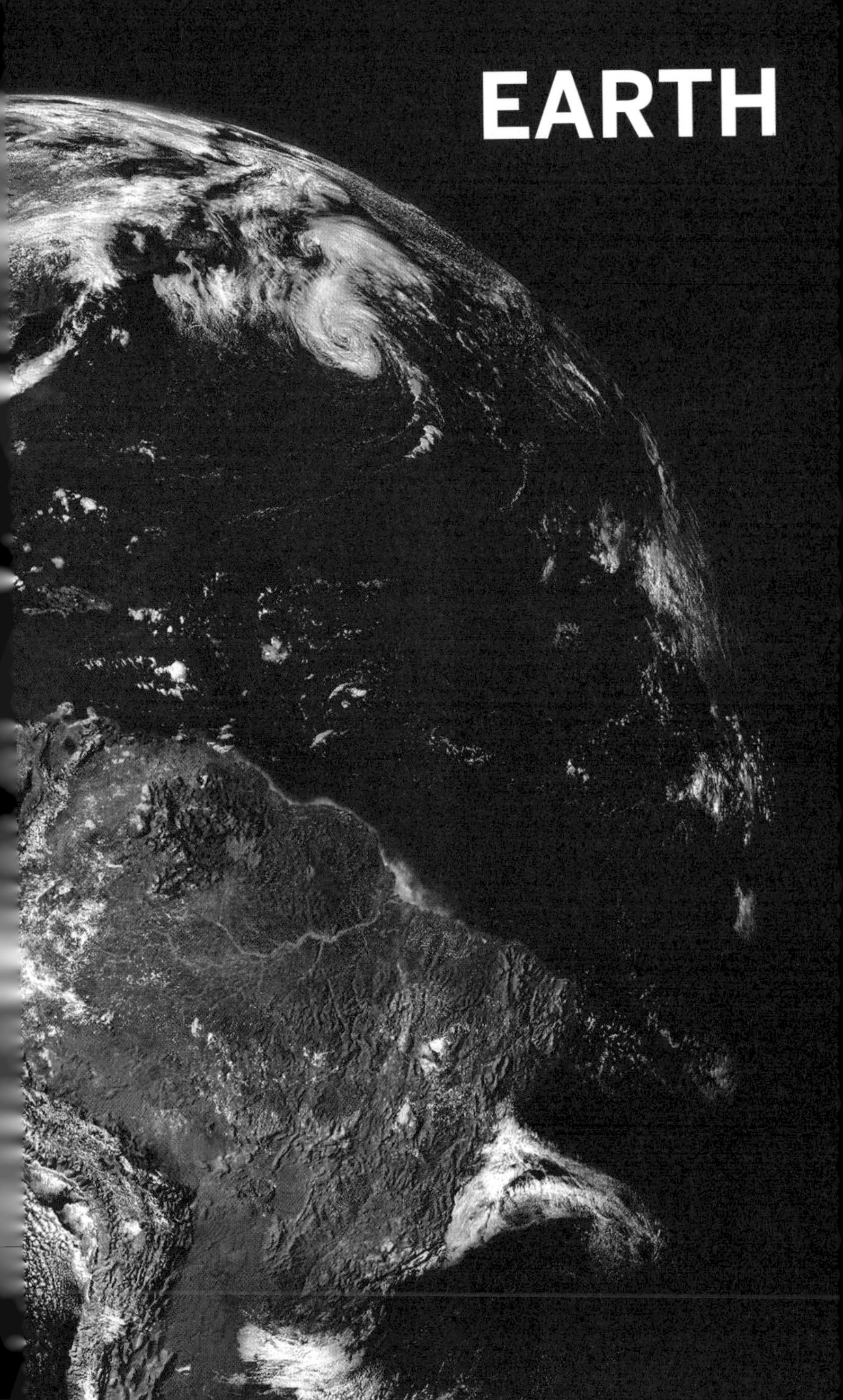
EARTH

Earth as seen from space.

Earth at a Glance

Despite the number of planets, not to mention universes, which astrophysicists and astronomers now believe might exist, one fact remains: Earth is the only planet we know of that sustains life.

The word 'Earth' is at least 1000 years old, an amalgam of the Saxon 'ertha', the Dutch 'aerde' and the German 'erda' – all of which mean 'ground'. (Earth is the only planet not named after a Greek or Roman deity.) The third-closest planet to the Sun, our home is the fifth-largest planet in the solar system. If the Sun were the size of the average household door, the Earth would be the size of a nickel. The Earth orbits the Sun, which is in fact a star, at a distance of 150 million km (93 million mi) and one orbit takes 365 days. Earth is the only world in our solar system featuring liquid water on its surface. But, along with its fellow terrestrial planets, Earth is composed of a molten core, a rocky mantle and a solid crust. With one moon and no rings, the Earth is protected from incoming meteoroids by its atmosphere, which breaks up incoming debris.

The first Earthling to see its home from orbit was a terrier named Laika, who circled the planet in 1957 aboard Sputnik 2. Although she did not survive the trip, two subsequent Soviet space dogs – Belka and Strelka – became, in 1960, the first living creatures to return from orbit alive, paving the way for human explorers. Popular culture has generated countless alternative views of Earth, with the planet and its population governed by everything from apes to a stone monolith. But how much longer travellers, canine or otherwise, will be able to thrive on Earth is the subject of heated debate. Quite literally.

The fate of Earth is inextricably linked to that of the Sun. Models predict that, in around 5 billion years, the Sun will become a red giant. It will increase to 100 times its present size, reaching a luminosity 2000 times its current level. At that point it will vaporise the Earth, whose water will have already evaporated. But that leaves plenty of time to take in Earth's natural wonders: oceans, mountains, deserts and jungles – all teeming, exclusively in the entire known Universe, with an extravagant abundance of life.

Laika, the first dog in orbit, lives on in countless commemorations.

DISTANCE FROM SUN
1 AU

LIGHT-TIME TO THE SUN
8.25 minutes

LENGTH OF DAY
24 hours

LENGTH OF ORBITAL YEAR
365.25 days

ATMOSPHERE
Nitrogen, oxygen, trace gases

Top Tip

Visitors to Earth should plan their itinerary while there's still time. Venice is sinking, Machu Picchu is collapsing and the lush Congo Basin could be two-thirds gone by 2040, while experts say at least 27 species go extinct each day. Sobering facts, only partly offset by the regular new discoveries being made of countless new species in the Amazon and deep ocean.

Getting There & Away

With commercial space travel imminent, be sure you know what you're signing up for. The Earth's moon can be reached in about three days, while suborbital flights can pass in under an hour. But travelling to the former ninth planet from the Sun, Pluto, took New Horizons, launched in 2006 and the fastest probe ever to leave Earth, nine and a half years. For now, leaving our home planet is only for a rare few.

Earth's Seasons

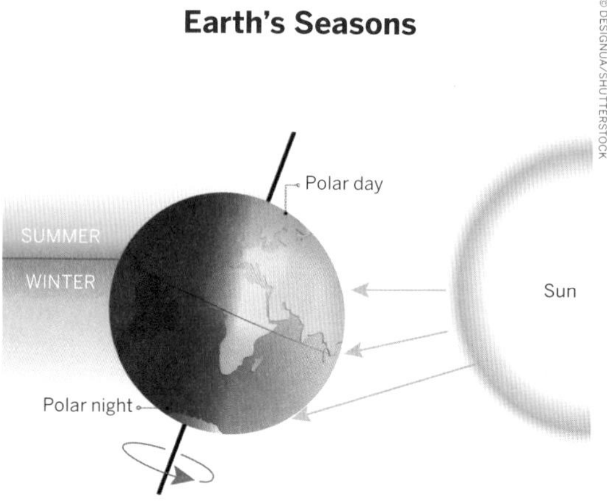

Earth's axis determines how the seasons change throughout the year.

Orientation

If there is anybody else out there, what would they see looking at Earth? A planet whose radius of 6371 km (3959 mi) makes it the biggest of the terrestrial planets and the fifth-largest planet overall. With an average distance of 150 million km (93 million mi), Earth is exactly one astronomical unit away from the Sun because one astronomical unit (abbreviated as AU), is the distance from the Sun to Earth. This unit provides an easy way to quickly compare other planets' relative distances from the Sun. It takes about eight minutes for light from the Sun to reach Earth.

As Earth orbits the Sun, it rotates once every 23.9 hours. It takes 365.25 days to complete one trip around the Sun. That extra quarter of a day presents a challenge to our calendar system, which counts one year as 365 days. To keep yearly calendars consistent with Earth's orbit around the Sun, every four years sees the addition of one extra day, a leap day, more commonly expressed by the year in which it is added – a leap year.

In fact, the length of Earth's day is increasing. When Earth was formed, 4.6 billion years ago, its day would have been roughly six hours long. Around 620 million years ago, this had increased to 21.9 hours. Today, the average day is 24 hours long, but its length is increasing by about 1.7 milliseconds every century. This is caused by the moon, whose gravity slows Earth's rotation through the tides it helps create. Earth's spin causes the position of its tidal ocean bulges to be pulled slightly ahead of the Moon-Earth axis, which creates a twisting force that in turn decreases the speed of Earth's rotation.

Earth *vs* the Planets

-- Radius --
11x
SMALLER
than Jupiter

-- Mass --
17x
LESS
than Neptune

-- Volume --
1321x
LESS
than Jupiter

-- Surface gravity --
2.5x
LESS
than Jupiter

-- Mean temperature -
-466°C (-808°F)
COLDER
than Venus

-- Surface area --
6.8x
MORE
than Mercury

-- Surface pressure --
92x
LESS
than Venus

-- Density --
8x
DENSER
than Saturn

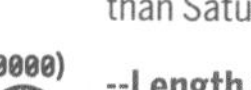

--Length of Day --
1.41x
LONGER
than Uranus

-- Orbit Period --
53%
SHORTER
than Mars

Earth's axis is an imaginary pole going right through the planet's centre from top to bottom. Earth spins around this pole, making one complete turn each day. That is why we have day and night, and why every part of Earth's surface gets some of each. When Earth was young, it is thought that something big hit Earth and knocked it off-kilter. So instead of rotating with its axis straight up and down, it leans over a bit. As Earth orbits the sun, its tilted axis always points in the same direction. This means that throughout the year, different parts of Earth get the sun's direct rays. This tilt causes the yearly cycle of the seasons.

Roughly speaking, the northern hemisphere is tilted towards the Sun between the months of April and September, while the southern hemisphere is tilted away. With the Sun higher in the sky, direct solar heating is greater in the north, creating summer conditions. Conversely, reduced solar heating in the south induces winter. Six months later, the situation is reversed. At the beginning of their respective spring and fall seasons, both hemispheres receive roughly equal amounts of heat from the Sun. Today, the Earth's axis is tilted 23.5 degrees from the plane of its orbit around the sun. But this tilt changes. During a cycle that averages about 40,000 years, the tilt of the axis varies between 22.1 and 24.5 degrees. As the axis changes, the seasons as we know them can become exaggerated.

Distance from the Sun, however. doesn't impact our experience of the seasons. While the difference between perihelion (when Earth is closest to the Sun) and aphelion (our farthest point from the Sun) is over 4.8 million km (3 million mi), relative to our total distance from the Sun it isn't much, and has no appreciable impact on how Earth's weather changes throughout the year. The other major factor affecting the planet's climate and short-term local weather is Earth's own atmosphere.

Fun Fact: Guest from Above

In November 2018, NASA glaciologists discovered a prime example of just what can happen when the atmosphere is off its game: a large impact crater hiding beneath more than a half-mile of ice in northwest Greenland. The crater, under the Hiawatha Glacier, was created by a meteorite estimated to have struck at least 12,000 years ago. The crater is 300 m (1000 ft) deep and 13 km (19 mi) in diameter. NASA's Operation Icebridge discovered the crater's existence using radar data gathered on polar flights.

Hiawatha Glacier in Greenland, seen by NASA.

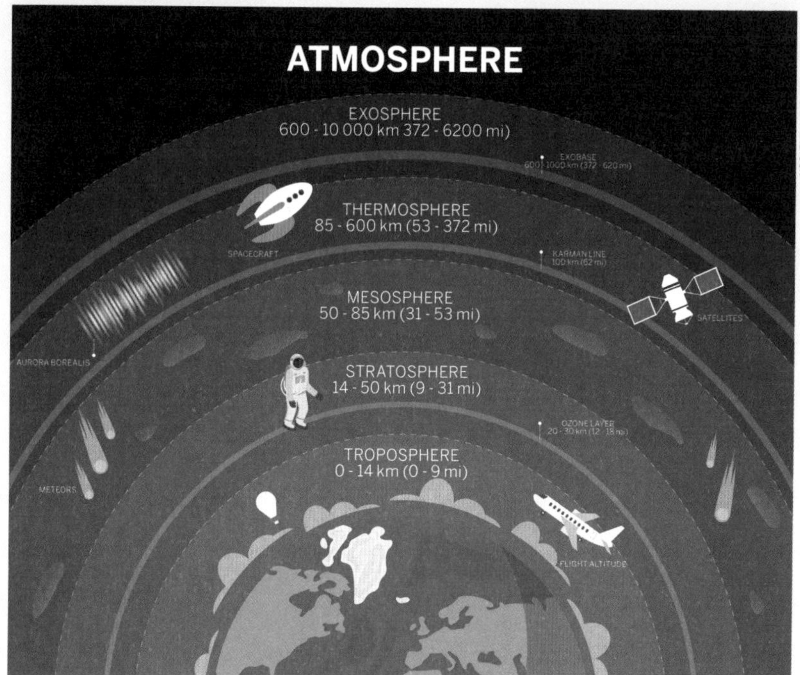

Layers of the Atmosphere

Troposphere

The densest part of the atmosphere, this region starts at the Earth's surface and extends for up to 14 km (9 mi). It is where the Earth's weather systems occur, the zone we experience on a daily basis.

Stratosphere

Above the troposphere, the stratosphere takes over up to 50 km (31 mi). Home to the crucial ozone layer, this atmospheric region absorbs and dissipates harmful solar ultraviolet radiation.

Mesosphere

This is the layer to thank for meteors not crashing to Earth more often. The meteoroid-burning mesosphere extends from the stratosphere's edge up to 85 km (53 mi).

Ionosphere

Abundant in electrons, ionized atoms and molecules, the ionosphere exists across all atmospheric layers above the stratosphere. It starts approximately 50 km (30 mi) above the Earth and extends to the edge of space, defined as about 600 km (372 mi). This dynamic part of the atmosphere expands or shrinks depending on solar conditions. It is critical to interactions with the solar wind and also makes radio communications possible, as the proliferation of ions and electrons 'reflect' radio signals from broadcaster to receiver.

Thermosphere

The region in which aurorae, aka the Northern and Southern Lights, occur, the thermosphere starts above the mesosphere and extends to 600 km (372 mi). It is also the layer in which satellites most commonly orbit the Earth.

Exosphere

The atmosphere's upper limit, the exosphere reaches to 10,000 km (6200 mi) above the surface of the Earth.

The Magnetosphere

The magnetosphere is formed by the magnetic field that fully covers the Earth. Other planets in the solar system have magnetospheres, but Earth has the strongest of all the rocky planets: our magnetosphere is a vast, comet-shaped bubble, which has played a crucial role in our planet's habitability. Constant bombardment by the solar wind compresses the sun-facing side of our magnetic field. The sun-facing side, or dayside, extends a distance of about six to ten times the radius of the Earth. The side of the magnetosphere facing away from the sun (the nightside) stretches out into an immense magnetotail, which fluctuates in length and can measure hundreds of Earth radii, well beyond the orbit of the moon.

The magnetosphere is created by the planet's rapid rotation and its molten nickel-iron core. Like the atmosphere, it protects the Earth, especially from the solar wind, a stream of charged particles continuously ejected from the Sun. Our magnetosphere deflects most particles away from the Earth. When charged particles do become trapped in Earth's magnetic field, they collide with air molecules above our planet's magnetic poles, which create the glowing aurorae.

The magnetic field is what causes compass needles to point to the north pole, no matter which way you're facing. But the magnetic polarity of Earth can change, flipping the direction of the magnetic field. The geologic record indicates that a magnetic reversal takes place on average about every 200,000 to 400,000 years, but the timing is irregular, and the impetus is still unknown.

As far as we know, such a magnetic reversal doesn't cause any harm to life on Earth, and the next polar reversal is unlikely to happen for at least another thousand years. When it does occur, compass needles will probably point in many different directions for a few centuries while the switch occurs. After the polar switch is completed, they will all point south, rather than north.

Invisible to us, the magnetosphere prevents cosmically charged particles from reaching Earth's surface.

History

Earth, the third-closest planet to the Sun, was originally formed by swirling gas and dust, pulled in by gravity about 4.5 billion years ago. Earth is composed of four main layers – starting with an inner core at the planet's centre, enveloped by an outer core, a rocky mantle and finally a solid crust.

The inner core is a solid sphere, about 1221 km (759 mi) in radius, and made of iron and nickel metals. The temperature is a spectacular 5400°C (9800°F). Surrounding the inner core is the outer core. This layer is about 2300 km (1400 mi) thick and is made of iron and nickel fluids. Between the outer core and crust is the thickest layer, the mantle. A hot, viscous mixture of molten rock about 2900 km (1800 mi) thick, it has the consistency of caramel.

Earth is relatively thin-skinned. Its outermost layer, the crust, runs to an average depth, on land, of about 30 km (19 mi). At the ocean bottom, the crust is thinner still, extending downward for about 5 km (3 mi), from the sea floor to the top of the mantle, named the asthenosphere.

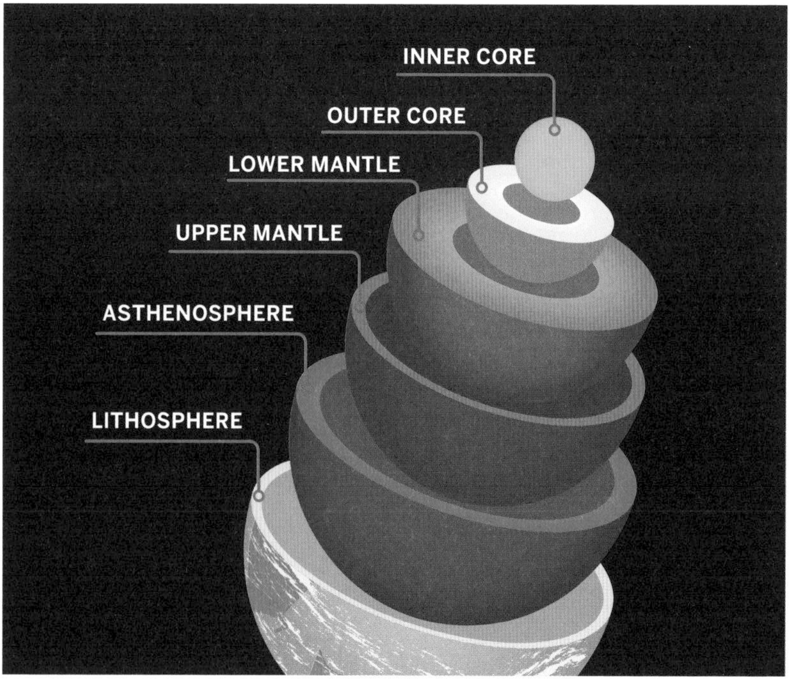

A diagram of Earth's inner mantle and core.

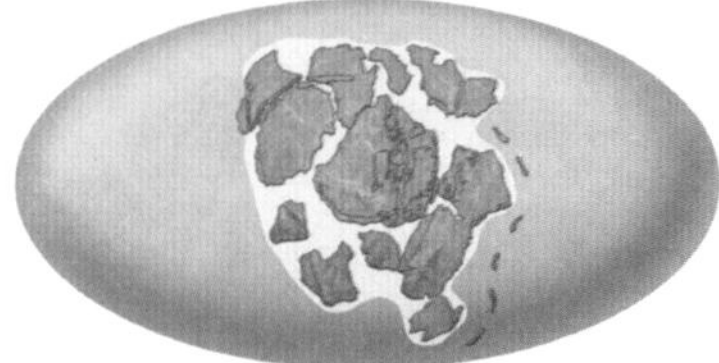

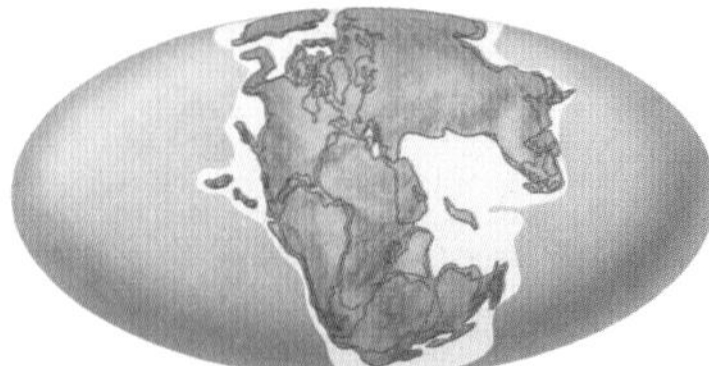

As Earth's tectonic plates have moved, the continents have changed shape.

Underwater Mountains

Earth's global ocean, which covers nearly 70% of the planet's surface, has an average depth of about 4 km (2.5 mi) and contains 97% of Earth's water. Almost all Earth's volcanoes are hidden beneath the sea. At over 10 km (6 mi), Hawaii's Mauna Kea volcano is taller, from base to summit, than Mount Everest (but most of it is beneath the water).

Earth's longest mountain range, the Mid-Ocean Ridge, is sub-aquatic too, at the bottom of the Arctic and Atlantic oceans. Spanning 65,000 km (40,389 mi), it is four times longer than the Andes, Rockies and Himalayas combined.

Like Mars and Venus, Earth also has volcanoes, mountains and valleys. Earth's lithosphere, which includes both the continental and oceanic crust and the upper mantle, is divided into huge constantly moving plates. The North American plate, for example, moves west over the Pacific Ocean basin, at a rate roughly equal to the growth of human fingernails. Earthquakes result when plates grind past or ride up over one another; mountain ranges occur when plates collide. In their travels, tectonic plates cycle through 100 million-year supercontinent cycles. The oldest one currently postulated, Kenorland, would date to over 2 billion years ago, followed by Nuna 1.8 billion years ago. Next was Rodinia, which formed 1 billion years ago and lasted until 800 million years previous. At the centre of Rodinia lay what is now North America. Scientists can only estimate when Rodinia broke apart, but some of the many pieces it split into re-collided some 250 million to 500 million years ago, creating the Appalachian Mountains in North America and the Ural Mountains in Russia and Kazakhstan. Around this time, the continents came together once again to form another supercontinent, Pangaea. Like Rodinia, it was surrounded by a single worldwide ocean. On Earth, it seems, breaking up isn't particularly hard to do. Fifty million years later, it was Pangaea's turn to fall to pieces: the separate land masses of Gondwanaland and Laurasia, which fragmented further into today's continents. In the long intervals between supercontinents, continents and smaller landmasses make their way across Earth's surface pell-mell, like errant puzzle pieces. Scientists can use detective work to put together an image of the past configuration by analysis of iron-bearing magnetic minerals, which show where rocks were when they first cooled. This is called their 'paleoaltitude,' or alignment to Earth's magnetic pole. In the future, one theory is that plate tectonics will erase the Caribbean and Arctic Oceans to make a supercontinent in the Northern Hemisphere: the new landmass of Amasia.

Snowball Earth

In the last 650,000 years, there have been seven cycles of glacial advance and retreat, and scientists even believe that, in some cases, the climate became so cold that Earth froze over completely. This is known as the 'Snowball Earth' theory (as opposed to 'Slushball Earth'). There may have been four such periods of alternate freezing and thawing, triggered by reductions in greenhouse gases such as methane and carbon dioxide. During these times, Earth would have been covered by glacial ice from pole to pole. Most of the Sun's energy would have been reflected back into space by the ice, and the planet's average temperature would have been about -50°C (-74°F).

The abrupt end of the last ice age, about 7000 years ago, marked the beginning of the modern climate era – and of human civilisation. Most of the previous climate changes have been attributed to small variations in Earth's orbit, which changed the amount of solar energy our planet received. The current warming trend is of particular significance because it is extremely likely, with greater than 95% probability, that most of it is the result of human activity since the mid-20th century. Over the course of just a few decades, this warming has proceeded at a rate greater than the increases previously spread out over millennia.

'Ice cores' drawn from Antarctica and Greenland contain records of historic atmospheric carbon dioxide (CO_2) levels. Other evidence can be found in tree rings, ocean sediments, coral reefs and layers of sedimentary rocks. This 'paleoclimate' data reveals current warming is occurring roughly ten times faster than the average rate of ice-age recovery.

The heat-trapping nature of CO_2 and other gases was demonstrated in the mid-19th century, and is known as the greenhouse effect. In fact, their ability to affect the transfer of infrared energy through the atmosphere is the scientific basis of many NASA instruments. The increase of carbon dioxide in the atmosphere may seem small, but it can drive big changes.

Since the late 19th century, Earth's average surface temperature has risen about 0.9°C (1.62°F), a change driven largely by increased CO_2 and other human-made emissions. Most of the warming occurred in the past 35 years; since 2010 we have experienced five of the warmest years on record, with 2016 the warmest year yet (likely to be surpassed).

Alaska's Yakutat Glacier seen in retreat as the climate warms.

The People Versus the Planet

Satellites orbiting the Earth, along with other technological advances, have enabled scientists to see the big picture, collecting evidence confirming how human activity has contributed to a rapidly warming climate on a global scale.

Earth and the moon seen from orbit.

1) Warming oceans

The oceans have absorbed much of the climate's increased heat. In 2019, research suggested ocean warming had outpaced IPCC estimates by 40%.

2) Shrinking ice sheets

Greenland lost an average of 286 billion tons of ice per year between 2002 and 2017. Antarctica lost about 127 billion tons per year during the same period. The rate in Antarctica's ice loss has tripled in the last decade.

3) Extreme weather events

Since 1950, the number of record-high-temperature events worldwide has been increasing, while the number of record-low-temperature events has been decreasing. Heat waves in Europe, Australia, Asia and the Americas all cause fatalities and lead to stronger wildfire seasons. The world has also witnessed increasing numbers of intense rainfall events, with drought years increased as well.

4) Sea-level rise

Global sea levels rose about 8 inches in the last century. The rate in the last two decades, however, is nearly double that and is accelerating every year.

5) Declining Arctic sea ice

Arctic sea ice has declined rapidly over the last few decades, both in extent and thickness. In 2012, the Arctic sea ice minimum, when the melting ice reaches its minimum annual extent (around September), was the lowest on record. An ice-free north pole during the summer may be in Earth's future this century, opening up entirely new areas for shipping and, potentially, commercialisation.

6) Glacial retreat

Monitored from space, glaciers can be seen retreating at a rapid pace from almost every mountain range in the world, including the Alps, Himalayas, Andes and Rockies, as well as from Mount Kilimanjaro.

7) Ocean acidification

Since the Industrial Revolution, ocean acidity has increased by about 30%, lowering pH levels in oceans worldwide. This is the result of increased carbon dioxide emissions into the atmosphere. The amount of carbon dioxide absorbed by the upper layer of the oceans is increasing by 2 billion tons per year. Acidification is one cause of coral reef bleaching, and it promotes invasive algae blooms as well.

© TANAKORN KHATIYASONTORN/SHUTTERSTOCK

The moon during a lunar eclipse, a phenomenon that is a clear sign of Earth's spherical nature.

The Myth of a Flat Earth

That our medieval ancestors once considered the Earth to be flat may be as big a myth as the idea itself. In fact, this vision of a flat Earth owes as much to art as to science: painters like Hieronymus Bosch, for example, whose 15th-century work *The Garden of Earthly Delights* depicts a striking flat earth floating inside a sphere.

In fact, we have known that Earth is round for more than 2000 years, on and off. The ancient Greeks measured shadows during the summer solstice and also calculated Earth's circumference. They used the positions of stars and constellations to estimate distances on the Earth's surface, and recorded seeing, during a lunar eclipse, the round shadow of the Earth against the moon. Today, scientists use 'geodesy' – the science of measuring Earth's shape, gravity and rotation. Geodesy provides accurate measurements that show Earth is spherical, and by using GPS, scientists can measure Earth's size and shape to within a centimetre. Pictures from space also show Earth is round, like its moon.

Thanks to social media among other factors, however, there are still plenty who doubt the science – the UK held its *first* Flat Earth Convention in 2018. The flat-Earthers have a point only inasmuch as the Earth has never been a perfect sphere. The planet bulges around the equator by an extra 0.3% as it rotates about its axis. Earth's diameter from the north pole to the south pole is 12,714 km (7900 mi), while through the equator it is 12,756 km (7926 mi). The difference is a mere 42.78 km (26.58 mi), about 1/300th the diameter of Earth. This variation is too tiny to be seen in pictures of Earth from space, so the planet appears round to the human eye. Recent research from NASA's Jet Propulsion Laboratory suggests that melting glaciers are also causing Earth's 'waistline' to expand.

The Shape of the Earth: A Brief Round-Up

1) Ancient Greece

From around the 6th century BC, Greek cosmologists were unanimous in the belief Earth was round. Aristotle noted stars seen from Egypt were different from those seen in Cyprus, suggesting a curved surface. Aristotle also claimed that heavy things, such as earth and water, were drawn to the centre of the sphere, while light objects – fire and air in particular – went up.

2) Roman Empire

By the time of Cicero and Pliny, around 50 BC, it was accepted the Earth was round. Roman geographers, such as Strabo, advanced theories based on evidence from seafarers, who noted high ground was visible from a distance when lower ground was not. This would only be possible, Strabo concluded, if the Earth curved.

3) India

By the 6th century AD, astronomers, such as the Indian scholar Aryhabatta, were getting close to accurate measurements of the planet. In his great work, modestly titled the *Aryabhatiya*, the circumference of the Earth is given as 4967 yojanas, or 39,968 km (24,834 mi). The actual circumference at the equator is 40,075 km (24,901 mi).

4) Middle Ages

Bishop – later Saint – Isidore of Seville preached that the Earth was round, although his tendency to use arcane Latin terminology led some to conclude that he in fact meant the planet was disk-shaped. Our medieval forebears may not have been flat-Earthers, yet they doubted it possible for humans to live in the southern hemisphere.

5) Islamic astronomy

To calculate the distance from any given point to Mecca, Islamic mathematicians pioneered the science of spherical trigonometry. In the 10th and 11th centuries, the scholar Abu Rayhan Al-Biruni developed principles of triangulation, allowing him to calculate the radius of the Earth to an almost exact number. The figure was not achieved in the West until the 16th century.

6) China

Relative latecomers to the round-Earth party, Chinese astronomers did not wholeheartedly acknowledge a spherical planet until the 17th century, when they observed that ships could return to their port of origin by effectively sailing in a straight line.

A flat-Earth concept illustration.

Satellite Earth: Ten NASA Missions Helping to Better Understand the Planet

NASA Earth Science is an extensive research programme that monitors the ways the planet is changing. Using highly sensitive instruments, NASA's observations provide an increasingly detailed understanding of the interaction between the Earth's oceans, air, land and life. NASA satellites help study and predict weather, drought, pollution, climate change, and many other phenomena. Here are some key ongoing missions, and several planned for the future.

1) Landsat

Launch date (as Earth Resources Technology Satellite): July 1972

Since its first launch, in 1972, data from Landsat, as the original satellite was renamed, has helped ecologists track deforestation, as well as the retreat of mountain glaciers. Water authorities have monitored irrigation of farmland in America's West, while population researchers have watched the growth of cities worldwide. A new Landsat, 9, is due to launch in December 2020.

2) Gravity Recovery and Climate Experiment (GRACE)

Launch date: March 2002

Tracking changes in Earth's water reservoirs – lakes, ice sheets, glaciers – GRACE monitored sea-level rises caused by the addition of water to the oceans. A 'follow-on', GRACE-FO, was launched in May 2018. Information from both is used to construct maps of Earth's gravity field, offering details of how mass, in particular water, moves around the planet.

3) Solar Radiation and Climate Experiment (SORCE)

Launch date: January 2003

SORCE measures electromagnetic radiation produced by the Sun, and its effects on the Earth's surface. One of SORCE's primary successes has been the daily record of Total Solar Irradiance (TSI), above Earth's atmosphere. TSI is a critical variable, as energy from the Sun affects Earth's climate and weather systems.

4) Suomi National Polar-orbiting Partnership (Suomi NPP)

Launch date: October 2011

Named after the meteorologist Verner Suomi, NPP maps land cover, as well as changes in vegetation productivity. NPP also tracks atmospheric ozone and aerosols; sea- and land-surface temperatures; and natural disasters, such as volcanic eruptions and floods.

Landsat orbiting Earth.

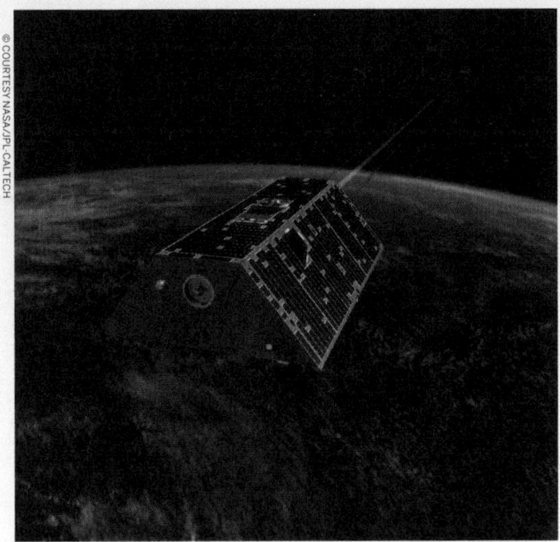

© COURTESY NASA/JPL-CALTECH

Artist's rendering of the twin spacecraft of the GRACE-FO mission.

5) Soil Moisture Active Passive SMAP

Launch date: January 2015

An orbiting observatory that measures water in the Earth's topsoil. SMAP's data on soil moisture have helped understand how water and carbon circulate. The amount of water that evaporates from the land into the atmosphere depends on soil moisture; data on soil moisture is key to understanding how water and heat energy impact the climate.

6) Ice, Cloud and Land Elevation Satellite-2 (ICESat-2)

Launch date: September 2018

ICESat-2's 'photon-counting laser altimeter' allows scientists to measure the elevation of ice sheets, glaciers and sea-ice. ICESat-2 investigates why, and how, the 'cryosphere', the frozen areas of the planet, are changing. ICESat-2 also measures temperate and tropical regions and vegetation levels.

7) Earth Observing System (EOS)

Launch dates: December 1999 (Terra); May 2002 (Aqua); July 2004 (Aura).

NASA's Earth Observing System (EOS) has a number of satellites working together to build a comprehensive picture of the Earth's climate: *Terra* helps map the impact of human activity and natural disasters on ecosystems, *Aqua* collects data on the water cycle and *Aura* measures ozone and gases in the atmosphere.

8) GPM (Global Precipitation Measurement)

Launch date: February 2014

A collaboration with the Japanese Aerospace Exploration Agency (JAXA), the GPM Core Observatory satellite is an international mission to provide next-generation observations of rain and snow worldwide. Relaying data every three hours, its mission is to unify global precipitation measurements through a network of partner satellites.

9) Surface Water and Ocean Topography (SWOT)

Launch date: September 2021

The SWOT mission unites US and French oceanographers and hydrologists in this satellite mission to make the first global survey of Earth's surface water, observe the fine details of the ocean's surface topography, and measure how water bodies change over time.

10) Tropospheric Emissions: Monitoring Pollution (TEMPO)

Launch date: TBC

The first space-based instrument to monitor major air pollutants across the North American continent will go into orbit above the equator. From here, TEMPO will monitor daily variations in ozone, nitrogen dioxide and other elements of air pollution.

Earth Highlights

From arid deserts to plains and waters teeming with life, we present the highs, lows and environmental wonders every visitor to Earth should add to their bucket list.

Mt Everest

1 At a towering 8848 m (29,029 ft), the summit of Mt Everest is Earth's highest point above sea level. The border between Tibet and Nepal straddles its peak.

Challenger Deep

2 Located in the Pacific Ocean, at the southern end of the Mariana Trench, Challenger Deep is the deepest known point in the planet's seabed, at 10,900 m (35,800 ft).

Atacama Desert

3 The Atacama Desert is the driest place on Earth. Located in northern Chile, it is found next to Earth's largest body of water, the Pacific.

Mauna Kea

4 A dormant Hawaiian volcano, Mauna Kea is 1 million years old. With its peak at 4207 m (13,802 ft) above sea level, and an oceanic base of more than 6000 m (16,685 ft), it is Earth's tallest mountain.

Chicxulub Crater

5 The largest surviving example of an impact crater yet found, this crater in Mexico was caused when an asteroid, estimated to be about 10 to 15 km across (6 to 9 mi), struck the Earth 66 million years ago.

Silfra

6 Silfra, in Iceland's Thingvellir National Park, is a rift between Earth's North American and Eurasian tectonic plates. It offers the clearest water on Earth, with visibility of up to more than 100 m (300 ft). Scuba heaven.

Death Valley

7 Located in eastern California's Mojave Desert, Death Valley holds the accolade for the hottest air temperature ever recorded on Earth, a blistering 56.7°C (134°F).

Inhospitable to most forms of life, Antarctica supports seven different species of penguin.

The sun sets over a massive crater with a salt deposit in the middle of eerie terrain at the Valle de Luna (Moon Valley).

Antarctica

8 With average winter temperatures of -50°C (-122°F), the coldest place on Earth has abundant wildlife and spectacular geological features, made more poignant given the threat to its climate.

Great Barrier Reef

9 At 2300 km long (1400 mi), Earth's largest coral reef is the biggest structure created by living organisms. It is even visible from space!

Amazon Rainforest

10 The 5,500,000 sq km (2 million plus sq mi) of Brazil's Amazon rainforest are home to 390 billion trees, one-fifth of all bird species, and 2.5 million species of insects. And more. Lots more.

Ngorongoro Conservation Area

11 Safari destination Ngorongoro is home to Africa's so-called big five: elephants, lions, rhinos, cape buffalo and leopards.

Great Wall of China

12 At over 21,000 km (13,048 mi) in length, the man-made Great Wall is visible from space, with magnification, that is.

Mount Everest from Kala Patthar on the way to Everest base camp, Sagarmatha National Park, Nepal.

Mt Everest

First climbed in 1953, by Edmund Hillary and Tenzing Norgay, Mt Everest is, at 8848m (29,029ft), Earth's most famous mountain. During peak season, April and May, climbers queue to summit.

Everest is the highest of Earth's 14 'eight-thousanders', peaks more than 8,000 m above sea-level, or over 26,247 ft. There are base camps in Nepal, on the south side, and Tibet, to the north, respective starting points for the two main routes up the mountain. Each year, more than 800 people attempt Everest, paying for the privilege. Though not regarded as the most technical climb out there – K2, for example, is more difficult – Everest still poses risks such as altitude sickness, rapidly changing weather, and avalanches. In Nepal, the main danger comes at the bottom, with the constantly shifting Khumbu Icefall; in Tibet it's near the top, on the long, exposed northeast ridge. The first attempts to climb Everest were made in the 1920s, from the Tibetan side, most famously by George Mallory and Sandy Irvine, who disappeared in June, 1924. Close to 300 people have died climbing Everest, with sherpas often bearing the brunt of the fatalities.

Sights

Tenzing-Hillary Airport (Nepal)

1 This is regarded as the world's most dangerous airport. The short flight from Kathmandu heightens nerves. But not as much as the landing on a short runway atop a sheer cliff drop.

Pang La

2 This is the last pass on Tibet's road to Everest. Translated as 'Meadow Pass', from here you can see five massive peaks rise up: Makalu, Cho Oyu, Shishapangma, Kanchenjunga and Everest.

Khumbu Icefall (Nepal)

3 This spectacular river of ice is riven with crevasses and prone to avalanche. The Icefall has accounted for 25% of deaths on Everest's Nepal side. A terrible beauty.

Namche Bazaar

4 Popular with trekkers, this mountain town is, at 3440 m (11,286 ft), a great base for acclimatising to high altitude. It's handy for last-minute mountaineering buys.

Rongbuk Monastery (Tibet)

5 Last stop before base camp on the Tibetan side, with the great mountain in full view. Of huge cultural significance, Rongbuk's monks still bless many Sherpas accompanying expeditions.

An aircraft on the runway of the Tenzing-Hillary airport in Nepal..

What's in a Name?

Everest is revered by the indigenous people of the Himalayas. To the Nepalese, it is *Sagarmatha*, to the Tibetans, *Chomolungma*. Varying translations mean, essentially, 'mother goddess'. It is best known by the moniker bestowed upon it in 1865, when the Royal Geographical Society named it after Sir George Everest, a British Surveyor General of India.

Measuring Contest

Everest provides the backdrop to one of Earth's more petty border disputes. The Himalayas are made of metamorphic rock and were created by the collision between the Indian and Asian tectonic plates. China, which owns half the summit, maintains that the highest rock, at 8844 m, is the vertical extent of the mountain; Nepal, which owns the other half of the summit, insists the mountain be measured by the extent of the snow, at an estimated 8848 m, leaving the officially acknowledged height in some confusion. Independent geologists have recorded vertical growth of about six cm a year from the upthrust of the colliding Asian and Indian continental plates, though the height can erode through weathering. Nepal has undertaken a new, more definitive survey of the height. At either altitude, the air atop Everest is only one-third as dense as the air at sea level.

These hydrothermal tube-worms thrive in the emissions of the Mariana region's underwater volcanoes.

Challenger Deep

Named after HMS Challenger*, which made the first soundings of its depth, in 1872. Modern submersibles have recorded depths of 10,898 m to 10,916 m (35,755 ft to 35,814 ft). Sonar tests suggest it could be deeper still.*

The Mariana Trench, which is home to the even more plunging Challenger Deep, was formed by subduction, a process in which two of the Earth's tectonic plates collide. When the lighter section is pulled beneath the heavier, a deep trench can form. Challenger Deep was created when the Pacific Plate struck the Philippine Sea Plate. The extent of its depth continues to fascinate scientists. The US National Oceanic and Atmospheric Administration (NOAA) measured it in 2010, using sonar, and found an unverified depth of 10,994 (36,069 ft). Director James Cameron undertook his own expedition, two years later, piloting a submersible to a depth of 10,898m (35,756 ft), and was convinced he could have gone deeper. In 2014, scientists at the University of New Hampshire published a sea-floor survey putting Challenger Deep at 10,984 m (36,036 ft) below sea level, even more than previously thought.

Top Trivia

Deep and Narrow

1 The Mariana Trench is 2542 km (1580 mi) long, more than five times the length of the Grand Canyon. However, the width of the trench is, on average, just 69 km (43 mi).

Spherical(ish)

2 Earth isn't a perfect sphere. It's technically an oblate spheroid, bulging somewhat at the equator and squashed near the poles. Thus the radius of Earth, at its poles, is 25 km (16 mi) shorter than its equator radius. This difference in the Earth's shape means the Arctic seabed is closer to the planet's centre than Challenger Deep, despite its having an average depth of just 1038 m (3406 ft). That's just over 1 km and less than a mile.

Pressure Cooker

3 The water pressure on the floor of Challenger Deep is 8 tons per square inch, more than 1000 times that at sea level and the equivalent of lifting 50 jumbo jets. For filmmaker and underwater explorer James Cameron to survive his March 2012 descent into the trench via submersible without succumbing to the pressure, a protective hull of steel 64 mm (2.5 in) thick was required.

White Smokers

4 Challenger Deep is home to examples of rare underwater volcanoes. The Eifuku 'submarine volcano', for example, emits liquid carbon dioxide at a temperature of 103°C (217°F). These bubbling vents are called 'white smokers'.

Everest Eclipsed

5 If Everest were placed at the bottom of Challenger Deep, there would still be a vertical mile of water between the mountain's summit and the ocean's surface.

NASA has considered sending a submersible similar to Challenger Deep to Titan.

Monumental Mariana

The Mariana Islands are a US Commonwealth, meaning the United States has jurisdiction over both the Mariana Trench and Challenger Deep. In 2009, the US established the Mariana Trench Marine National Monument, creating a protected marine reserve for the seafloor and its surrounding waters.

Aliens in the Ocean

We know more about some parts of space than we do about our deepest oceans. While humanity continues to harvest, and pollute, the oceans, they remain rich in life. Estimates suggest as much as 80% of life on Earth is contained in our seas and yet, to date, scientists have managed to identify a mere 225,000 species. This means an estimated 750,000 to 25 million marine species might still be out there awaiting discovery. Using mud samples, scientists have discovered more than 200 different micro-organisms just in Challenger Deep. In 2017, the elusive Mariana snailfish, which lives at depths of 8000 m (26,200 ft), was identified as the apex predator of the Mariana trench zone.

Geysers of Tatio in Atacama, Chile.

Atacama Desert

Occupying 105,000 sq km (40,500 sq mi), Chile's Atacama Desert offers an arid landscape of geysers, salt pans and haunting – some would say Mars-like – scenery.

The aridity of the Atacama, the driest place on Earth, is explained by its position between the mighty Andes and the Chilean Coast ranges. Both are of sufficient height to keep out almost all moisture, from both the Atlantic and Pacific Oceans. For the Atacama, this means a double-whammy - spending life in a two-sided rain shadow. In fact, research suggests that, when taken as a whole, the Atacama Desert may not have had any significant rainfall for more than 400 years. And some parts have had no rainfall. Ever. The annual average for the region is 1 mm, although better hydrated locations within Atacama - coastal cities, such as Arica and Iquique - enjoy up to 3mm (that's still not even an inch!). Dryness does have its upsides, among them the beautifully stark landscape and salt (lots of it). The Atacama region is home to Earth's largest natural supply of sodium nitrate, and also of lithium. Both are vital industries in the region.

Sights

Valle de la Luna (Valley of the Moon)

1 Part of the Reserva Nacional los Flamencos and the apogee of Atacama's otherworldly landscape. Home to wind sculptures so intricate they appear to have been carved by human hands.

Museo Arqueológico San Miguel de Azapa

2 The earliest Egyptian mummies date to 3000 BC; Atacama's Chinchorro mummies, on display here, predate them by 4000 years, conserved in no small part by the region's lack of moisture.

El Tatio

3 Located 4320 m (14,173 ft) above sea level, El Tatio is the largest geyser field in the southern hemisphere. Part of the Altiplano-Puna volcanic complex, it makes for an explosive day out.

Piedras Rojas

4 The red rocks that give the plateau its name are something to behold, although the 4000 m (13,000 ft) of elevation means you might want to take it easy hiking here.

Salar de Atacama

5 Chile's largest salt flat is home to the Laguna Cejar, a lake whose high salt concentration – as much as 28% – means swimmers float easily, as they do in the Dead Sea.

The Salar de Talar salt flats in the Chilean Andes.

Radio Earth

Its clear skies for 300-plus days a year, absence of light pollution and high altitude make Atacama one of the best places on Earth for stargazing. The perfect location, then, for the Atacama Large Millimeter/sub-millimeter Array. Otherwise known as ALMA, the facility studies the stars through its 66 radio telescopes, the world's largest ground telescope among them.

From Atacama to Mars

NASA studies of rocks and minerals suggest dry conditions have persisted in the Atacama for at least 10 million years. Coupled with strong ultraviolet radiation from the Sun, this means what little life there is in the Atacama exists in the form of microbes, living underground or inside rocks. Similarly, if life exists, or ever existed, on Mars, the planet's surface dryness and radiation exposure would likely drive it underground. That makes the Atacama a key location for Earth-based research into life on the Red Planet. In 2016, NASA used the arid core of the desert, near Estación Yungay, to test the KREX-2 prototype rover and various life-detection instruments in advance of a possible launch to Mars for a potential mission in search of microbes on the Red Planet.

Observation telescope at Mauna Kea.

Mauna Kea

Mauna Kea, or 'white mountain', takes its name from the snow that dusts the top of this colossal volcano, a strikingly unusual weather phenomenon in the tropical Pacific.

Located on Hawaii's Big Island, Mauna Kea rises directly from the sea floor, built by the same volcanic hotspot responsible for creating the entire Hawaiian archipelago. Though almost 4207 m (14,000 ft) of its elevation is visible above sea level, the majority of Earth's largest mountain dwells beneath the waves. Measuring 10,000 m (33,000 ft) from base to summit, Mauna Kea offers a dizzyingly diverse range of environments. The summit is effectively an Alpine desert inhabited by tiny predators, such as wēkiu bugs, whose blood contains a type of anti-freeze. The waters of Mauna Kea's foothills, in contrast, are home to mysterious deep-sea creatures, among them the cusk eel, a species living deeper than any other known fish. Between these two extremes, you'll find Montane forests inhabited by the Hoary Bat, Hawaii's only native mammal, as well as lowlands dense in shrubs, and numerous aquatic zones. Mauna Kea's slopes are frequented by birds and whales.

Sights

Summit and stars

1 Mauna Kea's official observatory, MKO, is world-renowned. The Mauna Kea Visitor Information Station (MKVIS) hosts local astronomers, who offer a free stargazing programme for tourists.

Whale-watching

2 The 'sunlight zone', a section of ocean that extends to 396 m (1300 ft) below the surface, is frequented by a wide range of aquatic species, including, from December to April, humpback whales.

Surf's up

3 Big Island might not see the four-story waves that surfers enjoy on Hawaii's northern coastlines, but reliable breakers of 4.5 m (15 ft) roll in during winter, including on Mauna Kea Beach.

Lake Waìau

4 Legend has it this mountain lake connects the heavens to the Earth. Native Hawaìians place their newborn's umbilical cord in Waìau's waters, to connect with the gods.

The world's hardest cycle road climb

5 From Waikoloa Beach, 4192 m (13,753 ft) of elevation awaits. This bike ride is 92 km (57 mi) long, almost exclusively uphill, with gradients of as much as 20%.

Extinct volcanic craters and an observatory atop the Mauna Kea summit.

Lava Life

Mauna Kea last erupted 4500 years ago, but the potential for renewed activity is high. The neighbouring Mauna Loa volcano, part of the same geothermic system, has erupted 33 times since 1843.

A Sacred Shield Mountain

Scientists estimate Mauna Kea broke the surface of the ocean 800,000 years ago, whereupon it began building a shield volcano, its explosive layers of lava settling in the shape of a warrior's shield. This volatile phase ended about 200,000 years ago, giving way to cinder cones on its flanks. Subsequent eruptions buried the central caldera. Mauna Kea continued to erupt, even when covered in ice, but is now under threat. Each year, the Pacific tectonic plate shifts Mauna Kea about three inches off the hotspot from which it draws its fuel. Eventually, with no lava to build, Mauna Kea could collapse into the sea. At present the highest mountain in the Hawaiian chain, it's one of five shield volcanoes that make up the Big Island (Hawai'i). No wonder that it's known to the native population as a sacred place, home to the gods and originally open only to priests and royalty. Though now called Mauna Kea, its name was once Mauna o Wakea, after the Sky Father Wākea who wed the earth mother Papahānaumoku.

© MARK GARLICK/GETTY IMAGES

An illustration of the Chicxulub asteroid impact.

Chicxulub Crater

At over 150 km (90 mi) wide, the Chicxulub impact crater is believed to have been caused by the asteroid that led to the end of the dinosaurs.

Located on Mexico's Yucatan peninsula, the Chicxulub Crater eluded detection until as recently as 1978. Hidden beneath a kilometre of younger rocks and sediment, it is seen by most scientists as evidence of a huge asteroid or comet that crashed into Earth's surface 65 million years ago, and estimated to have been between 11 and 81 km (7 to 50 mi) in diameter. The resulting instantaneous change to the climate brought about the extinction of more than 70% of all living species on the planet, including the dinosaurs. The true scale of the crater is only visible when seen from space: a string of sinkholes, or cenotes, are visible across the Yucatan's northern tip. Magnetic and gravity data from the area show a large circular structure, suggesting these cenotes resulted from fractures in the buried crater's rim.

Analysis of impact ejecta and melt spherules found in rocks and ocean sediments from as far away as Haiti helped convince the scientific world that the evidence shows Chicxulub Crater was the site of an impact which sent life on Earth in a new direction, from the age of the dinosaurs to our own age of mammals.

Snorkelling in Silfra.

Silfra

Enjoy the crystal-clear waters of Iceland's Thingvellir Park, as two of Earth's tectonic plates continue to go their separate ways.

Located in Iceland's Thingvellir National Park (found just outside of the capital, Reykjavik), Silfra is a fissure between the North American and Eurasian plates, which are drifting apart at a rate of about 2 cm a year. With a depth of 63 m (207 ft), this is among the deepest of Thingvellir's fissures. The lake in which Silfra sits is fed by natural springs at its base, and by meltwater from Langjökull, Iceland's second-largest glacier. Silfra attracts scuba divers from all over the planet, tempted not only by the rare opportunity of swimming beneath massive continental plates but also by the beautiful geology and superb visibility; the main dive sites rarely have visibility less than 70 m to 80 m (230 ft to 262 ft) and many divers report as much as 100 m (328 ft). Silfra offers three sites: the Hall, the Cathedral and the Lagoon. Of these, the Cathedral is the most spectacular, with visibility from one end to the other of this light-filled chasm. It's not every day that you can scuba between continental plates.

Sailing stones and their paths on the floor of the Race Track Playa in California's Death Valley National Park.

Death Valley

The hottest place on Earth is no forsaken wasteland. In fact, it's a 3.4-million-acre national park, and the main draw in UNESCO's Mojave and Colorado Deserts Biosphere Reserve.

Death Valley sits in the Badwater Basin, the lowest point in North America, at 86 m (282 ft) below sea level. While Death Valley's summer temperatures average 47°C (117°F), the day that brought it worldwide recognition was 10 July 1913, when the air temperature in the settlement of Furnace Creek was measured at 56.7°C (134°F). The ground here gets even hotter than the air. On 15 July 1972, the Creek's ground temperature was recorded as 93.9°C (201°F). And yet, life finds a way.

According to the National Park Service, who maintain the reserve, more than 1000 species of plants live here, including the iconic Joshua trees, creosote bushes and desert hollies. Bobcats, coyotes and eagles also proliferate. Visitors to Death Valley will find several campgrounds and at least one ultramarathon depending on the season. Winter brings snow to the peaks of the surrounding mountains, and rainfall, though rare, can cause wildflower meadows to burst into exuberant life.

Sights

The Badwater Ultramarathon

1 Under average temperatures of 54°C (130°F), July's 'Badwater 135' ultramarathon takes runners 217 km (135 mi) from Badwater Basin 86 m (282 ft) below sea level to Whitney Portal, at an elevation of 2500 m (8374 ft). Presumably via Hell.

Bighorn Gorge

2 For serious hikers, the Gorge offers trails that gain more than 1.5 km (5000 ft) in elevation. The distances may seem short, but – with no water available – early starts are a must.

West Side Road

3 One of the tougher routes available, the trail starts south of Highway 190. Ahead lays 64 km (40 mi) of mostly level road but with a rutted gravel surface. As ever, water is key.

Zabriskie Point / Dante's View

4 If you're after quintessential views of the Californian desert, choose either of these locations and you can't go wrong. Sunsets are special.

Scotty's Castle

5 An architectural curiosity located in the Grapevine Mountains, near Stovepipe Wells. The brainchild of theatrical impresario and conman Walter Scott, this Spanish colonial villa has a fascinating history.

Scotty's Castle facade in Death Valley National Park's Grapevine Mountains.

Lights, Camera, Action

Death Valley often stands in for otherworldly landscapes. Many of the desert scenes from the original *Star Wars* trilogy were filmed in Death Valley, in particular for *A New Hope* and *Return of the Jedi*. Luke Skywalker roamed here, before being attacked by the Tusken Raiders; so did the murderous cult leader Charles Manson, whose infamous hideout, Barker Ranch, was in a remote part of the area.

Climate Records

Not content with holding the record for the hottest day, in 2018 Death Valley raised the bar, not to mention the mercury, once again – eclipsing its own mark for the hottest month ever. During July, the average temperature was 42.28°C (108.1°F). That's almost half a degree hotter than the previous record, set in Death Valley in July 2017. Until that year, the record for the hottest month had stood for more than a century. In 2018, there were four consecutive days where the daytime high in Death Valley was 52.7°C (126.9°F) and where nighttime temperatures remained above 38°C (100°F). The standard for measuring air temperature is determined by the World Meteorological Organization (WMO) and must be taken from 1.5 m above the ground, outside of direct sunlight. Satellite measurements of temperature are believed to be less reliable.

A Zodiac explores Antarctica's icebergs.

Antarctica

Antarctica is the coldest, driest and windiest of Earth's continents. All but uninhabitable in winter, come summer it yields a mesmerising seascape, rich in marine life.

Almost 99% of Antarctica is covered with ice, with an average thickness of one mile. Antarctica's lowest-ever recorded temperature, at Vostok station in 1983, was -89.6°C (-128°F). The average winter temperature is a more clement -49°C (-56°F), although wind speeds can reach 320 km/h (199mph). The harshness of the climate is part of Antarctica's allure; to reach it you must navigate the Southern Ocean, the most treacherous on Earth. Although classified as a desert – its annual rainfall is 200 mm (not quite 8 inches) – Antarctica has abundant wildlife, including rare species of seals and penguins.

There is no permanent settlement as such, but 30 countries maintain a scientific presence. Antarctica is, along with its northern cousin the Arctic, under increasingly serious threat from climate change warming. Scientists continue to report dramatic reductions in the extent of its ice as a result of the greenhouse effect, with implications for wildlife and global sea-levels.

Sights

Tierra del Fuego

1 A city at the end of the world and the launchpad for most expeditions. Vessels depart via the Beagle Channel, named after Charles Darwin's ship on his 1831 voyage to South America.

South Georgia

2 One of the remote South Sandwich Islands and resting place of famed British explorer Sir Ernest Shackleton. King penguins and elephant seals offer a glimpse of the wildlife to come.

Ross Sea and Island

3 Glaciers abound on the ice shelf, and Shackleton's Hut, at Cape Royds, is a must on Ross Island. Only accessible for a few months each year, if the pack ice relents.

Drake Passage

4 Usually a two-day crossing, and a stomach-churning one for all but the most sea-hardy. A truly daunting stretch of ocean; distract yourself by watching magisterial giant albatross.

Antarctic archipelago

5 The numerous small islands off the main peninsula are packed with life: gentoo penguins, crabeater seals, brown skuas, humpback whales, orcas... simply too many to list here.

An orca in Antarctica, home to about 70% of the world's killer whale population.

Polar Bound

The south pole, the southernmost point on the planet, was one of the great prizes of the golden age of 19th- and 20th-century exploration of the extremes of Earth. The first to reach it was Roald Amundsen, the Norwegian explorer, whose expedition discovered their goal on 14 December 1911.

Iceberg Ahoy

In October 2018, NASA released the striking image of a notable iceberg photographed in Antarctica's Weddell Sea. Measuring roughly one mile across, it was perfectly rectangular. The geometric shape suggested the mass of ice had recently calved from an ice shelf. While the uniform nature of 'tabular' icebergs is not unheard of, to capture such a large example, and in such pristine condition, is extremely rare. As with all icebergs, the area visible above the ocean's surface represents a fraction of the object's total mass. NASA estimates that, in this case, about 90% of the iceberg was hidden underwater. Icebergs may collide with each other and with landmasses as they journey on water currents connecting to the Antarctic Circumpolar Current, the only current to entirely circumnavigate the globe. It was first discovered by the British astronomer Edmond Halley, after whom Britain's Antarctic Halley Research Station is named.

Stony Coral Colony and soldier fish on Australia's Great Barrier Reef.

Great Barrier Reef

Not so much lurking beneath the water as demanding your attention above it (the reef can even be seen from space), Australia's Great Barrier Reef is Earth's largest structure made by living organisms.

Situated off the coast of Queensland, the Great Barrier Reef is 2300 km (1400 mi) long and covers an area of 344,000 sq km (133,000 sq mi). Coral has been growing in this part of the planet for 25 million years, although Australian marine scientists put the age of the Barrier Reef at 20,000 years. Coral is created by polyps, small organisms that cluster together with the help of calcium carbonate to form an exoskeleton with a scaly texture, hence coral's reputation for cutting unwitting feet. Vital to coral's success is water shallow enough for sunlight to penetrate, thus allowing photosynthesis.

What makes the Barrier Reef such a draw is the marine life that calls its clear waters home: 1500 species of fish, 134 types of shark and 30 species of mammals, as well as rare turtles. Declared a World Heritage Site by UNESCO, the greater reef area includes 30 'bioregions', many of which are home to highly localised coral species.

Sights

Whitehaven Beach

1 Arriving by seaplane is popular, approaching via the Whitsunday Islands before landing on a crystal-blue lagoon. Stretch your legs on the expanse of pure-white sand, then grab your snorkel.

Yongala Shipwreck

2 One of the world's best dive sites, the wreck of the SS *Yongala* lies off the coast of Townsville. Home to sea snakes, giant trevally, groupers, rays, sharks and countless tropical fish.

Low Isles

3 This is among the most sheltered snorkelling destinations on the Reef. Boats leave daily from Port Douglas, including glass-bottomed vessels should you fancy taking it easy.

Turtle-watching at Mon Repos Island

4 From November to January, watch endangered loggerheads venture ashore nightly to lay eggs at Mon Repos beach. Between January and March, hatchlings leave the nests and race for the sea.

Swim with whales

5 Dwarf minke whales visit the Barrier Reef in June and July. Simply jump into the water, hold a surface rope and wait for them to approach. Friendly and curious, they almost certainly will.

Magnificent colours in the Great Barrier Reef.

Coral Bleaching

As with so many of Earth's natural treasures, the Great Barrier Reef is under threat from global warming. The optimum water temperature for this fragile ecosystem ranges from 23°C to 29°C (73°F to 84°F); if temperatures exceed this, the coral is vulnerable to 'bleaching'. In 2016, the Great Barrier Reef experienced its worst 'die-off' yet recorded, when almost half the corals were killed in a heat wave.

Gathering Data on Reef Conditions

Help is at hand. In 2016, NASA launched its Coral Reef Airborne Laboratory (CORAL), a three-year mission combining state-of-the-art aerial imaging technology with in-water research. The mission will provide critical data for analysing reef ecosystems. CORAL is in the process of gathering findings from a large sample of reefs, both across the Pacific and in Australia. There it will survey six sections of the Barrier Reef, from the Capricorn-Bunker Group of islands to the Torres Strait. Scientists will then use this data to look for trends corresponding to reef condition and examine the natural and human-produced factors, both biological and environmental, that affect reefs.

Aerial view of a river winding through the Amazon rainforest.

The Amazon rainforest

Earth's largest rainforest spans over 5.5 million sq km (2 million sq mi). The forest's ability to convert carbon dioxide into life-giving oxygen has earned it a reputation as the 'lungs' of the planet.

The majority of the Amazon rainforest, 60%, is located in Brazil – but eight other countries host portions of this raucous, verdant wonderland. The rainforest was formed 50 million years ago, during the Eocene period, when great swathes of the Earth would have been similarly cloaked in dense foliage. Today, due to deforestation, the Amazon is an important last redoubt, representing more than half of Earth's remaining rainforest. Situated in a vast basin, it is drained by the Amazon River.

The wet, warm, highly oxygenated climate makes it a perfect incubator for life. Earth's most biodiverse forest, it is home to an estimated one-third of all the planet's species – among them 2000 species of birds, mammals and reptiles. And, in the Amazon's 1100 tributaries, a similar number of fish species dwell. The diversity of plant species enjoys its greatest concentration here – the rainforest is home to 16,000 species of trees alone.

Sights

Aerial exploring

1 Buckle into a harness and pull yourself up to the forest canopy; at 10 stories high, it is an elevated hothouse of bromeliads and rare orchids, kept company by capuchin and howler monkeys.

Two rivers run through it

2 The region is home to not one but two mighty waterways. At Manaus, the Amazon is joined by its tributary the Rio Negro, whose dark waters blend with the creamy Amazon in floating whirlpools.

Anavilhanas archipelago

3 A network of 400-plus islands on the Rio Negro in Brazil, this area is home to pink dolphins and manatees. Birders come for the white-winged potoo and pompadour cotinga.

A walk in the dark

4 Night hikes are the best way to meet the forest's nocturnal species, such as tree frogs and giant crickets. Use a reliable operator if you want to find your way back.

Opera in the jungle

5 Take in an aria at the Teatro Amazonas. The luxurious opera house, a window on Manaus' faded colonial era glory, was built at the height of the rubber boom in the late 19th century.

A red howler monkey (alouatta seniculus) in Tambopata National Reserve, Peru.

New Species

A 2017 two-year study by the World Wildlife Fund found that, on average, one new plant or animal species is being discovered in the Amazon every other day. Between various research groups, 381 new species were counted in 24 months. These were broken down into 216 plants, 93 fish, 32 amphibians, 20 mammals, 19 reptiles and one new bird species.

Ticking Clock on Discovery

As elated as biologists are at finding new species, there is concern others may become extinct before they are found. Images from NASA's Landsat satellites have recorded the extent of deforestation since 1975. In one Amazon region, Rondonia, images have identified a 'fishbone' pattern. This is created by major roads being cut deep into the forest, with secondary roads branching off at right-angles. The forest between these secondary roads is then cleared, coming together to create a single deforested area. The World Wildlife Fund estimates that up to a quarter of the Amazon will be without trees by 2030.

The Ngorongoro Crater teeming with grazing wildlife.

Ngorongoro Conservation Area

Located in the Crater Highlands of Tanzania, between the Great Rift Valley and the plains of the Serengeti, the dramatic caldera of Ngorongoro is a game-spotter's paradise.

For pure wildlife drama, the Ngorongoro Conservation Area has few places on Earth to rival it. The wooded slopes of the Ngorongoro Crater drop 600 m (2000 ft) from rim to floor, where the profusion of game includes the big five – buffalo, rhino, lion, leopard and elephant – as well as numerous other species. The hyena population is particularly healthy, while the grasslands abound with zebra and antelope. The area is also part of the route of one of the great spectacles of the planet's natural world, the annual migration of more than 1.7 million wildebeest, from the Serengeti to Kenya's Masai Mara. And, of course, back again. The Crater is a perfect amphitheatre, and its lodges offer spectacular views as standard. The region also features soda lakes, habitat of the pink flamingo. The active volcano whose collapsed cone formed the crater 2.5 million years ago left a vital environment in its wake.

Sights

Lake Magadi

1 Rich in algae, this shallow soda lake in Kenya's Great Rift Valley attracts pink flamingos to its saline waters year-round.

The great migration

2 From December to March, this extraordinary gathering of Earth's ungulate species can be found clustered about Ngorongoro's Lake Ndutu as over a million wildebeest, Burchell's zebra and Thomson's gazelle make their seasonal trek.

Caldera nights

3 If the primary attraction here is wildlife, the accommodations can be a close second. Stay at a boutique lodge on the crater rim and take in the vastness.

Black rhino

4 Once on the brink of extinction, black rhino are recovering thanks to studious conservation work. Fifty of Earth's 5000-strong population live on the slopes of the crater.

Empakaai

5 Ngorongoro Crater undoubtedly gets busy with tourists. Its sister crater, Empakaai, is car-free and, with a lake of its own, is a haven for birds – from trogons to crowned eagles. Check out this alternate site to avoid the crowds.

A male lion resting in the Ngorongoro Crater.

Human Origins

The other species key to the region is, of course, humans. The Ngorongoro Conservation Area is where you'll find the Olduvai Gorge. It was here, in 1959, that paleoanthropologists Louis and Mary Leakey discovered the fossilised remains of *Homo habilis*, the 1.7-million-year-old primate species believed to be our oldest ancestor.

Large Landslides

While the Ngorongoro Crater is believed to have originally formed 2.5 million years ago, NASA's aerial reconnaissance of the Crater Highlands shows not just how far they rise above the adjacent savannas, but that geological change is ongoing. This is especially evident on the eastern flanks of the nearby Mt Loolmalasin, whose steep volcanic cone shows evidence of a much more recent phenomenon – a vast landslide. The deposits extend eastward for 10 km (6 mi), across the floor of the Rift Valley. On Earth, such a long run of landslide debris is unusual. And beyond its geological wonder, there's the view. Pick your own superlative: amazing, incredible, breathtaking... they all apply to the stunning, ethereal blue-green vistas of the Ngorongoro Crater, populated by teeming wildlife.

The Great Wall of China is estimated to be as many as 21,000 km (13,000 mi) long.

The Great Wall of China

Offering easy strolls as well as epic hikes, the longest construction ever built by human hands extends from Hushan, in eastern China, to the Jiayu Pass, in the west.

A grand total of 19 different constructions have laid claim to the title 'Great Wall of China'. As far back as the 7th century BC, the process of building then rebuilding this vast structure has been a vanity project for successive Imperial dynasties. The best known of these – at least within China – was the wall built by Qin Shai Hong, China's first emperor, around 200 BC. Located further north than the site of the current wall, not much of it remains, thanks in part to neglect, but also to light-fingered visitors helping themselves to its brickwork.

The current wall was built to protect the prosperous Ming dynasty from invaders from the north. At 15 m high and 9 m deep (50 ft high and 30 ft deep), the Wall houses a staggering 7000 watchtowers. It is arguably one of the most popular tourist destinations on Earth; if you're crowd-averse then opt for sections far from Beijing, such as in Jinshanling.

Sights

Simatai

1 Located just over an hour from Beijing, this steep and dramatic section enjoys the distinction of being the only stretch of the Wall you can visit at night.

The world's longest cemetery?

2 Of the one million labourers used, an estimated 400,000 died during the Wall's construction. Legend has it some are buried in the Wall, although no bones have been found.

Juyonggan

3 One of the best preserved Wall forts and home to the white-marble Cloud Platform. Genghis Khan used the nearby Juyong Pass during the Mongols' expansion into China, in the 12th century.

History takes a break

4 There are numerous gaps along the wall where natural obstacles, such as mountains or rivers, serve in place of bricks.

Disappearing treasure

5 Sections in the Ghobi are in bad shape due to desertification. Gansu and Ningxia provinces could see sections disappear altogether inside 20 years.

Stone, brick, tamped earth and wood comprise the Great Wall, seen here in Mutianyu.

Imperfect Defense

If the Wall was built to repel invaders, you could call it the world's largest failed construction project. It has been breached several times. The Ming dynasty, who built most of the current fortifications, were invaded by the Mongols in 1449 – and the walls succumbed again to the Manchus in 1644, when a treacherous Ming general opened a gate.

Visible from Space, or Not?

That it can be seen from space is probably the best-known fact about the Great Wall. Unless, that is, you happen to be an astronaut. While NASA radar images have occasionally picked up the Wall, no less an authority than Neil Armstrong has claimed it can't be seen by the naked eye. Armstrong's compatriot, William Pogue, a veteran of both the Apollo and Skylab programmes of the 1960s and 1970s respectively, did claim he was able to see the Wall – but it was only with the aid of binoculars, and then from a low-orbit distance of around 300 km (186 mi). The visible wall theory was further shaken after China's own astronaut, Yang Liwei, said he couldn't see the historic structure. The issue surfaced again after photos taken by Leroy Chiao with a 180mm lens and a digital camera from the ISS were determined to show small sections of the wall in Inner Mongolia about 320 km (200 mi) north of Beijing.

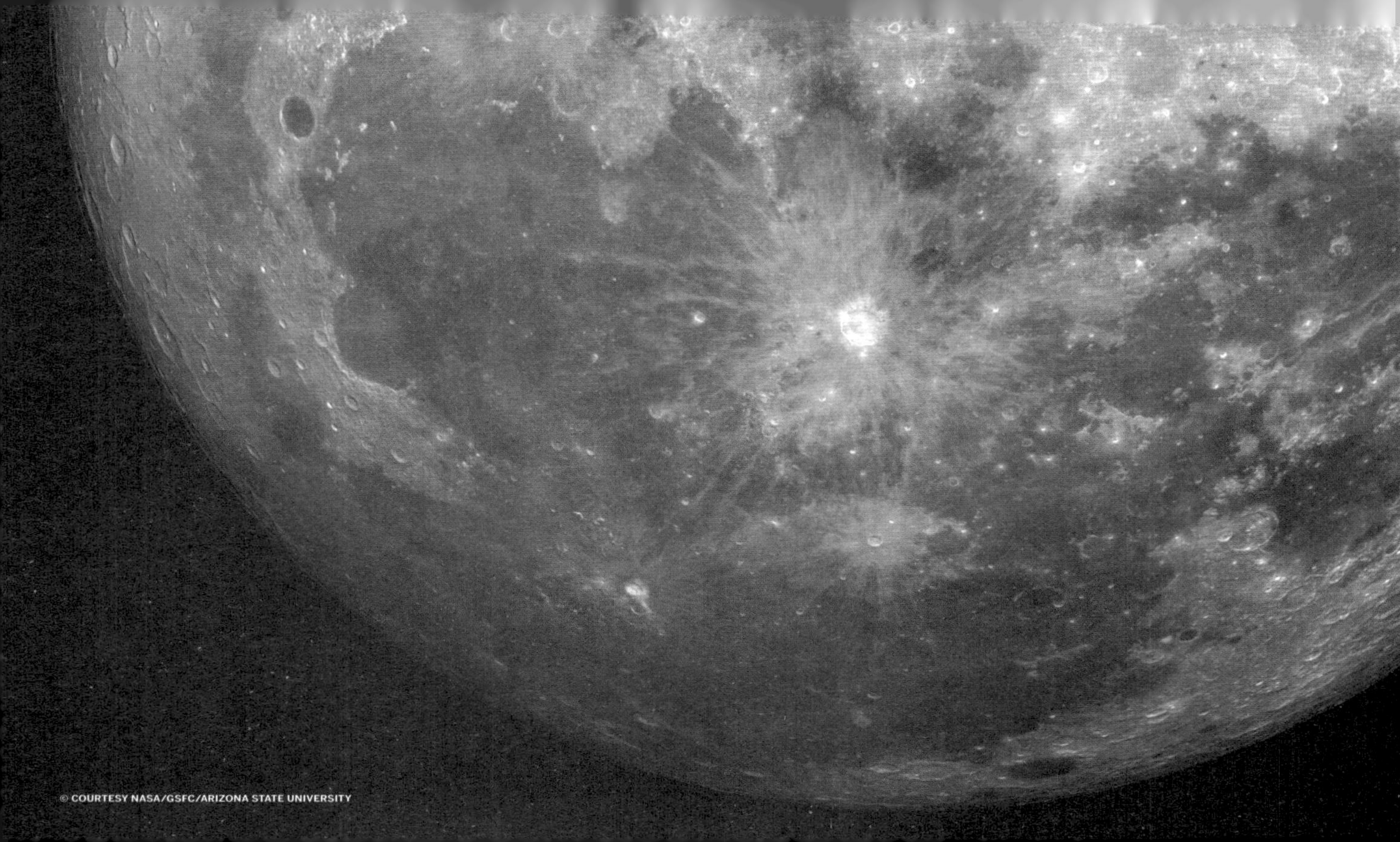

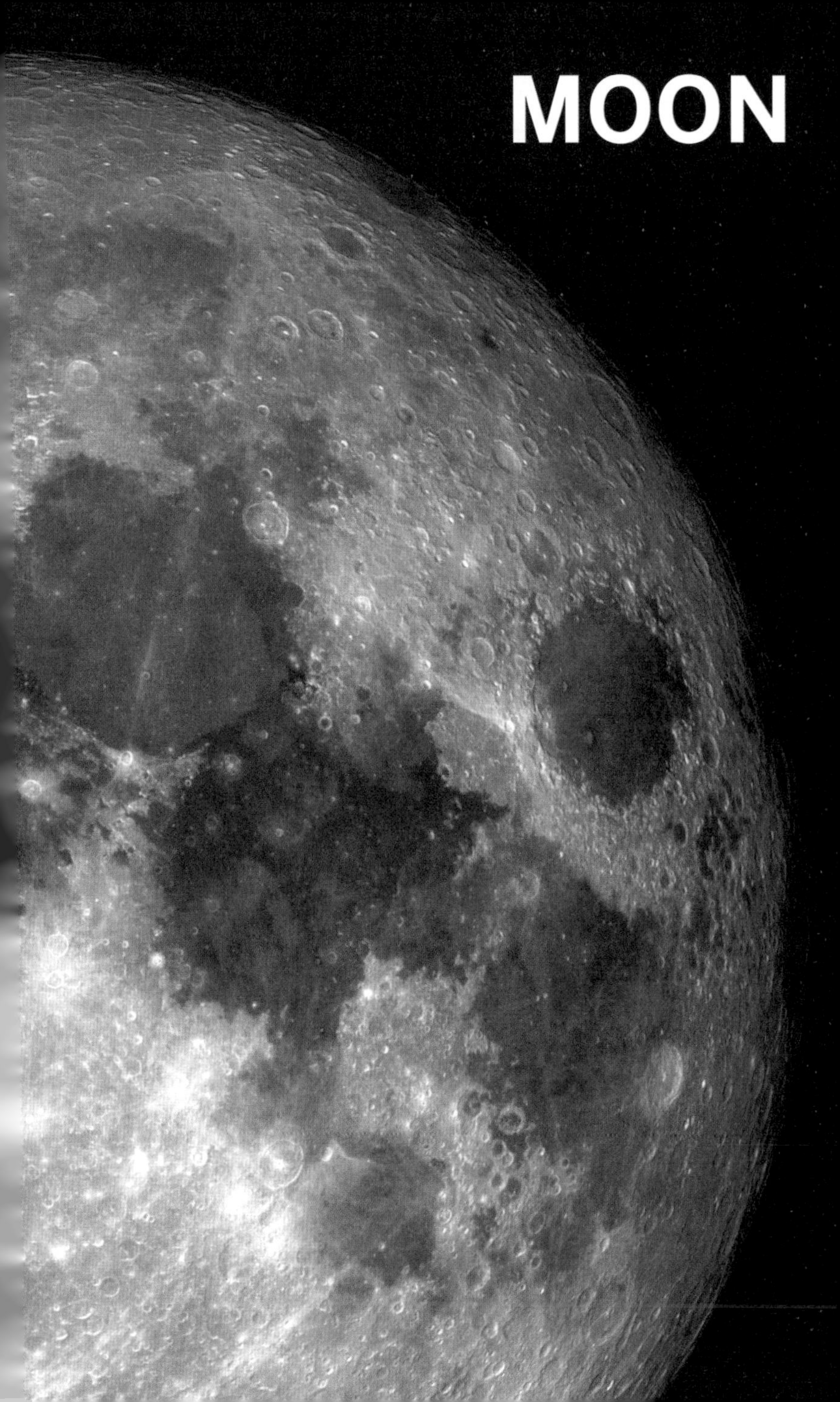
MOON

The moon's light and dark sides are divided by the so-called lunar terminator.

Moon at a Glance

The fifth-largest moon in the solar system, Earth's moon is the only place beyond Earth where humans have set foot.

Mankind's giant leap came on 20 July 1969, when the crew of Apollo 11 touched down on the lunar surface. 'Lunar', the adjective for all things moon-related, comes from the moon's Latin name, 'Luna'. More than 118 spacecraft have been launched to explore the moon, the only celestial body beyond Earth whose surface has been visited by humans. The astronauts of the Apollo missions brought back a combined total of 382-kg (842lbs) of lunar rocks and soil to Earth. So many samples, in fact, that NASA scientists are still studying them today.

Earth's moon is the fifth-largest moon in the solar system (after Ganymede, Titan, Callisto and Io). One of the reasons Earth's only natural satellite takes the definite article, *the* moon? Until Galileo Galilei discovered the first four moons orbiting Jupiter in 1610, the existence of other moons was unknown.

The brightest and largest object in Earth's night sky, our moon makes Earth a more liveable planet; by moderating our home planet's 'wobble' on its axis, the moon creates a relatively stable climate here. The moon's gravitational pull is also responsible for tides, particularly at the full moon, when they are at their strongest. While it's not also responsible for werewolves, as reputed, it does have an influence on the biological patterns of creatures on Earth in other ways large and small.

As Earth's constant companion, the moon has created a rhythm to life on Earth that has guided humans for thousands of years. The moon itself has a very thin, tenuous atmosphere, known as an exosphere. This weak atmosphere, along with the moon's lack of liquid water or breathable air, means it cannot support life. It's also part of what enabled astronauts on the moon's surface to bounce so far above the ground; its low mass and lack of atmosphere mean that your weight on the moon would be a mere 16.5% of your weight on Earth.

DISTANCE FROM SUN
1 AU

LIGHT-TIME TO THE SUN
8.3 minutes

LENGTH OF DAY
29.5 Earth days

LENGTH OF ORBITAL YEAR
27.3 days

ATMOSPHERE
Trace amounts of helium, argon and neon

An Apollo 17 astronaut on the moon, December 1972.

Top Tip

If you know where to look, you can still find pieces of NASA equipment on the moon: several American flags and even a camera left behind by astronauts. Note also, the moon's gravity is one-sixth of Earth's, which is why in that famous moonwalk footage – of Neil Armstrong, not Michael Jackson – Armstrong appears to bounce across the surface.

Getting There & Away 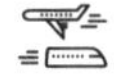

The moon is the most easy of all celestial bodies to visit from Earth, at least in relative terms. It takes just three days to get there, although if you want to land on the surface, you'll need to build in time to enter lunar orbit. Apollo 11 took 76 hours to reach lunar orbit, plus another day preparing its lander.

Orientation

It's largely thanks to the moon landings of the 1960s and '70s that many people assume the moon is relatively close. In fact, the moon is an average of 384,400 km (238,855 mi) away from Earth. That's enough room between Earth and the moon for 30 Earth-sized planets. Moreover, the moon is slowly moving away from Earth, the distance increasing by about an inch each year. In terms of size, with a radius of 1738 km (1080 mi), the moon is less than a third of the width of Earth.

The moon rotates at about the same rate it revolves around Earth, a phenomenon known as 'synchronous rotation'. This means the same hemisphere of the moon faces Earth at all times. The far, unseen hemisphere is sometimes referred to as the 'dark side' of the moon, inaccurately; the far side would be more correct. As the moon orbits Earth, different parts of its surface are in sunlight or darkness at different times. This changing illumination is why, from our perspective, the moon goes through phases. During a 'full moon', the hemisphere of the moon we see from Earth is fully illuminated by the Sun. A 'new moon' occurs when the far side of the moon is in sunlight and it is nighttime on the side facing Earth.

The moon makes a complete orbit around Earth in 27 Earth days, and rotates at that same rate. But Earth is moving as well, rotating on its own axis as it orbits the Sun. So, from our perspective, the moon appears to orbit Earth every 29 days.

Earth's moon is made up of a core, mantle and crust. The moon's core is smaller than those of other terres-

© ROBERT_S/SHUTTERSTOCK

The phases of the moon, waxing from new to full and waning back again.

Moon *vs* Earth

-- Radius --
27%
EARTH

-- Mass --
1.2%
EARTH

-- Volume --
2%
EARTH

-- Surface gravity --
16.3%
EARTH

-- Mean temperature --
-35°C
(-31°F)
COLDER than EARTH

-- Surface area --
7.4%
EARTH

-- Surface pressure --
0%
EARTH

-- Density --
60%
EARTH

-- Orbit velocity --
3.4%
EARTH

-- Orbit distance --
1/389th
EARTH

A comparative view of the sizes of Earth and the moon.

trial bodies. Its solid, iron-rich inner core is 240 km (149 mi) in radius (Earth's is five times as large). This inner core is surrounded by a liquid iron shell, 90 km (56 mi) thick. A partially molten layer, with a thickness of 150 km (93 mi), surrounds the iron core. The mantle extends from the top of the molten layer to the bottom of the moon's crust. Scientists believe it is made of minerals such as olivine and pyroxene, which are composed of atoms of magnesium, iron, silicon and oxygen.

The moon's crust is about 69 km (43 mi) thick on the Earth-side hemisphere, with a thickness of 150 km (93 mi) on the far side. The crust is made of oxygen, silicon, magnesium, iron, calcium and aluminium. It also features small amounts of titanium, uranium, thorium, potassium and hydrogen. Long ago, the moon had active volcanoes, but today they are dormant, having not erupted for millions of years. On the other side of the temperature spectrum, the moon does have its own remnants of water-ice around the poles.

The light areas visible on the moon are highlands. The dark features, called maria (Latin for 'seas'), are impact basins that were filled with lava sometime between 4.2 and 1.2 billion years ago. These light and dark areas represent rocks of different compositions and ages. They provide evidence for how the early crust may have crystallised, from what was once a lunar magma ocean. Although cooler than it once was, the surface temperature can still be a sweltering 127°C (260°F) when in full sunlight. In darkness, however, the temperature plummets, to about -173°C (-279°F).

Atmosphere

The moon has a very thin and weak atmosphere, called an exosphere. It does not provide any protection from the Sun's radiation, nor from impacts by meteoroids.

Magnetosphere

At an early stage in its evolution, the moon may have developed an internal dynamo, the mechanism for generating global magnetic fields for terrestrial planets. Today, however, the moon only has a very weak magnetic field, many thousands of times weaker than the magnetic field on Earth.

History

The discovery by Galileo Galilei of four moons orbiting the planet Jupiter, in 1611, fundamentally changed the way we view our own, natural satellite. Until then, it was assumed Earth was the only planet to be orbited by another celestial body. In the intervening years, scientists have learned that the origins of moons vary. Some of them, such as Phobos and Deimos, the twin moons of Mars, are little more than pieces of rock captured by the Red Planet's gravity. Earth's moon is different, most likely the result of a collision billions of years ago. When Earth was a young planet, a large chunk of rock, perhaps similar in size to Mars, smashed into it, displacing a portion of its interior. The resulting debris came together to form our moon.

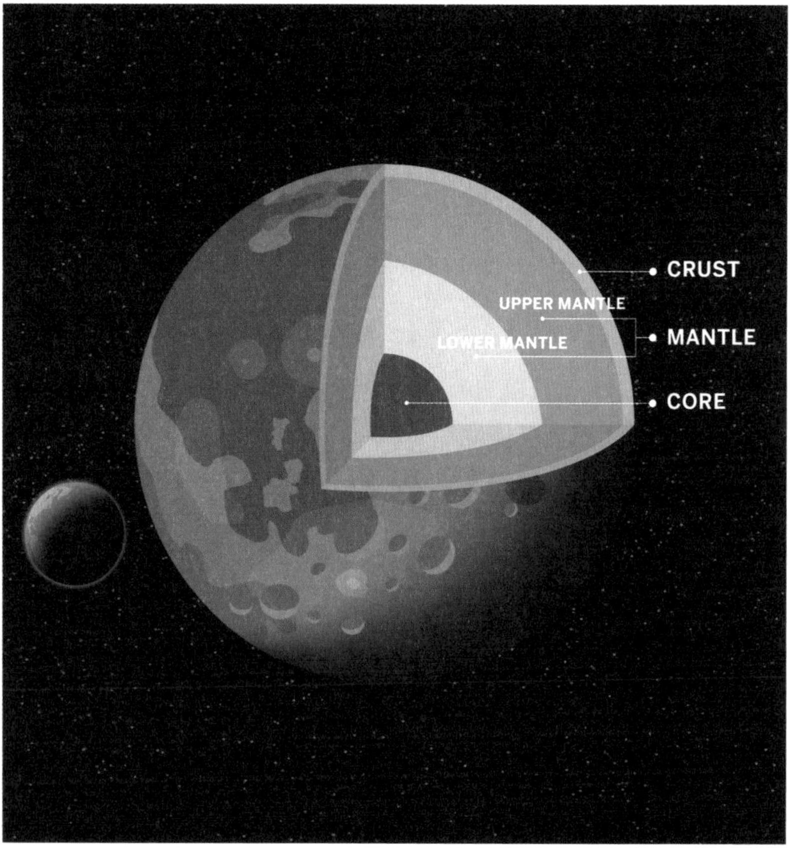

A cross-section of the moon's interior.

With too sparse an atmosphere to impede impacts, the surface of the moon is subject to a steady rain of asteroids, meteoroids and comets, whose strikes leave behind craters. Tycho Crater, to the south of the moon's Earth-facing hemisphere, is more than 83 km (52 mi) wide. At an estimated age of 108 million years, this relatively young crater is overlaid on top of previous impacts. The moon's surface has its own version of scar tissue, but rather than healing, it simply receives new marks to overwrite the old ones.

The impact record of the moon can provide important insights into the history of the Earth. Evidence of ancient impacts on Earth, such as crater structures, can be difficult to identify because of turnover at the planet's surface due to various geologic, chemical and physical processes. Because ancient craters on Earth get wiped away over time, scientists have turned to the moon in order to identify periods of time when impacts may have been common in the Earth-moon system.

Over billions of years, these impacts have ground the surface of the moon into fragments. These range in size from huge boulders to little more than deposits of powder. The vast majority of the moon's surface is covered by the lunar regolith, a mixture of rocky debris and charcoal-like dust. Beneath the regolith is a layer of fractured bedrock, known as the megaregolith.

Samples of lunar material brought back from the moon by Apollo astronauts led to development of the Late Heavy Bombardment (LHB) theory, positing that the Earth, and the entire inner solar system, suffered through an intense spike in asteroid bombardment roughly 4 billion years ago.

Since its development, the LHB theory has shaped our understanding of Earth's early evolution and, notably for astrobiology, concepts of when life originated on our planet. Due to the size and frequency of impact events during the LHB, scientists thought that water would have been vaporised at the surface and our planet, at least above ground, would have been rendered uninhabitable for life. If life originated in environments on the Earth's surface, it only could have happened after bombardment stopped.

Many astrobiologists have taken an interest in this topic, because it could have major implications for the timing of the origin(s) of life on Earth. However, in recent years, more and more scientists have begun to question the LHB theory. Some believe that the period of impact events was relatively short-lived. Others suggest that there may not have been a cataclysmic spike in impact events 3.9 billion years ago. Instead, bombardment of the moon may have occurred over a prolonged period lasting from roughly 4.2 to 3.4 billion years ago. The new interpretation of lunar data could affect our understanding of conditions on the ancient Earth at the time of life's origins. Its lack of plate tectonics makes it the perfect lab for testing the LHB and later Nice model, based on the instability of comet orbits.

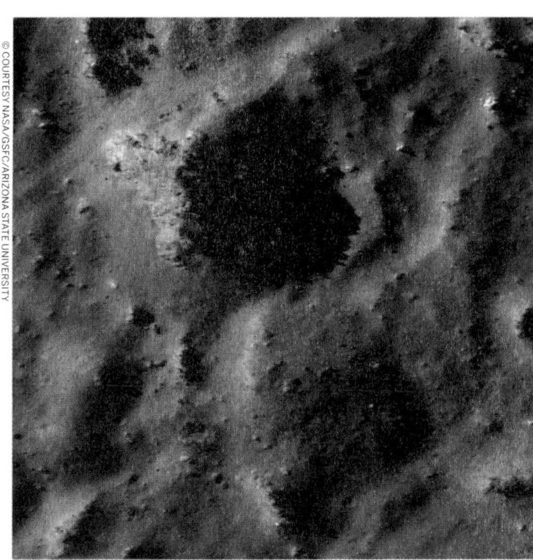

Depressions and reliefs in Tycho Crater are the result of impact melt.

Discovery Timeline

The US moon landing of Apollo 11 were far from humanity's first attempt to reach our celestial satellite by spacecraft. Its 1968 predecessor, Apollo 8, was the first manned mission to successfully travel beyond Earth's orbit. In total, there have been 118 various moon missions, from an increasing number of nations.

4 January 1959

The Soviet Union's Luna 1 makes a first successful flyby of the moon at a distance of 6000 km (3700 mi). Seven previous missions, launched by the Soviets and the US, all failed at launch.

13 September 1959

Luna 2 becomes the first human-made object to touch the surface of another world upon crashing into the moon.

7 October 1959

Luna 3 becomes first mission to photograph the far, or 'dark', side of the moon.

12 April 1961

Yuri Gagarin becomes the first human in space, aboard the Vostok spacecraft, setting the stage for later lunar approaches.

31 July 1964

Ranger 7 is the first US success in going to the moon. Following 13 consecutive failed missions, Ranger 7 is deliberately crashed into moon's surface. Fifteen minutes before impact, it begins to transmit spectacular footage of the surface.

24 March 1965

Plunging towards Alphonsus Crater, Ranger 9 broadcasts images of the moon's surface on live television.

2 June 1966

Robotic Surveyor 1 makes the first successful US soft landing on surface of the moon. It transmits more than 11,000 images, sending information on lunar soil, radar reflectivity and temperature.

24 August 1966

NASA's Lunar Orbiter 1 begins mapping the moon from space.

29 December 1968

Apollo 8 is the first mission to take humans to the moon. After swinging around the moon's far side, the astronauts are the first to witness an 'earthrise', the appearance of their home planet above the lunar horizon.

20 July 1969

Astronauts Neil Armstrong and Buzz Aldrin are the first humans to walk on the moon's surface. Ten more astronauts would

Astronaut James B Irwin works on Apollo 15's lunar module, Falcon, with Mount Hadley at back.

The unmanned Ranger spacecraft, designed to take the first close-up images of the moon.

enjoy the privilege before the end of the Apollo programme in 1972.

1970 & 1973

Soviet Union lands Lunokhod 1 on the moon. The first of two eight-wheeled robotic rovers, the second set a roving distance record only beaten in 2014, by the Mars rover Opportunity.

1994 & 1998–1999

The 'ice hunter' probes Lunar Prospector and Clementine eventually detect signs that water-ice may exist at the lunar poles.

2003–2006

SMART-1, a lunar orbiter sent by the European Space Agency (ESA), records key chemical elements.

2007–2008

Japan launches its second lunar spacecraft, Kaguya, and China its first, Chang'e 1. They are soon joined on their one-year missions by the Chandrayaan-1 orbiter, India's first mission to the moon.

2009

Dual launches of LRO and LCROSS mark NASA's return to lunar exploration. The following October, LCROSS is directed to crash into a permanently shadowed region of the moon, resulting in the confirmation of water-ice.

2011

The twin GRAIL spacecraft set off on their mission to map the moon, from crust to core. NASA's ARTEMIS also enters lunar orbit, a separate mission to study the moon's interior and surface composition.

2019

China's Chang'e 4 moon lander touches down in South Pole–Aitken Crater, on the far (or dark) side of the moon.

The Moon in Popular Culture

Our lunar neighbour has inspired stories since the first humans looked up at the sky and saw its grey, cratered face. Some observers are convinced they can identify human facial features, the so-called man in the moon. Hungrier observers have compared those craters to the holes found in cheese.

In literature, Jules Verne's 1865 novel, *From the Earth to the Moon*, is sometimes credited with inspiring real-life rocket pioneers, such as Robert H. Goddard and Hermann Oberth. The latter's innovative research was a mixed bag, helping to launch Hitler's V2 rockets across the English Channel but, later, America into space. Verne's novel is, of course, a work of science fiction – but certain details seem prescient.

When, in 1969, Apollo 11 became the first manned mission to land on the moon's surface, the size and shape of Columbia bore a close resemblance to the vehicle conceived by Verne. Verne also predicted the number of crew required to reach the moon would be three; step forward Buzz Aldrin, Neil Armstrong and Michael Collins. And just as Verne's spacecraft had used 'retro-rockets', thrusters designed to provide braking for his descending lander, similar technology assisted Neil Armstrong and his crewmates when they made their own historic flight.

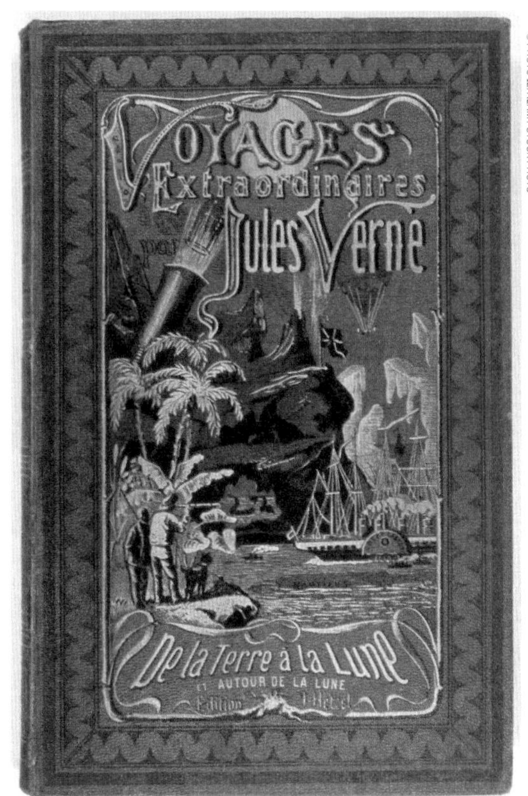

Early sci-fi luminary Jules Verne imagined a voyage to the moon well in advance.

Verne's prediction of weightlessness came close, albeit with no cigar: He believed it would only be experienced at the midpoint of the journey, when the moon's gravity was in balance with that of the Earth. Lastly, the Frenchman held that lunar explorers returning to Earth would splash down in the Pacific Ocean, precisely where Apollo 11 landed, 106 years after the publication of Verne's novel.

Verne's novel recounts how a telescope is able to track the progress of the author's moon mission. When an explosion took place aboard Apollo 13 in 1970, a telescope at Houston's Johnson Space Center monitored events as they unfolded more than 332,000 km (200,000 mi) from Earth.

The history of true moon exploration has spawned other cultural icons. The competition over who should host the Apollo 11 launch site was so hotly contested, between Florida and Texas, it could only be resolved by Congress. In the end, Florida's Kennedy Space Center was chosen as the launch site, with a certain Texan city providing

Mission Control. 'Houston, we've had a problem' first entered the popular imagination when uttered by Jack Swigert, during the Apollo 13 mission (and has often been misremembered in the present tense).

The moon made its cinematic debut in 1902, in the black-and-white, silent French film *Le Voyage dans la Lune* (*A Trip to the Moon*). The moon's most iconic movie role, however, came the year before astronauts walked on its surface. *2001: A Space Odyssey* imagines a colony of astronauts on a lunar outpost and en route to Jupiter. Decades later, Stanley Kubrick's masterpiece is still regarded as one of the best science-fiction movies ever, the malign intentions of HAL, the onboard computer, frequently invoked as a warning against unchecked faith in artificial intelligence.

And while we may not yet have a moon colony, there remains plenty of evidence of humanity's presence there. Astronauts have planted no fewer than six American flags on the surface, but that does not mean the United States has claimed it. In fact, it was written into international law, in 1967, that no single nation may claim ownership of planets, stars or any other natural celestial objects.

The moon, at least the 'dark' side, also gave Pink Floyd the name of their eighth album, and John Denver a cheesy song lyric, when the country singer told us of his desire to 'dance across the mountains on the moon'. The same name was once bestowed upon Uganda's Rwenzori Mountains, where the 'mountains of the moon' were once thought to be the source of the Nile. Always evocative, the moon may be our oldest inspiration.

An image from the 1902 French film *Le Voyage dans la Lune*.

Focus on Apollo 11

Buzz Aldrin walking on the moon near the Apollo 11 lunar module.

The primary objective of the Apollo 11 mission was set out by Presidential decree. On 25 May 1961, President John F. Kennedy tasked NASA with a genuine first. 'I believe that this nation should commit itself to achieving the goal, before this decade is out,' Kennedy famously said, 'of landing a man on the moon and returning him safely to the Earth.'

The Cold War was heating up. The Soviet Union's successful launch, in 1957, of Sputnik 1, the world's first artificial satellite, induced panic; many in the West felt Russia was on the verge of completing its intercontinental nuclear missile programme. The Sputnik crisis ushered in the Space Race and, in response, President Dwight D Eisenhower formed NASA in 1958.

The Apollo 11 mission was readied in the shadow of previous failures, not least the launchpad fire which claimed the lives of all three astronauts aboard Apollo 1, in 1967. The Apollo 11 spacecraft was made up of three parts: Columbia, a command module (CM) which contained a cabin for the three astronauts; a service module (SM) to provide propulsion, electrical power, oxygen and water; and lastly a lunar module (LM), dubbed Eagle, which would deliver men to the surface of the moon.

Apollo 11 launched from Cape Kennedy on 16 July 1969, powered by a giant Saturn V rocket. On board were Commander Neil Armstrong, Command Module Pilot Michael Collins and Lunar Module Pilot Edwin 'Buzz' Aldrin. From its initial Earth orbit, and clear of the Saturn V's third stage, Apollo 11 travelled for three days to reach lunar orbit. On 20 July, four days after launch, Armstrong and Aldrin climbed through the docking tunnel from Columbia to Eagle. Having made their final checks, 100 hours and 12 minutes into the flight, Eagle undocked from Columbia.

Behind the moon on what was now its 13th lunar orbit, Eagle fired its descent engine for 30 seconds, providing the retrograde thrust it would need for safe passage. Piloted manually in part, by Armstrong, Eagle arrived on the Sea of Tranquility at about 6.4 km (4 mi) from the predicted touchdown point.

Armstrong's words of confirmation – 'the Eagle has landed' – prompted hysteria back at Mission Control. Attached to the lander was a commemorative plaque, signed by President Richard Nixon and the three astronauts. It would be almost four hours before Armstrong emerged from Eagle for that historic transmission to Earth, perhaps the most significant television event in history. Around 109 hours and 42 minutes after launch, Armstrong stepped onto the moon. Aldrin followed 20 minutes later. An estimated 530 million people watched the images of Armstrong, listening in awe as he took 'one small step for man, one giant leap for mankind.' Half an hour later, President Nixon spoke with the astronauts by telephone link.

In total, Armstrong and Aldrin spent 21 hours and 36 minutes on the moon's surface, ranging up to 91 m (300 ft) from Eagle before the ascent-stage engine fired to take them back into lunar orbit. Docking with Columbia occurred on the command module's 27th revolution of the moon. During the time Aldrin and Armstrong had spent on the surface, Collins had piloted Columbia alone. Not everything was brought back to Columbia. In addition to the US flag, the astronauts also left a number of commemorative medallions, including three bearing the names of the Apollo 1 astronauts. Two others bore the names of Soviet cosmonauts, who had also died in accidents. A one-and-a-half-inch silicon disc also stayed behind, containing miniaturised goodwill messages from 73 different countries.

The command module of Apollo 11 made its re-entry on 24 July. After a total flight time of 195 hours, 18 minutes, 35 seconds – about 36 minutes longer than planned – Armstrong, Aldrin and Collins splashed down in the Pacific Ocean, 21 km (13 mi) from the recovery ship USS *Hornet*. Due to bad weather in the planned target area, the landing point had been changed by about 400 km (250 mi). For all three astronauts, Apollo 11 was their last spaceflight. It also signalled the end of the Space Race.

The Apollo 11 command module being recovered on the deck of the USS *Hornet*.

Focus on Orbital Gateway

In 2018, in collaboration with a number of international space agencies, NASA announced plans to return to the moon. Specifically, a new lunar space-station was proposed. The Lunar Orbital Platform-Gateway, or 'Gateway' for short, will allow astronauts to explore the lunar surface in extensive detail. It will also give them the opportunity to develop the science and human survival skills required for manned deep-space missions of the future. These may deal with radiation or with meteorites; missions to Mars are also of particular interest.

Unlike the International Space Station, orbiting a mere 400 km (250 mi) from the surface of the Earth, Gateway will operate hundreds of thousands of miles away – which is precisely the point. Once assembled, Gateway will, at the closest point of its approach, come within 1450 km (930 mi) of the moon's surface; the farthest it will be from the lunar surface is 70,000 km (43,500 mi), more than five times closer than the normal orbit distance between the moon and Earth.

With astronauts spending weeks, even months, aboard Gateway, conditions are likely to prove cramped. Aboard the two modules designed to house future lunar pioneers, Gateway will have just 55 cubic metres (1942 cubic feet) of habitable volume, compared with the palatial 388 cubic metres (13,696 cubic feet) enjoyed by guests at the International Space Station. The need to blast Gateway modules deep into space means they will need to be as light as possible. Once in space, they will be manoeuvred by ion propulsion rockets rather than chemical ones. An advantage in terms of fuel economy, but crucially at the expense of thrust. With that in mind, a new generation of ion propulsion rockets are in development.

In order to remain in constant contact with Earth, Gateway will remain clear of the moon's shadow during its planned six-day lunar orbit. NASA intends this orbit to serve as a stepping stone, either for deploying landers on the lunar surface, or for launching new sorties into deep space.

A vital part of Gateway-mission life will be the search for water on the moon's surface, believed to exist in significant quantities in the lunar polar regions. If it is located, this water could then be 'harvested' and used to create fuel for deep-space missions potentially, again, to Mars. The science behind this is, to a certain extent, already proven; broken down into water's constituent parts of hydrogen and oxygen, such propellants were used to fuel the space shuttle programme.

Astronomers are also keen to use Gateway as a deep-space observatory. From the vantage of lunar orbit, they aim to study the low-frequency radio waves they hope will provide insights into the creation of the universe 13.8 billion years ago. Attempts to study these signals on Earth are often compromised by radio traffic on our home planet.

An artist's conception of the updated Orbital Gateway.

Lunar Eclipses

A diagram showing Earth's shadow during a lunar eclipse.

A lunar eclipse occurs when the Earth passes between the Sun and the moon, blocking the sunlight from reaching the moon. There are two kinds of lunar eclipse: a total lunar eclipse occurs when the moon and Sun are on exact opposite sides of Earth, and a partial lunar eclipse happens when only part of the Earth's shadow covers the moon. This also means that an observer standing on the surface of the moon during a lunar eclipse would see all Earth's sunrises and sunsets at once. Pretty neat!

It's not often that we get a chance to see our own planet's shadow, but a lunar eclipse gives us a fleeting glimpse. During some stages of a lunar eclipse, which happens about twice a year, the moon can appear red in colour. This is because the only sunlight reaching the moon is that which escapes from around the edges of the Earth, and that sunlight is scattered through the Earth's atmosphere. The full moon rapidly darkens and then glows red as it enters the Earth's shadow. Evocatively, an eclipsed moon is called a blood moon for the red tint it takes on.

Occasionally a lunar eclipse will also be a so-called 'supermoon'. This is the phenomenon of the moon seeming particularly large in the sky, the result of its being full at perigee, when it is closest to Earth.

Lunar Eclipse Q&A

So why don't eclipses happen twice a month?

The reason is that, relative to the Earth's orbit around the Sun, the moon's orbit around the Earth is tilted.

If that's the case, why do eclipses happen at all?

Throughout the year, the moon's orbital tilt remains fixed with respect to the stars, but varies with respect to the Sun. About twice a year, this puts the moon in just the right position to pass through the Earth's shadow, causing a lunar eclipse.

As the moon passes into the central part of the Earth's shadow, known as the umbra, it darkens dramatically. Only when it is entirely within the umbra will the moon appear red.

You may have to stay up late to watch a lunar eclipse, but if you do you'll see the moon in rare form, and you'll catch a brief glimpse of the dramatic long shadow cast by our own planet.

Moon Highlights

Sea of Tranquility

1 Scene of one of the great achievements of science, engineering, even philosophy, leading to a greater understanding of our place, both on our home planet and within the wider universe: the moon landing.

South Pole–Aitken Crater

2 A vast monster of the dark side and the second-largest crater in the solar system. Takes its place between two Martian giants, Utopia Planitia and Hellas Crater.

Copernicus Crater

3 A light in the darkness, this bright spark of an impact crater is extremely visible, even when viewed from Earth with the naked eye.

Montes Apenninus

4 A moonwalker's lunar paradise, this rugged range is named for its earthly Italian counterpart and is home to the highest mountains on the moon.

Oceanus Procellarum

5 The only designated ocean on the moon, it may sound like a Harry Potter charm but it is in fact the biggest of all the lunar mare, and contains the moon's brightest feature, Aristarchus.

The terraced sides of the bright (high albedo) Aristarchus Crater in the Oceanus Procellarum.

A mosaic image of the Sea of Tranquility.

Sea of Tranquility

This historic location is where humans first set foot on a world beyond Earth. Look carefully and you just might find the tattered US flag planted by the icons of 1960s space exploration.

Perhaps the most famous of the lunar maria, the Mare Tranquillitatis (Sea of Tranquility) was the site of the Apollo 11 landing, where Neil Armstrong made his 'giant leap for mankind'. The landing site is noticeably blue in colour, most likely caused by the high metal content in the region's rocks. The mare is believed to be pre-Nectarian, making it around 3.9 billion years old. Unlike many other lunar maria, the Sea of Tranquility does not appear to have a mass concentration (abbreviated as mascon). 'Mascons' are anomalies which cause gravity at the centre of a mare to behave differently.

Despite its billing, those pioneering Apollo visitors to the Sea of Tranquility clearly did not find it tranquil enough. During their brief stay at 'Tranquility Base', the astronauts' schedule was supposed to include seven hours' sleep, to restore the pilots' strength; in the event, neither Aldrin nor Armstrong slept at all because they were too excited.

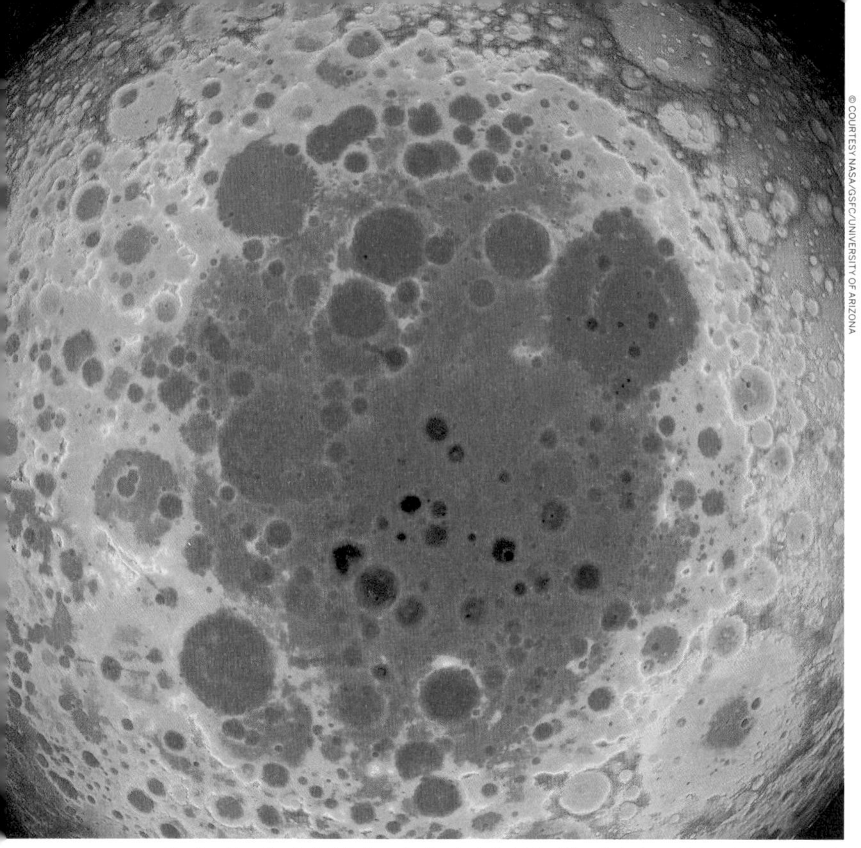

Elevation data of the South Pole shows the deep basin that is the Aitken Crater impact feature, in blue and purple.

South Pole–Aitken Crater

This mysterious dark basin recently made history of its own when, in January 2019, it saw the arrival of the Chinese lander, Chang'e 4, the first lander to make a soft touchdown on the moon's far side.

The largest and deepest impact basin on the moon, the South Pole–Aitken Crater is 2500 km (1600 mi) across and 13 km (8 mi) deep. It is the second-largest known crater in the solar system. Located on the far side of the moon, it takes its name from two features on different sides of its vast basin: the lunar south pole at one end, and the Aitken Crater – a depression on the basin's northern edge – at the other. South Pole–Aitken was first discovered in 1962.

Over the subsequent years, geologists began to establish its size and other characteristics. The South Pole–Aitken Crater has the lowest elevation found anywhere on the moon's surface, approximately 6000 m (19,685 ft) below the mean radius; it also features, on its northeastern rim, some of the moon's highest elevations, the 8000 m (26,247 ft) Leibnitz Mountains. At 4 billion years old, the South Pole–Aitken Crater is also the moon's oldest crater.

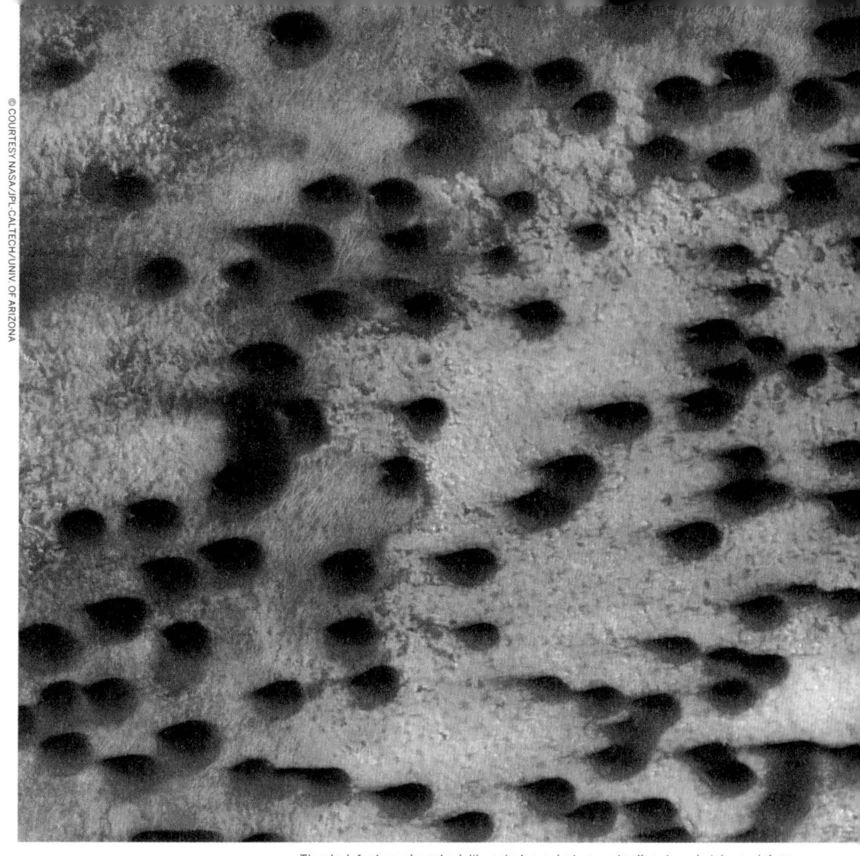

The dark features here look like raindrops, but are actually mineral-rich sand dunes.

Copernicus Crater

At just 800 million years old, this relative novice among lunar craters is nonetheless a standout feature – Copernicus gets itself noticed thanks to its marked contrast with surrounding mare.

At 107-km-wide (66-mi-wide), Copernicus is well known thanks to its extreme visibility. A bright spot surrounded by vast swathes of dark maria, Copernicus is easy to pick out from Earth using binoculars, and has been a moon-gazer's friend for generations. Like so many lunar features, Copernicus is an impact crater, located in the eastern part of Oceanus Procellarum, to the northwest of the moon's Earth-facing hemisphere. Named after the noted astronomer Nicolaus Copernicus, it is one of the more youthful lunar craters, formed during the moon's Copernican period – roughly 800 million years ago. Copernicus is typical of this era, in that it has a prominent ray system, lines of fine 'ejecta' material that radiate from the crater rim and resemble the spokes of a wheel. And unlike older craters, its floor has not been flooded by lava. The Apollo 12 mission landed just south of Copernicus, while Copernicus itself was named a possible site for proposed mission Apollo 20.

Mosaic photos from the LRO show the elevated peaks of the Montes Appenninus.

Montes Apenninus

Strap on your moon boots and head for the hills, amid soaring peaks whose number includes the tallest found anywhere on the lunar surface.

The moon's highest mountain range forms, simultaneously, the northwestern border of the Terra Nivium highlands and the southern boundary of the Mare Imbrium (Sea of Rains). An area certain to appeal to lunar hikers, it offers wide, expansive foothills, which give way to rugged views of the adjacent Mare Imbrium. The range is named after its Italian counterpart on Earth, although at 595 km (370 mi) in length, the lunar version is considerably shorter. Earth's namesake version is twice the length. What's more, the gentle slopes and lack of gravity – just one sixth of that found on Earth – means a fit moonwalker might plausibly hike the full range in a couple of days.

A number of the peaks in this range rise as high as 5 km (3 mi), among them Mons Huygens, at 5500 m (18,044 ft) the highest mountain on the moon. Other notable peaks include Mons Hadley and its neighbour, Mons Hadley Delta, whose flanks form the valley in which, in 1971, the Apollo 15 rover landed.

A mosaic of LRO images of the Oceanus Procellarum.

Oceanus Procellarum

Spanning more than 4 million sq km (1.5 million sq mi), Oceanus is the biggest of all the lunar mare. The landing site for Apollo 12, it is the only mare officially designated an ocean.

Located to the west of the moon's near side, the sheer size of the imposing 'Ocean of Storms' means scientists cannot say for certain if it was created by a single impact strike. For one thing, its 'shores' play host to numerous other bays and smaller seas, such as the maria Nubium and Humorum. If Procellarum was created by an ancient impact, the size of the initial strike would have been more than 3000 km (1864 mi) across, making it one of the largest in the solar system.

Another theory suggests such an impact would have created a smaller, secondary moon – on the far side, and about 1200 km (746 mi) in diameter. Tens of millions of years later, the two moons collided, creating a vast pile of material on the moon's far side. Oceanus Procellarum is also home to the 32-km-wide (19-mi-wide) Aristarchus crater, the brightest feature on the moon. Its albedo is almost double most lunar features, bright enough to be visible to the naked eye.

MARS

A mosaic of the Valles Marineris hemisphere on Mars.

Mars at a Glance

No planet fascinates more than Mars. The fourth-closest planet to the Sun, this terrestrial planet within the Sun's habitable zone shares Earth-like features such as polar ice caps, and once had a dense atmosphere.

The second smallest of the (currently accepted) planets, Mars has been known since humans first looked to the skies. The Red Planet earned its nickname due to its unique appearance, created by iron minerals in the Martian soil that oxidize, or rust, causing the soil and atmosphere to appear a warm, red colour. In fact, Mars is a cold, desert world with a very thin atmosphere. It is also a dynamic planet, with seasons, polar ice caps, extinct volcanoes, vast canyons and extremes of weather.

Named after the Roman God of war, Mars is located at an average distance from the Sun of about 228 million km (142 million mi) or 1.52 AU. Similar to Earth, one day on Mars takes a little over 24 hours. A year in Martian time, however, during which it makes one complete orbit

around the Sun, is 687 Earth days.

A rocky planet, Mars' surface has been altered by volcanoes, impacts from debris, and fierce winds, as well as crustal movement and chemical reactions. The atmosphere on Mars is thinner than Earth's, and is made up mostly of carbon dioxide, argon and nitrogen, as well as small amounts of oxygen and water vapour. Orbited by two moons, Phobos and Deimos, Mars has had many other visitors than its own satellites over the years: from flybys and orbiters to rovers on the surface. The first such successful mission to Mars was the 1965 Mariner 4 flyby.

Current NASA missions, including the InSight lander, are underway to determine Mars' potential for life. At present, Mars' surface cannot support life as we know it. What water exists does so only in icy dirt and thin clouds. However, there are signs of ancient floods on Mars, and, on some Martian hillsides, there is evidence of liquid, salty water in the ground. There is evidence that, billions of years ago, Mars was wetter and warmer, with a thicker atmosphere. Scientists are now investigating what happened to that atmosphere, and potentially exploring the planet as a possible outpost.

An artist's impression of the InSight lander probing the surface of Mars.

DISTANCE FROM SUN
1.524 AU

LIGHT-TIME TO THE SUN
12.6 minutes

LENGTH OF DAY
24.6 hours

LENGTH OF ORBITAL YEAR
687 days

ATMOSPHERE
Carbon dioxide, nitrogen, argon, trace gases

Top Tip

Mars' thin atmosphere means a space suit is essential. But you could also pack your Air Jordans. Mars' gravity is about 30% of the gravity of Earth, meaning pretty much anyone could slam dunk a basketball in a regulation height hoop.

Getting There & Away

On average, Earth and Mars are 225 million km (140 million mi) apart as they orbit the sun. The shortest distance between Earth and Mars, 54.6 million km (33.9 million mi), occurs every two years. In 1969, the Mariner 7 orbiter reached Mars in 128 days. The longest time taken by a probe, Viking 2 in 1975, was 333 days. The estimated time for manned flights is 250–300 days. Launches to Mars have to follow a parabolic trajectory in order to align with the planet's orbit.

© COURTESY NASA/JPL-CALTECH

Earth looms over Mars in this comparison of their relative heft, at twice the size.

Orientation

As large as Mars looms in our minds, it is significantly smaller than Earth. With a radius of 3390 km (2106 mi), Mars is about half the size of our home planet. Yet both planets have roughly the same amount of dry land area, since Mars has no oceans.

From an average distance of 228 million km (142 million mi), Mars is 1.5 AU (astronomical units, or the distance between the Sun and the Earth) away from the Sun. From this distance, it takes sunlight 13 minutes to travel from the Sun to Mars.

As Mars orbits the Sun, it completes one rotation every 24.6 hours, which is very similar to one day on Earth (at 23.9 hours). As any sci-fi fan can tell you, Martian days are called 'sols', short for 'solar day'. A year on Mars lasts 669.6 sols, which is the same as 687 Earth days, or almost twice the length of our year.

Mars' axis of rotation tilts as it orbits the Sun, by 25 degrees. This is another similarity it shares with Earth, whose 'axial tilt' is 23.4 degrees. And, like Earth, Mars has four distinct seasons. The seasons on Mars last longer than those on Earth because Mars is further from the Sun and its orbit is of greater duration.

Another difference is season length. While on Earth the seasons are evenly spaced with each of the four seasons lasting one quarter of a year (marked by the solstices and equinoxes), on Mars the seasons vary because of Mars' elliptical, egg-shaped orbit. A Martian spring, in the planet's northern hemisphere (autumn in the southern) is the longest season, at 194 sols. Autumn in the northern hemisphere (spring in the south) is the shortest season, at 142 days. In the middle-range are winters and summers: a northern winter/southern summer is 154 sols, while a northern summer/southern winter is 178 sols.

Mars *vs* Earth

-- Radius --
53% EARTH

-- Mass --
10.7% EARTH

-- Volume --
15% EARTH

-- Surface gravity --
38% EARTH

-- Mean temperature --
-74°C (-101°F)
COLDER than **EARTH**

-- Surface area --
28% EARTH

-- Surface pressure --
0.01% EARTH

-- Density --
71% EARTH

-- Orbit velocity --
81% EARTH

-- Orbit distance --
1.5x EARTH

Mapping Mars: From Seas to Craters

For the purposes of mapping, the United States Geographical Survey breaks Mars down into 30 sectors, or quadrangles. Some, such as Tharsis and Elysium Planitia, are well-known because they are home to prominent features; others remain relatively obscure. The first maps, such as the one created by Italian astronomer Giovanni Schiaparelli in the late 19th century, suggested a planet rich in water and vegetation. The advent of space reconnaissance, however, starting with the Mariner missions of the 1960s and accelerated significantly by the 10-year mission (1996 to 2006) of Mars Global Surveyor, has resulted in highly detailed topographical maps of the planet's surface. Since the first independent Mars rover landed in 1997, these maps have continued to be refined and updated.

The topography of Mars can roughly be divided into two types. Its northern plains are flat, the work of millennia of lava flows coursing over the surface. The more southerly quadrants are mountainous, and home to a greater profusion of impact craters. When describing an object in the solar system, space scientists often refer to its 'albedo', or whiteness; this indicates its ability to reflect light. On Mars, the albedo is of two very different kinds. First are the paler high-albedo plains, rich in iron oxides and once believed to be Martian continents, hence naming conventions such as Amazonis Planitia or Arabia Terra. The darker features, whose albedo is markedly lower, were thought to be Martian seas, such as Mare Erythraeum and Mare Sirenum. The largest of Mars' dark spots, Syrtis Major Planum, was first recorded in a sketch by Christiaan Huygens, the Dutch astronomer, in 1659.

The difference between Mars' highest and lowest points is around 30 km (19 mi). The highest point is the near-25-km (16 mi) summit of Olympus Mons, the gargantuan shield volcano located in the Tharsis region. The lowest point is the floor of the Hellas impact basin, which lies a little over 8 km (5 mi) below the planet's 'datum' (the comparable coordinate on Earth is sea-level). On Earth, a planet twice the size of Mars, that difference (between the summit of Mt Everest and the bottom of the Mariana Trench) is only 19 km (12 mi). While the surface of Earth is far from smooth, Mars is notably lumpier than our planet. This means the radius of Mars varies depending on whether you are measuring from a mountain or a crater.

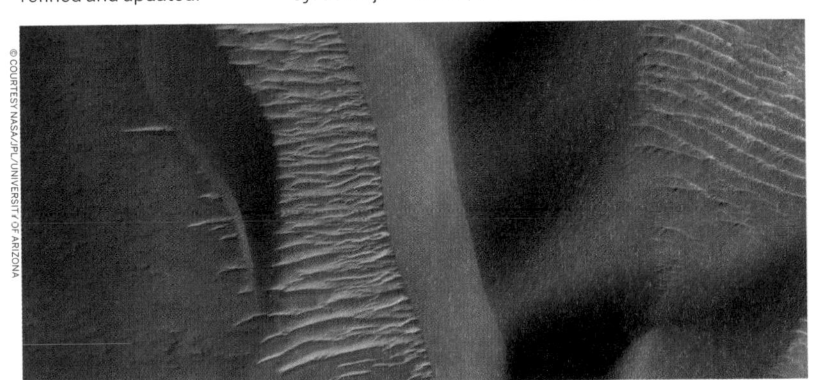

The steeply sloped sand dunes of Nectaris Montes.

Martian Moons: Phobos and Deimos

The American astronomer Asaph Hall was all but ready to give up his frustrating search for a Martian moon when, on the night of 16 August 1877, he was persuaded to continue by his wife, Angelina. He discovered the first of Mars' moons, Deimos, the very next night, and a second moon, Phobos, six nights later. Hall named the moons after the mythological sons of Ares, the Greek god of war. (Mars is Ares' Roman counterpart.) Phobos means fear, or panic, while Deimos translates as flight.

Mars' moons are among the smallest in the solar system. Phobos is slightly larger than Deimos and orbits a mere 5954 km (3700 mi) above the Martian surface. No known moon orbits closer to its planet. Phobos whips around Mars three times a day.

Ninety-four years after Asaph Hall's discovery, in 1971, NASA's Mariner 9 spacecraft got a much better look at the two moons during its orbit around Mars. The dominant feature on Phobos, Mariner discovered, was a crater 10 km wide (6 mi). That makes the crater nearly half the width of the moon itself, which is roughly 27 km (27 mi) long, 23 km (14 mi) wide and 18 km (11 mi) deep. Hall bestowed upon this feature his wife's maiden name; hence its title, Stickney Crater.

Asaph Hall observes Mars through the US Naval Observatory telescope.

Like Earth's Moon, Phobos and Deimos are tidally locked and hence always present the same face to their planet. To someone standing on the Mars-facing side of Phobos, Mars would take up a large part of the sky. Further out from Mars is smaller Deimos. At 14km (9 mi) long, 11 km (7 mi) wide and 10.9 km (6.8 mi) deep, Deimos is the junior sibling. Like Phobos, it bears the scars of numerous impact craters and has a thick topsoil, or regolith, of as much as 100 m (328ft) in depth.

Phobos and Deimos are among the darkest objects in the solar system, sheltered from the Sun by Mars. Scientists have even discussed the possibility of using one of the Martian moons as a base from which astronauts could observe the Red Planet, although no plans have been formalised. This would enable the astronauts to launch robots to Mars' surface while shielded from cosmic rays and solar radiation by miles of rock. Otherwise the toll on astronaut health would be extreme.

Phobos facts

- Phobos' gravity is 1/1000th that of Earth; on Phobos, a 68kg (150lb) person would weigh the equivalent of a mere 68g (2.4oz)!
- This Martian moon is gradually spiralling inward, drawing closer to Mars by about 1.8m (5.9ft) every 100 years. Within 50 million years, Phobos will either crash into Mars, or break up and form a ring around the planet.
- The moon of Phobos orbits faster than Mars rotates; this means that, after rising in the west and setting in the east, Phobos then does it all over again before the Martian sol, or day, ends.

Deimos facts

- Deimos has the far longer orbit; at 30.3 hours, it takes more than 3.5 times as long to travel around Mars.
- Neither Deimos nor Phobos are spherical; Deimos, in particular, might be named after a demigod but looks more like a potato.
- Deimos appears to be composed of carbon-rich rock, as does its larger sibling. Both moons may in fact be captured asteroids.
- Both Deimos and Phobos feature (albeit not by name) in Jonathan Swift's 1726 novel *Gulliver's Travels*.
- Deimos is being considered as a staging-post for exploration on Mars.

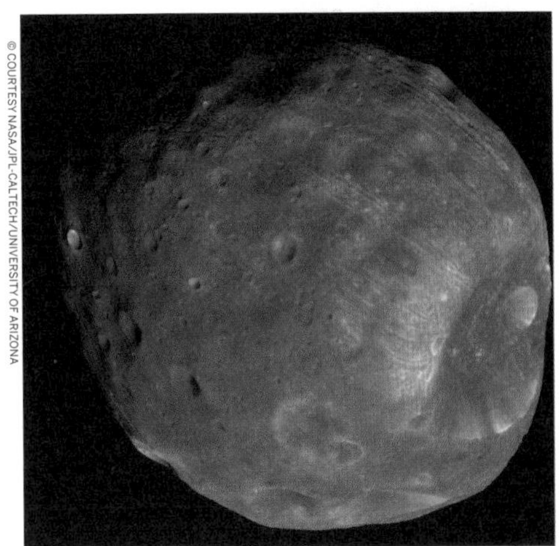

Crater-pockmarked Phobos is the larger of Mars' two moons.

Phobos is the inner moon and Deimos the outer.

In the absence of oceans and continents, cartographers of Mars have had to use those reference points visible from Earth: highlands and lowlands, as well as impact craters, plains and volcanoes. Planetary features are named following the conventions set down by the International Astronomical Union (IAU) and must be Greek or Roman. Landforms and plains are prefixed with *terra*, *planitia* or *planum*; areas once thought to be seas with *mare*. Mountains take the suffix *mons*, but craters – however impactful – stay just that.

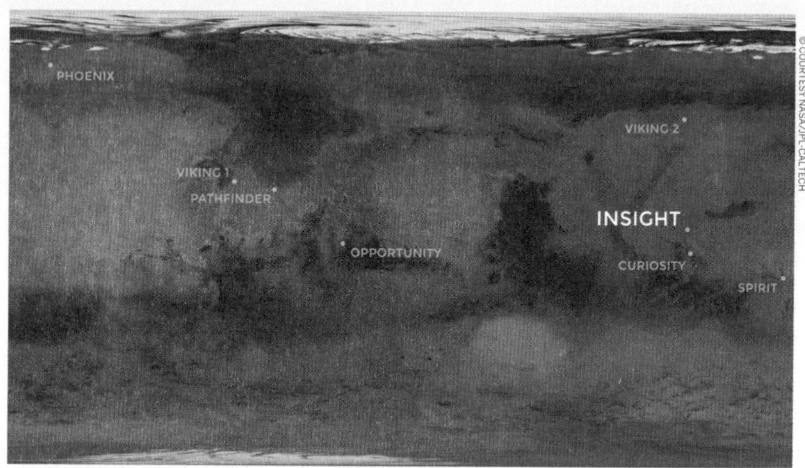

The landing sites of the Mars rovers are shown between the planet's two poles.

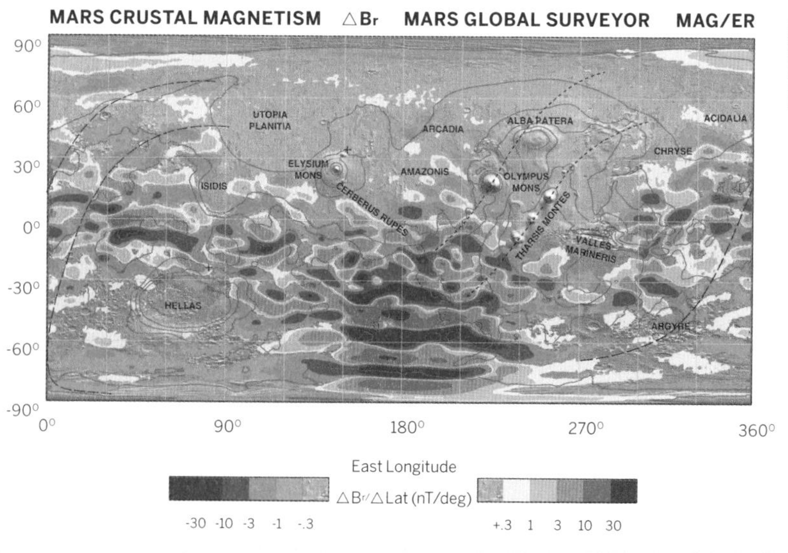

Though Mars lacks a magnetic field, the southern hemisphere is strongly magnetised.

Atmosphere/Magnetosphere

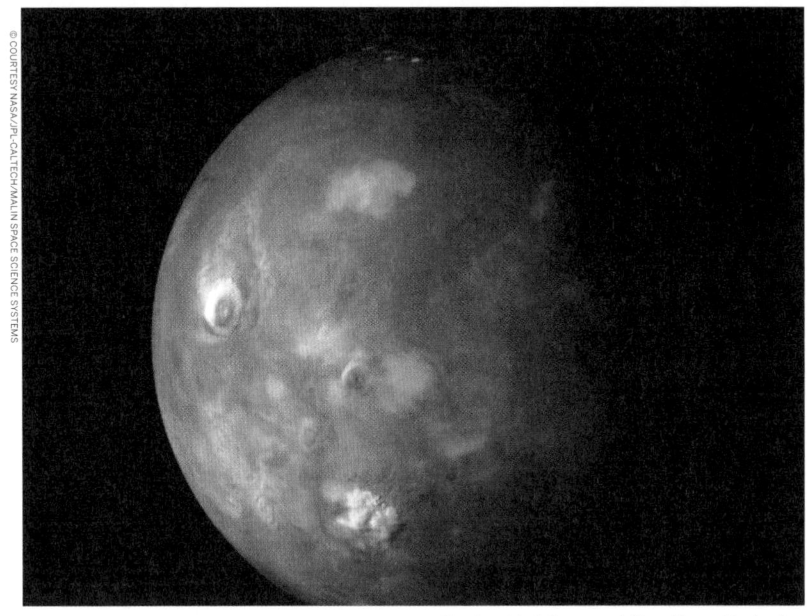

Clouds and dust storms are visible swirling on Mars' surface.

Mars has a thin atmosphere, made up mostly of carbon dioxide, nitrogen and argon gases. If viewed from the planet's surface, the Martian sky would be hazy and red – unlike the familiar blue tint we see on Earth – because of dust suspended in Mars' atmosphere. In fact, winds on Mars are strong enough to create dust storms that cover much of the planet. After such storms, it can be months before the dust fully settles.

The thin atmosphere also explains the countless craters on we see on Mars. The Red Planet has taken a pounding over the millennia; as of 2017, Martian craters accounted for 21% of all 5211 named craters in the solar system. Unlike Earth's atmosphere, Mars' atmosphere provides insufficient resistance to break up incoming debris, and so offers relatively little protection from objects such as meteorites, asteroids and comets. Scientists think Mars may have had a thicker atmosphere early in its history and that it has been reduced over time by the solar wind.

In addition to dodging missiles while peering through dust storms, another vital challenge for human inhabitants of Mars would be coping with wildly varying temperatures. Its thin atmosphere provides little regulation between highs and lows. The temperature on the Red Planet can be as high as 20°C (70°F), or as low as about -153°C (-225°F). And because the atmosphere is so thin, very little heat from the Sun is trapped at the planet's surface. If you were to stand on the surface of Mars on the equator at noon, it would feel like a balmy 24°C (75°F) at your feet, while at your head the temperature would be a chilly 0°C (32°F)!

As for finding your way around, you wouldn't be able to rely on a compass, as Mars has no magnetic field. Areas of the Martian crust in the southern hemisphere, however, are highly magnetized, indicating traces of a magnetic field that might have existed around 4 billion years ago.

History

Like the other terrestrial planets in our solar system, Mars was formed around 4.5 million years ago by the pull of gravity on dust and gases. Today it consists of a central core, a rocky mantle and a solid crust. Mars' central core is particularly dense, measuring between 1500 km to 2100 km (930 and 1300 mi) in radius, and is made of iron, nickel and sulphur. Surrounding this core is a rocky mantle between 1240 km to 1880 km (770 and 1170 mi) thick and, above that, a crust made of iron, magnesium, aluminium, calcium and potassium. This crust is between 10 km to 50 km (6 and 30 mi) deep.

The reason Mars appears red is due to the oxidization of iron. This process occurs in the iron-rich rocks, regolith (the Martian 'soil') and dust of Mars. When this dust gets kicked up into the atmosphere, it makes the planet appear mostly red. In fact, the Red Planet is many colours – at the surface you can see a rich assortment of browns, gold and tan.

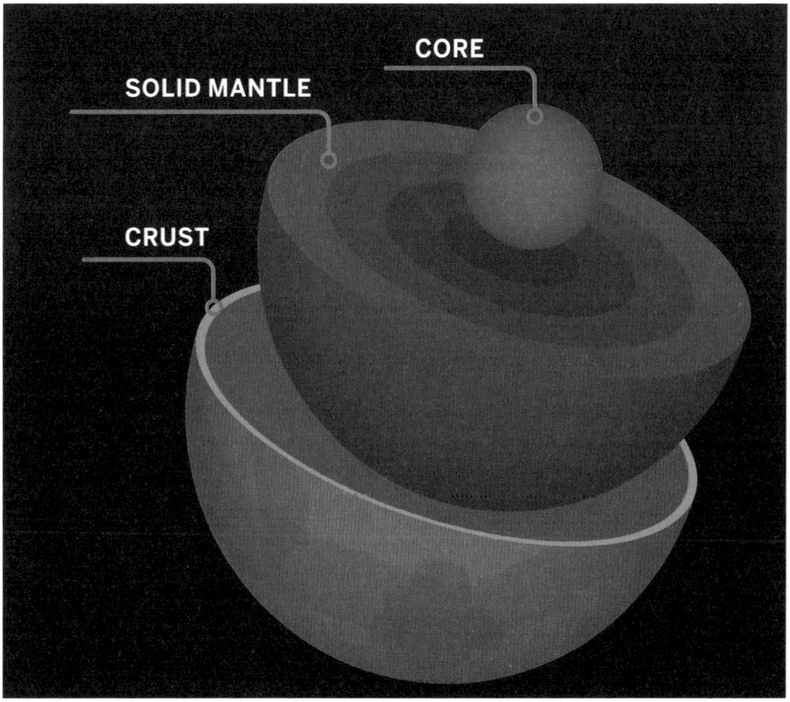

CORE
SOLID MANTLE
CRUST

Mars' core has a lower density than Earth's, indicating it contains more lighter elements.

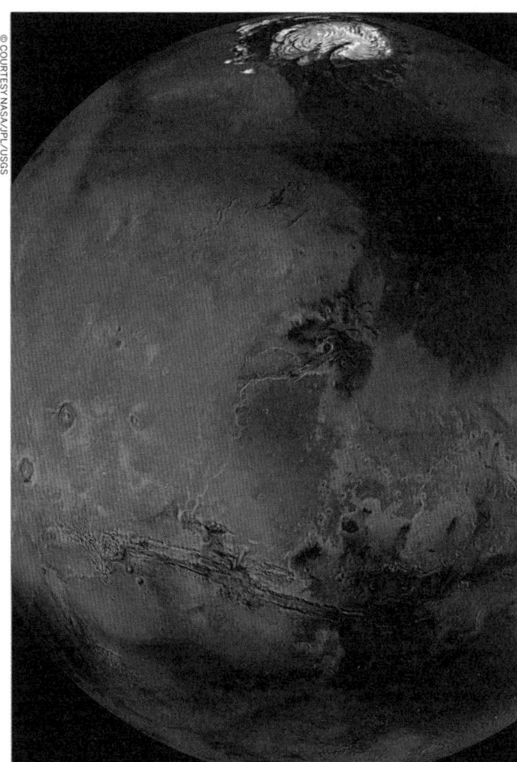

The polar ice caps of Mars present a striking visual contrast from its red surface.

Its volcanoes, impact craters, crustal movement and atmospheric conditions (such as dust storms) have altered the landscape of Mars over many years, creating some of the solar system's most interesting topographical features. Although Mars is half the size of Earth, these features are far bigger than their counterparts on our home planet. This means that while Mars is about half the diameter of Earth, its surface has nearly the same area as Earth's dry land.

A large canyon system, called Valles Marineris, is more than 4800 km (3000 mi) long. With that length, on Earth it would stretch from California to New York. This vast Martian canyon is also 320 km (200 mi) across at its widest point, and 7 km (4.3 mi) deep. That's about 10 times the size of the Grand Canyon.

Mars is also home to the largest volcano in the solar system, Olympus Mons. At close to 25 km (15 mi) high, it's three times taller than Mount Everest, and 60% taller than Mauna Kea, Earth's largest mountain. This Martian monster has a base the size of Italy. Yet because it's spread out, it would be less steep to climb than most Earth peaks.

Mars' Watery Past

» The belief that water once existed on Mars can be traced back to the late 19th century, when the Italian astronomer Giovanni Schiaparelli claimed, in 1877, to have seen a system of channels criss-crossing the planet's surface. Schiaparelli's term for these features, 'canali', was translated into English as 'canals', prompting unwarranted speculation that they had been built by intelligent life.

» Mars does appear to have had a watery past, however, with ancient river valley networks, deltas and lake beds, as well as rocks and minerals on the surface that could only have formed in liquid water. Some features suggest that Mars experienced huge floods about 3.5 billion years ago.

» There is water on Mars today, but the Martian atmosphere is too thin for liquid water to exist for long on the surface. Water is found in the form of ice just under the surface in the polar regions, as well as in briny water which flows, seasonally, down hillsides and crater walls.

Discovery Timeline

No planet other than Earth has been studied as intensely as Mars. Observations of the Red Planet can be traced back more than 4000 years, to Ancient Egypt, whose astrologers named it *Har Decher* – the 'red one'. Studies of Mars have continued ever since, and some key moments include:

1659

Christiaan Huygens, the celebrated Dutch scientist who ranks alongside Isaac Newton and Albert Einstein, is the first to sketch Syrtis Major Planum, a shield volcano whose dark colouring appears in striking contrast to Mars' red surface.

1877

Asaph Hall discovers Phobos and Deimos, the two moons of Mars, which are now believed to be asteroids brought within their home planet's orbit.

1965

Launched the previous year from Cape Canaveral, NASA's unmanned Mariner 4 travels for eight months to reach Mars; the 22 images it sends back are the first close-up photographs of any planet besides Earth.

1971

Mariner 9 becomes the first spacecraft to successfully orbit Mars. Two Soviet spacecraft, Mars 2 and Mars 3, arrive just three weeks later. Following several months of dust storms, Mariner finally sends back clear images, helping to further map the planet's surface.

1976

Tasked with searching for signs of life, Viking 1 makes a successful landing on the surface of Mars. Its record mission of 6.33 years, or 2245 Martian sols, was only surpassed in 2006 by the Mars Global Surveyor.

1996

Launched November 7, 1996, Mars Global Surveyor became the first successful mission to the red planet in two decades, ending communications on 2 November 2006.

Christiaan Huygens made observations of Syrtis Major Planum.

Curiosity took this selfie during a Martian dust storm on the surface.

1997

With perfect timing, on the 4th of July, Mars Pathfinder lands in the Chryse Planitia region. Its subsequent dispatch of Sojourner marks the first wheeled rover to explore the surface of another planet.

2002

Launched in 2001, Mars Odyssey begins its mission to make global observations and find buried water-ice on Mars. In addition to finding extensive reserves of water-ice, in 2010 it breaks the Mars Global Surveyor's record for the longest-working NASA instrument on Mars.

2004

Twin Mars Exploration Rovers, named Spirit and Opportunity, find strong evidence that Mars once had long-term liquid water on its surface.

2006

The Mars Reconnaissance Orbiter begins returning high-resolution images as it studies the history of water on Mars. More recent images helped select a landing site for the InSight mission.

2008

Phoenix finds signs of possible habitability; in addition to the occasional presence of liquid water, some soil chemistry appears favourable.

2012

NASA's Mars rover Curiosity lands in Gale Crater and finds conditions once suited for ancient microbial life.

2018

InSight rover lands and begins its mission.

© GRAHAM PRENTICE/ALAMY STOCK PHOTO

Surrey's Crown Square features a sculpture of the invaders from *The War of the Worlds*.

Life on Mars: The Red Planet in Popular Culture

No other planet has captured our collective imagination quite like Mars. Long before David Bowie posed the question, filmmakers and writers offered their own versions of life on Mars, creating delight and alarm in equal measure. Edgar Rice Burroughs' *Barsoom*, a comic serial set on the Red Planet, began as early as 1912. In the concluding storyline, in 1940, the eponymous hero *John Carter of Mars* is sent to the planet by 'astral projection'. Upon his arrival he discovers that Mars' weak gravity has given him superpowers. Interplanetary travel between Earth and Mars has, of course, been two-way traffic. During the 1938 radio adaptation of HG Wells' *The War of the Worlds*, so legend has it, so many listeners believed the story to be real news coverage of a Martian invasion that it caused widespread panic. Wells' novel was even converted into a rock opera by Jeff Wayne in 1978. By the age of the B-movies of the 1950s, Martians were ever-present, making their way onto the big screens of Earth, the reds above our heads during the era of McCarthyism. Martians also teamed up, in 1953, with comedy team Abbott and Costello in the movie *Abbott and Costello Go to Mars*. Decades later, Tim Burton's 1996 comedy, *Mars Attacks*, sees Earth-dwellers vaporised by cartoonish aliens from Mars whose brains are exposed at the tops of their heads.

The master of science fiction, Isaac Asimov, has a Martian impact crater named after him, while Mars lends its name, in

© JOHN THYS/GETTY IMAGES

Though oft linked to the planet, the Mars Bar was named after the maker's last name.

The War of the Worlds was one of the first big sci-fi successes.

turn, to arguably the world's most famous chocolate bar.

Yet many of the more recent appearances of Mars in pop culture reveal a deeper curiosity of when and how people may eventually populate the planet. The real challenges that will face humans on Mars, namely the lack of air, food and water, were addressed in 1990 with Arnold Schwarzenegger's *Total Recall*, themes that were repeated in a 2012 remake.

Perhaps the most plausible creation of an actual Martian experience is more recent still. Andy Weir's 2014 novel *The Martian* became a bestseller, and the subsequent movie starring Matt Damon, released in 2015, has the Hollywood star playing botanist Mark Watney. Stranded alone on the planet as he awaits a rescue mission, Watney famously manages to grow potatoes. The ISS astronauts likely approved.

Space-age Technology

The scientific know-how required to grow crops on Mars already exists, but what else would humans need to reach and survive on Mars? Here are NASA technologies already being used, or being tested, in preparation for humans visiting the Red Planet:

1) Ion propulsion

Gases such as argon or xenon are electrically charged and the ions then pushed out at speeds of 320,000 km/h (200,000 mph). The spacecraft experiences a force no greater than a gentle breeze but can continuously accelerate for several years. This technology has allowed NASA's Dawn spacecraft to minimize fuel consumption and complete more than five years of continuous acceleration; for a total velocity change around 40,000 km/h (25,000mph).

2) Habitation

At NASA's Johnson Space Center, crews train for long-duration, deep-space missions in the Human Exploration Research Analog (HERA). A self-contained environment that simulates a deep-space habit, the two-story structure features living quarters, workspaces, a hygiene module and simulated airlock. Within the module, test subjects conduct operational tasks, living together for 14 to 45 days (planned to increase to up to 60 days).

3) Solar power

By the 6th century AD, astronomers, such as the Indian scholar Aryabhata, were getting close to accurate measurements of the planet. In his great work, modestly titled the *Aryabhatiya*, the circumference of the Earth is given as 4967 yojanas, or 39,968 km (24,835 mi). The actual circumference at the equator almost exactly matches his prediction.

4) Plant farms

'Veggie' is a deployable fresh-food production system currently being trialled aboard the International Space Station. Using red, blue and green light, Veggie helps plants grow in small bags, whose wicking surfaces contains fertiliser. Astronauts have used the system to grow lettuce. Some crops are harvested for consumption on board, while the rest gets sent back to Earth for analysis.

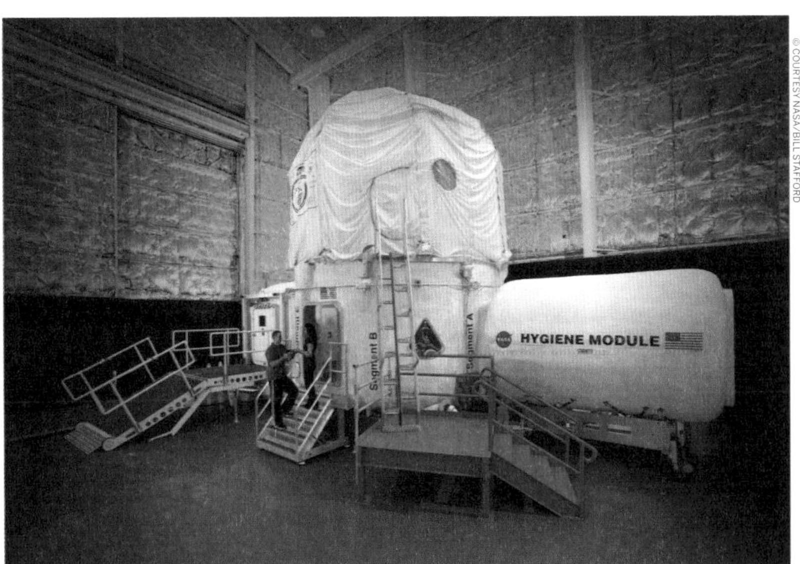

NASA's HERA Lab at Johnson Space Center, a three-story habitat designed to serve as an analog for remote conditions in space.

5) RTG

NASA has used Radioisotope Thermoelectric Generators, or RTGs, for forty years. RTGs helped power Apollo missions to the Moon, as well as the Mars rover Curiosity. These 'space batteries' convert heat from the natural, radioactive decay of plutonium into reliable electrical power. The RTG on Curiosity generates 110 watts of power, slightly more than the average light bulb.

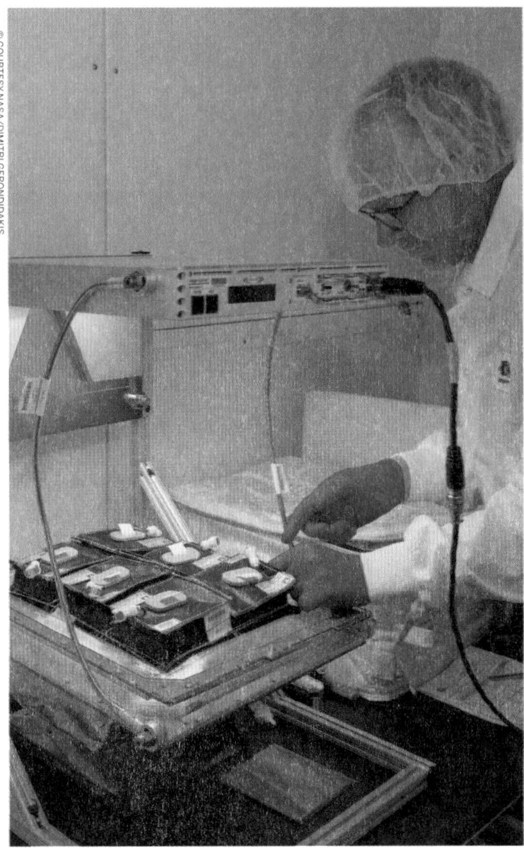

The Veggie lab grows vegetables (what else?) on the ISS.

6) Water recovery

The International Space Station recovers and recycles water from everywhere, including urine, hand-washing and oral hygiene. Its Water Recovery System (WRS) reclaims and filters water to get it ready for consumption. Liquids can present problems in space as they behave differently in a microgravity environment. The part of the WRS that processes urine uses a centrifuge for distillation, since gases and liquids do not separate like they do on Earth.

7) Oxygen generation

NASA's Oxygen Generation System (OGS), reprocesses the atmosphere of the spacecraft to continuously provide breathable air efficiently and sustainably. The system produces oxygen through electrolysis, splitting water molecules into their component oxygen and hydrogen atoms. The oxygen is released into the atmosphere, while the hydrogen is either discarded into space or fed into the water recovery system.

8) Spacesuits

On the Martian surface, a spacesuit is vital for survival. NASA's Z-2 suit and Prototype eXploration Suit will address the challenges of walking on Mars, including dealing with dust. The red soil on Mars could affect both astronauts and systems inside a habitation unit. New designs will feature a 'suitport', enabling astronauts to quickly hop into a spacecraft while the outer layer of their spacesuit remains outside.

9) Next-gen rovers

The Multi-Mission Space Exploration Vehicle (MMSEV) will support missions to asteroids, Mars and its moons, as well as many other missions in future. The MMSEV is built to address such issues as extended range, rapid entry or exit, and radiation protection. Some versions of the MMSEV have six pivoting wheels for improved manoeuvrability. In the event of a flat tyre, the vehicle simply lifts up the bad wheel and keeps on rolling.

4.5-billion-year-old Martian meteorite ALH84001 was found in Antarctica.

Investigating Mars

Understanding the evolution of Mars, its geological processes and whether or not it has ever hosted life helps us understand Earth by determining similarities between early Earth and early Mars and what, if anything, caused the development of their respective climates to diverge.

Astrobiology is a useful tool in the search. This exciting branch of science unites astronomers, biologists, geologists and physicists, and owes much to the discovery of a lump of rock the size of a small melon. ALH84001 is the oldest Martian meteorite to have landed on Earth and was found in Allan Hills, Antarctica, in 1984. Formed about 4.1 billion years ago and composed of a rock known as orthopyroxenite, the six-inch block contains, in traces of water, mineral deposits that may just be fossilized organisms. Finding fossils preserved from early Mars might tell us that life once flourished on the Red Planet. Evidence of cells preserved in Martian rocks might be found at an incredibly small scale. Known as 'biosignatures', these molecular fossils give an indication of the organisms that created them. Unfortunately, over hundreds of millions of years, any Martian biosignatures may well have been destroyed or transformed to a point where they may no longer be identifiable.

In an effort to find evidence of past life, and the conditions it might require in the future, an international fleet of robotic spacecraft is currently studying Mars from all angles, both in orbit and on the surface.

1) Mars Odyssey

Launch date: 7 April 2001

Still in orbit around Mars, NASA's 2001 Mars Odyssey spacecraft has collected more than 130,000 images and continues to send information to Earth about Martian geology, climate and mineralogy. Odyssey also relayed communications from surface landers such as Spirit and Opportunity.

2) Mars Express

Launch date: 2 June 2003

Express has discovered minerals that form only in the presence of liquid water, confirming Mars was once much wetter. It detected sufficient water-ice in the polar caps to create a global ocean 11m (36ft) deep; also located, at 100 km (62 mi), were the highest clouds ever seen above any planetary surface. The Express mission found indications of methane on Mars, which on Earth is attributed to volcanism and biochemical processes.

3) Opportunity Rover

Launch date: 8 July 2003

Having notched up over 5000 sols on Mars, Opportunity returned dramatic evidence of historically wet conditions that might have sustained microbial life. Scientists believe that Opportunity's Meridiani Planum landing site was once the shoreline of a salty sea. In July 2014, Opportunity set the 'off-Earth' roving-distance record, having clocked 40 km (25 mi) of driving. The mission finally ended in February 2019, after months without contact.

NASA's Mars Reconnaissance Orbiter in a concept illustration.

4) Mars Reconnaissance Orbiter (MRO)

Launch date: 12 Aug. 2005

The MRO revealed that Mars is more dynamic and diverse than had previously been realized. The data this mission gathered on 'warm-season flows' is the strongest evidence to date of liquid water on Mars, and MRO returns more data about Mars every week than the weekly total from all six other active Mars missions combined.

5) Mars Atmospheric and Volatile EvolutioN (MAVEN)

Launch date: 18 Nov. 2013

During its primary mission, to study Mars' atmospheric conditions, MAVEN carried out 10 months of observations, including four 'deep-dip' campaigns into the Martian atmosphere. Equally notable for surviving an encounter with a comet, named Siding Spring.

6) ExoMars Trace Gas Orbiter

Launch date: 14 March 2016

ExoMars is searching for methane and other trace gases in the Martian atmosphere. Organisms on Earth release methane during digestion, and geological processes, such as the oxidation of minerals, also produce the gas. ExoMars monitors seasonal changes in the Martian atmosphere and looks for evidence of water-ice beneath the surface.

Focus on Curiosity

© COURTESY NASA/JPL-CALTECH/MSSS

Curiosity prepares to drill down into a rock target on Mt Sharp.

Launch date: 26 November 2011

The Curiosity rover is NASA's current six-wheeled ambassador on the Red Planet. It arrived, on 5 August 2012, aboard the Mars Science Laboratory, the first planetary mission to use precision-landing techniques. In steering itself towards the Martian surface, the host spacecraft mimicked the way a space shuttle controls entry through Earth's upper atmosphere. The rover's spacecraft flew to a desired location above Mars' surface before deploying a parachute and 'retro rockets' to slow it down. It then lowered the rover, much like the way helicopters move large objects. A new technique for NASA, it enabled Curiosity to use a landing area, or ellipsis, of 4 miles by 12 miles – about one-third the size used by earlier rovers Spirit and Opportunity in 2004. NASA selected this ellipsis, in Gale Crater, based on highly detailed images sent by the aptly-named Mars Reconnaissance Orbiter.

At 3 m (9ft) long and weighing 899 kg (1982lb), Curiosity is twice as long and three times as heavy as previous rovers, helping to provide stability as it navigates up and down steep terrain near its landing site. The Curiosity rover carries a radioisotope power system that generates electricity from the radioactive decay of plutonium. This both provides greater mobility and allows the rover to explore a much larger range of latitudes and altitudes than was possible on previous missions.

Like Spirit and Opportunity, Curiosity has six wheels, as well as cameras mounted on a mast. Unlike the twin rovers, Curiosity carries a laser to vaporise thin layers from the surface of rocks and analyse the elemental composition of the underlying material. By drilling down, NASA scientists can test soil samples from a range of locales, including some possible dried streambeds with signs of water erosion.

Since its arrival, Curiosity has continually collected samples, relaying them to its on-board test chambers for chemical analysis. Curiosity carries a suite of instruments for identifying compounds, known as organic molecules. Crucially, these contain one or more carbon atoms, bound to hydrogen. They can exist without life, but life as we know it cannot exist without these building blocks – their presence would be an important plus for future human habitation. Curiosity can also check for other chemical elements important for life, such as nitrogen, phosphorus, sulphur and oxygen.

A view from the 'Kimberley' formation on Mars, taken by Curiosity.

Curiosity took this self-portrait at the 'Big Sky' drilling site.

Curiosity's key findings so far:

» During the trip to Mars, the Mars Science Laboratory and Curiosity encountered radiation levels that could pose potential health risks to human astronauts.

» A crucial part of the mission was the testing of landing equipment and techniques never before used on Mars. Thus Curiosity's safe delivery to the Martian surface gives NASA scientists significantly more freedom – i.e., weight – in developing instrumentation.

» During the 'Noachian' period on ancient Mars, between 4.1 billion to 3.7 billion years ago, evidence suggests the planet might have had the right chemical conditions to be a suitable home for microorganisms.

» Curiosity has found evidence of an ancient streambed, where water once flowed knee-deep. The landing site, in Gale Crater, is home to a range of different environments. Again, this indicates the presence of significant levels of water at some time in Mars' past.

Focus on InSight: The Latest NASA Mission to Mars

Launch date:
5 May 2018

Arrival on surface:
26 November 2018

Launch phase

On 5 May 2018, InSight took off from Vandenberg Air Force Base aboard the hugely powerful Atlas V-401 rocket, an expendable launch vehicle. The mission that launched InSight also bid a hopeful sendoff to a separate technological experiment. InSight was followed by two cuboid spacecraft, or CubeSats, each no bigger than a briefcase. Their goal is to test new, miniaturised deep-space-communication equipment. Upon their arrival at Mars, the twin MarCOs (Mars CubeSat One) successfully relayed back data from InSight as it entered the Martian atmosphere and landed. The first successful test of miniaturized CubeSat technology in orbit around another planet, researchers hope it will offer future missions increased communication capabilities going forward.

By examining physical features, such as canyons and volcanoes, previous missions to Mars have investigated only the surface of the Red Planet. Signatures of the planet's formation can only be found by sensing and studying far below the surface, checking the planet's 'vital signs'. InSight (short for Interior Exploration using Seismic Investigations, Geodesy and Heat Transport) is a Mars lander designed to give the Red Planet its first thorough checkup since it was formed, 4.5 billion years ago. It is the first robotic explorer to study the 'inner space' of Mars: the planet's crust, mantle and core. Examining Mars' interior will help answer key questions about the early formation of other rocky planets in our inner solar system – Mercury, Venus and Earth – as well as how rocky exoplanets formed.

The record of Mars' formation is preserved in its rocks and soil. The presence of a 25 km-high (16 mi) volcano on the Martian surface is a reliable indicator that Mars experienced extreme geological activity in the past. A lander such as InSight is helping to reveal the extent of tectonic activity on Mars today, which seems to have slowed down but not yet ceased entirely.

A view of InSight's deck.

Unlike some spacecraft, InSight did not spin around its axis as it travelled to the Red Planet. Instead, sensors attached to the spacecraft told it which way was up, down, left, right and so on. This is known as '3-axis stabilisation'. InSight was powered by eight thrusters, fired intermittently. Four larger thrusters helped turn the craft in the right direction; four smaller ones kept the spacecraft stable. InSight's flight path was carefully planned to minimise travel time. By late July 2018, more than two months into InSight's journey, Mars was at the point in its orbit where it is closest to Earth. This happens roughly every 26 months and is known as Mars Close Approach. The InSight launch was timed so that the spacecraft travelled less than halfway around the Sun before it reached Mars.

Cruise phase

InSight left Earth at a speed of 10,000 km/h (6200 mph). The journey to Mars is about 485 million km (301 million mi) and, after launch, mission navigators tracked InSight continuously. The navigation team adjusted InSight's flight path several times during the 'cruise' phase, and the spacecraft had several tools to help it.

1) Star tracker

The star tracker logged InSight's position against the stars in the night sky to tell navigators back on Earth how the spacecraft was oriented, similarly to how ancient navigators looked to the stars to chart their own courses while at sea.

2) Inertial measurement unit

Working in tandem with an onboard gyroscope, this device provided information on which direction InSight was travelling, and how fast.

3) Sun-sensors

Sensors helped maintain its position relative to the Sun.

An artist's conception shows the InSight lander deploying its seismometer and heat probe.

© COURTESY NASA/JPL-CALTECH

InSight has weather-monitoring instruments it can deploy on Mars.

Landing on Mars

On 26 November 2018, protected by its heat shield, the InSight lander plunged through the thin Martian atmosphere. As Curiosity had done six years previously, it used a parachute to slow down before firing its retro rockets for the final descent onto the smooth ground of Elysium Planitia. The chosen site needed to meet several criteria. Firstly, because the spacecraft relied on the atmosphere for deceleration, only a landing site with a low elevation would have sufficient atmosphere above it. Secondly, it had to be close to Mars' equator, both to ensure that the lander's solar array could provide power year-round, and that the lander could stay warm enough to function.

InSight's success depended on landing on a flat area. It lacks the mobility of a rover; a steep slope might prevent its robotic arm from being able to reach enough terrain. An incline in the wrong direction might also jeopardize how much power the solar arrays could produce, while a large rock might prevent the solar arrays from opening.

InSight's heat-flow probe also needed to penetrate the ground. Designed to burrow into soil (not rock), to a depth of 3 to 5 m (10 to 16 ft), the Thermal Imaging System (THEMIS) on NASA's Mars Odyssey orbiter provided key images. Armed with the knowledge that solid rock changes temperature more slowly than soft ground, THEMIS was able to pinpoint the most appropriate locations to target.

Checking a Planet's Health

The goals of the InSight mission are very different from the Mars landers or rovers that have gone before. In some ways, InSight's activities are more of a marathon than a sprint. This means its instruments are different, too. InSight has three specific instruments, each designed to check a different aspect of Mars' health. The lander's instruments will delve deep beneath the surface and seek the fingerprints of the processes that formed the terrestrial planets. It does so by measuring the 'vital signs' of the planet: its 'pulse' (seismology), 'temperature' (heat flow), and 'reflexes' (precision tracking).

1) Taking Mars' pulse

SEIS, the Seismic Experiment for Interior Structure, is InSight's seismometer. Created to check the heartbeat of the planet, it is a dome-shaped instrument that sits on the Martian surface and records seismic vibrations. SEIS' measurements provide a glimpse into the planet's internal activity. Seismic waves are produced by marsquakes (see what they did, there?) and meteorite impacts. A suite of sensors – for wind, pressure, temperature and magnetic fields – fine-tune the seismometer's measurements. SEIS can also sense surface vibrations generated by weather systems, such as dust storms, or by turbulence in the atmosphere: phenomena such as dust devils can also generate seismic waves. In revealing what lies beneath, SEIS may even be able to tell us if there are liquid water deposits or plumes of active volcanoes under the Martian surface.

2) Taking Mars' temperature

To see how hot Mars is under the hood, NASA created the Heat Flow and Physical Properties Probe, or HP3 for short. HP3 can burrow to almost 5 m (16ft) – deeper than any previous instrument. A 'mole' at the end of the probe burrows into the ground, generating heat-pulses as it descends. The probe measures the soil's 'thermal conductivity' by monitoring how quickly, or

An artist's rendering of the Mars Reconnaissance Orbiter.

slowly, these pulses heat up the surrounding soil. As the probe makes its way down, a hammering motion generates vibrations. Scientists back on Earth can then use the seismometer to detect 'reflections' – such as lava flows – as the vibrational waves bounce off layers of Mars' subsurface, helping to build a cross-sectional picture of the planet's interior. In calculating how much heat is being generated, and what its source might be, scientists can determine whether Mars formed from the same stuff as Earth and the Moon.

3) Checking Mars' reflexes

As Mars orbits the Sun, InSight's Rotation and Interior Structure Experiment (RISE), a set of radio antennae, track the location of the lander to determine how much Mars' north pole moves. Each day, RISE sends X-band radio signals back and forth with Earth for an hour or so each day. These observations provide detailed information on the size of Mars' iron-rich core, by tracing how the planet wobbles. They will help determine whether the core is liquid, and which other elements, besides iron, may be present.

4) Talking to the physician

InSight normally talks to Earth once every Martian day (or sol), with its transmissions relayed via the Mars Reconnaissance Orbiter (MRO), the Mars Odyssey, MAVEN and ESA's Trace Gas Orbiter. During the first critical few weeks after landing, it was in touch twice daily. The MRO functions a bit like an airplane's black box, especially during the key landing period. Data travels by satellite relay.

Travelling to Mars: Your Chariot Awaits

The Orion Multi-Purpose Crew Vehicle (MPCV) is the NASA spacecraft that will take astronauts into space and beyond Earth's orbit. The agency began testing Orion in December 2014, and crewed missions are scheduled to depart as soon as the mid-2020s.

Similar in shape to the Apollo spacecraft, Orion is designed to carry up to six astronauts to destinations such as the Moon or Mars. Orion is a significant upgrade from Apollo – the spacecraft is much larger than Apollo, and features electronics decades more advanced than what Apollo's astronauts used to fly to the moon.

Orion was originally conceived for NASA's Constellation, a programme that was intended to bring humans to the International Space Station and, ultimately, Mars. It is now one of the key developments in NASA's new plan. Dubbed Moon to Mars, scientists aim to use Earth's Moon as a springboard for sorties into deep space.

Orion has a modular construction, made up of a distinctive teardrop-shaped astronaut capsule along with an additional service capsule containing various life-support equipment. If you're tall and struggle with legroom on modern airlines, you might find Orion a bit snug. The habitable space inside Orion is just 9 cubic metres (318 cu ft), although that's roughly one-and-a-half times more than the space inside Apollo. The service module is stocked with solar panels to generate electricity and oxygen equipment for breathing, as well as thrusters to propel the capsule. One of Orion's first destinations will almost certainly be NASA's Lunar Orbital Platform-Gateway, a sort of lunar resort for those who prefer heavy science to light reading. Commercial aerospace companies want to create modular 'hotels' in orbit around Earth, too.

ULA's Delta 4 rocket launches off Cape Canaveral with NASA's Orion on board.

Mars Highlights

A mosaic of Curiosity's images from Yellowknife Bay.

Challenging conditions will ensure Mars is only for the intrepid – but if you're after majestic plains, gargantuan peaks and mesmerising dunes, the Red Planet will be worth the wait to get there.

Polar Ice Caps

1 Mars is home to two permanent polar ice caps: Planum Boreum, at the northern extreme of the planet, and Planum Australe to the south.

Tharsis Montes

2 Towering above a vast, mountainous plateau stand some of Mars' most celebrated attractions – the largest volcanoes found anywhere in the solar system.

Olympus Mons

3 The undisputed heavy weight champion of Martian volcanoes rises 25 km (16 miles) above the planet's floor. Its 602-km (374-mi) diameter is roughly equivalent to that of the US state of Arizona.

Valles Marineris

4 What likely started life, around 1 billion years ago, as just a small crack is today a canyon some 4000 km (2500 mi) long, 500 km (370 mi) wide and 6 km (4 mi) deep.

Hellas Planitia

5 The asteroid that created this 2253 km (1400 mi) depression was without doubt a behemoth. The result, the vast plains of the Hellas basin, is one of the largest visible impact craters on Mars.

Bagnold Dune Field

6 Located by the Curiosity rover in 2015, these intriguing, crescent-shaped dunes are known as 'barchans'. A notable feature is their convex shape in the lee of the wind.

Gale Crater

7 First observed in the 19th century by Walter Frederick Gale, subsequent observations of carved outflow channels have led scientists to speculate this geologically rich area was once a lake.

Elysium Planitia

8 The second-largest volcanic region on Mars might not quite match the monsters of Tharsis but its three principal peaks – Hecates Tholus, Albor Tholus and Elysium Mons – are no shrinking violets.

Syrtis Major Planum

9 The largest of Mars' famous dark spots, the visibly dense colouring of the region has been attributed to its relative lack of dust and dark basaltic rocks.

Utopia Planitia

10 Less clearly defined than Hellas, the largest impact crater in the solar system features 'scalloped topography' – terrain that appears to have been scooped out with a spoon.

Vastitas Borealis

11 Located on the dramatic northern lowlands of the Red Planet, the broad swathes of this atmospheric wasteland may once have been home to a deep ocean.

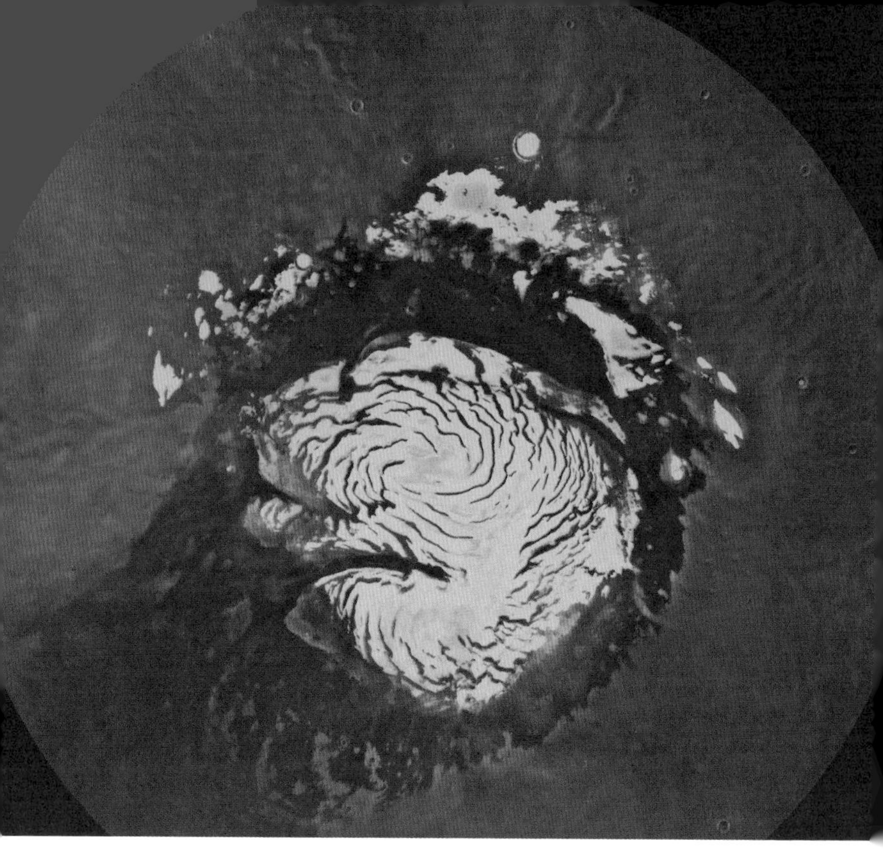

The Mars Boerum, or northern polar ice cap.

Polar Caps

With their mysterious weather systems and unique geological formations, Mars' north and south poles are as bleak as they are beautiful.

Covering an area roughly 1.5 times that of Texas, Mars' northern polar cap, Planum Boreum, is mainly water-ice. NASA's MRO, however, has observed a thick seasonal covering of carbon-dioxide-ice over gullies during winter. Otherwise known as dry ice, this does not exist naturally on Earth, but is plentifully available on Mars. The first reports of active gully-formation on Mars in 2000 generated excitement because they suggested the presence of liquid water on the Red Planet, the eroding action of which forms gullies here on Earth. Mars has water vapour and plenty of frozen water, but the presence of liquid water here has not been confirmed yet.

The southern cap, Planum Australe, is of similar size to the northern ice cap, although it stands about four miles above sea level. It is also made up of water-ice but – unlike its northern counterpart – retains its carbon-dioxide ice, even in summer. Over hundreds of thousands of years, variations in the tilt of Mars and the shape of its orbit have had significant effects on the planet's climate, including a number of ice ages. Data so far suggests the most recent Martian ice age ended 400,000 years ago.

Sights

Planum Boreum

Chasma Boreale

1 The most prominent feature at Mars' south pole, this canyon is 100 km (62 mi) wide and 1.9 km (1.2 mi) high. It is home to striking red cliffs formed, in part, by particles scoured into their faces by dust storms.

Olympia Undae

2 The largest of the 'sand seas' that ring Mars' northern polar cap, Olympia Undae offers rippling dunes as far as the eye can see.

Annular Cloud

3 The striking annular cloud will appeal to storm-chasing types. This circular cloud occurs over Planum Boreum in the morning and dissipates by the afternoon. Roughly 1600 km (1000 mi) across and featuring a 320-km-wide (200-mi-wide) 'eye', it resembles a cyclonic storm but, oddly, does not rotate.

Planum Australe

Martian Geysers

4 During the spring thaw, jets of carbon dioxide erupt through the polar ice. The dark material these jets carry drifts on the wind, creating mysterious, spider-like geological formations.

Perfect Powder

5 Mars' western hemisphere is the location for two huge impact basins, Hellas Planitia and Argyre Planitia. Together, they create a permanent area of low pressure over Planum Australe, which in turn results in a year-round covering of possible glaciers. It's not exactly Aspen, but beggars can't be choosers. For a colder planet than Earth with taller mountains, frosty peaks are within reason.

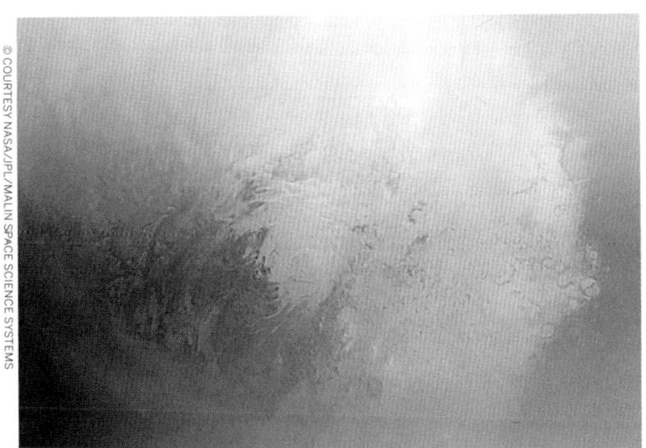

An image of Planum Australe taken by the Mars Global Surveyor, showing the ice cap.

Got Ice?

The end of the last Martian ice age saw the polar caps steadily thicken. NASA scientists estimate that Martian polar ice reached a maximum thickness of 320 m (1050 ft)– enough to cover the entire planet in a layer 60 cm (24 in) thick.

Frozen Planet

On Earth, our ice ages occur when the higher latitudes of the planet, including its polar regions, cool below average temperatures, for periods extending thousands of years. This causes glaciers to grow towards the milder latitudes, typically the middle of the planet. Martian ice ages have seemingly gone in the opposite direction. Thanks to the planet's increased tilt, Mars' poles have often become warmer than its mid-latitudes. During these periods, the polar caps retreat and water vapour migrates towards the equator, forming ice and glaciers. Ice formations contain information on these historical climate conditions on the Red Planet. 'Of all the solar system planets, Mars has the climate most like that of Earth. Both are sensitive to small changes in orbital parameters', said planetary scientist Dr. James Head of Brown University, Providence, RI. 'Now we're seeing that Mars, like Earth, is in a period between ice ages.'

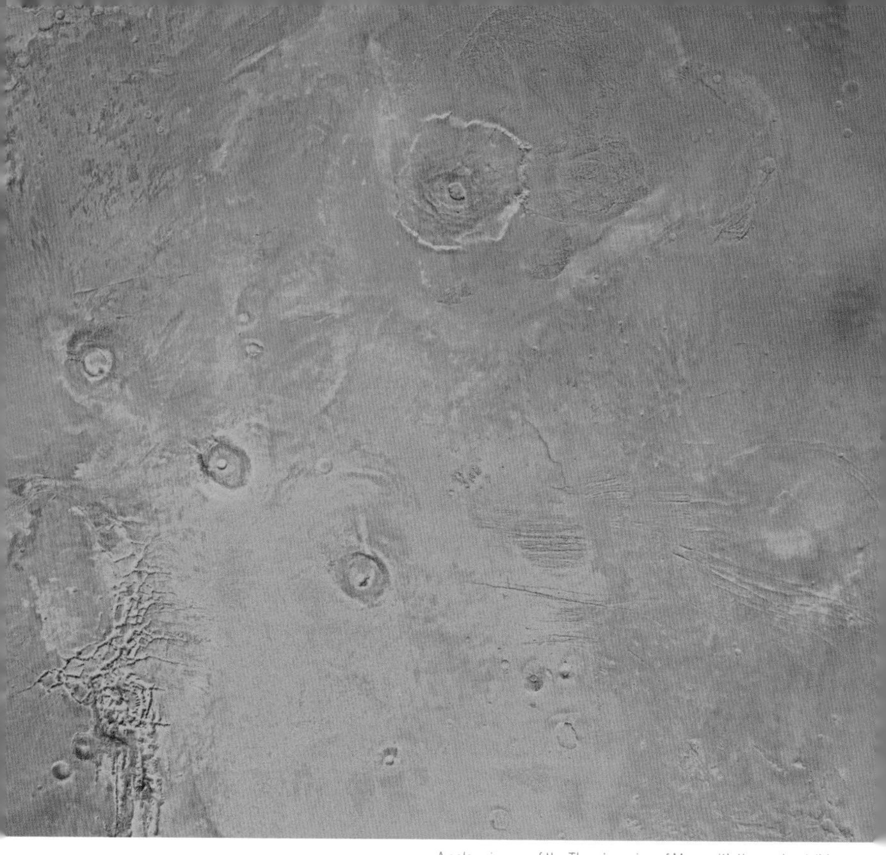

A colour image of the Tharsis region of Mars, with the peaks visible.

Tharsis Montes

Tharsis Montes is the name given to the solar system's titans of volcanology: Ascraeus Mons, Pavonis Mons and Arsia Mons, a trio of massive Martian shield volcanoes.

The three volcanoes of Tharsis Montes are located on a bulge in Mars' crust, at an average of 10 km (6 mi) above the mean elevation of the planet. Their summits are about the same elevation as that of Olympus Mons, the largest volcano in the Tharsis region. Tharsis Montes was first observed by the Mariner 9 spacecraft, in 1971, among the few surface features visible as the spacecraft entered orbit during a global dust storm. Running southwest to northeast, the peaks of Tharsis Montes are more or less equidistant, at about 692 km (430 mi) apart.

While the origins behind this formation are uncertain, it's unlikely to be a coincidence. Martian shield volcanoes form over single volcanic hotspots, much like the Hawaiian Islands on Earth. Unlike Earth's shield volcanoes, their Martian counterparts do not depend on tectonic movement, or lack thereof, for their longevity, and thus have been able to grow to quite enormous elevations.

Sights

Ascraeus Mons

1 The second-highest shield volcano on Mars was named after Ascra, birthplace of the ancient Greek poet Hesiod. The mountain took on its official moniker in 1973.

Pavonis Mons

2 Evidence of moraine deposits suggests glaciers may once have existed on the flanks of the mountain – and may still exist today.

Arsia Mons

3 Arsia is another of the planet's tallest volcanoes at 17.7 km (12 mi) high. Its central caldera, at 120 km (75 mi) wide, makes it the widest on Mars.

A wide-angle view of Ascraeus Mons, the northernmost Tharsis Montes volcano.

Pop Culture Quiz

Tharsis has featured in the cultural imagination for as long as humans have stared at the heavens, or at their console screens. Tharsis is the name of an angel in the Bible, and less sacredly, a PlayStation game. Pavonis Mons features in the title of a song by The Flaming Lips as well.

Super-sized

The main difference between the volcanoes on Mars and Earth is their size; volcanoes in the Tharsis region of Mars are 10 to 100 times larger than those anywhere on Earth. The lava flows on the Martian surface are observed to be much longer, probably a result of higher eruption rates and lower surface gravity. Another reason why the volcanoes on Mars are so massive is because the crust on Mars doesn't move the way it does on Earth. On Earth, the hot spots remain stationary but crustal plates are moving above them. The Hawaiian islands result from the northwesterly movement of the Pacific plate over a stationary hotspot producing lava. As the plate moves over the hotspot, new volcanoes are formed and the existing ones become extinct. This distributes the total volume of lava among many volcanoes rather than one large volcano. On Mars, the crust remains stationary and the lava piles up in one, very large volcano.

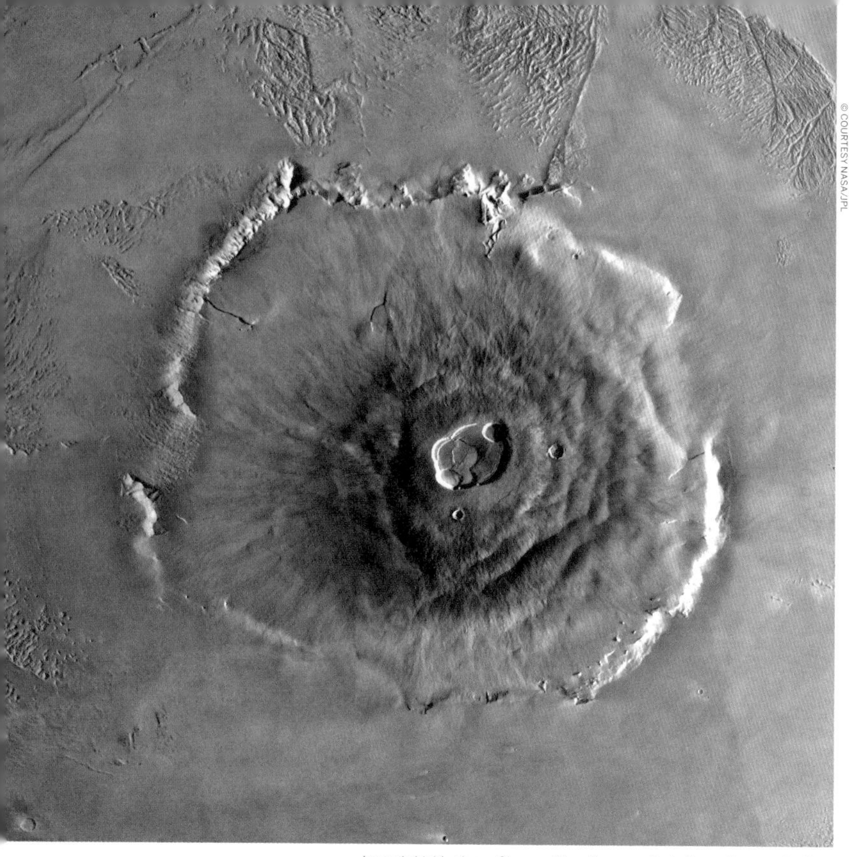

Around shield volcano Olympus Mons is a scarp that sits next to a moat of lava.

Olympus Mons

Formerly known as Nix Olympica ('snow of Olympus'), the largest mountain in the solar system covers an area approximately the size of Italy.

A beast of a volcano even by Martian standards, and first seen from Earth in the early 19th century, the shield volcano Olympus Mons rises 25 km (16 mi) into the sky. It has something of a lopsided profile, often likened to a gently sloping circus tent whose pole is off-centre. It's 624 km (374 mi) in diameter, about the size of the state of Arizona, with a huge 80-km-wide (50 mi) caldera at the top.

The youngest of Mars' shield volcanoes, the summit of Olympus Mons has no fewer than six collapsed craters, while the vast flanks of Olympus Mons are streaked with old lava flows. At 115 million years old, the lava flows on its northwestern slopes are, in geological terms, relatively new. This might indicate that Olympus Mons is still active. Among its most notable features are two impact craters, the Karzok and Pangboche, believed to be principal sources of what are called shergottites – the principal form of Martian meteorites ejected by the Red Planet before making their way to Earth.

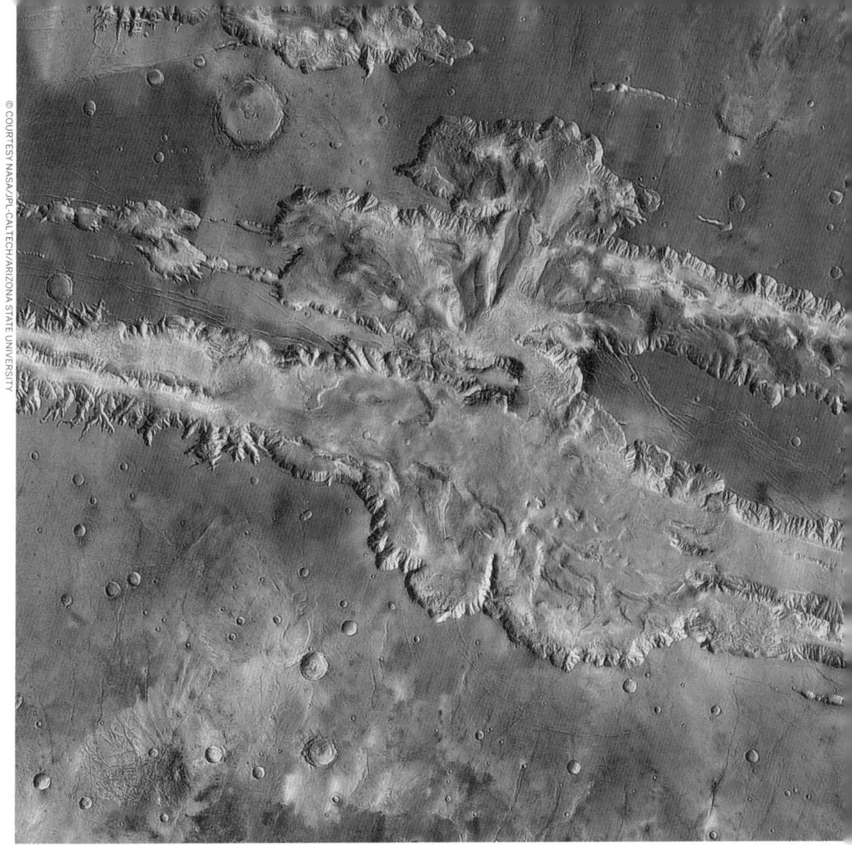

The Valles Marineris rift dwarfs Earth's Grand Canyon in every respect, and would cross America coast-to-coast.

Valles Marineris

Formed in a similar fashion to east Africa's Rift Valley, the immense channel of the Valles Marineris was first observed, in 1972, by NASA's Mariner 9 spacecraft.

Valles Marineris (Mariner Valley) is a vast canyon system that runs along the Martian equator just east of the Tharsis region. At 4000 km (2500 mi) long, and reaching depths of 7 km (4 mi), it dwarfs terrestrial rivals such as Earth's Grand Canyon; the Arizona chasm, by comparison, is a mere 800 km (500 mi) in length and just 1.6 km (1 mi) deep. Valles Marineris is as long as the United States and spans about 20% of the distance around Mars' mid-latitudes. The canyon extends from the so-called Noctis Labyrinthus region in the west to the rough terrain of the east. Most researchers seem to agree that Valles Marineris is a large tectonic crack, created when the planet cooled. The rising crust in the Tharsis region, as well as subsequent erosion, shaped the canyon we see today. Its widest section, the Melas chasma, is home to 'wall dune fields', formed by sands along the canyon's slopes. Common on Earth, these types of dunes are rare on Mars.

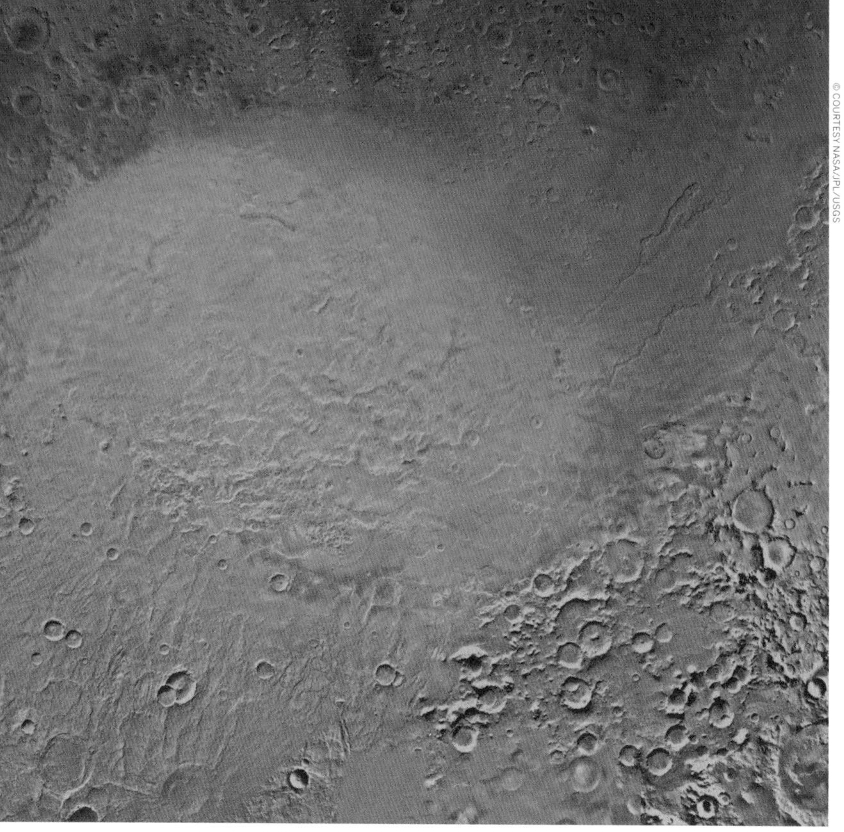

The Hellas Planitia is a wide plain within an ancient impact basin, the largest known on Mars.

Hellas Planitia

Measuring approximately 2250 km (1400 mi) across, this vast impact crater and its undulating plains was formed when a large projectile (an asteroid, comet, or meteor) hit the surface of the planet.

Named in honour of ancient Greece, by the astronomer Giovanni Schiaparelli, the broad outline of Hellas Planitia was the first Martian feature to be observed, by telescope, from Earth. The largest basin on Mars was formed when a large projectile struck Mars' surface during what astronomers call the Late Heavy Bombardment, a period 4 million years ago when space debris was fizzing about all over the place. The exact diameter of the basin is difficult to determine because large portions of the rim, to the northeast and southwest, are missing. In addition, several large patera or 'low volcanoes' – Tyrrhena Mons, Hadriacus Mons and Amphitrites Patera – occur near the rim and their lava flows have partially buried older impact deposits. The basin is the site of frequent dust storms; when these clear, they reveal a spectacular honeycomb topography, indicating landscape formed, in part, by glaciation.

Bagnold Dune Field sits on lower Mount Sharp, ripe for exploration by Curiosity.

Bagnold Dune Field

Another delight from the geological emporium of Gale Crater, the ethereal sands of the constantly shifting Bagnold Dune Field are at once alien and familiar.

Observing the changing nature of sand dunes on Mars allows scientists to monitor wind strength and direction. Those found in the Bagnold Dune Field – the sight is named after the British desert explorer Ralph Alger Bagnold – are known as 'barchan' dunes. It may sound like something from a *Star Wars* primer, but the word 'barchan' is Russian in origin, named for the desert region of Turkistan where the phenomenon was first observed, in 1881. On Earth, barchan dunes are found in numerous deserts, most notably in the Namib, in southern Africa. Typified by a gentle slope on their windward face and a steep, convex slope in the lee of the wind, Earth's barchans tend be formed by wind blowing from one direction. The Bagnold Dune Field, however, is formed by winds from a number of directions, giving it a beautiful, cross-hatched appearance. Martian dunes are dark, thanks to the high basalt content found in volcanic rock.

A colour-enhanced picture of Gale Crater.

Gale Crater

The site of extensive, and ongoing, exploration by NASA's Curiosity rover, the diverse terrain of the Gale Crater offers unique insights into Martian history.

Situated just below the Martian equator and 154 km (96 mi) in diameter, Gale Crater is believed to have once been home to a vast lake. A site of intense interest for NASA scientists, the wealth of discernible layers of sediment offers clues to the Red Planet's geological evolution. The crater has a special moat-like formation; a continuous depression around its outer rim rises towards the centre to create a three-and-a-half-mile-high mound of layered deposits known as Mount Sharp (in honour of US geologist Robert Sharp). Studies from orbit have revealed that the layers have different minerals depending on their height. It forms a natural repository for Martian geologic history. Near the bottom of the mound are clay minerals. Above these clay-bearing layers are others containing sulphur and minerals containing oxygen. These all form the stratigraphic columns used by geologists to study sedimentary rocks, and help to show the history of the features investigated by Curiosity. The original crater, similar in size to the US state of Connecticut, was formed by a meteor strike between 3.5 and 3.8 million years ago.

Sights

Peace Vallis

1 Gale Crater is home to water deltas, or 'fans', that provide information about historical lake levels. One of the most prominent, Peace Vallis, contains deposits consistent with an outflowing river.

Aeolis Palus

2 A windswept plain on the northern rim of the crater floor, it was here that the rover first uncovered evidence of an ancient freshwater lake.

Yellowknife Bay

3 The sedimentary rocks here were exposed by erosion about 70 million years ago. These rocks record superimposed ancient lake and stream deposits that offered past environmental conditions favourable for microbial life.

Aeolis Mons

4 This is Mount Sharp's official title, following the conventions of the International Astronomical Union. Winds scoured the crater's edges then carried material inwards to create an extraordinary central showpiece.

Confidence Hills

5 This historically significant site was the location of Curiosity's first drilling on the Red Planet. The rover reached the site, at the base of Mount Sharp, in September 2014.

A patch of windblown sand and dust dubbed 'Rocknest' near Curiosity's landing site.

RIP to Ray

The location of Curiosity's touchdown was later renamed 'Bradbury Landing' by NASA, in recognition of the work of the celebrated science-fiction writer. The author, whose *Martian Chronicles* (1950) tells of a lost human civilisation on the Red Planet, died in June 2012, aged 92.

Hidden Waters?

The choice of Gale Crater as the landing site, in August 2012, for NASA's Curiosity rover was a wise one. The rover has detected numerous signs that water was once present here; Gale Crater features both clays and sulphates, both of which form in water. Rock samples taken by the rover, on and around Mount Sharp, also suggest the Red Planet may once have been a habitat for microbes, one of the key building blocks for living organisms. So far, determining that the environments are microbe-friendly is about as far as Curiosity has gotten. It has, however, discovered some organic molecules (life's carbon-based buildings blocks), and noted seasonal variations in the methane concentrations in Mars' atmosphere, which may indicate the gas seeps out from underground reservoirs. On Earth, methane is primarily produced by living organisms. The mystery goes on, with new data slowly giving tantalising but inconclusive hints of liquid water.

Cerberus Fossae making tracks in Elysium Planitia.

Elysium Planitia

The volcanic region that served as the arrival lounge for the InSight lander in 2018. A broad plateau, it was from here the lander transmitted the first recordings of Martian winds.

Elysium Planitia is the second-largest volcanic region on Mars. It is 1700 km by 2400 km (1050 mi x 1500 mi) in size and, like the Tharsis region, is located on an area of crustal uplift. Elysium's three largest volcanoes – Hecates Tholus, Albor Tholus and Elysium Mons – might be smaller than those found in Tharsis, but they are still mightily impressive. As the name suggests, Elysium Planitia is relatively flat, precisely the reason it was chosen for the InSight landing. But you will also find both volcanic and impact craters here.

Unlike volcanic craters, impact craters are characterised by an identifiable rim, with fragments of the collision, known as 'ejecta', scattered around. In larger craters, typically those 10 km (6.2 mi) across or more, you might also find a peak. This is formed from the original surface material, which effectively 'rebounds' when an incoming projectile strikes the ground with great force.

Sights

Elysium Mons

1 Discovered in 1972, Elysium Mons is the largest volcano in this region. It measures 692 km (430 mi) across and rises 12 km (8 mi) above the surrounding plains.

Eddie Crater

2 At 89 km (55 mi) wide, the largest of three major impact craters in the region. The feature was named after South African astronomer Lindsay Eddie.

Cerberus Fossae

3 Running more than 1200 km (750 mi), this striking series of fissures was likely formed by faults in Mars' crust. Crustal movement may have put pressure on water underground, rupturing the crust further.

Orcus Patera

4 The origins of this elongated, elliptical depression remain a mystery. NASA's best guess is that it will eventually be confirmed as an impact crater.

Athabasca Valles

5 A now dry river valley, Athabasca Valles is part of a wider network of outflow channels located in Elysium Planitia. Scientists believe their courses begin beneath the Martian surface, where water once burst out of fissures at a rate that surpassed the flow of the Mississippi River.

InSight deploys its instruments while investigating Elysium Planitia.

Ice in the Valley

In 2005, Mars Express imaged an area of Elysium Planitia that appeared to show a huge volume of ice. At 800 km wide (500 mi), and of a similar length, estimates put the depth of the ice at around 45 m (147ft). Though volcanic ash has made precise measurements difficult, if verified this would be a similar volume of water to that found in Europe's North Sea.

Flood Watch...for Lava?

Orbital reconnaissance of Elysium Planitia has identified the presence of 'flood basalts'. These are created by fissures in the surface, which allow the slow release of huge quantities of basalt lava – and occasionally water – and which smooth over the surface to create, effectively, a lava floodplain. Flood basalts occur on Earth, too, with famous examples in Washington state and India. In 2018, scientists from NASA's Goddard Space Flight Center travelled to Iceland to study the Holuhraun lava field, a flood basalt that displays similar characteristics to those observed in Elysium Planitia. Analogs like this are being discovered with increasing frequency, as Mars is the terrestrial planet most similar to Earth.

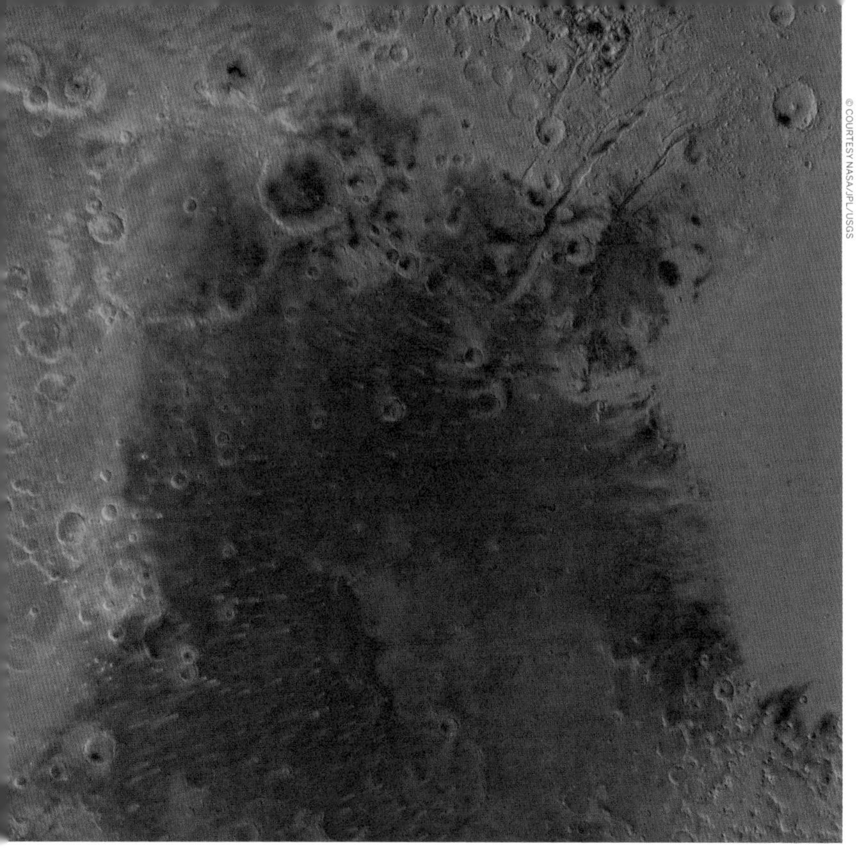

Syrtis Major shows up as a darker spot in images of Mars due to its basaltic rock.

Syrtis Major Planum

The first Martian feature to be recorded on Earth, Syrtis Major Planum is the best known of Mars' 'dark spots'. It was once thought to be a landlocked ocean.

Sketched in 1659 by Dutch astronomer Christiaan Huygens, Syrtis Major Planum is located between Mars' northern lowlands and the planet's southern highlands. The feature extends some 1500 km (930 mi) north from the planet's equator, and spans 1000 km (620 mi), west to east. The name Syrtis Major was conferred, like so many others, by Giovanni Schiaparelli, whose tentative map of the planet was made during the Mars Close Approach of 1877. Its name is derived from Syrtis maior, the ancient Roman name for the Gulf of Sirte, off the coast of modern-day Libya. What gives this 1450-km (900-mi) spot its dark colouring is basaltic volcanic rock allied to relatively clear air; a distinction which made it a potential landing site for future Mars missions.

Before orbital missions to Mars, Syrtis Major Planum was mistaken for water: the noted French astronomer Camille Flammarion dubbed it the *Mer du Sablier*, the Hourglass Sea. Though Flammarion was proved wrong, it's still thought to have potential scientific significance as an ancient fluvial basin that was once volcanically active.

The gully channels shown here were most likely formed by water action.

Utopia Planitia

With an estimated diameter of 3300 km (2050 mi), the haunting wilderness of this vast impact crater boasts spectacular landforms and watery secrets yet to be unlocked.

The largest impact crater in the solar system, it was on the icy wastes of Utopia Planitia that, in September 1976, Viking 2 successfully touched down on the Martian surface. What it encountered was perhaps the best example of 'scalloped topography', a type of geological formation commonly found at Mars' mid-latitudes. The rocks here seem to rest on, or just above, the ground, as if the regolith beneath them has been pared away. Scientists believe this occurs through the partial thawing and refreezing of Utopia Planitia's permafrost; the process leaves behind distinctive hollows, much like scallop shells in appearance. Ice is present here year-round. In 2016, NASA detected a vast frozen lake, with a volume of water similar to that of Lake Victoria in east Africa.

This Martian region has also enjoyed a different 15 minutes of fame. In the *Star Trek* franchise, Utopia Planitia is the name of an intergalactic auto shop.

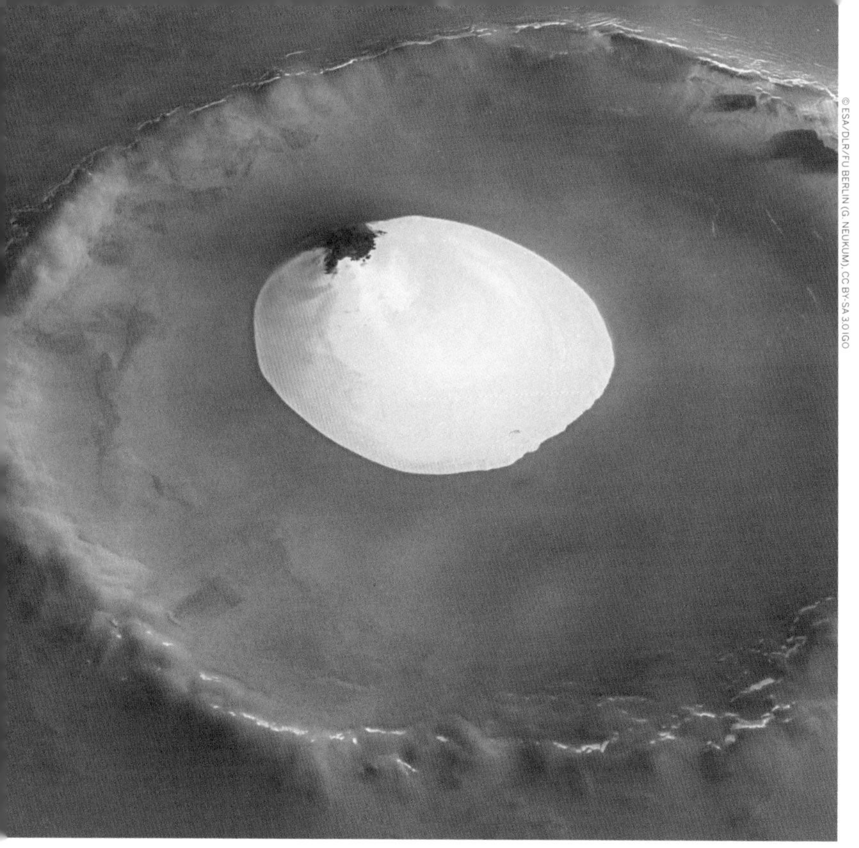

Residual water-ice is seen within the Vastitas Borealis crater.

Vastitas Borealis

Covering a vast area beneath much of Mars' northern polar region, this stark landscape may once have been home to an ancient Martian ocean.

Of Mars' various northern lowlands, the largest is the dramatic Vastitas Borealis ('northern wasteland'). As well as offering a stark, icy landscape, the region also lies at the heart of an ongoing debate regarding the 'Mars ocean hypothesis'. By measuring atmospheric water near the Martian poles, scientists have calculated that, at some point in history, Mars' polar ice caps lost around 20 million cubic km (5 million cu mi) of water. The ocean hypothesis proposes that, about 4.3 billion years ago, potentially even before Tharsis Montes arose, Mars would have had enough water to cover its entire surface in a liquid layer, as much as 137m (450ft) deep, though potentially shallower.

More likely, however, is that due to its low-lying nature, this water would have formed an ocean occupying almost half of Mars' northern hemisphere, and Vastitas Borealis in particular. Reaching depths of 1.6 km (1 mi), this ocean, named Arabia (and a subsequent one called Deuteronilus), would have covered 19% of the planet's surface. By comparison, the Atlantic Ocean presently occupies 17% of the surface of the Earth.

Bluish-white water-ice clouds can be seen in Mars' atmosphere in this image from the Mars Global Surveyor.

JUPITER

Jupiter's stripes and swirls are actually cold, windy clouds of ammonia and water.

Jupiter at a Glance

When it comes to size, no other planet in the solar system compares to the ginormous, gargantuan gas giant of Jupiter.

It's far and away the biggest planet in our solar system: a whopping eleven times the size of Earth, more than twice as massive as all the other planets combined. To put that in perspective, if Earth were the size of a grape, Jupiter would be approximately the size of a basketball. That's big.

First found in history records as early as 350 BC by Babylonian astronomers, it's named after the King of the Gods – a fitting moniker for this most kingly of planets. Seen from afar, Jupiter is marked by distinct patterns of stripes and swirls, like a gigantic gobstopper floating in space. These striations are actually cold, windy clouds of ammonia and water, floating in an atmospheric soup that's approximately nine-tenths hydrogen, one-tenth helium; pretty much the precise ingredients needed to make a star. In another parallel universe, Jupiter may perhaps have grown massive enough to ignite, and maybe even eclipse our Sun – but in this one, it remained stuck at mere planetary status.

Jupiter's most famous feature is its Great Red Spot, a storm of truly epic proportions – twice the size of Earth. Like many of Jupiter's storms, it's thought to have raged for hundreds of years. In 1979, the Voyager mission also discovered that Jupiter has its own faint ring system, but unlike the ones that encircle Saturn, Jupiter's rings are very faint, and made of dust, not ice.

For some scientists, however, the most interesting things about Jupiter are the bodies in orbit around it: its many moons. At the last count, there were 79, more than any other planet in our solar system. The nearest contender, Saturn, only has 62. Even more intriguingly, scientists believe that a few of them may have the atmospheric and chemical conditions necessary to support life. The main focus at present is on Europa, where there is evidence of a vast ocean hidden just beneath its icy crust – a prime location for possibly finding life.

This image captures the intense jets and vortices in the North Temperate Belt.

DISTANCE FROM SUN
5.2 AU

LIGHT-TIME TO THE SUN
43 minutes

LENGTH OF DAY
9.9 hours

LENGTH OF ORBITAL YEAR
4333 Earth days (about 12 Earth years)

ATMOSPHERE
Hydrogen and helium

Top Tip

On average, the surface of Jupiter is about twice as cold as the South Pole in the middle of winter – so you might want to pack a few extra warm garments.

Getting There & Away

Jupiter orbits the Sun at a distance of around 778 million km (484 million mi), or 5.2 Astronomical Units (Earth is at one AU). Travelling nonstop at the average speed of a jumbo jet, it would take over 60 years to reach it. The Galileo spacecraft took over six years to reach the planet, covering a lengthy course of 4.6 billion km (2.8 billion mi) from launch to impact. Reaching the planet for a flyby can take as little as 390 days, or about thirteen months.

Orientation

Jupiter is the fifth furthest planet from the Sun, positioned between Mars and Saturn, and the closest gas giant to the Sun (as well as the largest planet). When viewed from Earth, Jupiter is usually the second brightest planet in the night sky, after Venus.

Like the Sun, it is mostly composed of hydrogen and helium, and has a thick, cloudy atmosphere, sitting above a vast ocean made of liquid hydrogen. Jupiter's equator is tilted by just 3 degrees, which means the planet spins nearly upright and does not have seasons as extreme as other planets do. It is surrounded by four large moons, first discovered by Galileo Galilei in 1610, and many other small moons, like a solar system in miniature. There are so many Jovian moons that they haven't all even been given names yet!

An approximate comparison of the scale of Earth, Jupiter, and the Great Red Spot.

Jupiter *vs* Earth

-- Radius --
11.2x EARTH

-- Mass --
317.8x EARTH

-- Volume --
1321x EARTH

-- Surface gravity --
2.5x EARTH

-- Mean temperature
-161°C (-257°F)
COLDER than **EARTH**

-- Surface area --
120x EARTH

-- Surface pressure -
Unknown

-- Density --
0.24x EARTH

-- Orbit velocity --
44% EARTH

-- Orbit distance --
5.2x EARTH

Swirling clouds on Jupiter.

Atmosphere

What exactly does it mean to be a gas giant and have an atmosphere, when there's no surface to mark its clear beginning? We are most familiar with Jupiter's dramatic cloud tops, visible beneath a transparent, mostly hydrogen upper atmosphere. Beneath the cloud layers, the clear atmosphere is denser and warmer and slowly transforms from a gas to a liquid without a sharp boundary to mark the change. The interior liquid hydrogen oceans are certainly not part of the atmosphere, but the line isn't as rigid as it is on a planet with a crust.

Then there's the question of what the atmosphere is made of. While the atmosphere of Earth is about 78% nitrogen and 21% oxygen, and Mars and Venus have a high concentration of carbon dioxide, Jupiter's atmosphere is almost 90% hydrogen and 10% helium. The gas giants and terrestrial planets, in other words, share no common elements in their respective atmospheres, except as trace elements.

Because nitrogen and oxygen are heavier gases than hydrogen and helium, the lower-mass Earth can hold onto them with its gravitational pull, while the lighter gases escape into space. Appropriately, this process is called 'atmospheric escape', and each molecule has its own escape velocity: the speed it needs to reach to break away from the gravitational pull of a planet or moon. Based on the molecule's own mass and the mass of the planet, it's a principle that applies to rockets as well as to gases. Inner planets, warmer than those in the outer solar system, have a lower escape velocity due to their temperature as well as their lower mass. This is how Titan, smaller than Earth but more distant from the Sun as well, has retained its atmosphere.

Gas giants, however, are so weighty that they exert a stronger pull on the molecules in their atmosphere, making these planets capable of holding even the lighter gases in their respective atmospheres. The same principle is in operation for Saturn and the other gas giants as well. So-called hot Jupiters, high-mass exoplanets orbiting very close to their stars, might have a different atmospheric composition due to the impact of evaporation. Due to their large mass, these were many of the first exoplanets found. Increasing study of hot Jupiters may tell us more about how they differ from the atmosphere on our planetary neighbour.

History

Jupiter is believed to be the first of the planets to form, perhaps within a few million years (or less) after the solar system's creation. Jupiter took most of the mass left over after the formation of the Sun, as gravity pulled swirling gas and dust in to become this gas giant with more than twice the combined material of the other bodies in the solar system. About 4 billion years ago, Jupiter settled into its current position in the outer solar system, where it is the fifth planet from the Sun.

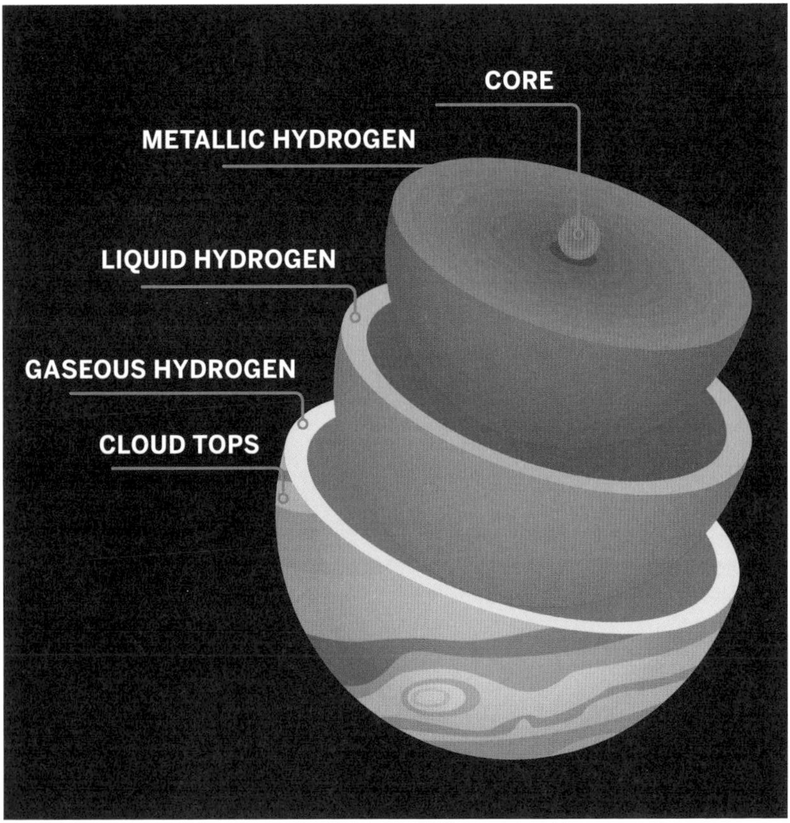

CORE

METALLIC HYDROGEN

LIQUID HYDROGEN

GASEOUS HYDROGEN

CLOUD TOPS

© MEVAN/SHUTTERSTOCK

Jupiter's core is estimated to be about the size of Earth, though scientists are seeking more information about its interior.

Exploring Jupiter

While Jupiter has been known since ancient times, the first detailed observations of this planet were made by Galileo Galilei in 1610 with a small telescope. More recently, this planet has been visited by several spacecraft, orbiters and probes.

Nine spacecraft have visited Jupiter so far. Seven flew by and two have orbited the gas giant. Pioneer 10 and 11 and Voyager 1 and 2 were the first to fly by Jupiter in the 1970s, and since then we've sent Galileo to orbit the gas giant and drop a probe into its atmosphere. Cassini took detailed photos of Jupiter on its way to neighbouring Saturn, as did New Horizons on its quest to observe Pluto and the Kuiper Belt. NASA's Juno spacecraft, which arrived in the Jovian system in July 2016, is currently studying the giant planet from orbit. Its magnetic field, storms, and atmosphere are all rich targets; then there are the many moons to consider. Jupiter's large gravitational field acts like a magnet for capturing space debris.

Discovery Timeline

1610

Galileo Galilei makes the first detailed observations of Jupiter and its largest moons, which had never before been seen.

1830

Observation of the Great Red Spot is first confirmed.

1972

Pioneer 10 becomes the first spacecraft to cross the asteroid belt and fly past Jupiter.

1979

Voyager 1 and 2 are the first to discover Jupiter's faint rings, t new moons, and the volcanic activity on Io's surface.

1992

Ulysses swings by Jupiter on 8 February 1992. The giant planet's gravity bends the spacecraft's flight path southward and away from the ecliptic plane, putting the probe into a final orbit that takes it over the Sun's south and north poles.

1994

Astronomers observe as pieces of comet Shoemaker-Levy 9 collide with Jupiter's southern hemisphere.

1995–2003

The Galileo spacecraft drops a probe into Jupiter's atmosphere and conducts extended observations of Jupiter and its moons and rings.

2000

Cassini makes its closest approach to Jupiter at a distance of approximately 10 million km (6.2 million mi), taking a highly detailed true colour mosaic photo of the gas giant.

2007

Images taken by NASA's New Horizons spacecraft, on the way to Pluto, show new perspectives on Jupiter's atmospheric storms, the rings, volcanic Io and icy Europa.

2009

On 20 July, almost exactly 15 years after fragments of comet Shoemaker-Levy slammed into Jupiter, a comet or asteroid crashes into the giant planet's southern hemisphere.

2011

NASA's Juno spacecraft launches to examine Jupiter's chemistry, atmosphere, interior structure and magnetosphere.

2016

Juno arrives at Jupiter, conducting an in-depth investigation of the planet's atmosphere, deep structure and magnetosphere for clues to its origin and evolution.

Time on Jupiter

Jupiter has the shortest day in the solar system. One day on Jupiter takes only about 10 hours (the time it takes for Jupiter to rotate or spin around once), and Jupiter makes a complete orbit around the Sun (a year in Jovian time) in just over 12 Earth years (4333 Earth days).

To Jupiter & Beyond

There are no rockets powerful enough to hurl a spacecraft into the outer solar system and beyond, but in 1962, scientists calculated how to use Jupiter's intense gravity to hurl spacecraft into the farthest regions of the solar system. We've been travelling farther and faster ever since.

A Shrinking Jupiter

Jupiter is so massive that its own gravity caused it to contract – in fact, 4.5 billion years after its formation, Jupiter is still shrinking. As Jupiter contracts, all the matter inside squishes against itself and churns, causing friction and heat. Due to this, Jupiter radiates more heat than it absorbs from the Sun.

Some of the Jovian moons can be seen in orbit around Jupiter in this 3D rendering.

Jupiter's New Moons

On 17 July 2018, scientists announced they had discovered 12 new moons orbiting Jupiter. That raised Jupiter's total number of moons to 79 – the most of any planet in the solar system. The team first spotted the moons in the spring of 2017 while they were looking for very distant solar system objects as part of a hunt for a possible massive planet far beyond Pluto. 'Our other discovery is a real oddball and has an orbit like no other known Jovian moon', team leader Scott Sheppard said. 'It's also likely Jupiter's smallest known moon, being less than one kilometre in diameter.'

Jupiter in the Movies

The biggest planet in our solar system, Jupiter has a large presence in pop culture, including many movies, TV shows, video games and comics. Jupiter is a notable destination in the Wachowski Brothers' science fiction spectacle *Jupiter Ascending*, while various Jovian moons provide settings for *Cloud Atlas*, *Futurama*, *Power Rangers* and *Halo*, among many others. In *Men in Black* when Agent J (played by Will Smith) mentions he thought one of his childhood teachers was from Venus, Agent K (played by Tommy Lee Jones) replies that she is actually from one of Jupiter's moons.

Jupiter Highlights

The Great Red Spot

1 This planet-sized monster makes our worst hurricane look like a storm in a teacup.

The Jovian Ring System

2 Jupiter is encircled by a system of three vast rings, thought to be made mainly of dust.

Surface of Jupiter

3 As a gas giant, Jupiter lacks an Earth-like surface; instead it has something more intriguing.

Jupiter's Clouds

4 Jupiter likely has three distinct cloud layers that, taken together, span about 50 to 70 km (30 to 50 mi).

Jupiter's Oceans

5 Hydrogen gas compressed into a liquid gives Jupiter the largest 'ocean' in the solar system.

Jupiter's Magnetosphere

6 After the sun, this is the biggest magnetosphere in our solar system, stretching almost 15 times the width of the sun.

The Juno Mission

7 This pioneering probe arrived in 2016, and aims to unlock Jupiter's secrets.

Europa

8 Cold and icy though it may be, this Jovian moon may perhaps be capable of supporting life.

Io

9 Unlike its icy sisters, this moon is a bubbling cauldron of volcanic activity.

Callisto

10 Jupiter's second-largest moon is roughly the size of Mercury.

Ganymede

11 Icy and pockmarked, this is the biggest moon discovered so far in our solar system.

An artist's rendering of Juno soaring above Jupiter's south pole.

Observations of the Great Red Spot show a changing environment.

The Great Red Spot

Jupiter's gargantuan Great Red Spot, a swirling storm twice as wide as Earth, is thought to have been raging for at least 150 years – with no (immediate) end in sight.

Jupiter is an extremely stormy world, swept by over a dozen prevailing winds, some of which reach astonishing speeds, up to 539 km/h (335mph) at the equator. By comparison, the strongest wind speed ever recorded on Earth was 408 km/h (253mph). Jupiter's ferocious winds swirl the planet's clouds into thick coloured bands and create storms of truly terrifying proportions. The Great Red Spot is one such storm – the largest ever observed by human eyes. With no solid surface to slow them down, Jupiter's storms, or 'spots', can persist for many decades, and sometimes – as in the case of the Great Red Spot – even centuries. Recent findings indicate that the Great Red Spot has started to drift westward faster than before. The storm always stays at the same latitude, held there by jet streams to the north and south, but it circles the globe in the opposite direction relative to the planet's eastward rotation. It remains big enough to accommodate all of Earth.

5 Facts About The Great Red Spot

1 The Great Red Spot has officially been monitored since 1830, but is thought to have first been observed by telescope two centuries before that.

2 The oval-shaped storm rotates anti-clockwise.

3 The Spot's clouds are composed mainly of ammonia, ammonium hydrosulphide and water.

4 Wind speeds in the Great Red Spot are thought to peak at about 640 km/h (400mph), twice as fast as the fastest hurricane winds ever recorded on Earth.

5 The Spot's colour varies from dark scarlet to pale pink, but no one is sure what chemical process causes the Spot's reddish colour.

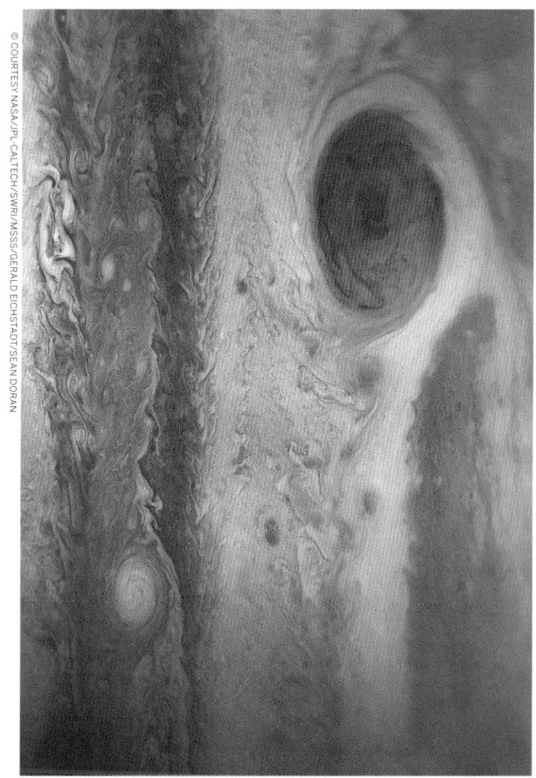

© COURTESY NASA/JPL-CALTECH/SWRI/MSSS/GERALD EICHSTADT/SEAN DORAN

Juno captured this image of the Great Red Spot and its surrounding area.

The Little Red Spot

More recently, three smaller ovals merged to form the Little Red Spot, about half the size of its larger cousin. Scientists do not yet know if these ovals and planet-circling bands are shallow or deeply rooted to the interior.

The Shrinking Spot

Observations of Jupiter date back centuries, but the first confirmed sighting of the Great Red Spot was in 1831. (Researchers aren't certain whether earlier observers who saw a red spot on Jupiter were looking at the same storm.) Keen observers have long been able to measure the size and drift of the Great Red Spot by fitting their telescopes with an eyepiece scored with crosshairs. Though once big enough to swallow three Earths with room to spare, Jupiter's Great Red Spot has actually been shrinking for a century and a half. Nobody is sure how long the storm will continue to contract or whether it will disappear altogether. A new study also suggests the storm seems to have increased in area briefly during the 1920s, but has since been getting both smaller and taller, like clay being stretched on a potter's wheel. Its colour has been deepening too, becoming more intensely orange since 2014.

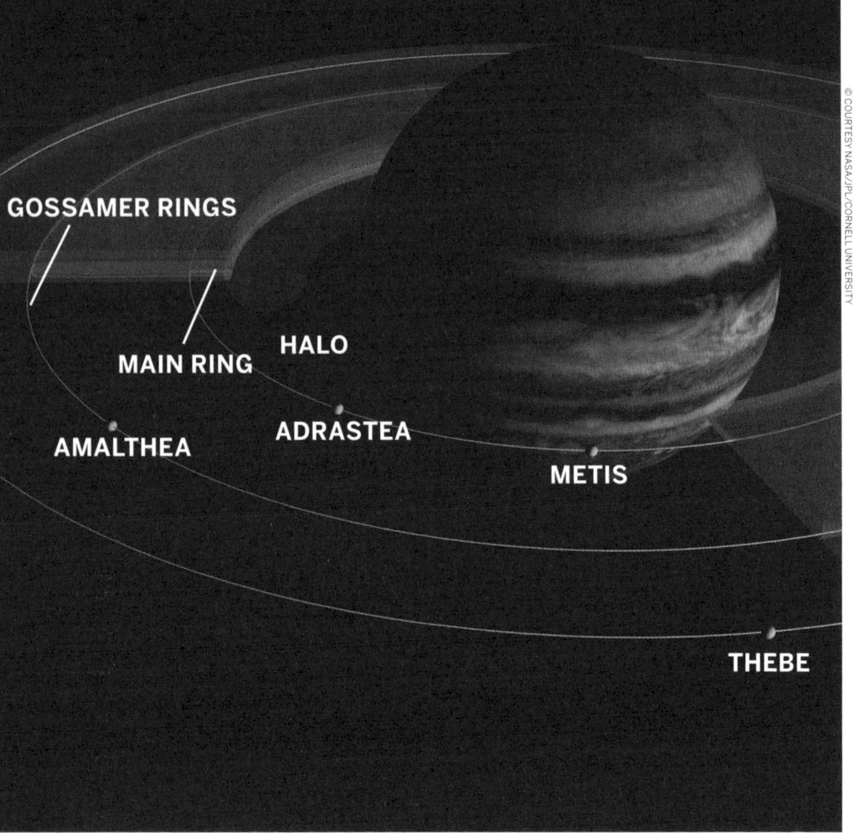

Jupiter's rings were only recently discovered and are still under study.

Jovian Ring System

Jupiter's three rings were first discovered by the Voyager 1 spacecraft in 1979. They are so faint, they are only visible when backlit by the Sun or when viewed in infrared.

The ring system is composed of three parts: a flat main ring, a toroidal 'halo' inside the main ring, and the 'gossamer ring', a vast, diffuse band which stretches out into space for thousands of miles from the main ring.

Unlikely Saturn's icy rings, which are full of large chunks of ice and rock, Jupiter's rings are composed of small dust particles, which scientists think are hurled up by meteor impacts on Jupiter's inner moons and then swirled up into orbit. These meteor impacts must be frequent and ongoing, as the rings must constantly be replenished with new dust to continue to exist. This theory was first suggested by the Galileo spacecraft, which made observations that the dust coincided with small moon locations: the two Gossamer rings near the small moons of Amalthea and Thebe, and the main ring, near Adrastea and Metis.

Juno captured this image of Jupiter's southern hemisphere.

The Surface of Jupiter

As a gas giant, Jupiter doesn't have a true surface, and is instead composed mostly of swirling gases and liquids. Below the cloud surface, likely a minimum of 50 km (30 mi) thick, remains mostly a tempting mystery.

Spacecraft attempting to land on Jupiter would run into trouble. As a planet without a large rocky core, its outer gas layers compress into a liquid, electrically charged and without solid ground. But finding somewhere to land would actually be the least of a spacecraft's problems. Chances are it wouldn't even make it as far as the surface, wherever that may be: the extreme pressures and temperatures deep inside the planet would crush, melt and vaporise any spacecraft trying to fly towards the surface, squashing it like a tin can. It is still unknown, however, what happens towards Jupiter's super-dense core: some scientists think it could be composed of solid material, or maybe a thick, dense soup made mostly of iron and silicate minerals (similar to quartz). One thing is certain – it will be unimaginably hot down there, perhaps up to 50,000°C (90,032°F).

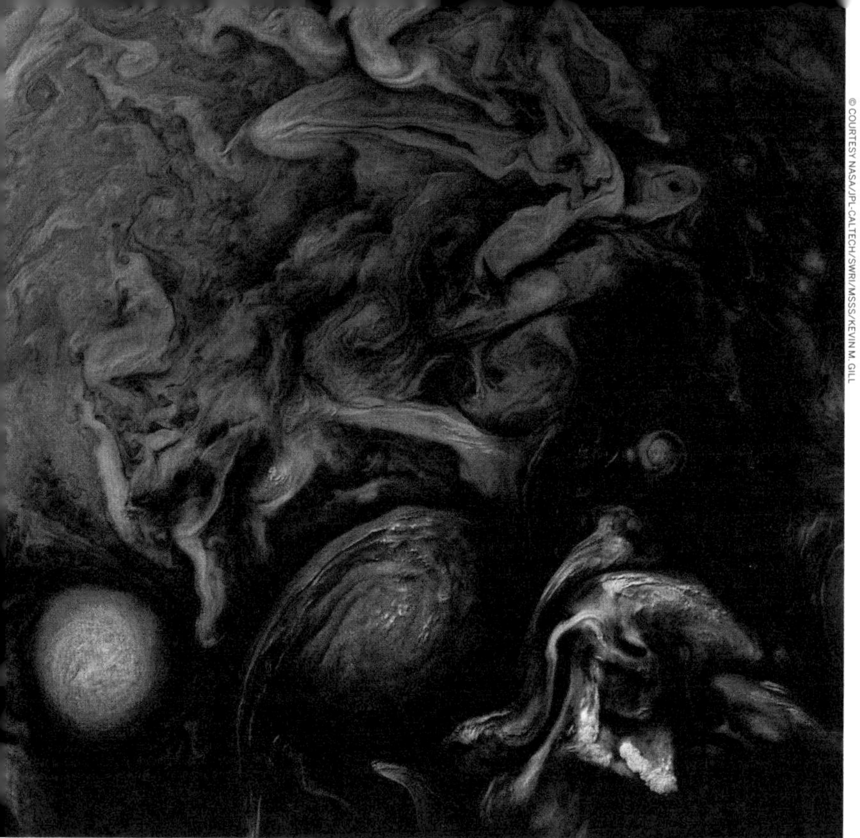

© COURTESY NASA/JPL-CALTECH/SWRI/MSSS/KEVIN M. GILL

Jupiter's intricate cloud pattens and formations are under intense study from Juno.

Jupiter's Clouds

In addition to its 'spots', Jupiter's appearance is also marked by a tapestry of colourful, ever-changing cloud bands, ranging from pale pinks to ochre browns and fiery reds.

These colourful bands of cloud are created by immensely strong winds, whirled into existence by the planet's rapid rotation, spinning once every 10 hours. There are believed to be three distinct cloud layers in Jupiter's 'skies' that, taken together, span about 71 km (44 mi), divided into dark belts and brighter zones.

The top cloud is probably made of ammonia ice, while the middle layer is likely made of ammonium hydrosulphide crystals. The innermost layer may be made of water ice and vapour. The vivid colours of the bands may be plumes of sulphur and phosphorus-containing gases rising from the planet's warmer interior. They shouldn't be understood as directly analogous to Earth's clouds and weather; Juno findings indicate that Jupiter's bands, though derived from powerful winds, may arise from the inner planet's convection system and not from a jet stream–like force as on Earth. Jupiter's air currents may have deep wellsprings.

The inner liquid metallic hydrogen oceans of Jupiter generate its strong dipole magnetic field.

Jupiter's Oceans

Jupiter has the largest ocean in the solar system – but it's quite different from the oceans we know on Earth. On this mighty gas giant, the ocean is made of liquid hydrogen instead of water.

Due to the immense pressure and temperature increase near Jupiter's surface, hydrogen gas condenses into a liquid here. In fact, scientists think that, at depths perhaps halfway to the planet's centre, the pressure becomes so great that electrons are squeezed off the hydrogen atoms, making the liquid capable of conducting electricity – just like metal. This is expected to be a key to Jupiter's intense magnetic field. Studying these liquid hydrogen oceans and their impact may reveal why Jupiter's magnetosphere seems less binary than other known planetary magnetic fields, with a less orderly north pole. Convection currents, created by residual heat from the planet's original formation, stir the liquid metallic hydrogen oceans. More still needs to be unearthed about how these movements and the general makeup of Jupiter's interior work. Though a dense mixture of compressed liquid gas, the inner 96% of the planet rotates like a solid body.

Hubble captured this stunning image of Jupiter's polar aurora.

Jupiter's Magnetosphere

The vast Jovian magnetosphere is the region of space influenced by Jupiter's powerful magnetic field – 16 to 54 times more powerful than Earth's, making it the largest and strongest in the solar system except for the Sun's.

Jupiter's strange type of liquid hydrogen that can even conduct electricity creates a churning, conducting fluid. It is likely the engine behind Jupiter's vast magnetic field, which balloons out between 1 and 3 million km (0.6 and 1.9 million mi) towards the Sun, and more than a billion km (620 million mi) toward Saturn.

As Jupiter spins, once every 10 hours, it drags its magnetic field around with it, creating strong electric currents. Because the charged particles in Jupiter's belt fly around extremely fast, this region is filled with intense radiation that can destroy a spacecraft. The large Galilean moons interact with Jupiter's rotating magnetic field as well.

This incredible magnetic field also causes some of the solar system's most spectacular aurorae, ribbons of glowing, electrified gas, located at the planet's poles where the magnetic field is strongest (similarly to Earth). Talk about a bucket list item.

Is There Water on Jupiter?

Gordon L Bjoraker, an astrophysicist at NASA's Goddard Space Flight Centre in Greenbelt, Maryland, is aiming to answer the question of whether there is water on Jupiter by peering deep into the Great Red Spot. His team's infrared telescopes captured and examined thermal radiation leaking from the storm, and detected the chemical signatures of water above the clouds. It's an exciting breakthrough: the Great Red Spot is full of dense clouds, which makes it hard for electromagnetic energy to escape and teach us about the chemistry within.

Their search also provides new information to clarify the depth of each of Jupiter's layers of cloud, which all have their own chemical composition. The pressure of the water and the amount of carbon monoxide in the atmosphere imply that Jupiter has two to nine times more oxygen than the Sun. That suggests there must also be abundant water on Jupiter, mostly made up of oxygen and molecular hydrogen.

Answering this question once and for all is one of the most important tasks of the Juno Mission. It's known that many of the Jovian moons have water-ice, meaning that the neighbourhood of Jupiter certainly has water, and chemical signals so far seem to indicate it's not the neighbourhood alone.

Jupiter & the Beginning of the Solar System

Four and a half billion years ago, a giant cloud of gas and dust, called a nebula, collapsed to form our solar system. Composed mainly of hydrogen gas, most of the nebula became the star we know as the Sun. The rest of the swirling cloud would condense to form Earth and the other planets, asteroids and comets. Since Jupiter contains a lot of the same light gases that the Sun is made of – hydrogen and helium – it is very likely to have been one of the first planets to form, and as such may hold important clues regarding the origin of the rest of the solar system.

Juno is depicted in orbit of the Great Red Spot.

Juno reached Jupiter's orbit only after a five-year journey from Earth.

Juno Mission

Launched in August 2011 from the Cape Canaveral Air Force Station, the pioneering Juno spacecraft has four main aims: to study Jupiter's origins, interior, atmosphere and magnetic field.

After orbiting Earth, the Juno spacecraft cruised across the inner solar system, finally arriving in July 2016. Powered by solar panels, and packed with delicate equipment, the spacecraft is in orbit over Jupiter's poles, traversing the planet in an elliptical north-south direction. After completing the mission's more than 30 planned orbits, Juno will have covered the entire surface of Jupiter, giving scientists a better understanding of how Jupiter formed, what lies within Jupiter's swirling clouds, what Jupiter is made of, and how much of it is water. The mission was originally scheduled to end in February 2018, with a deliberate crash into Jupiter's atmosphere, but NASA has confirmed it will continue for another three years until 2021, sending back data all along the way. Its nine instruments include a sensitive spectroscope, which separates white light into signatures of the elements.

5 Facts About The Juno Mission

1. Juno was launched on top of a 5-stage Atlas V 551 launcher, one of the most powerful rockets ever built.

2. Juno takes 53 days to complete a full revolution around Jupiter.

3. Juno only has about as much hard-drive space as an average laptop, 256MB of flash memory and 128MB of DRAM.

4. Juno is named after the wife of the King of the Gods – who peered down through the clouds to expose her husband's wrongdoings.

5. On each orbit, Juno comes within a mere 5000 km (3100 mi) of the planet's cloud tops.

Juno taking off on an Atlas V launcher.

Juno's Solar Panels

Juno is powered by three huge, 9m-long solar panels, generating about 400 watts of power. Because Juno relies entirely on solar power, the orbit has to keep the spacecraft in sunlight for its entire mission.

Jovian Radiation

In a given year, the amount of radiation expected to bombard Juno is equivalent to tens of millions of dental X-rays. The radiation is so destructive that two of Juno's instruments – the Jovian infrared auroral mapper and the camera, JunoCam – are only planned to last through eight of the total number of planned orbits. The microwave radiometer is designed to last through 11. A radiation vault surrounds the instruments to give them added protection. The radiation vault has titanium walls to protect the spacecraft's electronic brain and heart from Jupiter's harsh radiation environment. This vault will dramatically slow the aging effect radiation has on the electronics for the duration of the mission. Each titanium wall measures nearly a square meter (nearly 10 square feet) in area and about 1 centimetre (a third of an inch) in thickness. With more than 20 electronic assemblies inside, the whole vault weighs about 200 kg (440lbs).

Volcanic calderas and even active lava flows dot the surface of Io.

Io

Jupiter's moon Io is the most volcanically active world in the solar system, with hundreds of volcanoes spewing out lava fountains dozens of kilometres high.

Discovered on 8 January 1610 by Galileo Galilei, Io is a little bit larger than Earth's moon, the third largest of Jupiter's moons, and the fifth furthest from the planet. Its fiery volcanic activity is the result of a cataclysmic tug of war between the powerful gravity of Jupiter and two neighbouring moons, Europa and Ganymede. These forces cause Io's surface to bulge in and out by as much as 100m (328ft), and generate a tremendous amount of heat that keeps much of the moon's crust liquid. This gushes up onto the surface through giant volcanoes and fissures, forming massive lava lakes and floodplains of liquid rock. Although it's not known exactly what Io's lava is made of, it's believed to be mainly molten sulphur and silicate rock, while the thin atmosphere is mostly sulphur dioxide.

Io's volcanic nature means it's not a place where you'd like to hang around too long. Inside some of the volcanoes, temperatures reach more than 1600°C (over 2900°F), and yet at its surface, temperatures average -130°C (-202°F) – more than twice as cold as the south pole. This really is a world of fire and ice.

5 Facts About Io

1. Io's very thin atmosphere is primarily sulphur dioxide, which on Earth is sometimes used to preserve dried food.

2. Io's volcanoes are at times so powerful that they can be seen from Earth with large telescopes.

3. Over 1.8 Earth days, Io rotates once on its axis and completes one orbit of Jupiter, causing the same side of Io to always face Jupiter.

4. Data from the Galileo spacecraft indicates that an iron core may form Io's centre, thus giving Io its own magnetic field.

5. Io's orbit cuts across the planet's magnetic lines of force, turning it into a giant electric generator capable of developing 400,000 volts.

Io floats before Jupiter in this rendering.

The Mythology of Io

In classical mythology, Io is a mortal woman transformed into a cow during a dispute between the Greek god Zeus (Jupiter in Roman mythology) and his wife, Hera (Juno to the Romans).

Io on Film

Io's powerful volcanoes have captured imaginations since their discovery decades ago. The moon's most memorable role was arguably in *2010: The Year We Make Contact*, the sequel to Stanley Kubrick's 1968 cult classic *2001: A Space Odyssey*. Released in 1984 and directed by Peter Hyams, the film features a sequence in which astronauts make a spacewalk above Io's volcanoes to board an abandoned spacecraft. The Jovian moons, so much more similar to our terrestrial planet than to Jupiter, are fodder for dreams of exploration by scientists as well as writers, at least by instruments if not by humans (though sending astronauts to explore the Galilean moons has been floated by NASA).

This view of Europa, created from Galileo data, shows its cracked-seeming surface.

Europa

Beneath its icy surface, Europa is believed to conceal a global ocean of salty liquid up to twice the volume of Earth's oceans. This ocean is thought to be one of the most promising places to look for life in our solar system.

Slightly smaller than Earth's own moon, Europa has a shell of ice that is likely 15 km to 25 km (9 to 15 mi) thick, criss-crossed by long, linear fractures. These cracks are caused by the moon's elliptical orbit and the gravitational pull of Jupiter, which create tidal forces that flex and stretch the ice on the moon's surface. The same process also creates heat, and perhaps volcanic activity, as on the neighbouring moon of Io.

Beneath the icy crust, the moon's salt-water ocean, if it exists, may be between 60 km and 150 km (37 and 93 mi) deep. Scientists believe there may be volcanic or hydrothermal vents on the seafloor, thus supplying the three ingredients necessary for life as we know it: abundant liquid water, energy and chemistry. Who knows what lurks amid these elements. As yet, though, it's just an intriguing theory which won't be tested until the Europa Clipper mission launches in the 2020s.

5 Facts About Europa

1. Europa is slightly smaller than Earth's Moon and barely one-fourth the diameter of Earth.

2. Every 3.5 Earth days Europa rotates once on its axis and completes one orbit of Jupiter, so the same side of Europa always faces Jupiter.

3. Europa has an extremely thin oxygen atmosphere – far too thin for humans to breathe.

4. Europa has been visited by several spacecraft, including Galileo, which made repeated visits to Europa while in orbit around Jupiter.

5. An unknown reddish-brown material has been observed along Europa's fractures, and in splotchy patterns across its surface.

Plumes of water shoot into space from Europa's surface in this rendering.

The Mythology of Europa

Europa is named for the daughter of Agenor, who was abducted by Zeus (the Greek equivalent of the Roman god Jupiter), who had disguised himself as a spotless white bull. Zeus rode away with her to the island of Crete, where Europa bore him many children, including notorious Minos.

The Clipper Mission

The Europa Clipper spacecraft is designed to find out whether the icy moon could harbour conditions suitable for life. The radiation-tolerant spacecraft will carry cameras and spectrometers that can produce high-resolution images of Europa's surface and determine its composition, while an ice-penetrating radar will determine the thickness of the moon's icy shell and search for subsurface lakes similar to those beneath Antarctica. The probe will perform 45 flybys of Europa at altitudes varying from 2700 km to 25 km (1677 mi to 16 mi). NASA's Clipper is currently scheduled for launch sometime around the mid-2020s.

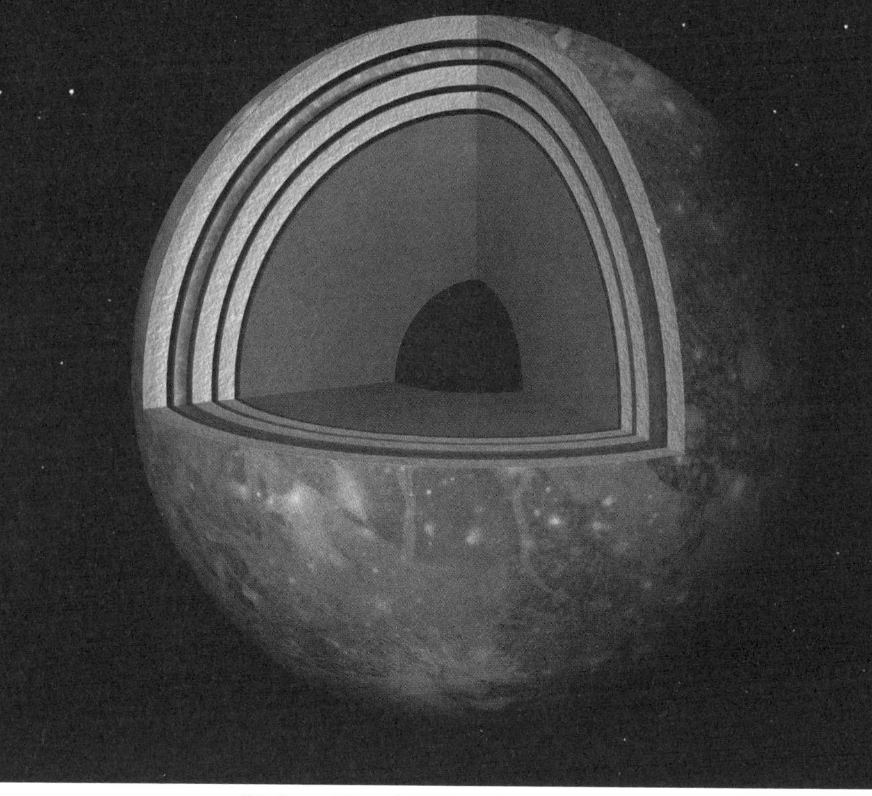

This diagram of Ganymede's interior reveals the model of alternating layers of ocean and icy crust.

Ganymede

Ganymede is the largest satellite anywhere in our solar system. It is larger than both Mercury and the dwarf planet Pluto, and three-quarters the size of Mars.

Ganymede is thought to have three main layers. A sphere of metallic iron at the centre (the core, which generates a magnetic field), surrounded by a spherical shell of rock (mantle), and another spherical shell of ice. Scientists also believe there must be a fair amount of rock in the icy surface; the moon has many ridges and grooves (known as 'sulcus') that suggest its surface experienced dramatic upheavals in the distant past. There is also evidence that there may be an underground ocean.

Ganymede is the only moon in the solar system (that we know of) with its own magnetic field, which causes Ganymede to have its stunning aurorae around the moon's north and south poles when buffeted by the solar wind. Other Jovian satellites may have magnetic fields induced by Jupiter's own powerful magnetosphere as it rotates.

Astronomers using the Hubble Space Telescope found evidence of an oxygen atmosphere in 1996, but it is far too thin to support life.

Ganymede in Numbers

1. Ganymede is 2.4 times smaller than Earth.

2. The moon was discovered by Galileo in 1610.

3. The moon's icy crust may be around 800 km (497 mi) thick.

4. Some of Ganymede's ridges run for thousands of kilometres, and may be as high as 700 m (2296 ft).

5. Many large, flat craters have been observed, ranging from (50 km to 400 km (31 to 248 mi) in diameter.

Contrasting smooth bright terrain and fractured dark terrain reveal lengthy ridges.

The Mythology of Ganymede

In mythology, Ganymede (*gan*-uh-meed) was a beautiful young boy who was carried by Zeus to Olympus disguised as an eagle. Ganymede became the cupbearer of the Olympian gods.

The Rocks of Ganymede

In 2004, scientists discovered irregular lumps beneath the icy surface of Ganymede. The irregular masses may be rock formations, frozen into and supported by Ganymede's icy shell for billions of years. This tells scientists that the ice is probably strong enough, at least near the surface, to support these possible rock masses from sinking to the bottom of the ice. However, this anomaly could also be caused by piles of rock at the bottom of the ice. Interpreting data from such a distance, and without reams of context, is a mix of an art and science, in this mission as in many others. Scientists make their best hypothesis based on the evidence, and must be open to revising their conclusions as potential new information comes to light from later research.

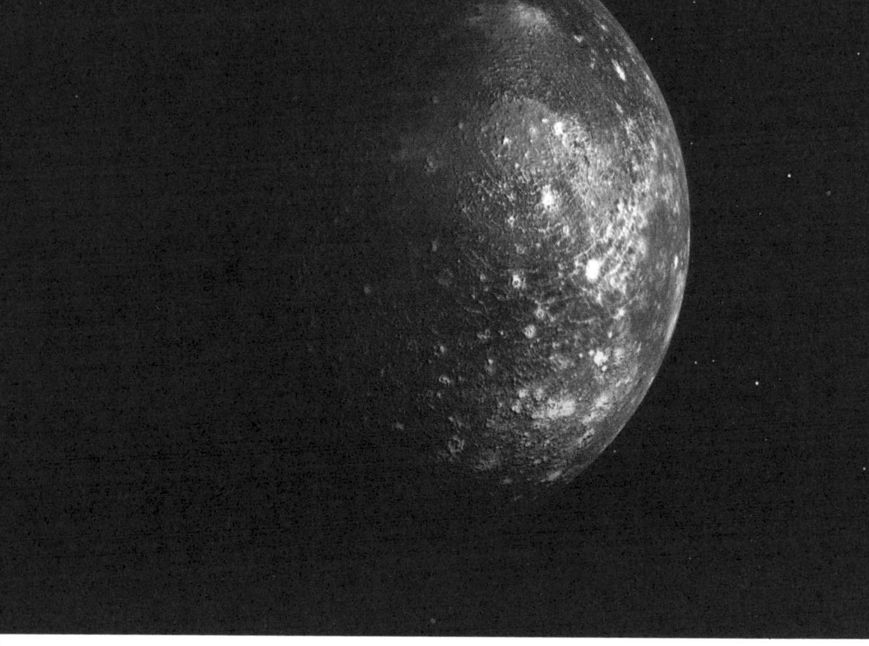

A rendering of Callisto, which may hold secret oceans under its icy crust.

Callisto

Second in size after Ganymede, the crater-covered moon Callisto is encased in a crust of rock and ice, and – like its sister moon Europa – may have a secret: a salty ocean beneath its surface.

Jupiter's second-largest moon and the third-largest in our solar system, Callisto is about the same size as Mercury. In the past, it was considered an 'ugly duckling moon' because it didn't seem to have much going on - no active volcanoes or shifting tectonic plates. But data from NASA's Galileo spacecraft in the 1990s revealed Callisto may have a salty ocean, placing the once-seemingly dead moon on the list of worlds that could possibly harbour life. More recent research reveals that this ocean may be located deeper beneath the surface than previously thought, around 250 km (155 mi) below the surface, or may not actually exist at all.

Callisto's rocky, icy surface is the oldest and most heavily cratered in our solar system. The surface is about 4 billion years old and it's been heavily pummelled by comets and asteroids.

5 Facts About Callisto

1. Callisto is 2.6 times smaller than Earth.

2. Callisto orbits about 1.9 million km (1.2 million mi) from Jupiter on average.

3. A day on Callisto lasts about 17 Earth days.

4. NASA's Galileo spacecraft detected a thin carbon dioxide atmosphere on the moon.

5. The NASA spacecraft to have observed Callisto are Pioneer, Voyager, Galileo, Cassini, Juno, New Horizons and Hubble.

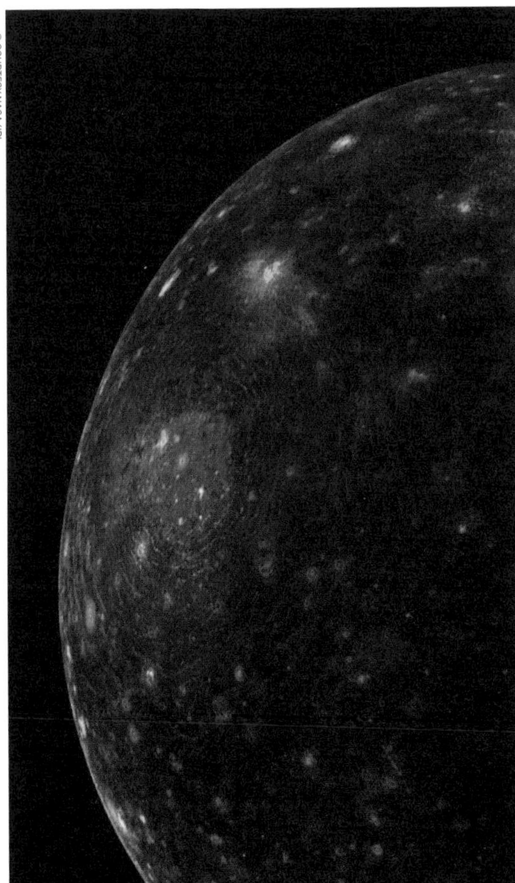

The prominent bright spot in this view of Callisto is the impact feature Valhalla.

The Mythology of Callisto

Callisto is named for a woman who was turned into a bear by Zeus in Greek mythology.

Callisto in Literature

Callisto has been a perennially popular location for science fiction writers. In the 1930s, Harl Vincent had Earth and Callisto at war in his novel *Callisto at War*, while Isaac Asimov's 1940 novel *The Callistan Menace* depicted Callisto as a deathtrap crawling with giant caterpillars. Perhaps the best-known writer to depict Callisto is Philip K. Dick, author of *Blade Runner* and *The Minority Report*, who wrote a short story in 1955 – called 'The Mold of Yancy' – about colonists living in a totalitarian society on this icy moon. Any future colonists would want to be sure they had a well-secured heat source! Callisto orbits just beyond Jupiter's main radiation belt, so at least radiation would be less of a concern. And the moon is believed to be a possible candidate for supporting life, with oxygen, hydrogen and water detected.

The Galilean four are just the start of Jupiter's many moons.

Jupiter's Other Moons

Jupiter's four largest satellites are now known as the Galilean moons, after their discoverer (see opposite page). Galileo himself originally called them the Medicean planets, after the powerful Italian Medici family. His observations of the moons were one of the things that first prompted to him to postulate that planets in our solar system orbit the Sun, not the Earth. But while they got the first attention (and basked in it for centuries), there are now 53 named moons known to orbit Jupiter, and 26 more awaiting their official names as of this book's printing. That's at least 79 moons total!

Some may have originated from passing asteroids, drawn into Jupiter's orbit by its immense gravitational field and 'captured' there. Considered irregular satellites, smaller and with eccentric orbits, they may not have the grandeur of the Galilean moons, which surpass even the dwarf planets in size, but they dot the neighbourhood in shocking proliferation. Metis, the closest moon to Jupiter, has a diameter of a mere 40 km (25 mi) and orbits Jupiter faster than Jupiter revolves on its axis, in .29 Earth days. Adrastea, Amalthea, Thebe and many more keep Metis company.

Galileo and the Medicean Planets

Jupiter's four largest moons – Io, Callisto, Ganymede and Europa – are called the Galilean satellites after Italian astronomer Galileo Galilei, who first observed them in 1610 (in fact, the German astronomer Simon Marius claimed to have seen the moons around the same time, but he did not publish his observations and so Galileo is generally given the credit for their discovery).

Galileo himself, however, originally named the newly discovered moons the Medicean planets, after the all-powerful Medici family, who ruled Florence during the 15th and 16th centuries (and also acted as his patrons). Galileo referred to the moons numerically as I, II, III, and IV, while the names of Io, Europa, Callisto, and Ganymede were provided by Simon Marius. In the 1800s, astronomers realised that the sheer number of Jupiter's newly discovered moons made a number-based naming system extremely confusing.

Galileo's discovery of these moons was revolutionary: it was the first time a moon had been observed orbiting a planet other than Earth. It led to Galileo's realisation that we live in a heliocentric solar system in which planets orbit the Sun, rather than the Earth, as under the long-standing geocentric model. Already posited by Copernicus, the heliocentric model was a profoundly controversial theory that received fierce opposition from the Catholic Church. Eventually, Galileo was forced to stand trial in defence of his theory in 1633. He was convicted of heresy, and put under house arrest until his death in 1642. The so-called Galileo Affair didn't officially resolve until the Vatican acknowledged his theory in 1992, but within the 17th century Kepler and Newton confirmed the scientific accuracy of a heliocentric solar system.

An 1891 depiction of Galileo at his studies.

Galileo's telescope.

SATURN

Saturn and its rings as seen by Cassini.

Saturn at a Glance

If there's one thing that pretty much everyone knows the planet of Saturn for, it's the rings – the vast, swirling discs of ice, dust and rock that encircle the planet like cosmic hula hoops.

We now know that it's not the only planet in our solar system that has rings (Jupiter and Uranus do too), but none are as spectacular or as complex as Saturn's: they're visible from Earth with even a half-decent telescope, a tantalising hint of the incredible and wondrous structures the universe is capable of constructing when it's in the mood. These billions of dust particles are likely cometary in origin.

The sixth-most-distant planet from our Sun, Saturn is another of the 'gas giants', mostly composed of swirling vortices of hydrogen and helium. The planet is blanketed with clouds that appear as faint stripes, jet streams and storms, tinted in different shades of yellow, brown and grey – giving Saturn the appearance of a gigantic planet-sized marble. Apart from Jupiter, it's by far the biggest world in our solar system – equivalent to nearly nine Earths positioned side by side (and that's not even including its famous rings).

After Jupiter, it's also the planet with the most moons: fifty-three confirmed so far, with another nine still to be 100% verified. Many of these moons are home to fascinating extraterrestrial landscapes quite unlike any we'd ever find on Earth. From the giant jets of water that gush up from Enceladus, to the methane lakes on smoggy Titan and the deep craters of Phoebe, the Saturn system is packed with the potential for scientific discovery, and holds many mysteries that are still waiting to be solved. And while planet Saturn is an unlikely place for living things to take hold, the same is not true of some of its satellites: moons like Enceladus and Titan, which are home to internal oceans, could conceivably support some kind of life.

DISTANCE FROM SUN
9.5AU

LIGHT-TIME TO THE SUN
79.34 minutes

LENGTH OF DAY
10.7 hours

LENGTH OF ORBITAL YEAR
10,759 Earth days
(29 Earth years)

ATMOSPHERE
Hydrogen and helium

Voyager 1 was only the second spacecraft to do a flyby of Saturn.

Top Tip

Saturn has the second-shortest day in the solar system (just over 10.5 hours) after Jupiter. That means a long night of sleep on Saturn could potentially turn into sleeping through an entire day!

Getting There & Away

At an average distance of 1.4 billion km (886 million mi), Saturn is 9.5 AU away from the Sun. From this distance, it takes sunlight around 80 minutes to travel from the Sun to Saturn. Pioneer 11, meanwhile, took six and a half years to arrive; Cassini took an extra three months on top of that, and Voyager 2 took four years. The speediest of all was Voyager 1, reaching Saturn after three years and two months of travel.

A comparison of Earth's size next to Saturn.

Orientation

The second-largest planet in our solar system, Saturn could fit over 750 of our planet inside its volume. However, Saturn's gravity is only 1.08 times the gravity on Earth because Saturn is gaseous rather than solid. In other words, an object weighing 100 kg on Earth would weigh 108 on Saturn.

Saturn is even less dense than water. If there were a body of water large enough to hold Saturn, it would float! No other planet in the solar system has this property. Its atmosphere is comprised of hydrogen (94%) and helium (6%). A spacecraft descending through the ammonia clouds of Saturn would encounter gases becoming hotter and denser until finally the spacecraft would be crushed and melted.

The planet rotates very rapidly - a day on Saturn is only 10 hours and 39 minutes long. This rapid rotation causes the planet to flatten at its poles and bulge at its equator. Saturn's equatorial atmosphere rotates around the planet faster than its inner layers and core. Imagine a set of nested cylinders, rotating at different speeds. Eventually, toward the centre of the planet, the layers move in synchrony and rotate together.

Notwithstanding its rapid rotation, Saturn takes a very long time to orbit the Sun, more than 29 Earth years. Since Saturn is so far from the Sun, it receives about 1% as much sunlight on its surface as does Earth. Saturn is far from its neighbours as well: Jupiter is a full 4.32 AU away and Uranus is 9.7 AU away.

Beautiful appearance aside, Saturn would not be a very fun place to live. Its seasons last a lengthy seven years and the average temperature at the cloud tops is an extremely cold -88 K. In addition, Saturn has a very high-velocity wind of 1800 km/h (1118mph).

Saturn *vs* Earth

-- Radius --
9.45x
EARTH

-- Mass --
95x
EARTH

-- Volume --
763x
EARTH

-- Surface gravity --
1.08x
EARTH

-- Mean temperature -
-155°C
(-247°F)

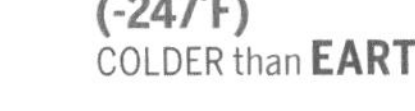

-- Surface area --
83.5x
EARTH

-- Surface pressure --
Unknown

-- Density --
12.5%
EARTH

-- Orbit velocity --
32.5%
EARTH

-- Orbit distance --
9.5x
EARTH

5 Facts About Saturn

1. Saturn is the furthest planet from Earth readily visible with the unaided human eye.

2. The planet is named for the Roman god of agriculture and wealth, who was also the father of Jupiter. Saturn shares its name with Saturday, arguably the best day of the week. Both are named after the same Roman god.

3. Saturn is tilted by 26.73 degrees, similar to Earth's 23.5-degree tilt. This means that, like Earth, Saturn experiences distinct seasons.

4. About two tons of Saturn's mass came from Earth: the Cassini spacecraft was intentionally vaporised in Saturn's atmosphere in 2017.

5. Saturn's rings are spread far into space, at generally only about 10m (30ft) thick!

Saturn surrounded by an array of its moons.

A Floating Saturn

It's hard to imagine, but Saturn is the only planet in our solar system whose average density is less than water. This means that the giant gas planet would float like a beach ball if it were dropped into a planet-sized bathtub – admittedly a rather unlikely scenario, but a fun fact nonetheless.

Saturn's North Pole

Saturn's north pole has an interesting atmospheric feature: a six-sided jet stream. This hexagon-shaped pattern was first noticed in images from the Voyager I spacecraft and has been more closely observed by the Cassini spacecraft since. Spanning about 30,000 km (20,000 mi) across, the hexagon is a wavy jet stream of winds (reaching 22km/h, or 200mph), with a massive, rotating storm at the centre. As far as we know, there is no weather feature like it anywhere else in the solar system.

History

Saturn took shape when the rest of the solar system formed about 4.5 billion years ago, when gravity pulled swirling gas and dust in to become this gas giant. About 4 billion years ago, Saturn settled into its current position in the outer solar system, where it is the sixth planet from the Sun. Like Jupiter, Saturn is mostly made of hydrogen and helium, the same two main components that make up the Sun. And its rings, constituted of ice and dust particles? Those may not just be relatively recent, but are destined to be short-lived. The rings are being pulled by gravity into Saturn as a dusty rain of ice particles under the influence of Saturn's magnetic field. They are believed to have less than 100 million years before disappearing, about the same span of time as they have existed so far.

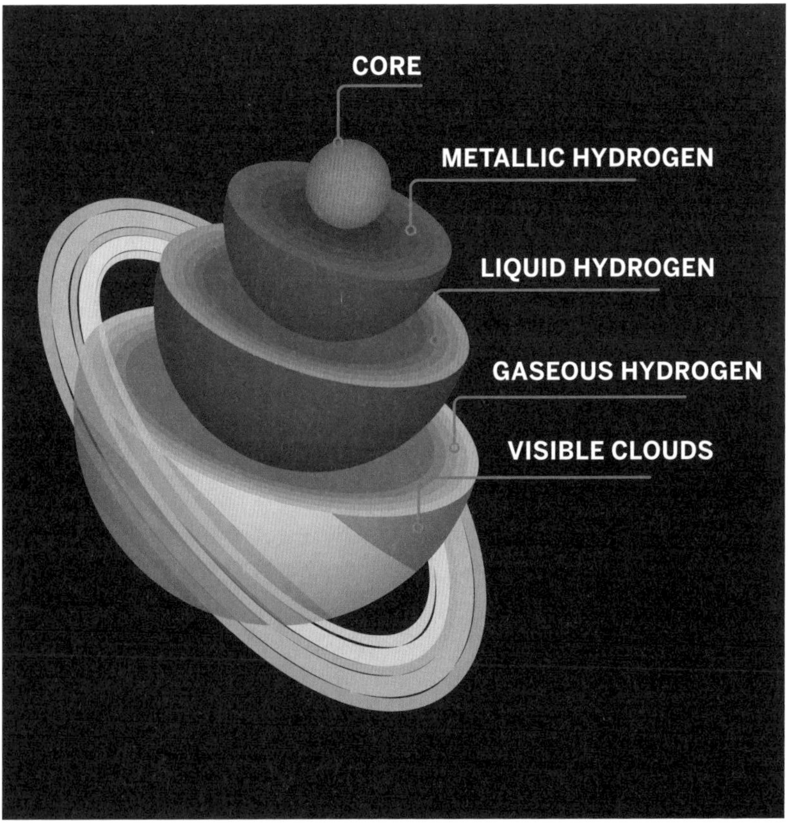

CORE

METALLIC HYDROGEN

LIQUID HYDROGEN

GASEOUS HYDROGEN

VISIBLE CLOUDS

Saturn's core is believed to be much smaller than Jupiter's, though surrounded by the same liquid hydrogen oceans.

Discovery Timeline

Galileo demonstrating his telescope.

~700 BCE

The oldest written records documenting Saturn, attributed to the Assyrians, describe the ringed planet as a sparkle in the night and name it 'Star of Ninib'.

~400 BCE

Ancient Greek astronomers name what they think is a wandering star in honour of Kronos, the god of agriculture. The Romans later change the name to Saturn, their god of agriculture.

July 1610

Galileo spots Saturn's rings through a telescope, but mistakes them for a 'triple planet'.

1655

Christiaan Huygens discovers Saturn's rings and its largest moon, Titan.

1675

Astronomer Jean-Dominique Cassini discovers a 'division' between what will later be called the A and B rings.

1979

Pioneer 11 is the first spacecraft to reach Saturn. Among its many discoveries are Saturn's F ring and a new moon.

1980 and 1981

In its 1980 flyby of Saturn, Voyager 1 reveals the intricate structure of the ring system, consisting of thousands of ringlets. Flying even closer to Saturn in 1981, Voyager 2 provides more detailed images and documents the thinness of some of the rings.

2004

NASA's Cassini spacecraft becomes the first to orbit Saturn, beginning a decade-long mission that revealed many secrets and surprises about Saturn and its system of rings and moons.

2005

ESA's Huygens probe is the first spacecraft to make a soft landing on the surface of another planet's moon – Saturn's giant moon Titan. The probe provides a detailed study of Titan's atmosphere during the over two hour descent and relays data and images from Titan's surface for about another hour and 10 minutes.

2006

Scientists discover a new ring, revealed using images obtained during the longest solar occultation of Cassini's four-year mission. During a solar occultation, the Sun passes directly behind Saturn, causing the rings to be brilliantly backlit. An occultation usually lasts only about an hour, but in this instance it lasts 12 hours. The new ring coincides with the orbits of Saturn's moons Janus and Epimetheus.

2009

The Spitzer Space Telescope reveals the presence of a gigantic, low-density ring associated with Saturn's distant moon Phoebe.

2017

Cassini ends a 13-year orbital mission with a spectacular, planned plunge into Saturn's atmosphere – sending science data back until the very last second. Cassini's final five orbits enabled scientists to directly sample Saturn's atmosphere for the first time.

Saturn on Screen

With its distinctive rings as calling card, Saturn is in the running for the most iconic of all the planets in our solar system. As such, it provides a popular backdrop for pop culture, from the giant sandworms inhabiting the planet in *Beetlejuice* to its proximity to the wormhole in *Interstellar*. You'll find it as a reference point in films such as *WALL-E*, *2001: A Space Odyssey*, *Star Trek*, and *Final Fantasy VII* among many more.

Exploring Saturn

Saturn was the most distant of the five planets known to the ancients. In 1610, Italian astronomer Galileo was the first to gaze at Saturn through a telescope. To his surprise, he saw a pair of objects on either side of the planet. He sketched them as separate circles, thinking that Saturn was triple-bodied. In 1655, Dutch astronomer Christiaan Huygens, using a more powerful telescope than Galileo's, observed that Saturn was surrounded by a thin, flat ring. He proposed the correct explanation for his observations, and for the regular disappearances and reappearances of Saturn's ring, in 1659 in *Systema Saturnium*.

In modern times, Saturn has been revealed in great detail by Hubble Space Telescope observations. Pioneer 10 made the first close observations of Saturn, followed by more detailed flybys by both Voyager 1 and 2. But no spacecraft has shown us more about Saturn, its rings and its moons than the Cassini orbiter and the Huygens probe, which landed on Titan in 2005.

An artist's conception of the landing area for ESA's Huygens probe to Titan.

A striking view of Saturn and its rings.

Saturn Highlights

The Cassini Mission

1 Before sacrificing itself, Cassini returned vast troves of data.

Saturn's Rings

2 Sail through Saturn's mighty rings, arguably the most impressive sight in the solar system.

Saturn's Magnetosphere

3 Aurora and a 'true north' polar alignment make this magnetosphere hum, with help from the planet's moons.

Saturn's Surface

4 This light gas giant is raked by gale-force winds (and then some).

Titan

5 Explore this massive moon, which has a huge hidden ocean and a methane cycle not too dissimilar from Earth's own water cycle.

Enceladus

6 Put on your sunglasses when you visit Enceladus – it's blindingly bright, and may also harbour life.

Rhea, Dione & Tethys

7 These so-called sister moons are tidally locked 'dirty snowballs'.

Iapetus

8 Two starkly opposite hemispheres, light and dark, divide Iapetus.

Mimas

9 Feel the force on the crater-pocked moon dubbed the 'Death Star' by astronomers.

Phoebe

10 Dark Phoebe is likely a captured asteroid, also known as a Centaur.

An artist's concept of floating carbon ice on a liquid hydrogen sea.

© COURTESY NASA/JPL

An artist's image of Cassini inserting itself into orbit around Saturn.

The Cassini Mission

For more than a decade, NASA's Cassini spacecraft shared the wonders of Saturn and its family of icy moons, taking us to astounding worlds.

The Voyager and Pioneer flybys of the 1970s and 1980s provided the first looks at Saturn and its satellites, but the secrets of the planet were only fully revealed by the Cassini mission during the spacecraft's many years in Saturn orbit. Cassini revealed in great detail the true wonders of Saturn, a giant world ruled by raging storms and delicate harmonies of gravity. In total, Cassini orbited Saturn 294 times from 2004 to 2017. Along the way, it looked at, listened to, sniffed and even tasted Saturn's moons.

During its mission, Cassini discovered many secrets about Saturn, its rings and its moons that have fascinated scientists. Saturn's rings were found to be active and dynamic, a laboratory for how planets or moons form. And while the larger moons are spherical, others are shaped like a sweet potato (Prometheus), a regular potato (Pandora), a meatball (Janus) and even a sponge (Hyperion). Some have a gnarled, irregular shape and texture like a dirty ice ball (Epimetheus). One object observed in the rings (unofficially called Peggy) may be a moon forming or disintegrating, or it might not be a moon at all. Most intriguingly of all, Cassini revealed that while the Saturn system lies far outside the 'habitable zone' of the solar system, several moons have sub-surface oceans which could potentially harbour forms of life.

An illustration of Saturn's rings lit by the Sun.

Saturn's Rings

The most recognisable and unique feature of this gas giant is doubtless it rings, spanning an immense distance with billions of particles in orbit.

Saturn's iconic rings are thought to be pieces of comets, asteroids or shattered moons, torn apart by Saturn's gravity. Swirling out around the planet at a width of 400,000 km (240,000 mi), Saturn's main rings never reach a thickness of more than about 10m (30ft). There are seven main rings, named alphabetically in the order they were discovered. Starting at Saturn and moving outward, these are the D ring, C ring, B ring, Cassini Division, A ring, F ring, G ring, and finally, the E ring. Much farther out, the very faint Phoebe ring orbits its namesake moon. Most of the rings are relatively close to each other, with the exception of Rings A and B, which are separated by a gap measuring 4700 km (2920 mi) wide called the Cassini Division.

The rings are composed of billions of small chunks of ice and rock coated with another material such as dust. The particles range from tiny, dust-sized icy grains to chunks as big as a house, and a few of the particles are even as large as mountains. Viewed from the cloud tops of Saturn, the rings would look mostly white – and each one orbits at a different speed.

Intriguingly, data from Cassini's final orbits suggests that the rings may be relatively young – probably having formed between 10 million and 100 million years ago, compared to 4.5 billion years for Saturn itself. This seems to support theories that the rings formed from a comet that wandered too close and was torn apart by Saturn's gravity – or perhaps by an event that broke up an earlier generation of icy moons. The rings they've left behind are so wide that it would take a car a week to drive across some of them. It is not the only planet to have rings, but none are as spectacular or as complex as Saturn's.

As Cassini headed for its grand finale, it passed between Saturn's rings.

Now You See Them, Now You Don't

Once about every 15 years, Saturn's rings appear to vanish into space. In fact, this is an optical illusion, caused by the fact that we cannot see Saturn's rings when they are viewed edge-on from Earth; they are only barely visible through the most powerful telescopes.

Saturn's Disappearing Rings

Still quite young by planetary standards, recent research has confirmed that Saturn is slowly losing its iconic rings. 'We estimate that this 'ring rain' of particles falling into the planet drains an amount of water products that could fill an Olympic-sized swimming pool from Saturn's rings each half an hour', according to James O'Donoghue of NASA's Goddard Space Flight Center. The first hints that ring rain existed came from Voyager observations of seemingly unrelated phenomena: peculiar variations in Saturn's electrically charged upper atmosphere (ionosphere), density variations in Saturn's rings, and a trio of narrow dark bands encircling the planet at northern mid-latitudes. These dark bands appeared in images of Saturn's hazy upper atmosphere (stratosphere) made by NASA's Voyager 2 mission in 1981.

A rendering of Saturn's magnetosphere.

Saturn's Magnetosphere

Saturn's magnetic field is smaller than Jupiter's, but it's still 578 times as powerful as Earth's, though the aurorae here operate by a different principle.

Saturn, its rings and many of its satellites lie totally within the planet's enormous magnetosphere, the region of space in which the behaviour of electrically charged particles is influenced more by Saturn's magnetic field than by the solar wind. Material blasted into space by the moon Enceladus feeds Saturn's giant E ring and is a major source of material (plasma) fuelling Saturn's magnetosphere. Saturn's magnetic field has north and south poles, as on a bar magnet, and the field rotates with the planet. On Jupiter and Earth, the magnetic fields are slightly tilted with respect to the planets' rotation axes – this tilt is the reason we say compass needles point to 'magnetic north' rather than true north. But Saturn's magnetic field is almost perfectly aligned with the planet's rotation.

This giant magnetosphere means that aurorae occur differently on Saturn. On Earth, these charged particles come from the solar wind, but at least some of Saturn's aurorae are like Jupiter's, caused by a combination of particles ejected from its moons, and the rotation of the planet's magnetic field.

Under its rings, Saturn's surface is a swirling mass of gases.

Saturn's Surface

As a gas giant, Saturn doesn't have a true surface: the planet is mostly swirling gases and liquids deeper down.

Like Jupiter, Saturn is made mostly of hydrogen and helium, with a dense core of metals like iron and nickel surrounded by rocky material and other compounds, solidified by the intense pressure and heat. It is enveloped by liquid metallic hydrogen inside a layer of liquid hydrogen – similar to Jupiter's core but considerably smaller. Saturn's density is the lowest in the solar system and its mass only 30% that of Jupiter, though it's about 84% Jupiter's diameter. Saturn's specific gravity (0.7) is less than that of water.

Like the other gas giants, the extreme pressures and temperatures deep inside Saturn would crush, melt and vaporise spacecraft trying to fly into the planet, which presents a challenge for researchers looking to study the planet! In fact, the pressure on Saturn is so powerful it squeezes gas into liquid within the planet.

Above the deep interior, winds in Saturn's upper atmosphere are around four times stronger than the most powerful hurricane-force winds on Earth: they can reach four times the speed of Earth's top winds in the equatorial region.

An artist's conception of a dust storm on Titan with Saturn in the background.

Titan

Principal among Saturn's moons is Titan – the second-largest moon in our solar system after Jupiter's Ganymede. But most thrilling? As far as we know, Titan is the only other place in the solar system where stable liquid can reliably be found on its surface.

Discovered in 1655 by the Dutch astronomer Christiaan Huygens, Titan is an icy world whose surface is completely obscured by a golden hazy atmosphere. It's much bigger than Earth's moon, and is even larger than the planet Mercury. It's the only moon in the solar system with a dense atmosphere, and the only world besides Earth that has standing bodies of liquid – including rivers, lakes and seas – on its surface.

Like Earth, Titan's atmosphere is primarily nitrogen, plus a small amount of methane. Scientists studying Cassini also found not only that liquid methane and ethane exist on Titan, but that they rain from the sky and fill liquid lakes. The grandest of these is larger than Lake Michigan–Huron, itself the largest freshwater lake on Earth. This means that Titan is the only other place in the solar system known to have an Earth-like water cycle, in which liquid methane rains from clouds, flows across its surface, fills lakes and seas, and evaporates back into the sky. Titan's air is dense enough that you could walk around without a spacesuit, but you'd need an oxygen mask and protection from the bitter cold.

Instruments on the Cassini-Huygens mission also found isotopes nitrogen-14 and nitrogen-15 in Titan's atmosphere. The results were similar to those found in comets from the Oort Cloud. The correspondence suggests Titan may have formed early in the solar system's history, perhaps in the same cold disc of gas and dust that formed the Sun, rather than the warmer disc of material that formed Saturn.

This composite image of Titan looks past the atmosphere, which is an orange colour.

5 Facts About Titan

1. Titan has a radius of about 2575 km (1600 mi), nearly 50% wider than Earth's moon.

2. Titan is about 1.2 million km (759,000 mi) from Saturn.

3. The moon takes 15 days and 22 hours to complete a full orbit of Saturn.

4. Surface pressure is about 60% higher than on Earth; temperatures can reach -179°C (-290°F).

5. Given the dense atmosphere and gravity roughly equivalent to Earth's Moon, a raindrop on Titan would fall about six times more slowly than on Earth.

Titan in Pop Culture

Many writers have set stories on Titan, including Arthur C Clarke, Philip K Dick, Isaac Asimov and Kurt Vonnegut. Titan also features in the 2009 film *Star Trek* – the USS *Enterprise* comes out of warp drive in Titan's atmosphere to sneak up on the Romulan ship attacking Earth. Other appearances include *Gattaca*, *Futurama*, *Eureka* and the anime series *Cowboy Bebop*.

The Seas of Titan

Titan's elements of clouds, rain, rivers, lakes and seas are composed of liquid hydrocarbons like methane and ethane. Theoretically at least, the largest seas could conceivably harbour forms of life that use different chemistry than we're used to. And even if they don't, beneath Titan's thick ice crust there is an ocean primarily composed of water rather than methane. This even could support life that's similar to that of the oceans of Earth, at least on a microscopic scale. Equally plausible is the possibility that Titan could be completely lifeless: it may be many years before we are able to answer the question for sure.

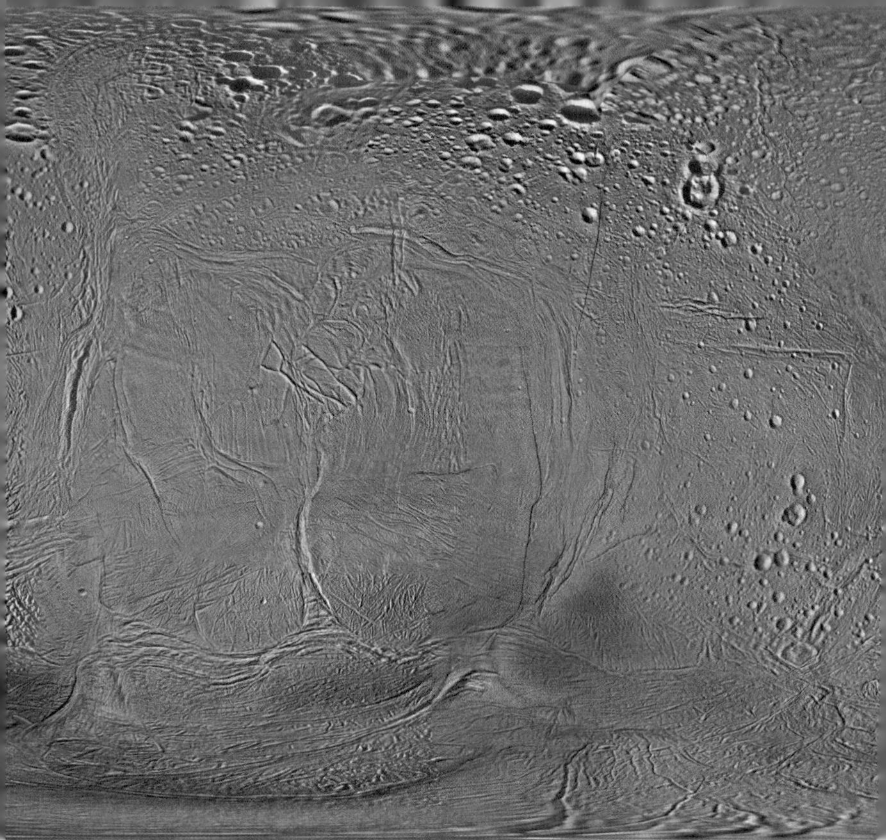

A global colour mosaic of Enceladus.

Enceladus

The brightest body in our solar system, Enceladus is coated in a crust of ice that reflects back the sunlight with dazzling intensity.

About as wide as Arizona, 500 km (310 mi) across, Enceladus orbits Saturn between the orbits of Mimas and Tethys. It has the whitest, most reflective surface (albedo) in the solar system, but for many years scientists had no idea why. It's now known to be a result of giant plumes that gush up from the subsurface saltwater ocean concealed beneath the moon's icy shell. These plumes were first detected by the Cassini spacecraft in 2005, spraying water vapour, ice particles and simple organic materials into space from the south polar region. The giant plumes can reach speeds of approximately 1300 km/h (800mph), leaving a trail of icy particles in the moon's wake. Some of these particles are carried into orbit and form Saturn's E ring. The remainder falls back like snow to the surface, which causes Enceladus' remarkable smoothness and brightness.

Enceladus' giant jets originate from warm fractures in the crust, which scientists call the 'tiger stripes'. Several gases, including water vapour, carbon dioxide, methane, perhaps a little ammonia and either carbon monoxide or nitrogen make up the gaseous envelope of the plume.

From these samples, scientists have determined that Enceladus has most of the chemical ingredients needed for life, and likely has hydrothermal vents spewing out hot, mineral-rich water into its ocean. With this global ocean, unique chemistry and internal heat, Enceladus has become a promising lead in our search for worlds where life could exist.

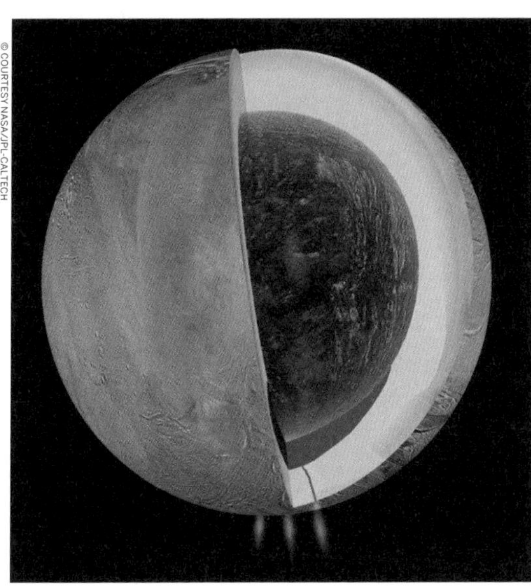

A diagram illustrating Enceladus' possible core with a regional water ocean above.

5 Facts About Enceladus

1. British astronomer William Herschel first spotted Enceladus orbiting Saturn on 28 August 1789.

2. Enceladus is named after a giant in Greek mythology.

3. Because Enceladus reflects so much sunlight, the surface temperature is extremely cold – about -201°C (-330°F).

4. Scientists think the average thickness of Enceladus' ice shell is about 20 to 25 km (12 to 16 mi), thinning to just 1 to 5 km (half a mile to 3 mi) at the south pole.

5. Parts of Enceladus show many large impact craters, while others have few, indicating major resurfacing events in the geologically recent past.

How Enceladus Got Its Name

William Herschel's son, John Herschel, suggested the name in his 1847 work *Results of Astronomical Observations*, in which he suggested names for the first seven Saturnian moons discovered. He chose these names in particular because Saturn, known in Greek mythology as Cronus, was the leader of the Titans.

Life on Enceladus

By taking samples from Saturn's E ring, which is largely formed by ice particles ejected from the moon's plumes, scientists have detected nanograins of silica, which can only be generated where liquid water and rock interact at temperatures above about 90°C (200°F). This suggests that hydrothermal vents may exist deep beneath the moon's icy shell, similar to the hydrothermal vents that dot Earth's ocean floor. And even if they don't exist, Enceladus' special combination of heat, organic chemistry and liquid water have made it an exciting potential site for life beyond Earth.

Sweet, Sweet Oxygen

In 2010, the Cassini spacecraft detected a very thin atmosphere known as an exosphere around Rhea, infused with molecules of oxygen and carbon dioxide – the first time an oxygen atmosphere has been discovered on a world other than Earth. The oxygen appears to arise when Saturn's magnetic field rotates over Rhea, decomposing the surface and releasing oxygen molecules (the origin of the carbon dioxide is less clear). Although the density of oxygen is estimated to be about 5 trillion times less than on Earth, these findings suggest complex chemistry could be happening on the surfaces of many icy bodies in the Universe.

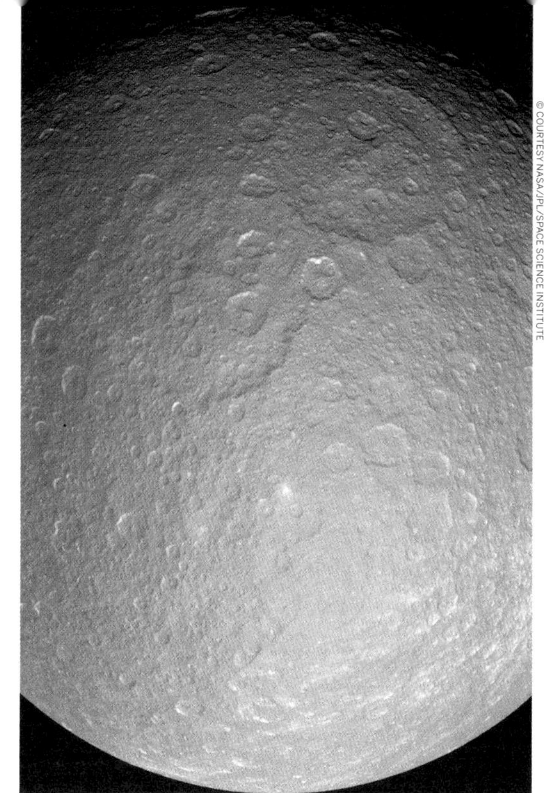

A mosaic of crater-scarred Rhea.

Rhea, Dione & Tethys

Rhea, Dione and Tethys share similar appearance and composition: discovered by Cassini, they're all small, cold, airless bodies that look rather like frozen, dirty snowballs.

Astronomer Giovanni Cassini discovered Rhea in 1672 and Dione and Tethys in 1684. All three sister moons are tidally locked in phase with their parent – one side always faces toward Saturn – as they complete a 2 to 4.5-Earth-day orbit around the planet. Surface temperatures on the moons are as warm as -174°C (-281°F) in sunlit areas, ranging down to -220°C (-364°F) in shaded areas. The moons have a high reflectivity (or geometric albedo) suggesting a surface composition largely of water-ice, which behaves like rock in this temperature range.

Rhea, at a distance of 527,000 km (327,500 mi), is farther from Saturn than Dione and Tethys, and because of this Rhea does not receive internal heating from Saturn's tidal variation. Dione and Tethys thus have more areas of smooth plains than Rhea. Such plains are probably areas where liquid water reached the surface and ponded in depressions, forming flat surfaces before refreezing and thus erasing existing craters. The lesser internal warmth at Rhea could have resulted in fewer erasures or there could have been more bombardment on Rhea.

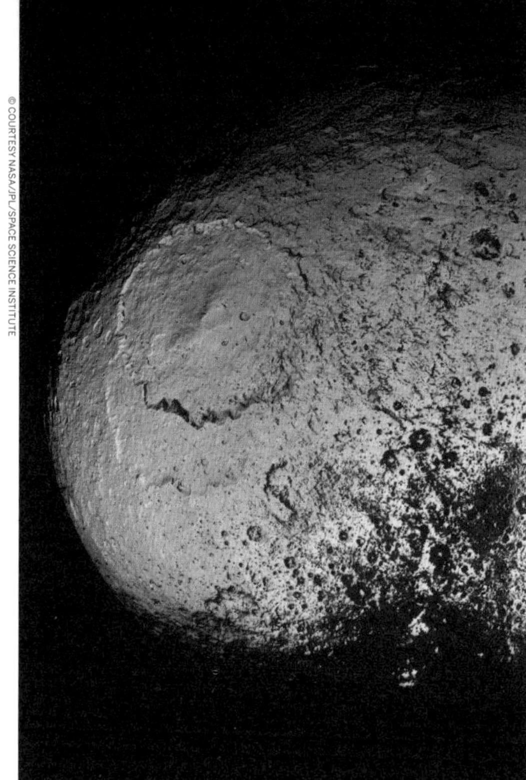

Cassini captured this first image of Iapetus' bright trailing hemisphere.

The Mysterious Ridge of Iapetus

The other most notable feature of Iapetus is its 'equatorial ridge', a chain of mountains 10 km (6 mi) high that girdles the moon's equator. On the anti-Saturnian side of Iapetus, the ridge appears to break up, and distinct semi-bright mountains can be seen; they were first spotted by the Voyager I and Voyager II probes, and are informally referred to as the Voyager Mountains. Some scientists think the ridge was formed at an earlier time when Iapetus rotated much faster than it does today; others think the ridge is made of material left from the collapse of a ring.

Iapetus

Iapetus has been called the yin and yang of the Saturn moons because it's divided into two distinct hemispheres, each with a contrasting albedo: one light, one dark.

With a mean radius of 736 km (457 mi), Iapetus is Saturn's third-largest moon. But it holds one of the bigger mysteries of the solar system in its stark albedo difference. Giovanni Cassini observed the dark-light divide when he discovered Iapetus in 1671. Scientists have long wondered about the cause: some think Iapetus may be sweeping up carbon particles ejected from the dark moon Phoebe, while an alternate theory is that ice volcanoes might be spewing darker material to the moon's surface. Cassini's flyby showed a third process, thermal segregation, may be the most likely cause of Iapetus' murky hemisphere: Iapetus' slow rotation of more than 79 days means that dark material can absorb large amounts of heat, sublimating out any lighter, colder material, accelerating the darkening process.

Iapetus orbits at 3,561,000 km (2,213,000 mi) from Saturn. The great distance from Saturn's tidal forces and most of the other moons and rings has left the Iapetus surface largely unaffected by any melting episodes that could have caused some smoothing or 'resurfacing' as on some of the moons closer to Saturn.

The Mimas Test

The fact that Mimas appears to be frozen solid is rather puzzling. The moon is closer to Saturn and has a much more eccentric (elongated) orbit than Enceladus, which should mean Mimas experiences more tidal heating. And yet its distinctive craters mean its frozen surface must be very old, and must have endured long enough to preserve them – in contrast to Enceladus, where the crust appears to have partially thawed. This paradox has prompted the 'Mimas Test', by which any theory that claims to explain the partially thawed water of Enceladus must also explain the entirely frozen water of Mimas.

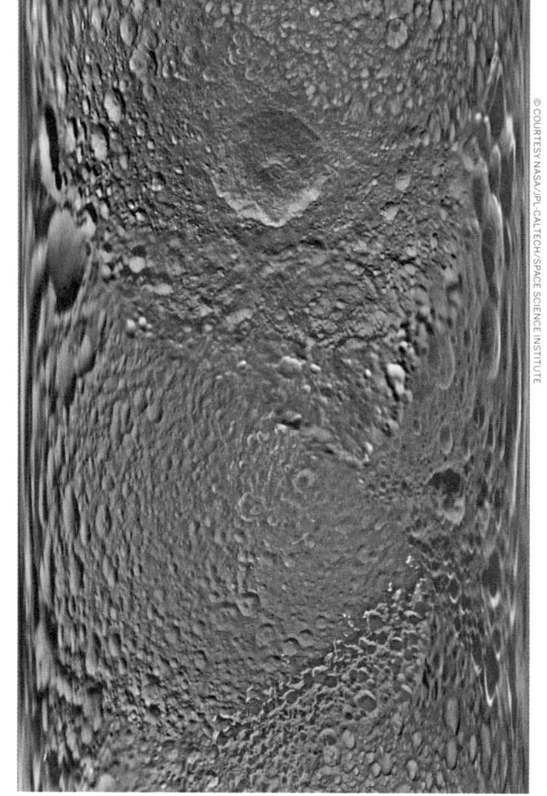

The depression of Hershel crater is visible on this global map of the Mimas surface.

Mimas

Crater-covered Mimas is the smallest and innermost of Saturn's major moons, and also bears a striking resemblance to the Death Star.

Less than 198 km (123 mi) across, Mimas's surface is pocked with impact craters ranging in size up to greater than 40 km (25 mi) in diameter. The largest is the Herschel Crater, named after the moon's discoverer, which stretches a third of the way across the face of the moon and makes it look eerily like Darth Vader's planet-destroying space station. This massive crater is 130 km (80 mi) across, with outer walls about 5 km (3 mi) high and a central peak 6 km (3.5 mi) high. The impact that blasted this crater out of Mimas probably came close to breaking the moon apart. Shock waves from the Herschel impact may have caused the fractures, also called chasmata, on the opposite side of Mimas. If it had splintered, its debris might have joined into Saturn's existing rings; torn-apart moons are one possible origin of the rings.

Mimas takes only 22 hours and 36 minutes to complete an orbit. Mimas is tidally locked: it keeps the same face toward Saturn as it flies around the planet, just as our Moon does with Earth. Ground-based astronomers could only see Mimas as little more than a dot until Voyagers I and II imaged it in 1980. The Cassini spacecraft later made several close approaches and provided detailed images of Mimas.

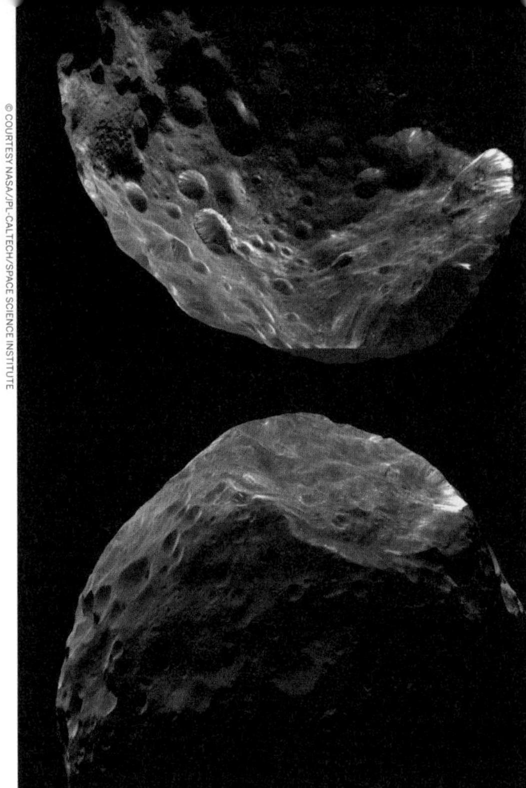

Two views of Phoebe, a possible captured Centaur.

Phoebe's Peculiarities

Phoebe is one of Saturn's most distant moons, orbiting almost 13 million km (over 8 million mi) from the planet, almost four times further out than its nearest neighbour, the satellite Iapetus. Phoebe's orbit is also peculiar: it's retrograde, which means it goes around Saturn in the opposite direction than most other moons, as well as most objects in the solar system. At about 106.5 km (66.2 mi), Phoebe's radius is roughly one-sixteenth that of Earth's own moon.

Phoebe

Unlike most of Saturn's moons, Phoebe is dark, reflecting only 6% of the sunlight that falls on it – which has led scientists to believe that it is a rare example of a 'captured object'.

A captured object is a celestial body that is trapped by the gravitational pull of a much bigger body, generally a planet. Phoebe's darkness, in particular, suggests that the small moon originates from the outer solar system, where there is plenty of dark material. Some scientists think Phoebe could be a captured Centaur – a primordial object dating from the early formation of the solar system that was never pulled into the formation of a planet. With its relatively small size, Phoebe might never have heated up enough to change its original chemical composition – meaning that it could be a rare fragment dating back to the very birth of the Milky Way itself.

Phoebe was discovered August 1898 by American astronomer William Pickering, making this moon the first body to be discovered using long-exposure photographic plates. The moon's name is an alternative for the Greek goddess Artemis (Diana to the Romans). She was the youthful goddess of Earth's moon, forests, wild animals and hunting. Sworn to chastity and independence, she never married and was closely identified with her brother Apollo.

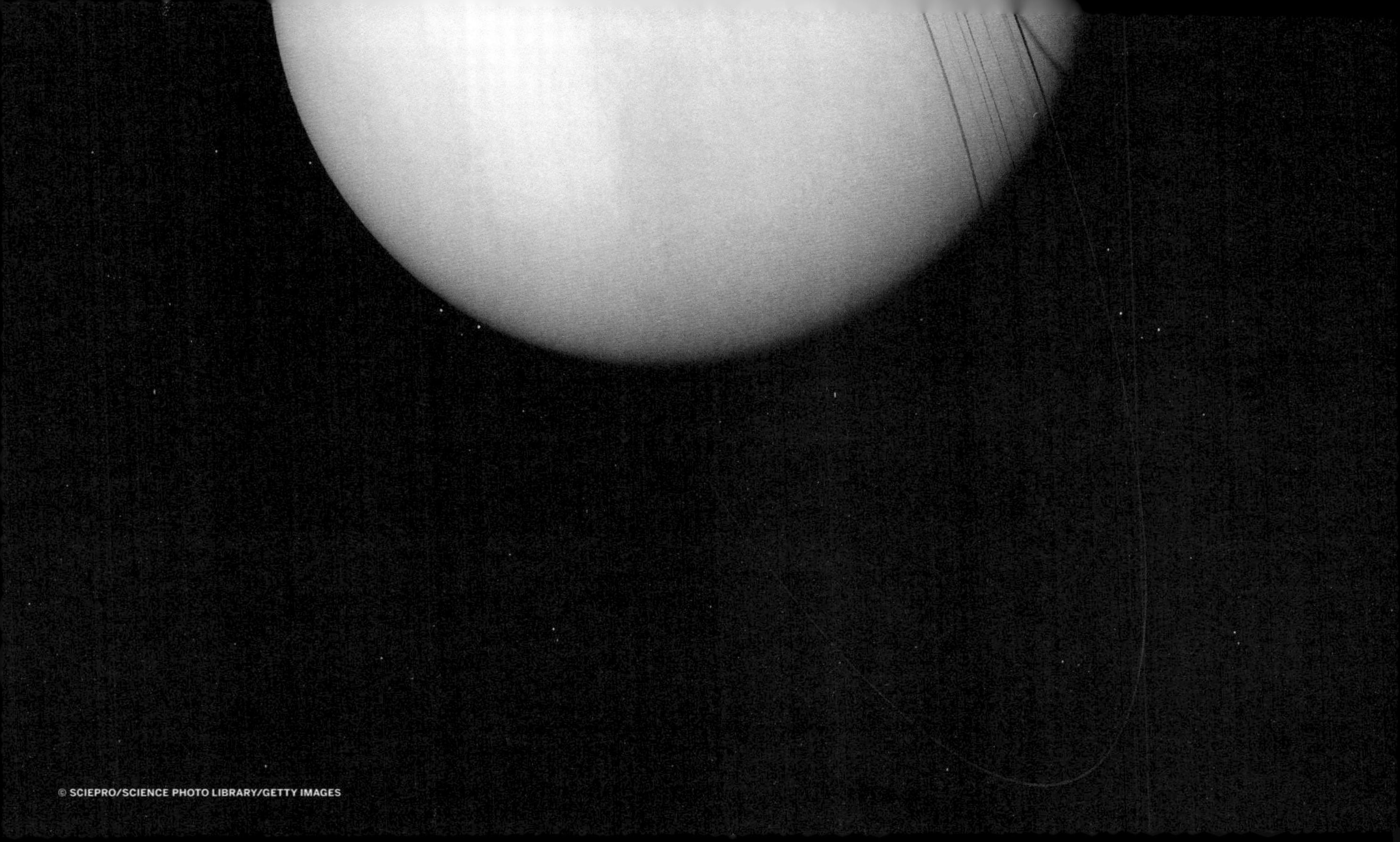

URANUS

PLANET TYPE
Ice Giant

NUMBER OF MOONS
27

SIZE COMPARED TO EARTH
4x

Uranus as seen by Voyager 2.

Uranus at a Glance

The most unfortunately named planet in the solar system has long been the butt of schoolyard jokes. FYI, the name relates to the Roman god of the sky, and it's pronounced yoor-*un-us, okay? So you can stop giggling at the back.*

Apart from its irresistibly snigger-worthy name, Uranus is fascinating for all kinds of scientific reasons. It's the third-largest planet in our solar system, and the seventh most distant from the Sun. Discovered in 1781 by astronomer William Herschel, Uranus is surrounded by 13 faint rings and 27 small moons. Its atmosphere is mainly composed of helium and hydrogen, with a bit of methane swirling in the mix which gives the planet its distinctive blue-green colour. Sunlight passes through the atmosphere and is reflected back out by the cloud tops. The methane in the atmosphere absorbs the red portion of the light, resulting in the planet's vivid blue-green tint.

Another thing that distinguishes Uranus is the way it spins. Like Venus, it rotates from east to west (the opposite of

Earth). Even more bizarrely, Uranus also spins nearly at a right angle to its orbit, which makes it appear to revolve on its side, like a rolling ball – the only planet in the solar system to do so.

This unique sideways rotation means the seasons on Uranus are extremely strange indeed: the planet's north pole experiences 21 years of nighttime in winter, 21 years of daytime in summer and 42 years of day and night in the spring and fall. That would really play havoc with the circadian rhythm of even the soundest of sleepers. And living on Uranus would also mess up your birthday – with an average orbit of around 84 Earth years, most Earthlings would count themselves lucky to enjoy a single birthday cake during their lifetime.

Not that you'd particularly want to spend your birthday on Uranus. With wind speeds that can reach 900km/h (560 mph) and an atmospheric temperature of -224°C (-371°F), you'd struggle to light the candles in the first place. And then there's the smell: recent research suggests that Uranus' upper atmosphere hydrogen sulfide clouds probably smell like rotten eggs, or, for the more scatologically minded, a particularly noxious bout of flatulence. So maybe Uranus' branding problem really does make sense after all.

It might, however, be able to tell us how the icy giants differ from their warmer brethren, information that may give new clues to the evolution of solar systems once another mission can reach its distant orbit.

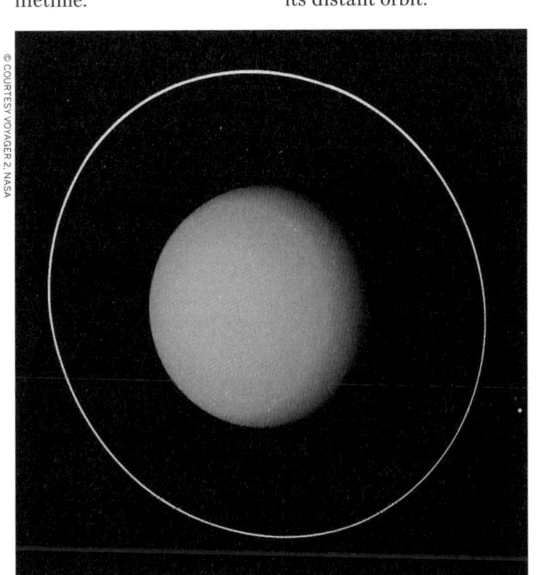

This Voyager 2 image also captured one of Uranus' newly discovered moons.

DISTANCE FROM SUN
19.8 AU

LIGHT-TIME TO THE SUN
165.07 minutes

LENGTH OF DAY
17 hours 14 minutes

LENGTH OF ORBITAL YEAR
30,687 Earth days (84 Earth years)

ATMOSPHERE
Hydrogen, helium and methane

Top Tip

Visitors might want to bring a gas mask to help ward off the smells of the hydrogen sulfide clouds.

Getting There & Away

With an average orbit distance of 2.9 billion km (1.8 billion mi) from the Sun, Uranus is 19.8 astronomical units away from the solar system's centre (nearly 20 times farther than Earth). At this distance, it takes a beam of sunlight 2 hours and 40 minutes to reach Uranus.

So far, Voyager 2 is the only spacecraft to fly by Uranus. No probes have orbited this distant planet to study it up close and at length.

Orientation

With a radius of 25,362 km (15,759 mi), Uranus is about four times wider than Earth. That means that if Earth were the size of a large apple, Uranus would be the size of a basketball. One day on Uranus (the time it takes for Uranus to rotate or spin once) lasts only about 17 hours, but because of its great distance from the Sun, Uranus makes a complete orbit around the Sun (a year in Uranian time) in about 84 Earth years (30,687 Earth days). 80% or more of the planet's mass is made up of a hot dense fluid of 'icy' materials – water, methane and ammonia – around a small rocky core. Near the core, it heats up to a relatively cool 4982°C (9000°F), making Uranus the only planet that doesn't give off more heat from its core than it receives from the Sun.

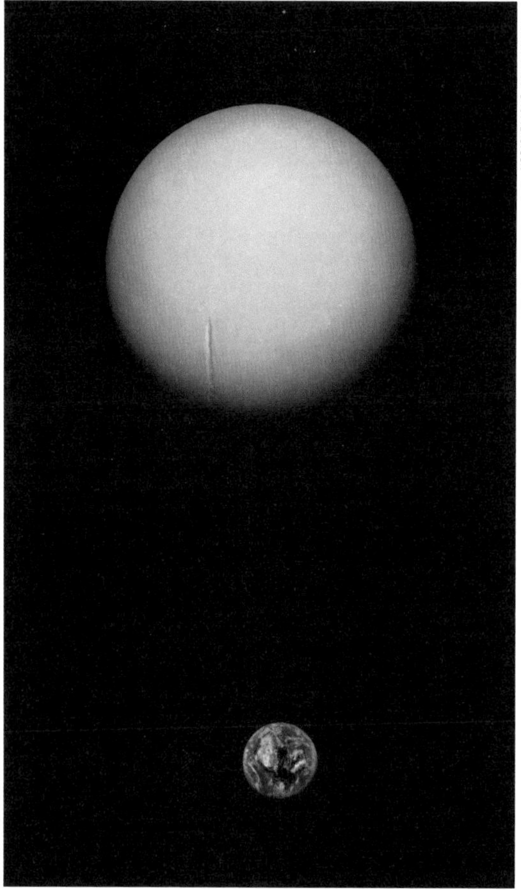

Earth is about four times smaller than our distant neighbour Uranus.

Uranus *vs* Earth

-- Radius --
4x
EARTH

-- Mass --
14.5x
EARTH

-- Volume --
63x
EARTH

-- Surface gravity --
90.5%
EARTH

-- Mean temperature --
-208°C
(-342°F)
COLDER than **EARTH**

-- Surface area --
16x
EARTH

-- Surface pressure --
Unknown

-- Density --
23%
EARTH

-- Orbit velocity --
22%
EARTH

-- Orbit distance --
19x
EARTH

Exploring Uranus

After travelling about 3 billion km (1.8 billion mi) in nine years, NASA's Voyager 2 flew past Uranus in 1986. The spacecraft gathered much of its critical information about the mysterious planet, including observations of its rings and moons, in just six hours. Voyager 2's images of the five largest moons around Uranus revealed complex surfaces indicative of a shifting geologic past. The cameras also detected 11 previously unseen moons. Several instruments studied the ring system, uncovering the fine detail of the previously known rings and two newly detected rings. Voyager data showed that the planet's rotation takes 17 hours, 14 minutes. The spacecraft also found a Uranian magnetic field that is both large and unusual. In addition, the temperature of the equatorial region, which receives less sunlight over a Uranian year, is nevertheless about the same as that at the poles.

The rest of what we know about Uranus comes from observations via the Hubble Space Telescope and several powerful ground-based telescopes.

Voyager 2 is still functioning today as it flies in deep space more than 120 AU from Earth, through the heliosphere and heliopause and toward interstellar space. Voyager 2 and its companion spacecraft, Voyager 1, returned data on how the influence of our Sun wanes outside the heliosphere. Now that both spacecraft have reached the interstellar medium, beyond the Sun's influence, their ability to gather and transmit data will eventually be hindered by failing power, but their journeys continue.

5 Facts About Uranus

1. Uranus is slightly larger in diameter than its neighbour Neptune, yet is smaller in mass.

2. It is the second least dense planet; Saturn is the least dense of all.

3. While wind speeds on Uranus can reach up to 900 km/h (560 mph), they're still less than half the maximum speed of winds on Neptune.

4. Winds at the equator blow in the opposite direction of the planet's rotation, but follow it closer to the poles.

5. Uranus' environment is not conducive to life. The temperatures, pressures and materials are too extreme.

Uranus on Screen

Uranus may be the butt of more than a few jokes and puns, but it's also a frequent destination in science fiction, including the video game *Mass Effect* and TV shows like *Doctor Who*.

Uranus Stinks

Despite decades of observations, Uranus held on to one critical secret – the composition of its clouds. Then, in April 2018, a global research team discovered they contained hydrogen sulphide, the odiferous, flammable gas that makes rotten eggs smell so bad. This is very different from the gas of giant planets located closer to the Sun, where the cloud tops are mainly thought to be composed of ammonia.

But according to one of the report's lead researchers, Patrick Irwin of Oxford University, bad smells would be the least of your worries if you were to find yourself teleported to Uranus. 'Suffocation and exposure in the -200°C atmosphere... would take its toll long before the smell.' Well, that's a relief then.

History

Uranus took shape as the rest of the solar system formed about 4.5 billion years ago, when gravity pulled swirling gas and dust in to create this ice giant. Like its neighbour Neptune, Uranus likely coalesced closer to the Sun and moved to the outer solar system about 4 billion years ago, leading to its present icy state.

It was the first planet discovered with the aid of a telescope. The English astronomer William Herschel spotted it in 1781; he originally thought it was either a comet or a star. Two years later the object was universally accepted as a new planet, partly because of observations by astronomer Johann Bode. William Herschel tried unsuccessfully to name his discovery Georgium Sidus, after King George III. Instead the planet was named for Uranus, the Greek god of the sky.

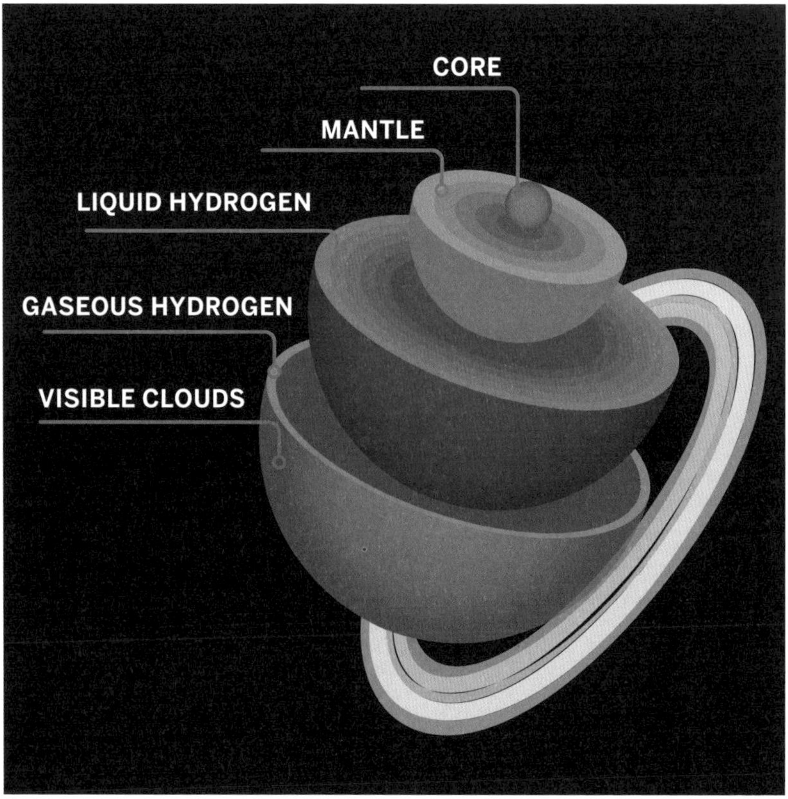

Uranus has the typical inner core of an ice giant, with faint rings in orbit.

Discovery Timeline

© ABBIE WARNOCK-MATTHEWS/SHUTTERSTOCK

Keck Observatory on the summit of Mauna Kea.

1781

British astronomer William Herschel discovers Uranus – the first new planet discovered since ancient times – while searching for faint stars. Despite being discovered by telescope, however, Uranus can be seen by a careful observer with the naked eye alone under the right conditions.

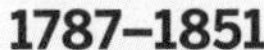

1787–1851

Four Uranian moons are discovered and named Titania, Oberon, Ariel and Umbriel.

1789

The radioactive element uranium is named after Uranus when it is discovered in 1789, eight years after the planet was discovered. Otherwise, there is no connection between the element and the planet.

1948

Another moon, Miranda, is discovered. The smallest, innermost of Uranus' larger round moons, like the rest it is thought to consist mostly of roughly equal amounts of water-ice and silicate rock.

1977

While observing Uranus' passing in front of a distant star (SAO 158687), scientists at the Kuiper Airborne Observatory and the Perth Observatory in Australia make a major discovery: Uranus, like Saturn, is encircled with rings.

1986

NASA's Voyager 2 makes the first – and so far the only – visit to Uranus. The spacecraft comes within 81,500 km (50,600 mi) of the planet's cloud tops. Voyager discovers 10 new moons, two new rings and a magnetic field stronger than that of Saturn.

2005

NASA announces the discovery of a new pair of rings around Uranus and two new, small moons (Mab and Cupid) orbiting the planet – from photographs taken by the Hubble Space Telescope. The largest ring discovered by Hubble is twice the diameter of the planet's previously known rings.

2006

Observations made at the Keck Observatory and by the Hubble Space Telescope show that Uranus' outer ring is coloured blue while the new inner ring is reddish.

Dec. 2007

Uranus is seen as it reaches equinox. Equinox is when the planet is fully illuminated as the Sun passes over its equator. Equinox also brings a ring-plane crossing, when Uranus' rings appear to get narrower as they pass through, appearing edge-on, and then widen again as seen from Earth.

March 2011

New Horizons passes the orbit of Uranus on its way to Pluto, becoming the first spacecraft to journey beyond Uranus' orbit since Voyager 2. However, Uranus was not near the crossing point, so no new data was gathered.

November 2011

This month, NASA's Hubble Space Telescope captures the first photographic evidence taken from Earth of aurorae on Uranus.

Uranus Highlights

Uranus' Surface & Atmosphere

1 Windy Uranus has dynamic weather systems.

The Aurorae of Uranus

2 Unrestricted to the poles, Uranus' newly-spotted aurorae can appear across the planet.

A Sideways Planet

3 The axis of Uranus is set at almost a right angle against its orbit.

Ring Systems

4 Faint, particle-strewn and extremely narrow rings surround Uranus.

Uranus' Moons

5 Five larger, mostly round moons share Uranus' orbit with many smaller satellites.

Miranda

6 Innermost Miranda has an icy subsurface.

Ariel

7 The young surface of Ariel may show micro-meteorite strikes.

Umbriel

8 Umbriel's strikingly low albedo reflects little light.

Oberon

9 Craters and a mountain adorn Oberon.

Titania

10 Titania is marked by fault valleys and frost.

Uranus' Shepherd Moons

11 Shepherd moons keep the thin rings of Neptune in line.

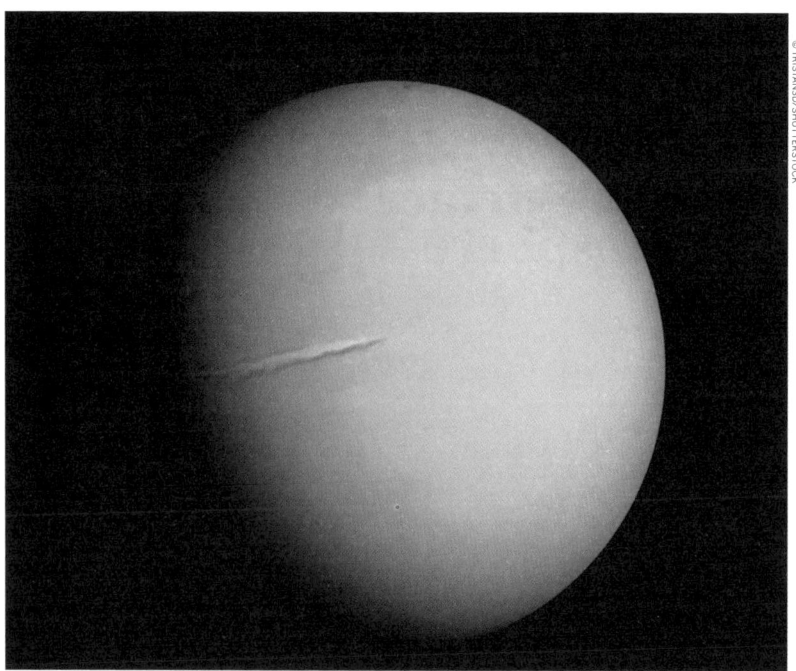

A rendering of Uranus with a visible weather system.

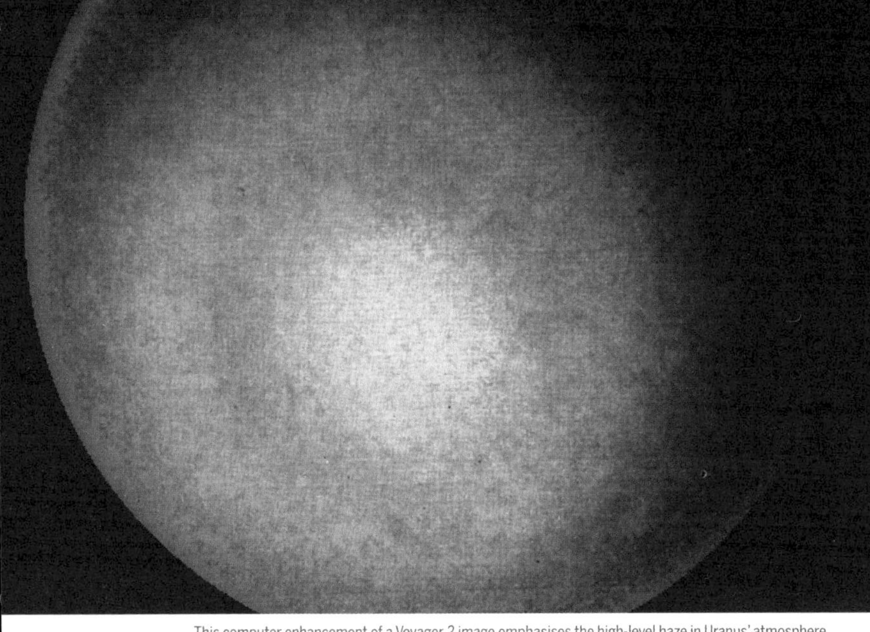

This computer enhancement of a Voyager 2 image emphasises the high-level haze in Uranus' atmosphere.

Uranus' Surface & Atmosphere

Uranus is one of two ice giants in the outer solar system (the other is Neptune). The planet has no true surface: it's a swirling, dense fluid made of 'icy' materials, mainly water, methane and ammonia.

As an ice giant with no firm surface, Uranus would have nowhere for a spacecraft to land, but it wouldn't be able to fly through Uranus' atmosphere unscathed anyway – the extreme pressures and temperatures would likely destroy it before it could get anywhere close. The Voyager 2 spacecraft recorded the first detailed view of windy Uranus' clouds, and more recent observations by Hubble have given new information on a dark spot in the northern hemisphere, which lasted for at least two months; not quite giving Jupiter a run for its money, but a glimpse at new complexity in the planet's weather systems and atmosphere. Observations reveal that Uranus exhibits dynamic clouds as it approaches equinox, including rapidly changing bright features; these would be in line with the extreme variations in sunlight over the duration of Uranus' orbit.

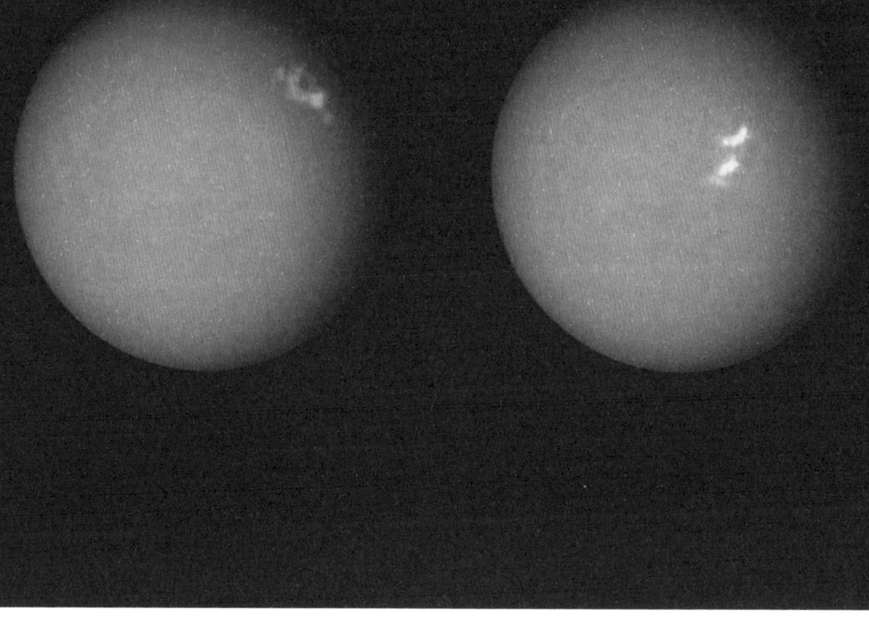

The aurora captured on Uranus.

The Aurorae of Uranus

Uranus has its own aurorae, just like the northern lights – but scientists think they could be way more spectacular than any you'd see on Earth.

In 2011, the NASA/ESA Hubble Space Telescope became the first Earth-based telescope to snap an image of the aurorae on Uranus. In 2012 and 2014, a team led by an astronomer from Paris Observatory took a second look at the aurorae using the ultraviolet capabilities of the Space Telescope Imaging Spectrograph (STIS) installed on Hubble.

They tracked the interplanetary shocks caused by two powerful bursts of solar wind travelling from the Sun to Uranus, then used Hubble to capture their effect on Uranus' aurorae – and found themselves observing the most intense aurorae ever seen on the planet. By watching the aurorae over time, they collected the first direct evidence that these powerful shimmering regions rotate with the planet. Unlike on Earth, Jupiter and Saturn, however, Uranus' aurorae are not in line with the poles, due to the planet's lopsided magnetic field. The magnetosphere that gives rise to the aurorae is believed to be generated at a relatively shallow depth under the planet's surface, and may have an almost strobe-like function, opening and closing daily.

4 Facts About the Magnetosphere

1. Uranus' dipole magnetic field is tilted nearly 60 degrees from the planet's rotational axis.

2. It is also offset from the planet's centre by a third of the planet's radius, meaning it doesn't sit equally between the poles.

3. The magnetosphere tail is 18 times wider than the planet itself.

4. The magnetic field is twisted by Uranus' sideways rotation into a long corkscrew shape.

The Science of Aurorae

Aurorae are caused by streams of charged particles like electrons that are caught in powerful magnetic fields and channelled into a planet's upper atmosphere – where their interactions with gas particles, such as oxygen or nitrogen, set off spectacular bursts of light.

An artist's rendering of Uranus and its ring systems.

Uranus orbits on an extreme tilt.

A Sideways Planet

Uranus is the only planet whose equator is nearly at a right angle to its orbit, with a tilt of 97.77 degrees – possibly as the result of a collision with a terrestrial planet-sized object long ago.

This planet's unique tilt causes the most extreme seasons in the solar system. For nearly a quarter of each Uranian year, the Sun shines directly over each pole, plunging the other half of the planet into a 21-year-long dark winter.

Recent research suggests that, around four billion years ago, a young proto-planet of rock and ice collided with Uranus, causing its extreme tilt. Using advanced computing techniques, scientists simulated the suspected impact, and determined that an object with more than the mass of Earth struck the young planet with a grazing blow. The collision was so strong it reshaped Uranus and pushed it onto its side, although it was not strong enough to blast the planet's atmosphere off into space, or significantly change its orbit around the Sun.

If it happened, the impact might have caused Uranus' tilted, off-centre magnetic field, too. Rock and ice thrown into orbit would have then clumped together to form the rings and moons around Uranus, now in its newly established rotation.

Tilt Time

1 In comparison to Uranus' 98-degree tilt, Jupiter tilts by just 3 degrees, and Earth by 23 degrees.

2 Some older simulations suggest rather than one single giant impact, the tilt may have been caused by at least two smaller collisions.

3 An alternative theory suggests the tilt may have been caused by 'orbital resonance' from Jupiter and Saturn, in which the planets' combined gravitational influence knocked Uranus off-axis.

Uranus & the Search for Exoplanets

Research into a massive collision early in Uranus' history have helped explain some of the planet's more peculiar features. It's also helped scientists with their search to understand other planets outside our solar system – known as exoplanets. Based on findings from the Kepler space telescope, the more common type of exoplanet is very similar to Uranus: a medium-size, gaseous planet with a rocky, icy core.

Information scientists have learned about this early impact suggests that similar collisions may have led to the formation of other planets, including the Uranus-like exoplanet 092LAb. An almost unimaginably distant 25,000 light-years away from our solar system, it was discovered by a team of scientists at Ohio State University and Warsaw University Observatory. With a mass four times that of Uranus, it orbits a binary star system in the constellation Sagittarius.

Similar to Uranus, 092LAb travels a wide orbit far from its stars, perhaps because the gravitational jostlings of the binary pair pushed it further away. Like all exoplanets with such a wide orbit, it was discovered by a technique called gravitational microlensing, which uses gravity's ability to bend light to detect objects which are otherwise too faint to view.

A rendering of exoplanet 092LAb, a Uranus-like planet at four times the size.

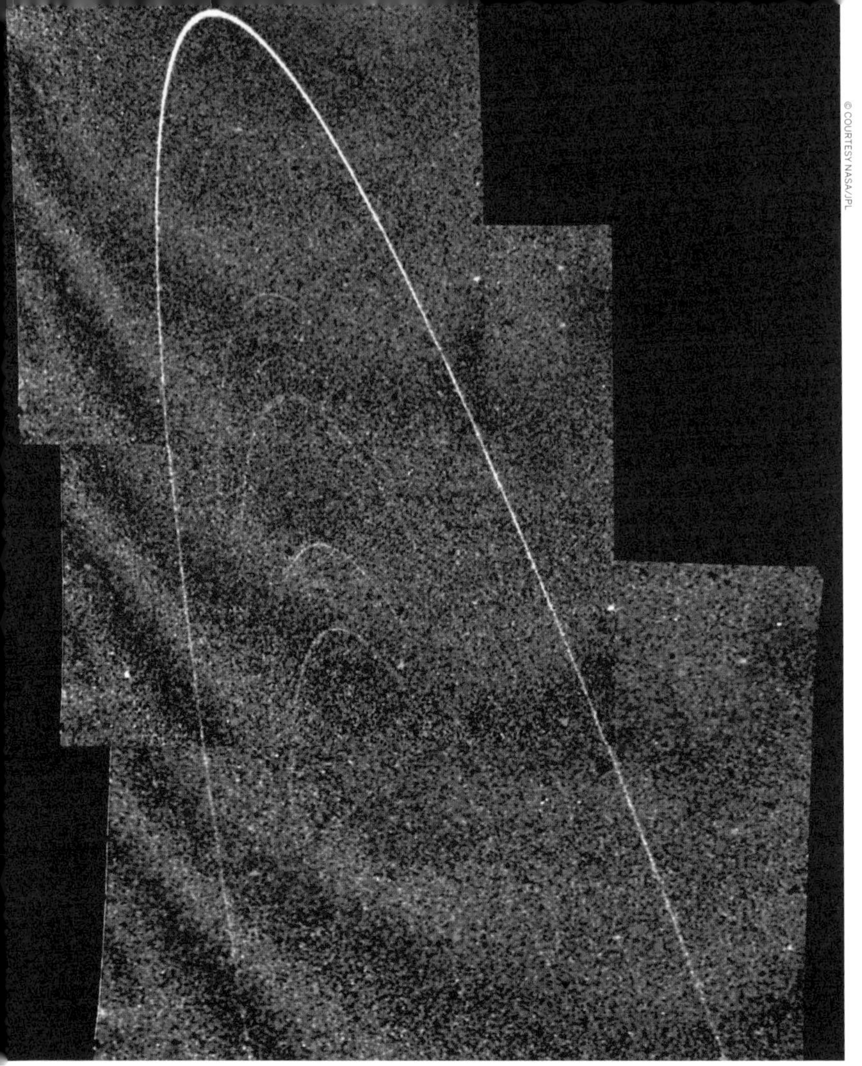

Uranus' rings are so faint they were only discovered very recently.

Ring Systems

Uranus has three distinct ring systems: an inner and two outer ring systems.

Uranus' inner system of nine rings consists mostly of narrow, dark grey bands. Then there are two outer ring systems: the innermost one is reddish, like dusty rings elsewhere in the solar system, and the outer ring is blue.

In total, Uranus has 13 known rings. In order of increasing distance from the planet, the rings are called Zeta, 6, 5, 4, Alpha, Beta, Eta, Gamma, Delta, Lambda, Epsilon, Nu and Mu. Some of the larger rings are surrounded by belts of fine dust.

The Citizen Astronomers: William & Caroline Herschel

Here's a good pub quiz question for you: which was the first planet found with the aid of a telescope?

Answer? Uranus. The planet was discovered in 1781 by the inveterate amateur astronomer Friedrich Wilhelm 'William' Herschel, although he thought it was either a comet or a star at first. When it was confirmed, Uranus became the first 'new' planet discovered since ancient times.

Herschel was originally a musician by training, but later discovered a passion for astronomy. He began constructing his own telescopes in the late 18th century (including a giant four-story refractor started in 1785). Sometimes he even made his own lenses, which he used to study Mars, double stars, star clusters and other deep sky objects, and to discover two moons of Saturn.

His companion in most of his work was his sister Caroline, who catalogued his observations, polished telescope mirrors, and organised his notes. Eventually she made astronomical discoveries of her own, especially comets, finding at least eight, along with nebulae. She also made an independent discovery of M110, a companion of the Andromeda galaxy, and became the first woman to be awarded a Gold Medal by the British Royal Astronomical Society.

William Herschel, meanwhile, became something of a celebrity and was Court Astronomer. Later, Herschel's name was immortalised on a space telescope mission, jointly funded by ESA and NASA.

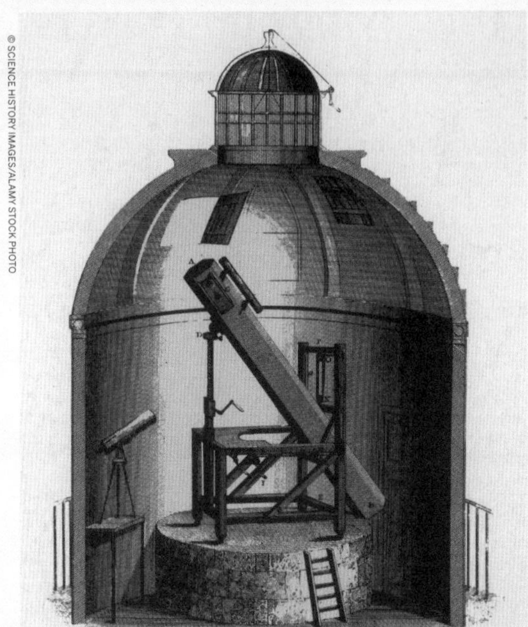

William Herschel's telescope.

Portrait of astronomer William Herschel.

A rendering of some of Uranus' 27 moons.

Uranus' Moons

While most of the solar system's satellites take their names from Greek or Roman mythology, Uranus' moons are unique in being named for characters from the works of William Shakespeare and Alexander Pope.

Oberon and Titania are the largest Uranian moons, and were the first to be discovered, by William Herschel in 1787. William Lassell, who had been first to see a moon orbiting Neptune, discovered the next two, Ariel and Umbriel. Nearly a century passed before Gerard Kuiper found Miranda in 1948. And that was it until a NASA robot made it to distant Uranus.

The Voyager 2 spacecraft visited the Uranian system in 1986 and found an additional 10 moons, all just 26 to 154 km (16 to 96 mi) in diameter: Juliet, Puck, Cordelia, Ophelia, Bianca, Desdemona, Portia, Rosalind, Cressida and Belinda.

Since then, astronomers using the Hubble Space Telescope and improved ground-based telescopes have raised the total to 27 known moons.

All of Uranus' inner moons (those observed by Voyager 2) appear to be roughly half water-ice and half rock. The composition of the moons outside the orbit of Oberon remains unknown, but they are likely captured asteroids, or perhaps even centaurs, whose orbits range between Neptune and Jupiter.

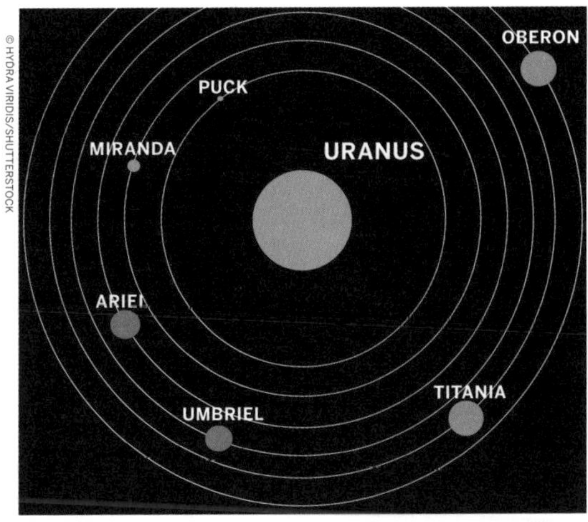

The major inner moons of Uranus.

Finding Uranus' Moons

Spotting the post-Voyager moons is an impressive feat. They're tiny – as little as 12–16 km (8–10 mi) across, and blacker than asphalt. And of course, they're about 2.9 billion km (1.8 billion mi) away from the Sun.

Uranus' Secret Moons

A study by University of Idaho researchers suggests there could be two tiny, previously undiscovered moonlets orbiting near Uranus' rings. Rob Chancia and Matt Hedman examined decades-old images of Uranus' rings taken by Voyager 2 in 1986, and compared it with data from NASA's Cassini spacecraft, currently orbiting Saturn. The scientists found a pattern in Uranus' rings similar to moon-related structures in Saturn's rings, called moonlet wakes. They estimate these moonlets to be around 4–14 km (2–9 mi) in diameter – although new images from future telescopes or spacecraft will be needed to fully confirm their existence.

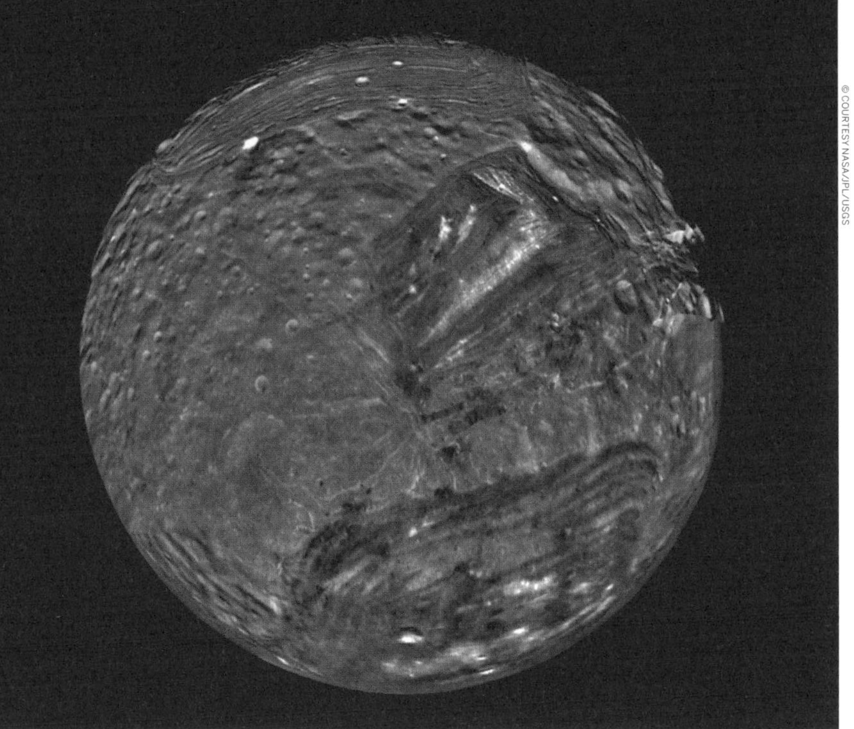

A mosaic of the surface of Uranus' moon Miranda.

Miranda

Named after Prospero's daughter in Shakespeare's The Tempest, *Miranda is the innermost and smallest of the five major satellites. Its fissured surface is split by giant canyons, including some that are 12 times deeper than the Grand Canyon.*

Like Frankenstein's monster, Miranda looks like it was pieced together from parts that didn't quite merge properly. At about 500 km (310 mi) in diameter, it's only one-seventh as large as Earth's moon, a size that seems unlikely to support much tectonic activity, and yet Miranda is notable for three large features known as 'coronae' – cratered areas of ridges and valleys, separated from the more heavily cratered (and presumably older) terrain by sharp boundaries, like mismatched patches on a moth-eaten coat.

Scientists disagree about what processes are responsible for these strange features. One possibility is that the moon may have been smashed apart in some colossal collision, and the pieces then haphazardly reassembled. Another, perhaps more likely, scenario is that the coronae are sites of large rocky or metallic meteorite strikes which partially melted the icy subsurface and resulted in episodic periods of slushy water rising to Miranda's surface and refreezing.

The moon was discovered in telescopic photos of the Uranian system by Gerard P Kuiper on 16 February 1948 at the McDonald Observatory in western Texas.

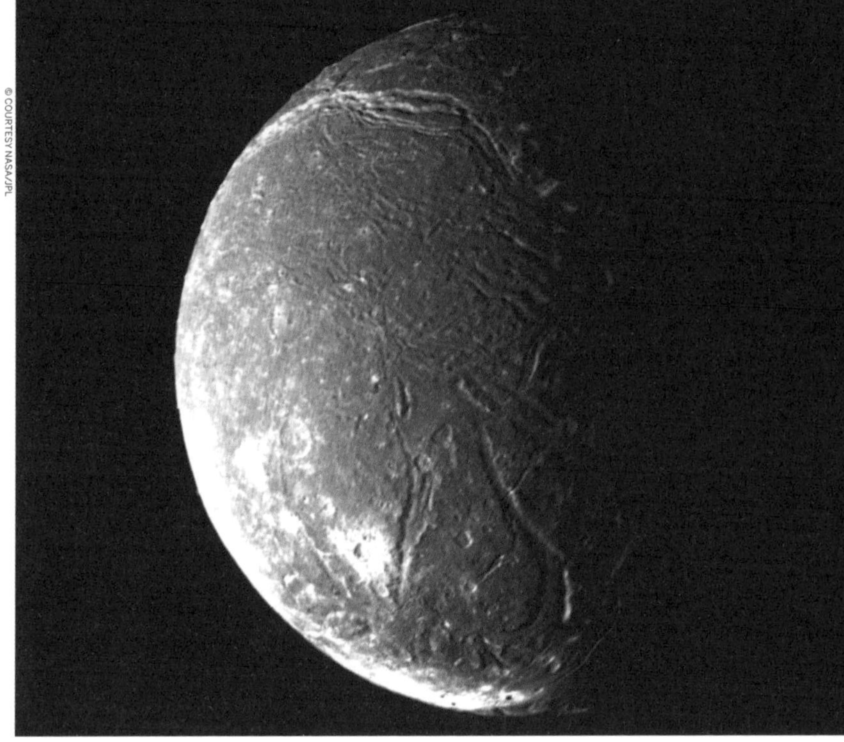

This is the most detailed mosaic of Ariel that exists, compiled from Voyager 2 images.

Ariel

Ariel's surface appears to be the youngest of all the moons of Uranus. It has few large craters but many small ones, indicating that recent collisions have probably wiped out the large craters that would have been left by much earlier, bigger strikes.

Ariel was discovered on 24 October 1851 by William Lassell, one of 19th-century England's grand amateur astronomers, who used the fortune he made in the brewery business to finance his telescopes. It has the brightest surface of the five largest Uranian moons, and is also thought to have had the most recent geologic activity. It is transected by grabens, or fault-bounded valleys.

Not much more is known about this distant moon, except for the fact that its brightness increases dramatically when it is in opposition – that is, when the observer is directly between it and the Sun. This indicates its surface is porous, casting shadows that reduce reflectivity when illuminated at other angles. Like Uranus' other moons, its surface reflects only a third of the sunlight that strikes it, suggesting the moon's surface has been darkened by a carbonaceous material – perhaps the result of eons of micrometeorite strikes tilling the soil.

Ariel is the name of a character in both Shakespeare's *The Tempest* and Pope's poem *The Rape of the Lock.*

Umbriel, one of Uranus' largest moons, is seen half in shadow here.

Umbriel

Umbriel is ancient, and by far the darkest of the five large moons. It has many old, large craters and sports a mysterious bright ring on one side. The moon is named for a malevolent spirit in Pope's poem The Rape of the Lock.

Along with Ariel, Umbriel was discovered on 24 October 1851 by English astronomer William Lassell. With a low albedo, it reflects only 16% of the light that strikes its surface, similar to the highland areas of Earth's moon. Other Uranian moons are much brighter, although the process by which Umbriel's ancient cratered surface has been darkened remains, for now, a mystery. Umbriel has a diameter of about 1200 km (750 mi), making it the third largest of Uranus' moons; Ariel has a very similar size. Images taken by Voyager 2 in 1986 revealed a curious bright ring about 140 km (90 mi) in diameter on the moon's dark surface. It is unclear what created the distinctive ring, although it may be frost deposits associated with an impact crater. Umbriel may have a rocky core composing about 50% of its mass, with the rest an icy outer shell. Spectographic analysis of the moon reveals that water and carbon dioxide are present, with possible methane deposits.

Voyager 2 captured this image of Oberon in its flyby.

Oberon

The outermost and second-largest of the major moons of Uranus was discovered on 11 January 1787 by William Herschel. Like all of Uranus' large moons, Oberon is roughly half ice and half rock, and has at least one large mountain that rises about 6 km (almost 4 mi) off the surface.

Little was known about this moon until Voyager 2 passed it during its flyby of Uranus in January 1986. Oberon is heavily cratered, similar to Umbriel – especially so when compared to three other major moons of Uranus: Ariel, Titania and Miranda. It is approximately 8.4 times smaller than Earth.

Oberon is old and shows little sign of internal activity, though tectonic activity was once likely based on the evidence of endogenic resurfacing, i.e., from below the surface. An unidentified dark material appears on the floors of many of its craters, while its (seemingly) lone mountain is among the largest in the entire solar system.

Keeping up the literary theme, the moon was named by Herschel's son, John, after the king of the fairies in Shakespeare's *A Midsummer Night's Dream*.

Voyager 2 captured this image of Titania. Further details on its surface will need to wait for another mission.

Titania

Appropriately enough for a moon named after the Queen of the Fairies in The Tempest, *Titania is the largest moon in orbit around Uranus. Like its sister moons, it was discovered by the ever-inquisitive astronomer William Herschel.*

Titania was discovered on 11 January 1787. It and Oberon were the first of Uranus' moons to be discovered, due to their size. Seeing objects as far away as Uranus, and on a much smaller scale, would be no easy feat if you didn't even know to look for them! Images taken by Voyager 2 almost 200 years after Titania's discovery revealed signs that the moon was geologically active. A prominent system of fault valleys, some nearly 1609 km (1000 mi) long, is visible near the moon's terminator, or shadow line. The troughs break the crust in two directions, an indication of some tectonic extension of Titania's crust. Deposits of highly reflective material, which may represent frost, can be seen along the Sun-facing valley walls. The moon is about 1600 km (1000 mi) in diameter. The neutral grey colour of Titania is typical of most of the significant Uranian moons.

An artist's conception of the view of Uranus from its satellite Puck.

Uranus' Shepherd Moons

Alongside the big five (Oberon, Titania, Ariel, Umbriel and Miranda), Uranus has another twenty-two moons, at least as far as we know.

Among the probable captured asteroids orbiting at a great distance, Uranus has some moons with a much more distinct purpose. Cordelia and Ophelia are two small shepherd moons that keep Uranus' thin, outermost 'epsilon' ring well defined through the power of their gravitational pull. Of the moons known to orbit Uranus, Cordelia is closest to the planet. Neither its size nor its albedo have been measured directly, but assuming an albedo of 0.07 like Puck, its surface probably consists of the dark, unprocessed, carbon-rich material found on the C-class of asteroids.

Between them and Miranda is a swarm of eight small satellites that are quite unlike any other system of planetary moons. In fact, this region is so crowded that astronomers don't yet understand how the little moons have managed to avoid crashing into each other. They may be shepherds for the planet's 10 narrow rings. Scientists think there must be still more moons to discover around Uranus, most likely orbiting inside the orbits of any of the moons discovered so far. These may confine the edges of Uranus' inner rings.

NEPTUNE

A Voyager 2 image from the spacecraft's flyby.

Neptune at a Glance

When you're planning your jaunt around the solar system, Neptune is unlikely to figure very high on your 'must-visit' list, but its moons make a compelling case for stopping in.

The eighth and most distant planet in our solar system, icy Neptune is more than 30 times as far away from the Sun as Earth. So not only will it take an awfully long time to get there, it's also the kind of place that will more than likely make you wish you'd listened to the advice of the interstellar travel agent and stayed at home. First off, it's dark. Very dark. Neptune's distance from the Sun means that it's stuck in a permanent state of murk: light on Earth is roughly 900 times brighter than on Neptune, and high noon on the big blue planet would seem like a dim twilight to us. That means it's probably not a place you'd be wise to visit if you're a Sun-worshipper or a sufferer of Seasonal Affective Disorder.

Second, it's cold. Really, really cold. Average temperatures on Neptune are beyond bitter, barely exceeding much above -200°C (-392°F). Hardly a prime destination for sunbathing, in other words.

And third, it's windy. And we don't mean just a bit windy: supersonically, apocalyptically windy. Earth's strongest twisters are mere breezes in comparison to the winds on Neptune, which whip across the planet at speeds of more than 2000 km/h (1200mph) – nine times stronger than those found on Earth.

In fact, pretty much the best thing you can say about Neptune is that it's a striking colour: a bright, brilliant, cerulean blue. The planet's vibrant colour is caused by clouds of methane swirling around its atmosphere, which is also responsible for giving neighbouring Uranus its blue-green tint – although as yet, scientists aren't quite sure what's causing Neptune's distinctly brighter, clear-blue hue.

So while it looks very pretty in pictures, it's probably best experienced from afar, especially given its distance and the long delay before any next mission.

A glimpse of Neptune from Voyager 2.

DISTANCE FROM SUN
30 AU

LIGHT-TIME TO THE SUN
248.99 minutes

LENGTH OF DAY
16 hours

LENGTH OF ORBITAL YEAR
60,190 Earth days (165 Earth years)

ATMOSPHERE
Molecular hydrogen, atomic helium and methane

Top Tip

Neptune is the windiest world in our solar system, with gales of frozen methane speeding across the planet at close to the speed of the fastest fighter jet. So our advice: pack a windbreaker. Preferably one made of industrial-strength concrete.

Getting There & Away

With an average distance of 4.5 billion km (2.8 billion mi), Neptune is 30 astronomical units from the Sun. From this distance, it takes sunlight 4 hours to travel from the Sun to Neptune. By comparison, Voyager 2 travelled for 12 years at an average velocity of about 19 km/s (42,000 mph) to reach Neptune.

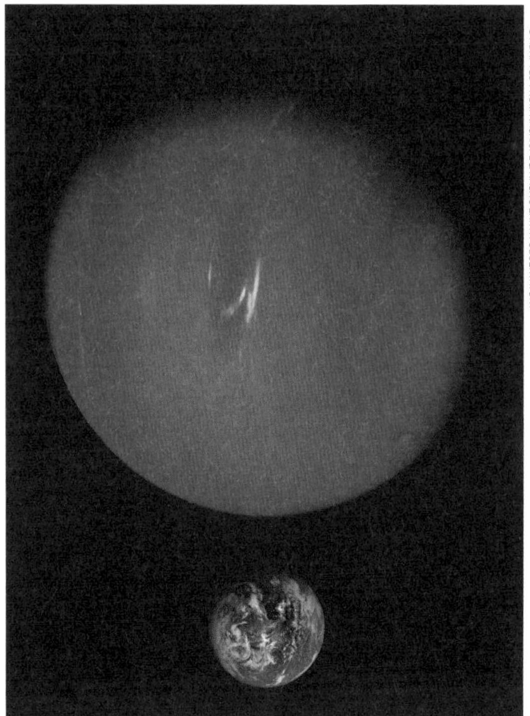

© ADAPTED FROM NASA/JPL & NASA'S EARTH OBSERVATORY

This comparative image shows Neptune and Earth side-by-side.

Neptune *vs* Earth

-- Radius --
4x EARTH

-- Mass --
17.15x EARTH

-- Volume --
57.7x EARTH

-- Surface gravity --
1.1x EARTH

-- Mean temperature --
-202°C (-334°F)
COLDER than **EARTH**

-- Surface area --
15x EARTH

-- Surface pressure --
Unknown

-- Density --
30% EARTH

-- Orbit velocity --
18% EARTH

-- Orbit distance --
30x EARTH

Orientation

Neptune is about four times wider than Earth. Due to its distance from the Sun, it has much less light to go by, however. The warm light we see on Earth is roughly 900 times as bright as the sunlight that reaches Neptune.

One day on Neptune takes about 16 hours, but one Neptunian year lasts about 165 Earth years (60,190 Earth days). In 2011 Neptune completed its first 165-year orbit since its discovery in 1846. Sometimes Neptune is even farther from the Sun than dwarf planet Pluto. Pluto's highly eccentric, oval-shaped orbit brings it inside Neptune's orbit for a 20-year period every 248 Earth years. Pluto can never crash into Neptune, though, because for every three laps Neptune takes around the Sun, Pluto makes two, a repeating pattern preventing close approaches of the two bodies. Such long orbits have their benefits when it comes to reducing collision risk.

Neptune's axis of rotation is tilted 28 degrees with respect to its orbit around the Sun, similar to the axial tilts of Mars and Earth. The tilt causes seasons on the planet, each of which lasts for over 40 years.

Magnetosphere

The main axis of Neptune's magnetic field is tipped over by about 47 degrees compared with the planet's rotation axis, similar to Uranus, whose magnetic axis is tilted about 60 degrees from the axis of rotation. This means that magnetic reconnection and auroras can appear all across the planet — not just close to the poles, like on Earth, Jupiter and Saturn.

Neptune's magnetosphere undergoes wild variations during each rotation because of this misalignment. According to scientific models, it's believed to have a strongly asymmetric shape, bulged out on one side. The magnetic field of Neptune is about 27 times more powerful than that of Earth. With potential missions to visit the gas giant in planning for the 2030s, there's still much to learn.

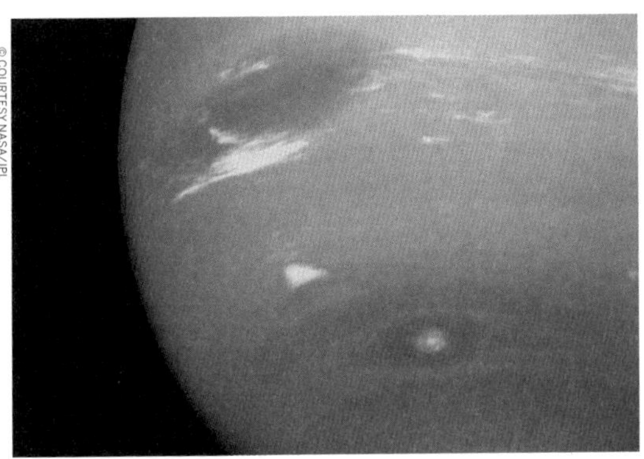

Neptune's storms rage over vast swaths of the planet before fading away.

5 Facts About Neptune

1. Neptune is named after the Roman god of the sea; appropriate given its blue colouring.

2. Neptune is the only planet in our solar system not visible to the naked eye.

3. Neptune's 165-year orbit spans roughly five generations of human life on Earth.

4. Neptune's magnetic field is about 27 times more powerful than Earth's.

5. Neptune has at least five rings, but they're very hard to see. Starting near the planet and moving outward, they are named Galle, Leverrier, Lassell, Arago and Adams. The rings are thought to be relatively young and short-lived.

The Neptunian Vortices

In 1989, along with the Great Dark Spot, Voyager 2 also took pictures of several other features on Neptune, including a bright smudge next to the Great Dark Spot; another fast-moving bright feature known to scientists as 'Scooter'; and a smaller dark spot (sometimes called Dark Spot 2) estimated to be about the size of the Earth's moon.

In 2016, the Hubble Space Telescope confirmed the presence of a new dark vortex in Neptune's atmosphere, the first seen in the 21st century. Like the original Great Dark Spot, such vortices are high-pressure systems, and are usually accompanied by bright 'companion clouds', formed when the flow of ambient air is diverted upward, causing gases to freeze into methane ice crystals.

The Gathering Storm

The 'Great Dark Spot' mentioned above was large enough to contain the entire Earth. That storm has since disappeared, but new ones have appeared on different parts of the planet. It's unclear how these storms form. But like Jupiter's Great Red Spot, the dark vortices swirl in an anti-cyclonic direction and seem to dredge up material from deeper levels in the ice giant's atmosphere. The vortices probably develop deeper in Neptune's atmosphere.

History

Neptune took shape around the same time as the rest of the solar system formed about 4.5 billion years ago, when gravity pulled swirling gas and dust in to form this ice giant. Like its neighbour Uranus, Neptune likely formed closer to the Sun and moved to the outer solar system about 4 billion years ago.

But just why did Neptune and the less-distant Uranus come into existence as such icy worlds? The answer is (mostly) as simple as it seems: further out in the solar system, it's much colder. While the planets, moons, comets and other bodies formed 4.5 billion years ago from the disc of gas and dust that surrounded our Sun, space closer in was hotter and drier than the space farther from the Sun, which was cold enough for water to condense. The dividing line, called the 'frost line', sat around Jupiter's present-day orbit. Even today, this is the approximate distance from the Sun at which the ice on most comets begins to melt and become 'active'. Objects that formed outside the frost line where water could condense and become solid had much more ice (variously composed of water, carbon dioxide, carbon, methane, and so on) and less rock and metals.

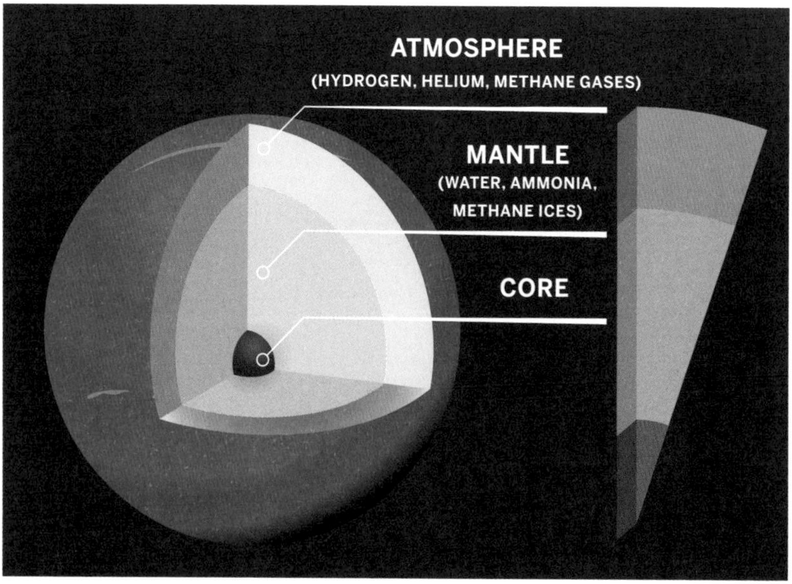

ATMOSPHERE
(HYDROGEN, HELIUM, METHANE GASES)
MANTLE
(WATER, AMMONIA, METHANE ICES)
CORE

Neptune's core is estimated to be about the same size as Earth.

Discovery Timeline

1612
Galileo incorrectly records Neptune as a fixed star during observations with his telescope.

1846
Using mathematical calculations, astronomers discover Neptune, increasing the number of known planets to eight. Neptune's largest moon, Triton, is found the same year.

1983
Pioneer 10 crosses the orbit of Neptune, the first human-made object to travel beyond the orbits of the planets.

1984
Astronomers find evidence for the existence of a ring system around Neptune.

1989
Voyager 2 becomes the first and only spacecraft to visit Neptune, passing about 4800 km (2983 mi) above the planet's north pole.

2002
Astronomers discover four new moons orbiting Neptune: Laomedeia, Neso, Sao and Halimede.

2003
Another moon, Psamathe, is discovered using ground-based telescopes.

2005
Scientists using the Keck Observatory find that some of the outer ring arcs have deteriorated.

2011
Neptune completes its first 165-year orbit of the Sun since its discovery in 1846.

2013
A previously unknown 14th moon of Neptune, Hippocamp, is discovered in Hubble images.

2016
Scientists using Hubble find a new dark spot on Neptune, a swirling vortex of high-speed winds and clouds.

Keck Observatory and Subaru Telescope on Mauna Kea.

Exploring Neptune

The ice giant Neptune was the first planet located through mathematical calculations. In fact, since Pluto's demotion, it's the only planet in the solar system that's not visible to the naked eye.

The first human to spot it, predictably, was Galileo; he recorded it as a fixed star during observations with his small telescope in 1612 and 1613. More than 200 years later, French mathematician Urbain Joseph Le Verrier proposed that a yet-unknown planet could be the cause of strange variations in the orbit of Uranus. Le Verrier sent his predictions to Johann Gottfried Galle at the Berlin Observatory, who found Neptune on his first night of searching in 1846. Seventeen days later, Neptune's largest moon, Triton, was discovered as well.

More than 140 years later, in 1989 Voyager 2 became the first – and only – spacecraft to study Neptune up close. Voyager returned a wealth of information about Neptune and its moons, and confirmed evidence the giant world had faint rings like the other gas planets. Scientists have also used the Hubble Space Telescope and powerful ground-based telescopes to gather information about this distant planet. Hubble uncovered a dark storm in September 2018 in Neptune's northern hemisphere. The storm is roughly 10,900 km (6800 mi) wide, accompanied by bright companion clouds.

Le Verrier being received by King Louise-Philippe after discovering Neptune.

A view of Hubble Space Telescope in orbit.

Neptune in Pop Culture

Even though Neptune is the furthest planet from our Sun, Neptune is a frequent stop-off in pop culture and fiction. The planet serves as the backdrop for the 1997 science fiction horror film *Event Horizon*, while in the cartoon series *Futurama*, the character Robot Santa Claus is based on Neptune's north pole. *Doctor Who* fans will remember that the episode 'Sleep No More' is set on a space station orbiting Neptune. And in the pilot episode of *Star Trek: Enterprise*, 'Broken Bow', viewers learn that at warp 4.5 speed, it is possible to fly to Neptune and back to Earth in six minutes (at least with superluminal spacecraft propulsion available to power your craft).

King Neptune presides over Virginia Beach in this sculpture by Paul DiPasquale.

Neptune Highlights

Neptune's Surface & Atmosphere

1 Fly down through clouds of hydrogen, helium and methane until it slowly merges into Neptune's swirling 'surface'.

Neptune's Rings

2 Delve into the dark rings of Neptune, so faint that they are almost invisible.

Proteus

3 Peer into the darkness of another of Neptune's moons, discovered by the Voyager 2 probe.

Triton

4 Venture onto the icy surface of Neptune's largest and coldest moon.

Nereid

5 Among the largest of Neptune's moons, its eccentric orbit suggests it may be a captured asteroid.

Neptune's Other Moons

6 Neptune's other satellites are shrouded in mystery; some have immense orbits and others are on collision courses.

An artist's impression of Neptune, seen from one of its moons.

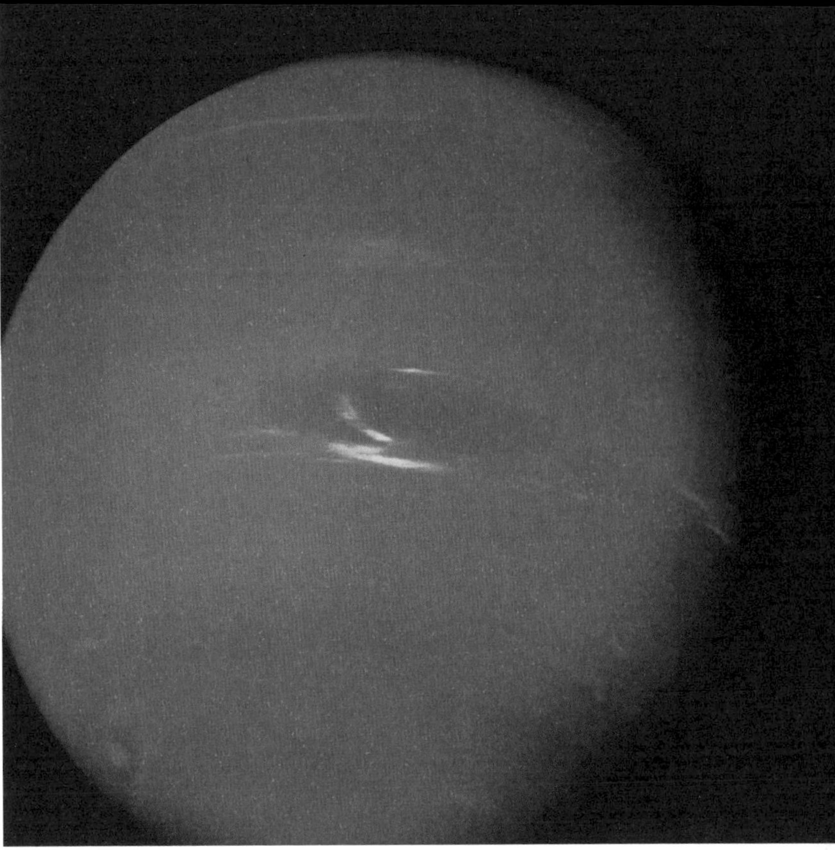

A view of Neptune captured by Voyager 2.

Neptune's Surface & Atmosphere

Similar to Uranus, Neptune is made of a thick soup of water, ammonia, and methane over an Earth-sized solid centre, while its atmosphere is mainly composed of hydrogen, helium and methane.

Along with Uranus, Neptune is one of two 'ice giants' in the outer solar system. More than 80% of the planet's mass is made up of a hot, dense fluid of 'icy' materials (water, methane and ammonia) swirling around a small, rocky core. Some scientists think that there may be an ocean of super hot water under Neptune's cold clouds, which doesn't boil away because of the planet's incredibly high pressure, keeping it locked tight inside.

While distinguishing between the surface, inner core or mantle and atmosphere is not a clear-cut delineation, as the ice giant's interior gradually intensifies pressure further in without having one solid surface, scientists do know that Neptune's atmosphere experiences powerful storms, similar to the Great Red Spot on Jupiter and driven by Neptune's powerful high-velocity winds. These whip across the planet at a speedy pace.

Neptune's arced rings.

Neptune's Rings

Like several other planets in the outer solar system, Neptune has its own system of rings, though you might have to squint to see them.

Starting near the planet and moving outward, the most prominent of these rings are named Galle, Leverrier, Lassell, Arago and Adams. The rings are thought to be relatively young and short-lived. Unlike the rings of Saturn, which contain some relatively large rock particles, Neptune's rings are believed to be composed of microscopic dust, which shows up as a dark material; it's hypothesised to be the same size as smoke particles, and gives the rings a strikingly dark appearance. They're more similar to Jupiter's rings than those of Saturn or Uranus, with a yet-unknown origin.

They also display peculiar clumps of dust called arcs. Four prominent arcs named Liberté (Liberty), Egalité (Equality), Fraternité (Fraternity) and Courage are in the outermost ring, Adams. The arcs are strange because the laws of motion would predict that they would spread out evenly rather than stay clumped together. Scientists now think the gravitational effects of Galatea, a moon just inward from the ring, stabilises these arcs.

Proteus in orbit around Neptune.

Proteus

Discovered in 1989 by Voyager 2, aptly named Proteus is more than six times smaller than its largest known sibling, Titan.

At just 420 km (260 mi) across, Proteus is eight times smaller than Earth's moon. It has an odd box-like shape, though if it had just a little more mass gravity would transform Proteus into a more spherical shape. The moon orbits Neptune about every 27 hours. Though heavily cratered, it shows no sign of geological modification from internal forces. Circling the planet in the same direction as Neptune rotates, tidally locked Proteus remains close to Neptune's equatorial plane.

Proteus is also one of the darkest objects in our solar system. Like Saturn's moon Phoebe, Proteus reflects only 6% of the sunlight that hits it. This is also referred to as a low albedo object.

Originally designated S/1989 N 1, Proteus was later named after the shape-changing sea god of Greek mythology. Proteus' discovery was unusual since a smaller moon, Nereid, had been discovered 33 years earlier using an Earth-based telescope. Proteus was most likely overlooked because it is so dark and the distance between Earth and Neptune is so great.

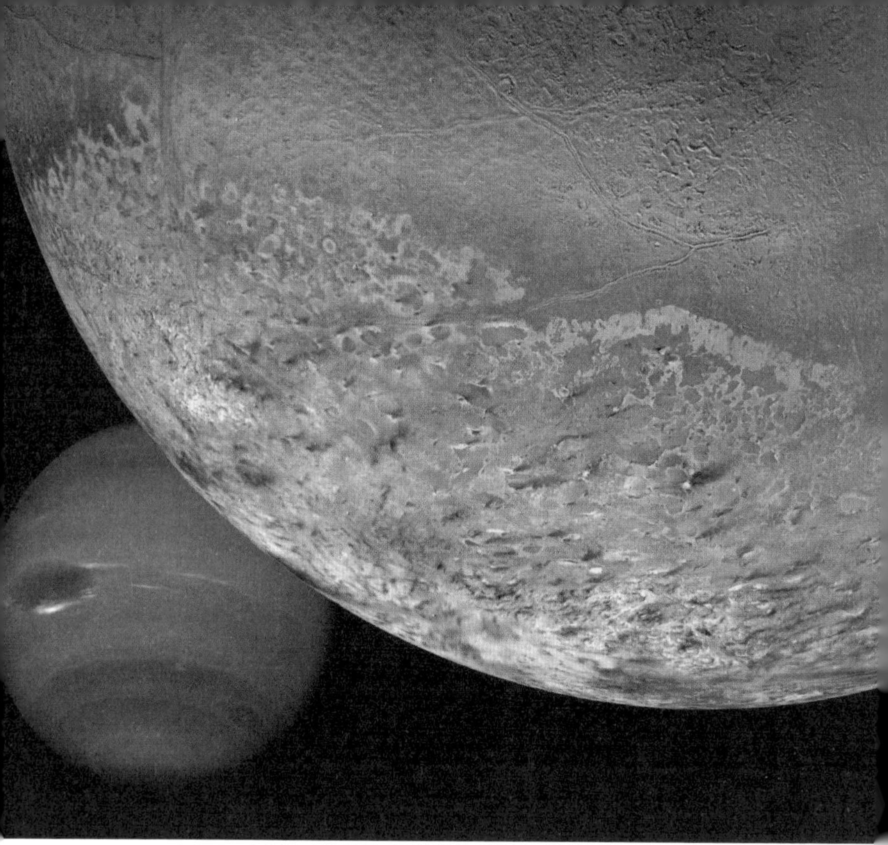

A computer-generated montage shows Neptune in the distance past Triton.

Triton

Neptune has 14 known moons, but one stands above them all. The largest of the moons is Triton, discovered on 10 October 1846 by William Lassell, just 17 days after Johann Gottfried Galle discovered the planet.

Triton is the only large moon in the solar system that has a retrograde orbit, which means it circles its planet in a direction opposite to the planet's rotation. This suggests that Triton may once have been an independent object captured by Neptune's orbit, most likely originating in the Kuiper Belt.

Triton is extremely cold, with surface temperatures around -235°C (-391°F). Despite this perpetual deep freeze, Voyager 2 discovered geysers spewing icy material upward more than 8 km (5 mi). Triton, Io and Venus are the only bodies in the solar system besides Earth that are known to be volcanically active at the present time.

Spacecraft images show the moon has a sparsely cratered surface with smooth volcanic plains, mounds and round pits formed by icy lava flows. Ice volcanoes spout a mixture of liquid nitrogen, methane and dust, which instantly freezes and then snows back down to the surface. One dramatic Voyager 2 image shows a frosty plume shooting 8 km (5 mi) into the sky and drifting 140 km (87 mi) downwind.

This volcanic activity is probably the origin of Triton's thin atmosphere. It has been detected from Earth several times since, and is growing warmer, although scientists do not yet know why. Even if warming continues, the planet will likely stay at quite a frosty temperature.

A colour photo of Triton.

5 Facts About Triton

1. Triton has a diameter of 2700 km (1680 mi).

2. The moon is named after the son of Poseidon (the Greek god comparable to the Roman Neptune).

3. Until the discovery of the second moon Nereid in 1949, Triton was commonly known as simply 'the satellite of Neptune'.

4. Triton has a density about twice that of water. It is one of the densest satellites in the solar system; only Europa and Io are more dense.

5. Like our own moon, Triton is locked in synchronous rotation with Neptune (one side faces the planet at all times). But because of its orbital inclination, both polar regions take turns facing the Sun.

Frozen Triton

Triton is one of the coldest objects in our solar system: Voyager 2 detected surface temperatures of -235°C (-391°F). It is so cold that most of Triton's nitrogen is condensed as frost, giving its surface an icy sheen that reflects 70% of the sunlight that hits it.

Finding Triton

We don't know with what beverage William Lassell may have celebrated his discovery of Neptune's moon Triton, but beer certainly made it possible. Lassell was one of 19th century England's grand amateur astronomers, and financed his telescopes with the fortune he made in the brewery business. He spotted Triton on 10 October 1846, just 17 days after a Berlin observatory discovered Neptune. Curiously, a week before he found the satellite, Lassell thought he saw a ring around the planet. That turned out to be a distortion caused by his telescope. But when NASA's Voyager 2 visited Neptune in 1989, it revealed that the gas giant does have rings, though they're far too faint to have been observed by Lassell.

© RON MILLER/STOCKTREK IMAGES

Nereid is so distant from Neptune that it's speculated to be a captured asteroid.

Nereid

Nereid is one of the outermost of Neptune's known moons, most notable for its bizarre and lengthy orbit.

Nereid has one of the most eccentric orbits of any moon in our solar system. It is so far away from Neptune that it requires 360 Earth days to make a single orbit – suggesting Nereid may be a captured asteroid or Kuiper Belt object, or perhaps that it was greatly disturbed during the capture of Neptune's largest moon, Triton.

The moon is named after the Nereids, or sea nymphs, which feature often in Greek mythology. It was discovered on 1 May 1949 by noted astronomer Gerard P Kuiper (of Kuiper Belt fame) with a ground-based telescope, the last satellite of Neptune to be discovered before Voyager 2's discoveries in its flyby four decades later. It is 37.5 times smaller than Earth, and around 10 times smaller than our moon.

Nearly everything that is known about Nereid is based on a series of images taken by Voyager 2 as it flew past between 20 April and 19 August 1989, with a closest distance of about 4.7 million km (2.9 million mi). The object is so small and irregular that it's hard to catch sight of from Earth.

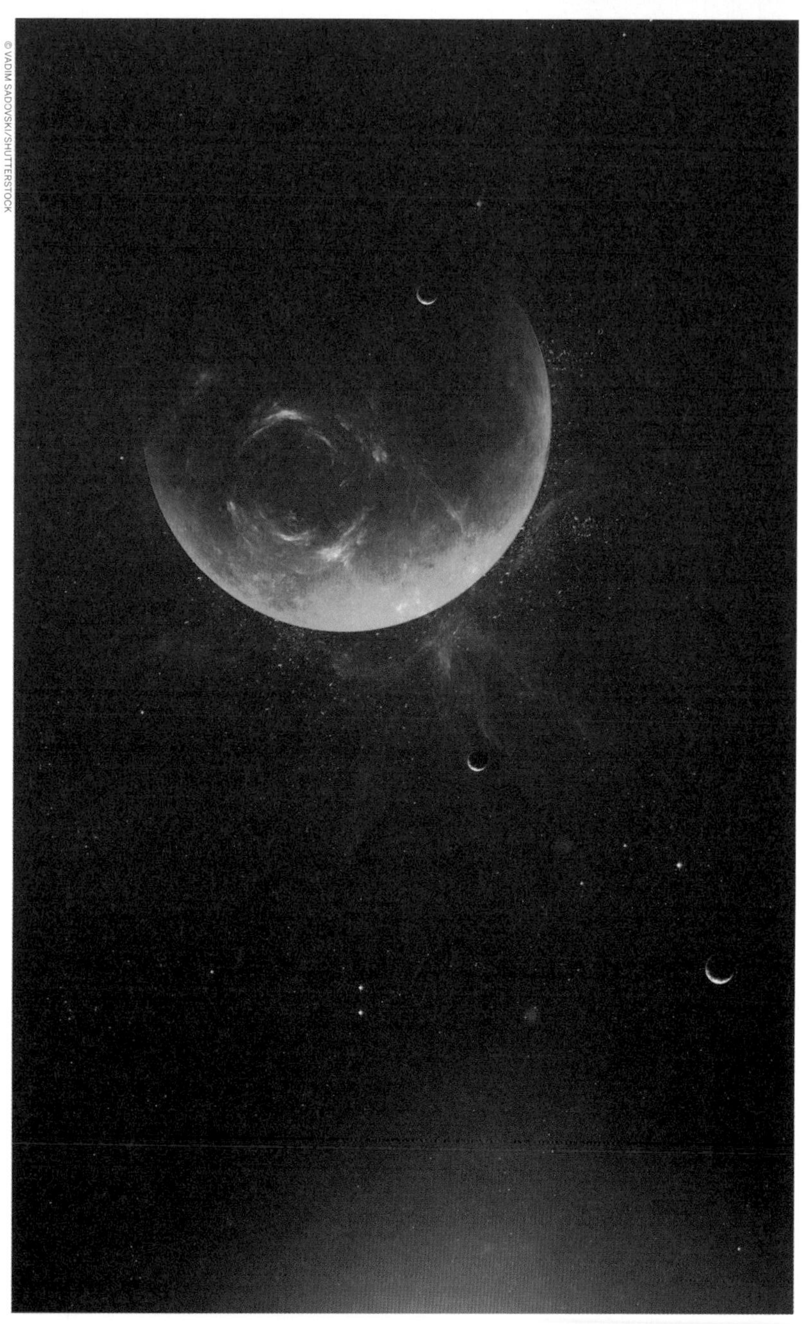

© VADIM SADOVSKI/SHUTTERSTOCK

Neptune surrounded by its satellites.

Cerro Tololo Observatory in Chile was the discovery site of small Neso.

Neptune's Other Moons

Several more moons are in orbit around Neptune, but most are so small and distant that they are hard to see from Earth, visible only to the largest and most powerful telescopes.

Several other moons were discovered by the Voyager 2 science team in 1989, including Galatea, Despina and Larissa. Small, irregularly shaped and heavily cratered, very little is known about these distant satellites other than their general orbits and their proximity to Neptune's faint ring system. Two other moons, Naiad and Thalassa, were also seen by Voyager; their close orbit means they may eventually crash into Neptune's atmosphere or be torn apart and form planetary rings, as others before them may have.

Additional moons have since been discovered from Earth using powerful telescopes. For example, tiny Neso, discovered in 2002 using the 4m Blanco telescope at Cerro Tololo Observatory in Chile. Or Psamathe, discovered by Hawaii's Mauna Kea Observatory in 2003. Both moons are known for their huge orbits; Neso is the moon that is furthest from its planet. Scientists think they may be fragments left over from the breakup of a larger moon billions of years ago.

Other moons discovered in this way include Halimede, Laomedeia and Sao – although it is possible there are more moons in orbit around Neptune that are still to be discovered, as suggested by the provisional discovery of S/2004 N1 in 2013. Only time will tell.

© MIKOLA/SHUTTERSTOCK

Neptune's New Moon

On 1 July 2013, astronomer Mark Showalter of the SETI Institute in Mountain View, California, analysed over 150 archival photographs of the Neptune system taken by the Hubble Space Telescope (HST) between 2004 and 2009. Showalter discovered a white dot which appeared over and over again, suggesting there was a new, as yet-undiscovered moon spinning around Neptune, somewhere between the orbits of Larissa and Proteus. With a mean radius of about 17 km (10 mi), it is much smaller than any of Neptune's previously known satellites, so small and dim that it is roughly 100 million times fainter than the faintest star that can be seen with the naked eye. Its tiny size means it was far below the detection threshold of the Voyager cameras sent there in 1989. After several years of analysis, a team of planetary scientists using NASA's Hubble Space Telescope at last came up with an explanation for a mysterious moon around Neptune that they discovered with Hubble in 2013.

The tiny moon, named Hippocamp, is unusually close to a much larger Neptunian moon called Proteus. Normally, a moon like Proteus should have gravitationally swept aside or swallowed the smaller moon while clearing out its orbital path.

So why does the tiny moon exist? Hippocamp is likely a chipped-off piece of the larger moon that resulted from a collision with a comet billions of years ago. The diminutive moon, only 34 km (20 mi) across, is one thousandth the mass of Proteus, which is about 418 km (260 mi) across. 'The first thing we realised was that you wouldn't expect to find such a tiny moon right next to Neptune's biggest inner moon', said Mark Showalter. 'In the distant past, given the slow migration outward of the larger moon, Proteus was once where Hippocamp is now.'

This scenario is supported by Voyager 2 images from 1989 that show a large impact crater on Proteus, almost large enough to have shattered the moon. 'In 1989, we thought the crater was the end of the story,' said Showalter. 'With Hubble, now we know that a little piece of Proteus got left behind and we see it today as Hippocamp.' The orbits of the two moons are now about 12,070 km (7500 mi) apart.

Neptune's satellite system has a violent and tortured history. Many billions of years ago, Neptune potentially captured the large moon Triton from the Kuiper Belt, a large region of icy and rocky objects beyond the orbit of Neptune. Triton's gravity would have torn up Neptune's original satellite system. Triton settled into a circular orbit and the debris from shattered Neptunian moons re-coalesced into a second generation of natural satellites. However, comet bombardment continued to tear things up, leading to the birth of Hippocamp, which might be called a third-generation satellite.

'Based on estimates of comet populations, we know that other moons in the outer solar system have been hit by comets, smashed apart, and re-accreted multiple times', noted Jack Lissauer of NASA's Ames Research Center in Silicon Valley, a coauthor on the new research. 'This pair of satellites provides a dramatic illustration that moons are sometimes broken apart by comets.'

Indeed, many moons of the solar system stem from violent collisions in the past, sometimes visible in their scarred surface as well as in their orbit paths.

An illustration of a meteroid entering Earth's atmosphere, debris from a passing comet or asteroid.

Asteroids, Dwarf Planets and Comets:

Non-Planetary Solar System Objects

At the very beginning of our solar system, a massive swirling cloud of dust and gas circled the young Sun. Not just the planets grew out of this debris; asteroids, dwarf planets and comets all derived from it as well.

The dust particles in this disc collided with each other and formed into larger bits of rock, and the process continued until they reached the size of boulders. Eventually, this process of accretion formed the planets of our solar system, accruing more mass and matter to become the eight planets we know today.

Yet billions of other objects in the solar system never aggregated to become planets under our current definition of the term. Some became smaller-than-planetary objects which behaved somewhat, but not entirely, like planets; others clustered together in smaller masses, orbiting the Sun on unique paths or following larger solar system objects on their orbits; others still began unique celestial journeys with a massively varied range of paths and behaviours. These dwarf planets, asteroids and comets comprise the remaining objects that we know exist in the solar system.

Amazingly, many of these mysterious objects have been altered very little in the 4.6 billion years since they first formed. Their relatively pristine state make these dwarf planets, comets and asteroids wonderful storytellers with much to share about what conditions were like in the early solar system. They can reveal secrets about our origins, chronicling the processes and events that led to the birth of our world. They might offer clues about where the water and raw materials that made life possible on Earth came from.

Top Highlights

Pluto

1 Though no longer classified as a planet, dwarf planet Pluto retains a special place of honour among solar system objects.

Kuiper Belt

2 Beyond the orbit of Neptune, the frigid Kuiper Belt contains a multitude of dwarf planets, icy comets and asteroids.

Comet Shoemaker-Levy 9

3 Shoemaker-Levy's impact with Jupiter was the first time that such a collision was witnessed.

Asteroid Belt

4 Stranded between Mars and Jupiter, this belt of rocky debris holds between one and two million objects over 1 km.

'Oumuamua

5 A traveller from interstellar space, small, speedy 'Oumuamua is just one of many objects of interstellar origin to pass unnoticed through the solar system each year.

Ceres

6 Ceres is the only dwarf planet located within the asteroid belt, and makes up 25% of the asteroid belt's overall mass.

Haumea

7 Dwarf planet Haumea in the Kuiper Belt is distinctive for its ring and moons.

Comet Hale-Bopp

8 A long-period comet, Hale–Bopp was visible to the naked eye for 18 months on its last flyby.

Eris

9 The discovery of this dwarf planet caused a reevaluation of how objects in the solar system were classified.

Comet ISON

10 Sungrazer comet ISON disintegrated in its path by the Sun, but not before giving astronomers a bevy of new information about comet behaviour.

Comet Chariklo

11 This comet is the first ever discovered to have its own ring system.

Oort Cloud

12 This speculative cloud of objects three light years from the Sun may be a source of long-period comets.

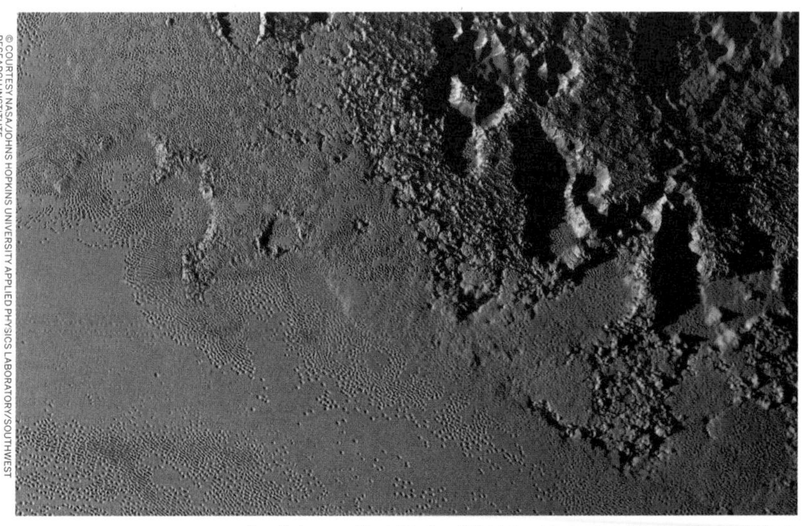

New Horizons sent back this view of Pluto's ice plains bordering the Krun Macula highlands.

The Asteroid Belt & Asteroids

Asteroids, sometimes called minor planets or planetoids, are the rocky remnants left over from the early days of our solar system. Many, but by no means all, dwell in the eponymous asteroid belt. They can range in size from Vesta – the largest at about 530 km (329 mi) in diameter – to bodies less than 10 m (33 ft) across. The total mass of all the asteroids combined is less than Earth's moon!

Most asteroids are irregularly shaped, though a few are nearly spherical, and they are often pitted or cratered. As they revolve around the Sun in elliptical orbits, the asteroids also rotate, sometimes quite erratically, tumbling as they go. More than 150 asteroids are known to have a small companion moon (some have two moons). There are also binary (double) asteroids, in which two rocky bodies of roughly equal size orbit each other, as well as triple asteroid systems.

The majority of known asteroids can be found orbiting the Sun within the asteroid belt between Mars and Jupiter, generally with rather normal elliptical orbits. The asteroid belt is estimated to contain between 1.1 and 1.9 million asteroids over than one kilometre in diameter, and millions of smaller ones. Early in the

history of the solar system, the gravity of newly formed Jupiter brought an end to the formation of planetary bodies in this region and caused the small bodies to collide with one another, fragmenting them into the asteroids we observe today.

Trojan asteroids are not located within the asteroid belt. Instead, these asteroids share an orbit with a larger planet, but do not collide with it because they gather around two special places in the orbit (called the L4 and L5 Lagrangian points). There, the gravitational pull from the Sun and the planet are balanced, overcoming a trojan's tendency to fly out of the orbit. The Jupiter trojans form the most significant population of trojan asteroids. It is thought that they are as numerous as the asteroids in the asteroid belt. Jupiter's massive gravity and occasional close encounters with another object changes the asteroids' orbits, knocking them out of the main belt and hurling them into space in all directions across the orbits of the other planets. Stray asteroids and asteroid fragments slammed into Earth and the other planets in the past, playing a major role in the geological history of the planets and in the evolution of life on Earth.

An artist's view of the early asteroid belt.

There are also Mars and Neptune trojans, and the first Earth trojan was discovered in 2011. Called 2010 TK7, it preceded Earth in our orbit around the Sun. Trojan 2010 TK7 has an extreme orbit that takes the asteroid far above and below the plane of Earth's orbit, in a motion referred to as an epicycle. In addition, the asteroid moves within Earth's orbit, circling horizontally around its stable point every 395 years.

Asteroids that actually cross Earth's orbital path are known as Earth-crossers. There are over 10,000 known near-Earth asteroids, and over 1,500 of them have been classified as potentially hazardous – those that could pose a threat to Earth. Scientists continuously monitor Earth-crossing asteroids and near-Earth asteroids that approach Earth's orbital distance to within about 45 million km (28 million mi), which may potentially pose an impact danger. Radar is a valuable tool in detecting and monitoring potential impact hazards. By reflecting transmitted radar signals off asteroids, scientists can learn a great deal about an asteroid's orbit, rotation, size, shape, and metal concentration – and be aware if their path ever poses a serious threat to Earth.

Small by space standards, Bennu is still taller than the Empire State Building and Eiffel Tower.

Bennu

Considered a potentially hazardous object because its orbit comes so close to Earth, asteroid Bennu will soon be one of the best-studied asteroids, helping us understand the history of the solar system.

Born from the rubble of a violent collision, hurled through space for millions of years and dismembered by the gravity of planets, asteroid Bennu had a tough life in a rough neighbourhood: the early solar system. Now, Bennu orbits the Sun between Mars and Earth, making scientists wary that it may someday get a bit too close for comfort. In the meantime, researchers want to learn all they can about this neighbouring asteroid, which carries a trove of data.

'We are going to Bennu because we want to know what it has witnessed over the course of its evolution', said Edward Beshore of the University of Arizona, Deputy Principal Investigator for the asteroid-sample-return mission OSIRIS-REx. Bennu was selected as the target because it is both old and well-preserved – but also close enough to Earth that a sample-return mission is possible within a few years. 'Bennu's experiences will tell us more about where our solar system came from and how it evolved. On planets like Earth, the original materials have been profoundly altered by geologic activity and chemical reactions with our atmosphere and water. We think Bennu may be relatively unchanged, so this asteroid is like a time capsule for us to examine', said Beshore.

Bennu may also harbour organic material from the young solar system. Organic material, composed of molecules containing primarily carbon and hydrogen atoms, is fundamental to terrestrial life. The analysis of any such material found on Bennu will give scientists a partial inventory of what was present at the beginning of the solar system that may have had a role in the origin of life.

RADIUS
246 m (807 ft)

MASS
6.0–7.8 × 10^{10} kg

COMPOSITION
Carbon, rocks and minerals

DISCOVERED
1999

ORBITAL PERIOD
1.2 years

Artist's image of OSIRIS-REx on the surface of Bennu.

OSIRIS-REx

The OSIRIS-REx (the Origins, Spectral Interpretation, Resource Identification, Security – Regolith Explorer) mission offers scientists a unique opportunity to study an asteroid and gain clues about the history of our solar system. The mission launched toward Bennu in late 2016, arrived at the asteroid in 2018 and will return a sample of Bennu's surface to Earth in 2023. Once it returns, the OSIRIS-REx team will be able to examine some of the most pristine asteroid regolith material that exists.

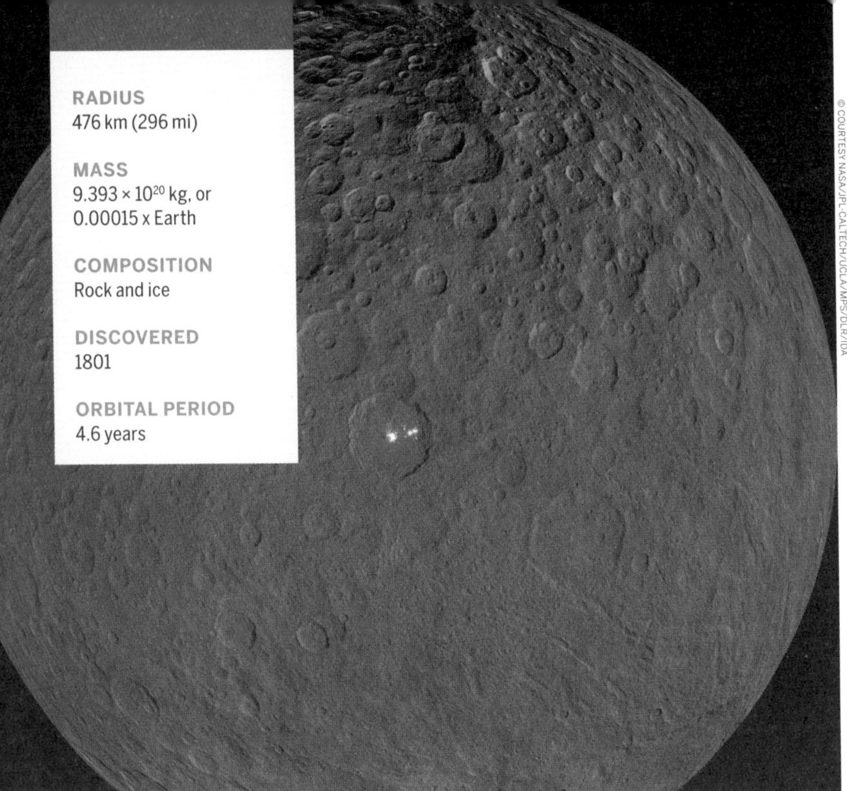

Ceres is a rare dwarf planet located within the asteroid belt.

Ceres

The largest object in the asteroid belt and the first to be discovered (in 1801), dwarf planet Ceres has delighted scientists with evidence of geologic activity and organic molecules.

Ceres may be the largest object in the main asteroid belt, but don't let its location fool you: this relatively large object is actually considered a dwarf planet. Yet though it comprises 25% of the asteroid belt's mass, this dwarf planet is still 14 times smaller than Pluto. In the centuries following its discovery, Ceres was considered an asteroid due to its composition and orbital path, but in 2006 it was reclassified among the first batch of dwarf planets along with Pluto and Eris. Technically, Ceres is still an asteroid, but calling it a dwarf planet is a more precise description.

In 2015, the Dawn spacecraft arrived at Ceres after wrapping up its observation of the asteroid Vesta elsewhere in the asteroid belt. Since then, Dawn has photographed Ceres extensively from multiple altitudes, revealing a dramatic landscape with geologic activity and features including volcanoes, water-ice and salt deposits that suggest that Ceres may once have had oceans. Though its mission ended in late 2018 having used the last of its fuel, Dawn will continue to orbit Ceres for decades, though it won't be able to transmit any additional data back to Earth.

RADIUS
302 km (108 mi)

MASS
Unknown

COMPOSITION
Unknown

DISCOVERED
1997

ORBITAL PERIOD
63 years

Chariklo's rings as seen in an artist's impression.

Chariklo

Stretching the definitions of an asteroid, Chariklo continues to surprise astronomers as they learn more about its unusual features – including the two smallest rings in the solar system.

Chariklo is an asteroid that lives outside the asteroid belt - but astronomers continue to debate whether it should be considered an asteroid at all. Chariklo is the largest of the 'centaurs', which are asteroids in the outer solar system. Some say that Chariklo may actually qualify as a dwarf planet.

Situated between Saturn and Uranus, Chariklo caught astronomers' attention in 2013 when they discovered it has at least two rings, similar to those of Saturn. Astronomers first discovered Chariklo's rings by examining dips in its brightness and used that data to determine that despite its relatively large size, Chariklo is the smallest known object to have rings. Before the discovery of rings around Chariklo, astronomers were unsure that small objects in the solar system were stable enough to have rings.

Planetary astronomers use computer simulations to investigate how Chariklo's unexpected ring system might have formed, how it survives and, given the asteroid's low mass and close passes of other small asteroids and the planet Uranus, how long it may last.

A Quarantids meteor streaks past a star trail field outside San Jose, California.

EH1

With its discovery in 2003, asteroid EH1 answered astronomers' questions about the source of one meteor shower and the lifecycle of comets.

For decades, astronomers pondered about the cause of the Quadrantids meteor shower (see sidebar), the first meteor shower of the year on the Gregorian calendar. With a brief but intense peak lasting only a few hours – rather than up to a few days as with other meteor showers – astronomers studied these meteors, which originate from a now-obsolete constellation named Quadrans Muralis between the constellations of Bootes and Draco, to try to determine what their parent object might be. Observable from a so-called circumpolar radiant, most northern hemisphere sites can view them, while points to the south cannot. Yet what causes them to streak across the sky each year?

The answer may be an asteroid that used to be a comet. In 2003, astronomers in the LONEOS program at Anderson Mesa Station near Flagstaff, Arizona, discovered a near-Earth object, and Peter Jenniskens of SETI identified the object as a likely candidate to be the Quadrantid's parent. Named EH1, this asteroid was quickly identified as the parent of the

Quadrantids based on its orbit and orbital period. Astronomers also posit, based on observations, that EH1 was likely formed on the breakup of comet C/1490 Y1, believed to have broken apart around 1490 based on Chinese records of an observed meteor shower at that time. EH1 is now considered an extinct comet, similar to Phaethon. Confused yet?

In asteroid terms, this means it was once a comet composed of a rocky core covered with ice. Over the millennia, EH1 shed its icy layers, leaving a trail of debris that causes the Quadrantids today – and making the object into an asteroid. The object's composition is the main factor in its definition as one or the other: comets are made of rocky material, ice and dust, while asteroids are made of rocky material and minerals, typically forming too close to the Sun for ice to be present. Both orbit the Sun; indeed, it's when comets approach the Sun that they gain their tail, made of vaporised ice and dust.

Armed with this knowledge, researchers now hypothesise that EH1 will continue its orbit as an asteroid, and over time – tens of thousands, if not millions, of years – the Quadrantids will eventually taper off. Tracing the origins and future path of meteor showers can be a complicated puzzle, but we know more now than ever before about what creates the awesome displays of meteor showers.

Flagstaff's Lowell Observatory at Anderson Mesa Station.

RADIUS
2 km (1.2 mi)

MASS
Unknown

COMPOSITION
Rocky

DISCOVERED
2003

ORBITAL PERIOD
5.52 years

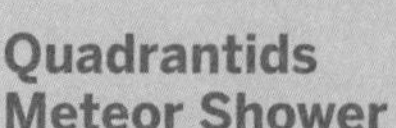

Quadrantids Meteor Shower

Many cultures celebrate the start of each Gregorian calendar year with fireworks, but did you know there's a show of celestial pyrotechnics you can watch instead? At their brief peak around 3 January each year, the Quadrantids (pronounced KAW-dran-tids) are as active as other meteor showers during the year, such as the Perseids and Geminids. During the peak window of the Quadrantids, you may see as many as 50–100 meteors per hour, spread over a few hours rather than across the two-day peak more common for meteor showers, making it the most intense annual shower. It's noted for its bright fireball meteors, persisting longer than an average meteor streak. Observed in the northern hemisphere since at least 1825, it was for a long time unknown what the parent body of the shower might be. The constellation that gave the shower its name, meanwhile, was made obsolete by the IAU in 1922.

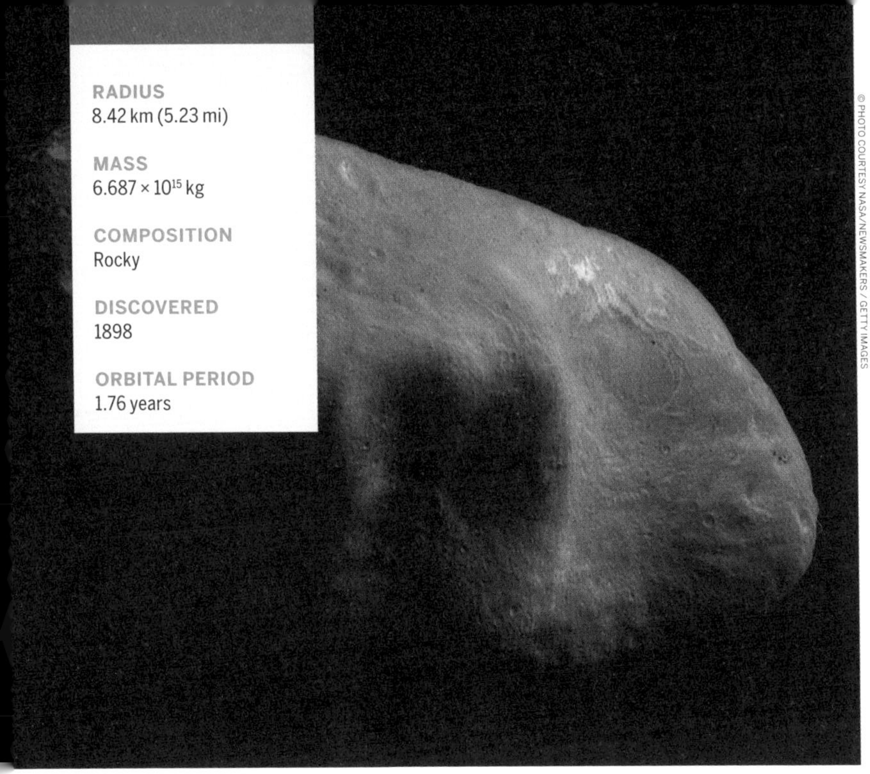

RADIUS
8.42 km (5.23 mi)

MASS
6.687×10^{15} kg

COMPOSITION
Rocky

DISCOVERED
1898

ORBITAL PERIOD
1.76 years

NEAR-Shoemaker captured this image of Eros, a near-earth asteroid in orbit near Mars.

Eros

The first near-Earth object discovered, at the turn of the 20th century, the asteroid Eros became the star of the show when a spacecraft orbited and landed there just over one hundred years later.

Perhaps because of its unusual shape, the astronomers who discovered Eros (formally 433 Eros) named it after the Greek god of love. Orbiting the Sun between Mars and Earth, Eros is considered a near-Earth object. It was the first ever discovered and the second largest yet found.

In both 1998 and 2000, the NEAR-Shoemaker spacecraft visited Eros as part of its mission to learn more about this unusually shaped rocky asteroid. On its second visit, NEAR began to orbit Eros, photographing it from every angle and allowing researchers an exceptionally detailed view of the asteroid. Over the course of its mission, NEAR helped scientists to discover that Eros is a single solid body, that its composition is nearly uniform, and that it formed during the early years of our solar system. Mysteries remain, however, including why some rocks on the surface have disintegrated.

On February 2001, the NEAR mission drew to a dramatic close as it was crash-landed onto Eros' surface. Unlike some missions that end on impact, NEAR survived well enough to return analysis of the composition of the surface regolith, a layer of loose, heterogeneous deposits on the surface, before the end of its mission.

Ida and its moon Dactyl, to scale.

Ida

Amid evidence of repeated cosmic collisions over millions of years, asteroid Ida surprised astronomers when its tiny moon was discovered during a Galileo flyby.

Ida, officially named 243 Ida, appears to be a relatively normal asteroid that orbits the Sun from a position in the outer asteroid belt. Part of a family of asteroids that astronomers believe formed roughly two billion years ago, Ida appears to have formed when two or more separate large objects collided softly and stuck together. As it turns out, Ida's shape was not the only unusual thing astronomers have discovered about the asteroid.

During a flyby en route to Jupiter in 1993, robot spacecraft Galileo encountered and photographed Ida. Those photos revealed a surprise: Ida has a tiny moon, which they named Dactyl. Dactyl is the first moon of an asteroid ever discovered and measures a mere 1.4 km (0.9 mi) in diameter, or about 1/20th Ida's size. Astronomers believe Dactyl was formed through one of the many cosmic collisions Ida has experienced over the eons.

Armed with the knowledge that asteroids could have moons, more than 150 'asteroid moons' have since been discovered by scientists. Some of these moons are partners to near-Earth objects we've been studying for centuries; there are even a few objects that have two moons.

A visualisation of JAXA's Hayabusa probe at 25143 Itokawa.

Itokawa

A mass of rocks and ice held together by gravity, asteroid Itokawa wanders through the inner solar system – and may someday cross paths with Earth.

Examine a photo of asteroid 25143 Itokawa, a small, sub-kilometre near-Earth object, and you'll be struck by a question: Where are its craters? The JAXA robot probe Hayabusa approached the Earth-crossing asteroid in 2005 and pictures showed a surface unlike any other solar system body yet photographed – one possibly devoid of craters.

The leading hypothesis for the lack of common circular indentations is that asteroid Itokawa is effectively a rubble pile: a bunch of rocks and ice chunks only loosely held together by a small amount of gravity. If so, craters might not form so easily, or are filled in whenever the asteroid gets jiggled by a nearby planet's gravity or struck by a massive meteor. Earth-based observations of asteroid Itokawa have shown that one part of the interior even has a higher average density than the other part, another unexpected discovery.

Soil samples returned to Earth by the Hayabusa mission show that while Itokawa is made of the same materials as other solar system objects, the physics that explain its formation and shape are relatively unique.

RADIUS
~2.9 km (1.8 mi)

MASS
Unknown

COMPOSITION
Rocky

DISCOVERED
1983

ORBITAL PERIOD
1.433 years

A star trail during the Geminids meteor shower.

Phaethon

Named for the mythological figure who was unable to drive the Sun's chariot, asteroid Phaethon's orbit volleys between a close Sun approach and travelling to the edge of the inner solar system.

Officially known as 3200 Phaethon, this is an asteroid with an unusual orbit: at times it passes within the orbit of Mercury and at other times travels beyond the orbit of Mars. With this kind of massively elliptical orbit, Phaethon behaves more like a comet than an asteroid. This orbit has led some astronomers to speculate that Phaethon may have once been a comet which shed its icy layers to form the rocky asteroid we see today. If this hypothesis proves true, it would explain Phaethon's unusual orbit and the fact that Phaethon leaves behind a trail of dust and pebbles, which is responsible for the Geminids meteor shower in December each year. The debris might be cometary vapour from heated water-ice and dust.

In late 2017, astronomers at Arecibo Observatory – which monitors and studies thousands of solar system objects – released new images that reveal more about Phaethon's shape. 'These new observations of Phaethon show it may be similar in shape to asteroid Bennu', said Patrick Taylor, a Universities Space Research Association (USRA) scientist and group leader for Planetary Radar at Arecibo. 'More than 1000 Bennus could fit inside of Phaethon'.

Geminids Meteor Shower

The last major meteor shower each year is one of the best. In fact, the Geminids meteor shower, which occurs from 4–17 December each year, is growing stronger and will likely outpace all other meteor showers in coming years. The Geminids typically peak on 14–15 December, and you can expect to see 100–200 meteors per hour during that time. The Geminids are one of only two meteor showers thought to be caused by an asteroid; the Quadrantids in January are caused by the asteroid EH 1.

Pictured is the possible Psyche mission.

Psyche

Possibly the exposed core of an early planet, or a unique metal mass orbiting in the asteroid belt, 16 Psyche is interesting enough that researchers have a mission planned to answer this question and others within the next decade.

One of the ten largest asteroids in the asteroid belt, that fact alone makes 16 Psyche interesting to astronomers. What makes it unique is that it appears to be the exposed nickel-iron core of an early planet, one of the building blocks of our solar system. Deep within rocky, terrestrial planets – including Earth – scientists infer the presence of metallic cores, but these lie unreachably far below the planets' rocky mantles and crusts. Because we cannot see or measure Earth's core directly, 16 Psyche offers a unique window into the violent history of collisions and accretion that created terrestrial planets.

NASA currently plans to launch the Psyche spacecraft to visit the asteroid which shares its name in 2022. The scientific goals of the Psyche mission are to understand the building blocks of planet formation and to explore first-hand a wholly new and unexplored type of world. The mission team seeks to determine whether Psyche is the core of an early planet, how old it is, whether it formed in similar ways to Earth's core, and what its surface is like. If the Psyche mission launches as planned, we will learn more about this asteroid in 2026–27 when the spacecraft is orbiting the asteroid.

RADIUS
312.5 km (194.2 mi)

MASS
2.59076 $\times 10^{20}$ kg

COMPOSITION
Rocky

DISCOVERED
1807

ORBITAL PERIOD
3.63 years

The asteroid belt's Vesta gives clues to ancient protoplanetary objects.

Vesta

The largest of the millions of main belt asteroids, and the only one bright enough to be visible to the naked eye, asteroid Vesta allows astronomers a view of the earliest stages of planetary formation.

Vesta is the largest protoplanet in the belt between Mars and Jupiter. The brightest object in the asteroid belt, it also makes up about 10% of the entire mass of the main asteroid belt. For more than two centuries after its discovery, this mysterious object appeared as little more than a fuzzy patch of light among the stars, and it was long believed to be an asteroid. Vesta beckoned, but its invitation was not answered until NASA's Dawn mission arrived in July 2011. During a 14-month orbit, Dawn examined Vesta from every angle, providing high-resolution photos and measurements that showed this purported asteroid was actually a leftover protoplanet.

So what does the surface of protoplanet Vesta look like? A small terrestrial world, rocky and dense like Mars and Mercury, Vesta has been battered since the birth of the solar system and is covered with craters, bulges, grooves and cliffs. Studying Vesta thus gives researchers insight into the formative years of our early solar system, as this unusual world may be one of the largest remaining protoplanets.

Kuiper Belt

© PAUL FLEET/GETTY IMAGES

The Kuiper Belt is a large region in the cold, outer reaches of our solar system beyond the orbit of Neptune, sometimes called the 'third zone' of the solar system. Astronomers think there are millions of small, icy objects in this region, including hundreds of thousands that are larger than 100 km (62 mi) wide. Some, including Pluto, are over 1000 km (620 mi) wide. In addition to rock and water-ice, objects in the Kuiper Belt can also contain a variety of other frozen compounds like ammonia and methane.

The region is named for astronomer Gerard Kuiper, who published a scientific paper in 1951 that speculated about objects beyond Pluto. Astronomer Kenneth Edgeworth also mentioned objects beyond Pluto in papers he published in the 1940s, and thus it's sometimes referred to as the Edgeworth-Kuiper Belt. Some researchers prefer to call it the Trans-Neptunian Region, and refer to Kuiper Belt objects (KBOs) as Trans-Neptunian objects, or TNOs. Whatever your preferred term is, the belt occupies an enormous volume in our planetary system, and the small worlds that inhabit it have a lot to tell us about the solar system's early history.

The Kuiper Belt is one of the largest structures in our solar system – others being the Oort Cloud, the Sun's heliosphere, and the magnetosphere of Jupiter. Its overall shape is like a puffed-up disc, or doughnut. Its inner edge begins at the orbit of Neptune, at about 30 AU from the Sun. (That's 30 times the distance of Earth.) The inner, main region of the Kuiper Belt extends to around 50 AU from the Sun. Overlapping the outer edge of the main part of the Belt is a second region called the scattered disc, which continues outward to nearly 1000 AU, with some bodies on irregular orbits that go even farther beyond the scattered disc.

So far, over 2000 trans-Neptunian objects have been catalogued by observers, representing only a tiny fraction of the total number of objects scientists think are out there. Despite this abundance of objects, the total mass of all the material in the Kuiper Belt is estimated to be no more than about 10% of the mass of Earth.

Astronomers think the icy objects of the Kuiper Belt are remnants left over from the formation of the solar system. Similar to the relationship between the main asteroid belt and Jupiter, the objects in the region might have come together to form a planet had Neptune not been there. Instead, Neptune's gravity stirred up this region of space so much that the small, icy objects there weren't able to coalesce into a large planet.

The amount of material in the Kuiper Belt today might be just a small fraction of what was originally there. According to one well-supported theory, the shifting orbits of the four giant planets (Jupiter, Saturn, Uranus and Neptune) could have caused most of the Kuiper Belt's original material – likely 7 to 10 times the mass of Earth – to be lost.

The basic idea is that early in the solar system's history, Uranus and Neptune were forced to orbit further from the Sun due to shifts in the orbits of Jupiter and Saturn. As they drifted further outward, they passed through the dense disc of small, icy bodies left over after the giant

A rendering of the Kuiper Belt's comets and KBOs in orbit.

planets formed. Neptune's orbit was the furthest out, and its gravity bent the paths of countless icy bodies inward toward the other giants. Jupiter ultimately catapulted most of these icy bodies either into extremely distant orbits (to form the Oort Cloud) or out of the solar system entirely. As Neptune tossed icy objects sunward, this caused its own orbit to drift even farther out, and its gravitational influence forced remaining icy objects into the range of locations where we find them in the Kuiper Belt.

Today, the Kuiper Belt is thought to be very slowly eroding itself away. Objects there occasionally collide, producing smaller KBOs (some of which may become comets), as well as dust that's blown out of the solar system by the solar wind. Over the eons, the rate of these collisions will become less and less frequent as the mass of objects in the Kuiper Belt slowly dwindles.

Fragments produced by colliding KBOs can be pushed by Neptune's gravity into orbits that send them sunward, where Jupiter further corrals them into short loops lasting 20 years or less; these are called short-period Jupiter-family comets. Given their frequent trips into the inner solar system, most tend to exhaust their volatile ices fairly quickly and eventually become dormant, or dead comets with little or no detectable activity. Researchers have found that some near-Earth asteroids, such as Phaethon or EH1 are most likely burned-out comets, and most of them would have started out in the Kuiper Belt. The Kuiper Belt is truly a frontier in space. It's a place we're still just beginning to explore and our understanding is still evolving.

Dwarf Planets

For a long time, our understanding of the solar system was simple. Large objects were either a planet, which orbited the Sun, or a moon, which orbited a planet. In 2003, this changed with the discovery of Eris, the first viable planetary candidate discovered in over 70 years, which broke prior molds. Astronomers grappled with the idea that there might be more planets than we had previously thought.

The International Astronomical Union (IAU) took on the challenge of classifying Eris. In 2006, the IAU passed a resolution that defined what a planet was – and by contrast, what a planet wasn't. This resolution established a new category of solar system object, called dwarf planets. Eris, then-planet Pluto, and Ceres, which had formerly been considered an

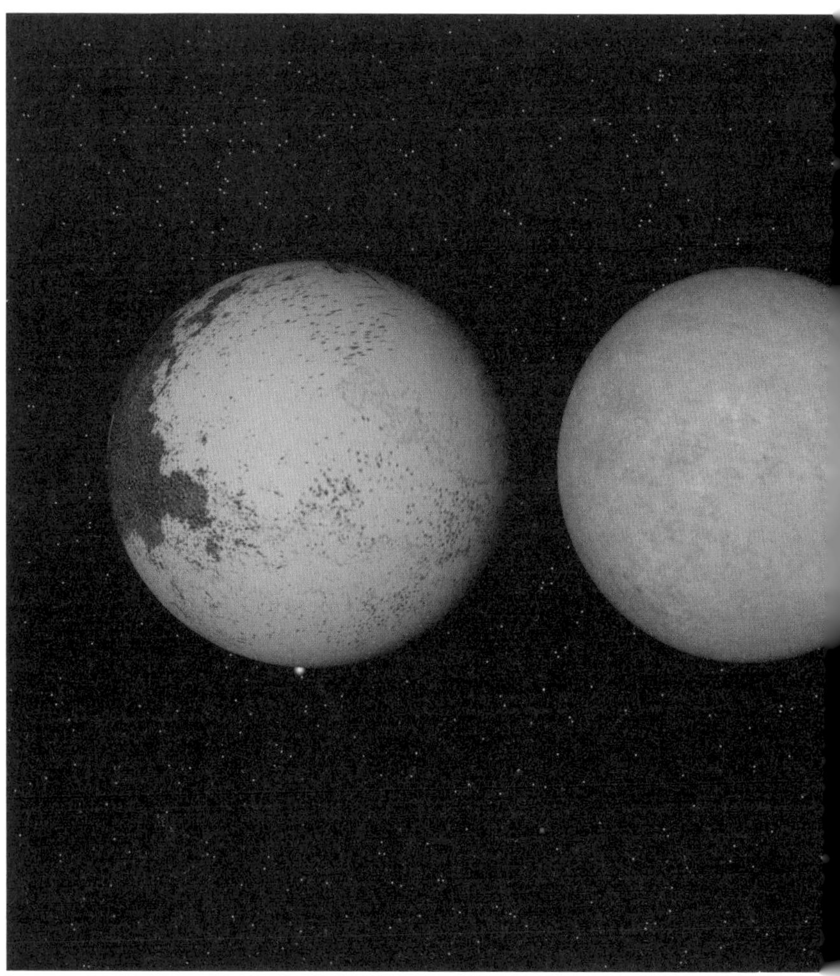

asteroid, became the first dwarf planets. Pluto was 'demoted' from its status as a planet in the process, but the dwarf planet classification better explained some of Pluto's unusual behaviours that had vexed astronomers since its discovery in 1930.

Haumea and Makemake later received the dwarf planet classification too, and round out the five dwarf planets currently recognised by the IAU. There may be another 100 dwarf planets in the solar system and hundreds more in and just outside the Kuiper Belt – and astronomers continue to discover more interesting, oddly-behaved objects in our solar system at increasingly mind-boggling distances. It turns out our solar system is a lot more interesting, and a lot bigger, than we had previously imagined.

Dwarf planets, from left: Pluto, Eris, Haumea, Makemake and Ceres.

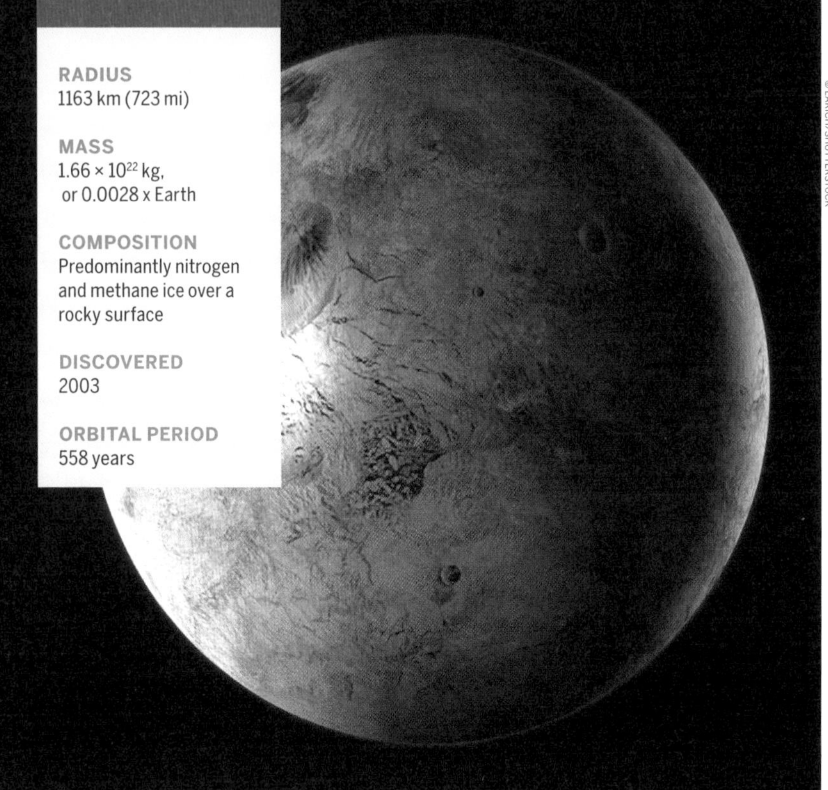

The discovery of Eris in the Kuiper Belt introduced an entirely new category of dwarf planets to astronomers.

Eris

Few newly discovered objects cause astronomers to reclassify existing ones. At its discovery, Eris became the first of two dwarf planets, redefining how we think about the objects in our solar system.

If you're looking for something to blame for Pluto's demotion to dwarf planet status, look to Eris. Discovered in 2003, astronomers measured Eris to have about 27% more mass and a larger radius than Pluto. Confronted with this knowledge, astronomers had a choice: announce the tenth planet in the solar system, or consider the unique characteristics of Pluto and Eris' and reclassify both objects something new. Thus the definition of dwarf planets was born.

Eris is on average more than three times farther from the Sun than Pluto. It is so cold out there that the dwarf planet's atmosphere has frozen onto the surface as a frosty glaze. The coating gleams brightly, reflecting as much sunlight as freshly fallen snow. The path Eris takes around the Sun is shaped like an oval rather than a circle, similar to Pluto. In about 275 years, Eris will move close enough to the Sun to thaw partially. Its icy veneer will melt away revealing a rocky, speckled landscape similar to Pluto's.

RADIUS
250 km (155 mi)

MASS
Unknown

COMPOSITION
Unknown

DISCOVERED
2018

ORBITAL PERIOD
Unknown

The Sun is just a faint light from the perspective of distant Farout.

Farout (2018 VG18)

What do you call the most distant object ever observed that is still part of our solar system? Farout, obviously.

As one of the most distant objects astronomers have found which can be considered part of our solar system, 2018 VG18, discovered by a team of astronomers using the Suburu telescope at Mauna Kea in 2018, was nicknamed Farout early on to emphasise just how far away this object is. At an average orbital distance of 120 AU, Farout is over three times farther from the Sun than Pluto on average – and over twice as far as Eris. Farout may eventually join these two objects and be classified as a dwarf planet, as astronomers believe it is roughly 500 km (310 mi) across in diameter.

While Farout is currently considered to be the most distant object observed that is part of our solar system, it is not necessarily the most distant at all times. That object is 2014 FE72, which is estimated to orbit possibly as far as 3660 AU during its 69,000-year orbit. While astronomers continue to study Farout to determine its average orbital distance and orbital period, it may not retain its title as the most distant object in the solar system. For now, it continues to be the most distant at the time of observation, but it will most likely have company soon.

An artist's conception of a distant Solar System Planet X, a possible force on the orbits of extremely distant objects like The Goblin.

The Goblin

The Goblin was nicknamed for the time of year it was discovered rather than its appearance – in fact, we still don't have a good view of The Goblin's visage, as it is one of the far distant objects in the solar system.

While 2015 TG387 may be its official name, most astronomers also know it as 'The Goblin' – a play on 'TG' and the fact that this unusual Trans-Neptunian object was discovered near Halloween. Very little has been discovered about The Goblin since astronomers in Mauna Kea, Hawaii first spotted it in 2015; it hasn't yet been classified as a dwarf planet, though it may well be big enough to earn that title.

The Goblin is one of three solar system objects considered a 'sednoid'. A sednoid is an object which is no less than 50 AU from the Sun at its closest approach (or perihelion), while half of its major axis is more than 150 AU. In short, it is an object that never gets close to the Sun, and orbits very far from it, possibly even under the influence of the gravity of an unknown planet from a distant solar system.

Only three sednoid objects are currently classified: The Goblin, minor planet Sedna and 2012 VP113, also known as Biden. The three sednoids are often grouped with other 'detached objects' in the solar system, which never interact with the planets or dwarf planets.

RADIUS
385 km (239 mi)

MASS
4.006×10^{21} kg,
or 0.0007 x Earth

COMPOSITION
Rock covered with a thin layer of ice

DISCOVERED
2003

ORBITAL PERIOD
285 years

Haumea with its ring and moons.

Haumea

The result of a solar system collision that left it spinning, oddly shaped, and wonderfully perplexing, Haumea continues to surprise astronomers.

Almost everything about Haumea is unusual. It is the fifth designated dwarf planet, after Pluto, Ceres, Eris, and Makemake, an exclusive class of solar system objects that haven't let a new member join the club since 2008 when Haumea and Makemake were added to the group.

Haumea's oblong shape also makes it quite unique. Along one axis, Haumea is significantly longer than Pluto; in another axis Haumea has an extent very similar to Pluto; while in the third axis, it is much smaller. You can imagine the puzzle this presented astronomers as they discovered and observed Haumea, before proposing the non-spherical shape we now believe it has.

But wait, there's even more that makes Haumea special: Haumea's orbit sometimes brings it closer to the Sun than Pluto, though usually it is farther away. Haumea also has two small moons discovered in 2005, named Hiʻiaka and Namaka for daughters of the Hawaiian goddess after whom this dwarf planet is named. In the past few years, astronomers have discovered that Haumea even has a ring, similar to the gas giant planets. The only question remains: what other secrets does Haumea hold?

RADIUS
357 km (222 mi)

MASS
<4.4 × 1021 kg, or
<0.0007 x Earth

COMPOSITION
Methane and possibly nitrogen ice

DISCOVERED
2005

ORBITAL PERIOD
307.5 years

An artists's rendering of dwarf planet Makemake.

Makemake

Bright, icy and incredibly distant from the Sun, Makemake was the latest – and perhaps final – object in the solar system to receive classification as a dwarf planet.

Makemake is one of the largest objects known in the outer solar system, and is one of the few objects officially designated as a dwarf planet by the IAU. It is a Kuiper belt object that is about two-thirds the size of Pluto, orbits the Sun only slightly further out than Pluto, and appears only slightly dimmer than Pluto. However, Makemake has an orbit much more tilted to the ecliptic plane of the planets than Pluto does.

Makemake is named for the creator of humanity in the Rapa Nui mythology of Easter Island. It is known to be a world somewhat red in appearance, with colours indicating that it is likely covered with patchy areas of frozen methane. Careful monitoring of the brightness drop of a distant star recently eclipsed by Makemake indicates that the dwarf planet has little atmosphere.

For several years, astronomers hypothesized that Makemake had a moon as other dwarf planets do. In 2016, they confirmed the existence of MK2, which is more than 1300 times fainter than Makemake. MK2 was first spotted orbiting approximately 21,000 km (13,049 mi) from the dwarf planet, and its diameter is estimated to be 175 km (110 mi) across.

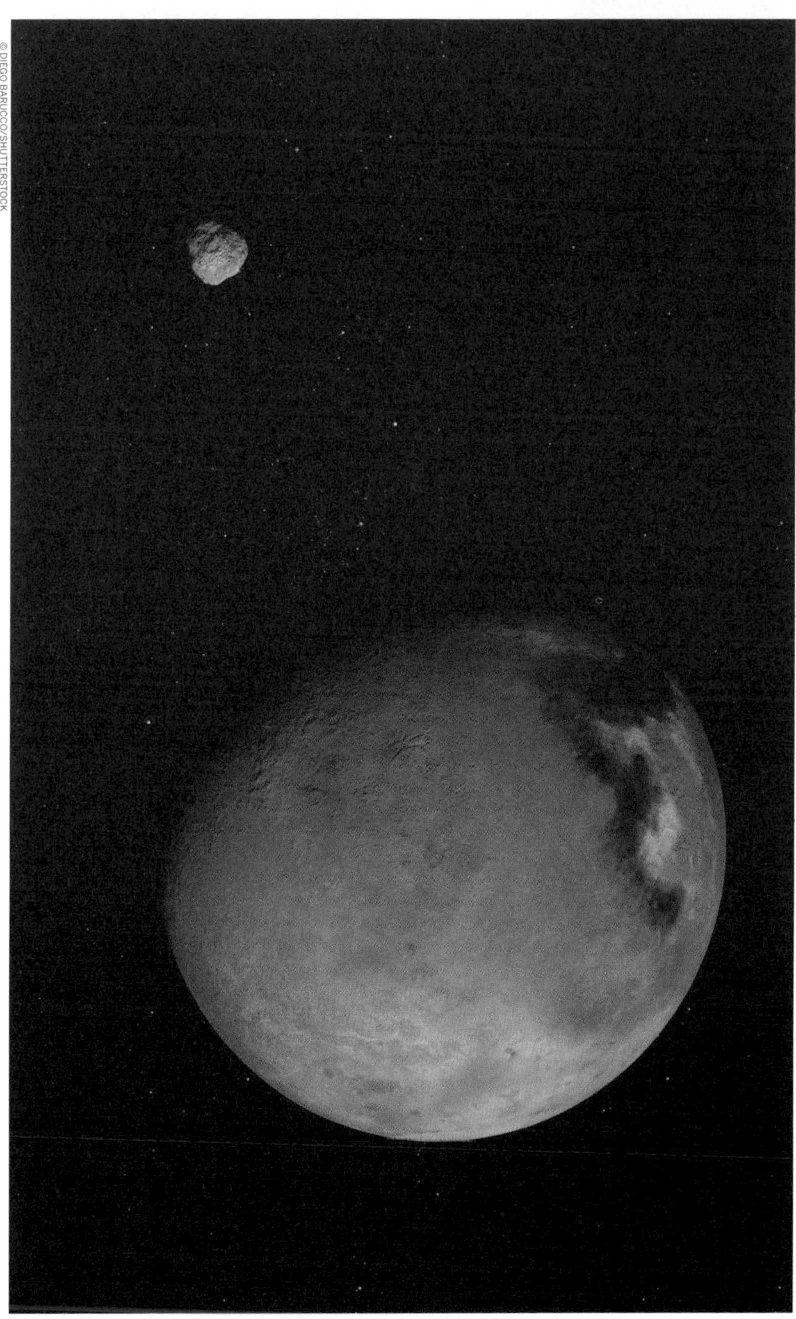

An artist's conception of Makemake's small moon, MK2.

Frozen canyons of Pluto's north pole.

Pluto

Wherever you fall on the 'is Pluto a planet?' debate, it's hard to deny that Pluto's unusual characteristics put it in an astronomical class of its own.

Pluto *used* to be the ninth planet from the Sun, but now it goes by a different designation. In 2006, astronomers reclassified Pluto as a dwarf planet. A dwarf planet must orbit the Sun just like other planets, and has sufficient mass to have a round shape, but the key deciding factor is that a planet has cleared the area surrounding its object. Though Pluto is among the biggest dwarf planets we've discovered (Eris holds the title for the largest based on most measurements), it is still small. It's about half the width of the United States and smaller than Earth's moon.

On average, Pluto is a distance of 39.5 astronomical units (AU) from the Sun, but because of its elliptical orbit, it does not remain the same distance from the Sun all the time. Its furthest point away from the Sun is 49.3 AU and at its closest, 29.7 AU, Pluto is actually nearer to the Sun than Neptune (30 AU distant).

Pluto has five known moons. The largest, named Charon, is about half the size of Pluto. Astronomers discovered Pluto's four other, smaller moons – Nix, Hydra, Kerberos and Styx – using the Hubble Space Telescope.

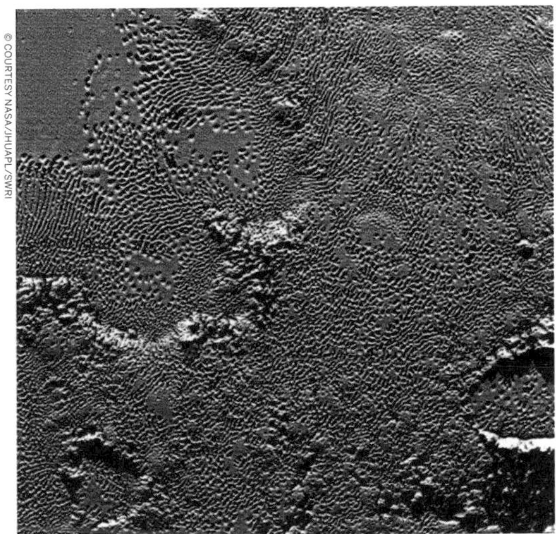

A zoomed-in view of Pluto's pitted surface.

RADIUS
1151 km (715 mi)

MASS
1.303×10^{22} kg,
or 0.00218 x Earth

COMPOSITION
Predominantly nitrogen ice, methane and carbon monoxide over a rocky surface

DISCOVERED
1930

ORBITAL PERIOD
248 years

Object Seeking Permanent Status

Is Pluto a planet, or not? When Pluto was discovered in 1930, astronomers made the assumption that this far distant body of the solar system was a planet, despite its small size and somewhat unusual orbital behaviour. However, by 2006, astronomers began to realise that the aspects that made Pluto unique were shared in common with other objects like distant Eris, Haumea, and Makemake. To better describe Pluto and these other Kuiper Belt Objects, the International Astronomical Union created the classification of dwarf planets. Pluto was among the first named a dwarf planet in 2006. However, the debate rages on: as late as 2018, the same astronomers continued to debate the definition of a planet and whether Pluto should regain its planetary status.

Comets

In the distant past, people were both awed and alarmed by comets, perceiving them as long-haired stars that appeared in the sky unannounced and unpredictably. Chinese astronomers kept extensive records for centuries, including illustrations of characteristic types of comet tails, times of cometary appearances and disappearances, and celestial positions. These historic comet annals have proven to be a valuable resource for later astronomers.

Scientists now know that comets are leftovers from the dawn of our solar system around 4.6 billion years ago, and may yield important clues about its formation. Comets are made mostly of ice coated with dark organic material; they have been referred to as 'dirty snowballs'. Comets may have brought water and organic compounds, the building blocks of life, to the early Earth and other parts of the solar system. They may be either periodic, on a known orbit

An artist's rendering of a comet passing near Earth.

(either long or short), or single-apparition, passing through our solar system only once.

Many arrive from the Kuiper Belt, a disc-like region of icy bodies beyond Neptune, where a population of dark comets orbits the Sun in the realm of Pluto. These icy objects, occasionally pushed by gravity into orbits bringing them closer to the Sun, become the so-called short-period comets. Taking less than 200 years to orbit the Sun, in many cases their appearance in our skies is predictable because they have passed by before. Less predictable are long-period comets, many of which arrive from a region about 100,000 AUs from the Sun called the Oort Cloud. These Oort Cloud comets can take as long as 30 million years to complete one trip around the Sun.

Each comet has a tiny frozen core, called a nucleus, often no larger than a few kilometres across. The nucleus contains icy chunks consisting of frozen gases with bits of embedded dust. A comet warms up as it nears the Sun and develops an atmosphere, or coma. The Sun's heat causes the comet's ices to change to gases which are ejected from the comet as they warm, so the coma gets larger and may extend to hundreds of thousands of kilometres. The pressure of sunlight and high-speed solar particles (solar wind) can blow the coma dust and gas away from the Sun, sometimes forming a long, bright tail. Comets actually have two tails – a dust tail and an ion (gas) tail.

Scientists have long wanted to study comets in greater detail, tantalised by the images taken of the Halley Comet's nucleus during its flyby in 1986. At the turn of the 21st century, NASA began sending spacecraft missions, orbiters and landers to well-documented comets in earnest.

This began in 1998 with the Deep Space 1 spacecraft, which flew by Comet Borrelly in 2001 and photographed its nucleus. The Stardust mission successfully flew within 236 km of the nucleus of Comet Wild 2 in early 2004, collecting cometary particles and interstellar dust for a sample return to Earth in 2006. Another NASA mission, Deep Impact, consisted of a flyby spacecraft and an 'impactor', a piece of hardware designed to crash on the comet and allow scientists to study the results. In July 2005, the impactor was released into the path of the nucleus of Comet Tempel 1 in a planned collision, which vaporised the impactor and ejected massive amounts of fine, powdery material from beneath the comet's surface. En route to the collision, the impactor camera photographed the comet in increasing detail. Two cameras and a spectrometer recorded the dramatic excavation, which helped determine the interior composition and structure of the nucleus.

These missions are some of the earliest attempts researchers on Earth have made to study the itinerant comets in our solar system. Future missions will examine comets in greater detail, and new comets are discovered regularly as they pass near enough to the Sun and Earth that we can spot their hazy tails behind them.

RADIUS
2.4 km (1.5 mi)

MASS
2×10^{13} kg

COMPOSITION
Rock and ice

DISCOVERED
1904

ORBITAL PERIOD
6.8 years

An artist's conception of Deep Space 1 flying by Borrelly.

Borrelly

A celestial bowling pin of rock and ice, Borrelly was one of the best-photographed comets in the early 21st century.

Comet Borrelly, officially designated 19P/Borrelly, is one of many short-period comets in the solar system. Borrelly orbits the Sun within the asteroid belt and is known as a Jupiter-family comet. These are defined as having an orbital period of fewer than 20 years and one that has been modified by close passages by the gas giant. It takes only 6.85 years for this comet to orbit the Sun once. Borrelly last reached perihelion (which activates the comet's tail) in 2015 before beginning its journey back to the asteroid belt, to return in 2022.

Resembling a chicken leg or bowling pin, the nucleus of comet 19P/Borrelly is small. In 1998, the Deep Space 1 mission was launched to test new engineering capabilities and to try and visit Comet Borrelly and an asteroid. Deep Space 1 flew by Comet Borrelly in September 2001, capturing the highest-resolution pictures of any comet's nucleus at the time. The technology on Deep Space 1 informed the design of the Dawn spacecraft, which made visits to the asteroid Vesta and the dwarf planet Ceres in the asteroid belt.

RADIUS
Unknown

MASS
Unknown

COMPOSITION
Rock and ice

DISCOVERED
1861

ORBITAL PERIOD
415 years

The Lyrids meteor shower above California's Sierra Nevada mountains.

C/1861 G1 Thatcher

As a solar system object that passes Earth once every five generations, Comet Thatcher might seem irrelevant – except that it's responsible for an annual astronomical event we can all enjoy.

C/1861 G1 Thatcher, sometimes called Comet Thatcher or just Thatcher after the astronomer who discovered it, is one of several comets we know are responsible for annual meteor showers. Comet Thatcher is considered a long-period comet – rather than short-period comets like Borrelly or Tempel 1 – since its orbital period is greater than 200 years. It takes 415.5 years for Thatcher to orbit the Sun once. Comet Thatcher last reached perihelion all the way back in 1861, and won't return again until 2276!

When comets like Thatcher come around the Sun, the dust they emit gradually spreads into a trail around their orbits. Every year, the Earth passes through these debris trails, which causes the bits to collide with our atmosphere, where they disintegrate and create fiery and colourful streaks in the sky – meteor showers. In the case of Comet Thatcher, we call the trail of debris the Lyrids meteor shower, which takes place each April (see below). Each shower has its own particular features based on the path and nature of its parent object.

The Lyrids

The Lyrids meteor shower in mid- and late April each year has the longest recorded history, dating back to 687 BC by Chinese astronomers. While the Lyrids may not be as active as other meteor showers, peaking at an average of 15–20 meteors per hour on 22–23 April each year, they make up for it in other ways. In a phenomenon known as a 'Lyrid fireball', some meteors are bright enough to cast shadows and leave trails of smoky debris in the sky for several minutes.

RADIUS
Varied

MASS
9.982×10^{12} kg

COMPOSITION
Rock and ice

DISCOVERED
1969

ORBITAL PERIOD
6.45 years

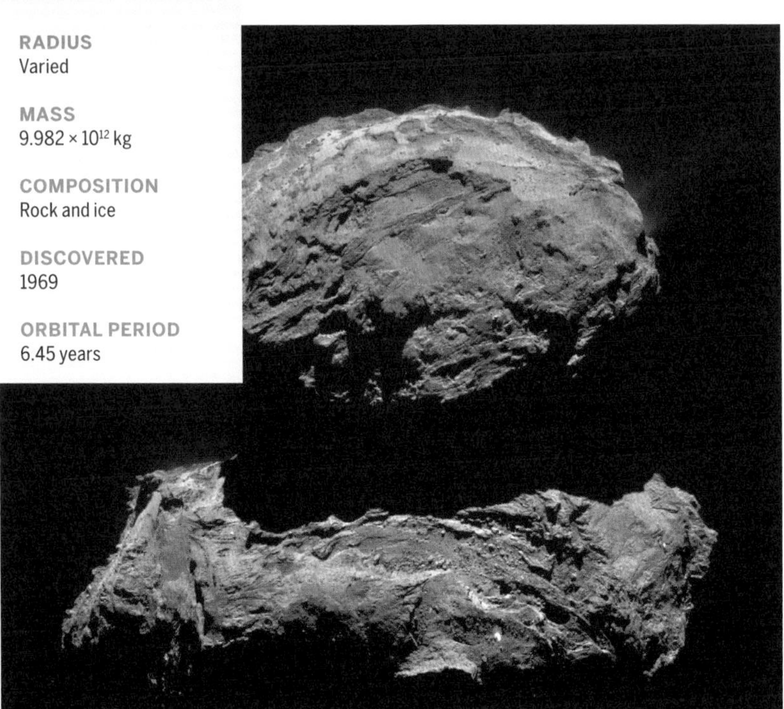

Image of Churyumov-Gersaimenko taken using the OSIRIS instrument aboard the ESA's Rosetta spacecraft.

Churyumov-Gerasimenko

The Rosetta mission to comet Churyumov-Gerasimenko shows that even when spacecraft don't stick the landing, the mission can still provide valuable knowledge – and amazing photographs.

Comet 67P, also called Churyumov-Gerasimenko (it's a mouthful), loops around the Sun in an orbit that crosses those of Jupiter and Mars, approaching but not reaching Earth's orbit. Named after its Soviet discoverers, the '67P' in its title comes from the fact that it was the 67th periodic comet ever found. Like most Jupiter-family comets, it is thought to have fallen from the Kuiper Belt as a result of one or more collisions or gravitational tugs from the gas giant.

Churyumov-Gerasimenko made history as the first comet to be orbited and landed upon by robots from Earth. In 2014, the Rosetta spacecraft and its payload, the Philae lander, reached Churyumov-Gerasimenko and the lander was sent to the surface of the comet. Unfortunately, Philae did not land well, and had limited access to sunlight to generate solar power for its mission. In 2016, the Rosetta mission ended when the spacecraft crashed into another part of Churyumov-Gerasimenko. During the two years Rosetta and Philae spent at Churyumov-Gerasimenko, both robots were able to take high-resolution photos of the comet, including one infamous photo of 'snow' formed by dust and ice blowing on the surface.

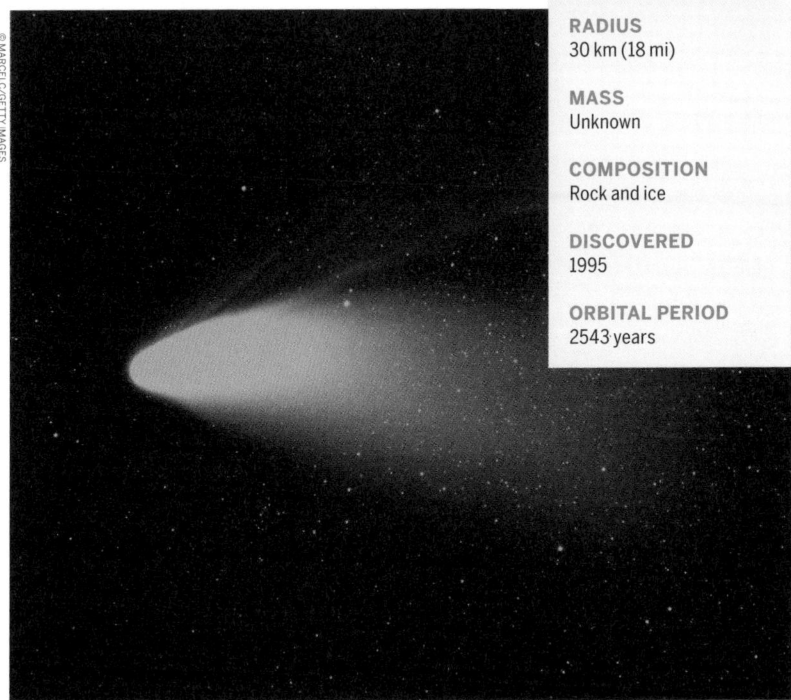

RADIUS
30 km (18 mi)

MASS
Unknown

COMPOSITION
Rock and ice

DISCOVERED
1995

ORBITAL PERIOD
2543 years

Hale-Bopp's blue ion tail is blown by the solar wind, while the white tail is caused by radiation pressure on dust particles.

Hale-Bopp

Many readers will remember the smudge of long-period comet Hale-Bopp in the night sky as it sailed past Earth in 1997, cementing its nickname as the Great Comet of 1997.

Comet C/1995 O1, also called the Great Comet of 1997 or just Hale-Bopp, is about five times the size of the object hypothesised to have led to the demise of the dinosaurs. Due to its large size, this comet was visible to the naked eye for 18 months in 1996 and 1997. Hale-Bopp was even visible to casual sky-watchers in light-polluted cities around the globe, leading researchers to speculate that it was the most viewed – and most photographed – comet in history.

As it passed, Hale-Bopp was visually memorable for its two tails: a whitish dust tail and a blue ion tail. The ion tail was created when fast-moving particles from the solar wind struck expelled ions from the comet's nucleus. The white dust tail comprises larger particles of dust and ice expelled by the nucleus, which orbit behind the comet.

As a long-period comet, it takes over 2500 years for Hale-Bopp to orbit the Sun. The comet last reached perihelion in early April 1997, making those who saw it extremely fortunate. When it returns to pass Earth in the 46th century, it's exciting (or dizzying) to imagine the world that might see it pass by above.

RADIUS
~5.5 km (3.7 mi)

MASS
2.2 x 10^{14} kg

COMPOSITION
Rock and ice

DISCOVERED
Known to Antiquity

ORBITAL PERIOD
75–76 years

The Orionids meteor shower in 2016.

Halley

Halley's Comet has astonished sky-watchers every few generations. Today, we look forward to its next visit to the inner solar system – in several decades.

Comet 1P/Halley is perhaps the most famous comet – it has been sighted for millennia. The first observations of comet Halley are lost in time, more than 2200 years ago, and it is even featured in the Bayeux Tapestry, which chronicles the Battle of Hastings in 1066. At the time, it wasn't known that periodic comets travelled on a predictable orbit. But in 1705, when Edmond Halley was studying the orbits of previously observed comets, he noted some that seemed to re-appear every 75–76 years. Based on the similarity of the orbits, he suggested these were the same comet and correctly predicted its next return in 1758. It was quite a feat of prognostication.

The last time Halley was seen from Earth was in 1986 and it will not enter the inner solar system again until 2061. Each time that Halley returns to the inner solar system, its nucleus sprays ice and rock into space under the warming influence of the Sun. This debris stream results in two weak meteor showers each year: the Eta Aquarids in May and the Orionids in October (see below).

The Orionids

Of the two meteor showers caused by Halley's comet, the Orionids – which occur in October each year – are the more active and impressive. The Orionids are a more concentrated meteor shower, with specific peak days that are optimal for viewing; during the typical peak in late October, you can sometimes see up to 25 meteors per hour. During the Eta Aquarids, which typically peak in late April or early May, you can expect to see 10–20 meteors per hour.

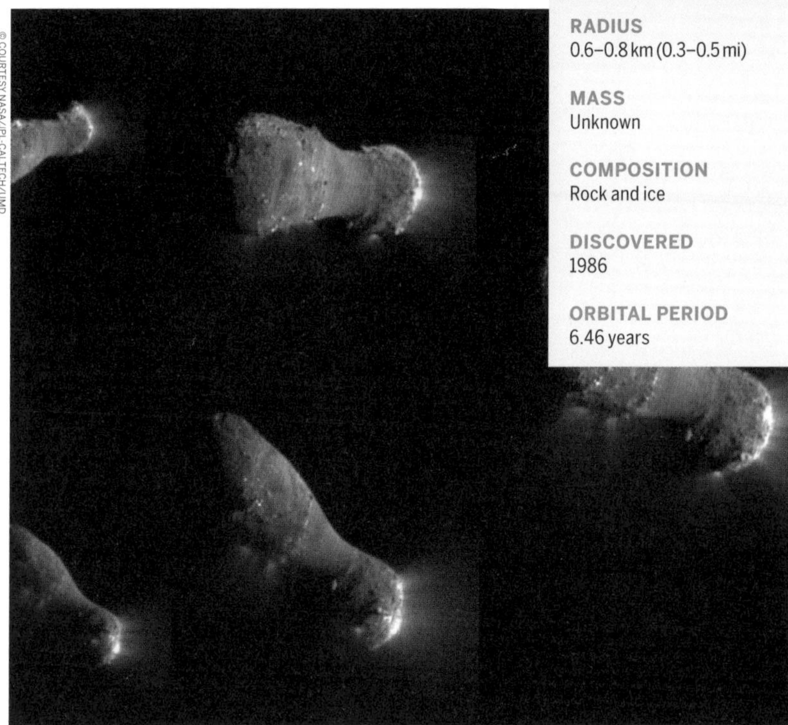

RADIUS
0.6–0.8 km (0.3–0.5 mi)

MASS
Unknown

COMPOSITION
Rock and ice

DISCOVERED
1986

ORBITAL PERIOD
6.46 years

Hartley's movements are partly driven by its carbon dioxide jets.

Hartley 2

Tumbling through space between Jupiter and Mars, Hartley 2 was visited by a robotic Earth ambassador in 2010, which observed the gas emissions that cause this comet's activity.

Comet 103P/Hartley (Hartley 2) is a small, peanut-shaped periodic comet, the third of three comets discovered by astronomer Malcolm Hartley in the late 1980s. Hartley 2 orbits the Sun within the asteroid belt, which lies between the orbits of Mars and Jupiter. As such, Hartley 2 is classified as a Jupiter-family comet, defined as comets having an orbital period of fewer than 20 years and that have been modified by close passages with the gas giant.

The Deep Impact (EPOXI) spacecraft performed a flyby of Hartley 2 in 2010. Hartley 2 was the fifth comet to be visited by man-made spacecraft and was the second encounter made by this particular spacecraft. Deep Impact had previously visited Comet 9P/Tempel 1 in 2005. Called 'hyperactive' by the EPOXI mission, Hartley 2 spins around one axis while rolling around another one. Comet Hartley's core is also not uniform but is made up of water-ice with methanol, carbon dioxide and possibly ethane. The release of these gases occurs at different locations on the comet, and the resulting carbon dioxide driven jets perpetuate its unpredictable celestial tumbles.

RADIUS
n/a

MASS
n/a

COMPOSITION
n/a

DISCOVERED
2012

ORBITAL PERIOD
n/a

Though hardly a spectacular object in the sky, ISON taught us an enormous amount about long-period comets.

ISON

One of the most studied comets in history, ISON offered scientists a spectacular and intriguing finale as it rounded the Sun.

Comet C/2012 S1 (ISON) was the subject of the most coordinated comet observation campaign in history. Russian astronomers Vitali Nevski and Artyom Novichonok discovered the comet in 2012 with the International Scientific Optical Network (ISON) telescope in Kislovodsk, Russia. Estimates of its size ranged from 5 km (3 mi) to as little as 0.8 km (0.5 mi) in diameter. Based on ISON's orbit, astronomers think the comet was making its first-ever trip through the inner solar system. Over the course of a year, more than a dozen spacecraft and numerous ground-based observers collected what is believed to be the largest ever single cometary dataset as sungrazing comet ISON plunged toward its close encounter with our star.

Just prior to its closest approach to the Sun on 28 November 2013, Comet ISON went through a major heating event and was torn to pieces. What remained of the comet appeared to brighten and spread out, then fade. Continuing its history of surprising behaviour, material from the Comet ISON soon appeared on the other side of the Sun. Within a short span of time, that remaining material disintegrated, suggesting that the long-period comet, which had journeyed to the inner solar system from the Oort Cloud, had made its final journey. Scientists have seen comets come close to the Sun and disintegrate and disappear before, like Comets Lovejoy and Elenin in 2012.

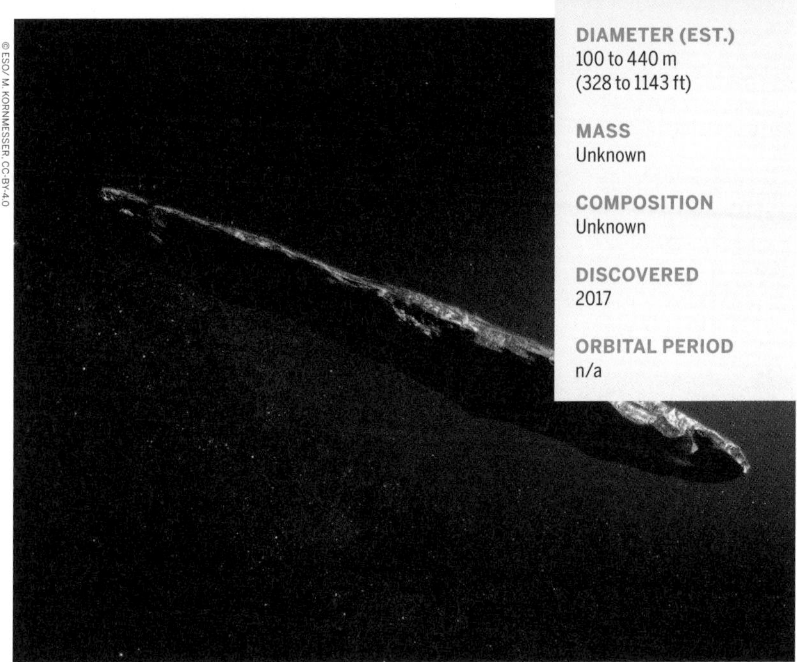

DIAMETER (EST.)
100 to 440 m
(328 to 1143 ft)

MASS
Unknown

COMPOSITION
Unknown

DISCOVERED
2017

ORBITAL PERIOD
n/a

An artist's rendering of 'Oumuamua as it hurtled through the solar system.

'Oumuamua

Discovered in 2017 as it approached the Sun, 'Oumuamua had been on an interstellar journey for hundreds of millions of years before its chance encounter with our star system.

The first confirmed object from another star to visit our solar system, this interstellar interloper appears to be a rocky, cigar-shaped object with a somewhat reddish hue. The object may be highly-elongated – perhaps 10 times as long as it is wide. Dubbed 'Oumuamua by its discoverers, its name means 'a visitor from a great distance' in Hawaiian. Its aspect ratio is greater than that of any asteroid or comet observed in our solar system to date. While its elongated shape is quite surprising, and unlike objects seen in our solar system, it may provide new clues into how other solar systems formed.

Observations suggest that this unusual object had been wandering through the Milky Way, unattached to any star system, for hundreds of millions of years before its chance encounter with our star system. 'For decades we've theorised that such interstellar objects are out there, and now – for the first time – we have direct evidence they exist', said Thomas Zurbuchen of NASA's Science Mission Directorate.

After looping around the Sun in September 2017, 'Oumuamua began its journey back out of the solar system. In May 2018 it passed outside Jupiter's orbit and by 2022 it will pass beyond Neptune's orbit and begin the long journey across the outer solar system through the Kuiper Belt and Oort Cloud, never to return.

RADIUS
n/a

MASS
n/a

COMPOSITION
Unknown

DISCOVERED
1993

ORBITAL PERIOD
n/a

The fragments of comet Shoemaker-Levy 9 approaching Jupiter on its collision course.

Shoemaker-Levy 9

Comet Shoemaker-Levy 9 enthralled astronomers when it crashed into Jupiter in the mid-1990s, proving that our solar system is a dynamic system with objects that interact – and sometimes collide.

Comet Shoemaker-Levy 9 was captured by the gravity of Jupiter, torn apart and then crashed in fragments into the giant planet in July 1994. When the comet was discovered in 1993 by Carolyn and Eugene Shoemaker and David Levy, it had already been torn into more than 20 pieces travelling around the planet in a two-year orbit. Further observations revealed that the comet (believed to be a single body at the time) had made a close approach to Jupiter in July 1992 and been torn apart by the tidal forces of the planet's powerful gravity. The comet is thought to have been orbiting Jupiter for about a decade before its demise.

Luckily, NASA had spacecraft in position to watch a collision between two bodies in the solar system for the first time in history. The Galileo orbiter (then still en route to Jupiter) captured unprecedented direct views as the string of fragments labelled A through W smashed into Jupiter's cloud tops. The impacts started on 16 July 1994 and ended on 22 July 1994. Observatories including the Hubble Space Telescope, Ulysses, and Voyager 2, also studied the impact and its aftermath.

The Perseids against the Milky Way.

RADIUS
26 km (16 mi)

MASS
Unknown

COMPOSITION
Rock and ice

DISCOVERED
1862

ORBITAL PERIOD
133 years

Swift-Tuttle

While many amateur stargazers may have never heard of Comet Swift-Tuttle, you've likely seen its signature in the night sky: the Perseids meteor shower which occurs each August.

Comet 109P, also called Swift-Tuttle after the two astronomers who independently discovered it in 1862, is a large comet, about twice the size of the object whose impact is hypothesised to have led to the demise of the dinosaurs. There's nothing to fear from Swift-Tuttle, though: while it crosses the Earth's orbit on its 133-year journey around the Sun, scientists predict that the closest approach it will make to our planet will be in 3044 AD at a distance of over 1.6 million km (994,000 mi).

A periodic comet, Swift-Tuttle last reached perihelion in 1992 and will not return again until distant 2125. Yet each year, we witness pieces of Swift-Tuttle left behind from its celestial journey. When comets come around the Sun, the dust and water vapour they emit gradually spreads into a trail around their orbits, a process called 'outgassing'. Every year, the Earth passes through these debris trails, which allows the bits to collide with our atmosphere, where they disintegrate and create fiery and colourful streaks in the sky. The pieces of space debris that interact with our atmosphere to form the popular Perseids meteor shower originate from Swift-Tuttle. It was Giovanni Schiaparelli who realised in 1865 that this comet was the source of the Perseids (see below for more).

The Perseids

The Perseids form most people's introduction to meteor viewing, and are a centrepiece of many young stargazer's memories. The Perseids have been observed dating back to the 1st century AD by the Chinese, but the connection between the meteors and Swift-Tuttle was not fully understood until the comet made its flyby in the mid-19th century. Most years, the Perseids peak at a rate of 60–70 meteors per hour in mid-August. During a 'meteor storm' or 'outburst year', the Perseids peak at anywhere between 100–200 meteors per hour, visible from the northern hemisphere down to the mid-southern latitudes.

RADIUS
6 km (3.73 mi)

MASS
Unknown

COMPOSITION
Rock and ice

DISCOVERED
1867

ORBITAL PERIOD
5.56 years

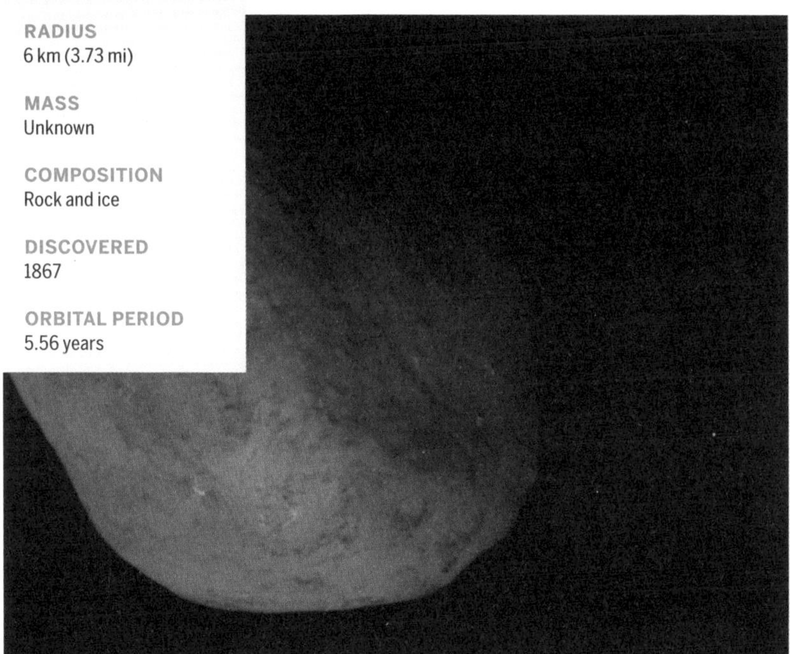

© COURTESY NASA/JPL-CALTECH/CORNELL

Comet Tempel as seen by NASA's Stardust.

Tempel 1

An otherwise normal object living in the asteroid belt, Comet Tempel 1 is the only comet to be visited by two different spacecraft – and the collision from the first was measured by the second.

Comet 9P, also called Tempel 1, is a periodic comet (denoted by the 'P' in its official designation) which orbits the Sun within the asteroid belt between the orbits of Mars and Jupiter. Tempel 1 is a Jupiter-family comet, one of those defined as having an orbital period of less than 20 years and one that has been modified by close passages by the gas giant. Tempel 1 offers great evidence of Jupiter's ability to change the orbit of comets: it takes about five years for Tempel 1 to circle the Sun once, but that period is changing slowly over time. When Tempel 1 was first discovered, its orbit measured 5.68 years.

Two missions have encountered Comet Tempel 1: Deep Impact in 2005 and Stardust-NExT in 2011. During the Deep Impact mission, scientists sent an impactor probe on a collision course with Tempel 1 to learn more about its composition, structure, and other physical properties. When the Stardust spacecraft visited Tempel 1 in 2011, it marked the first time a comet had been visited twice. Stardust was even able to photograph the crater caused by the Deep Impact probe six years earlier.

The Milky Way seen behind a Leonid meteor.

RADIUS
1.8 km (1.1 mi)

MASS
1.2×10^{13} kg

COMPOSITION
Rock and ice

DISCOVERED
First observed in 1699, discovered as a comet in 1865

ORBITAL PERIOD
33 years

Tempel-Tuttle

On its celestial journey, Comet Tempel-Tuttle passes astonishingly close to Earth (on a galactic scale), and as a result, produces a spectacular meteor storm every 33 years.

Comet 55P, or Tempel-Tuttle, is a small comet that, on the scale of the universe, gets very close to the Earth during its 33-year orbit. At its closest, Tempel-Tuttle passes within 1.2 million km (745,645 mi) of Earth. Luckily scientists don't predict that it will ever come closer than that.

Because of the close distance at which Tempel-Tuttle passes Earth, the stream of debris this comet leaves behind creates a unique experience. Most years, the meteor shower caused by Tempel-Tuttle debris – the Leonids in November (see below) – is relatively weak. But, every 33 years or so, the Leonids meteor shower becomes a meteor storm. A meteor storm – versus a shower – is defined as having at least 1000 meteors per hour. Tempel-Tuttle creates a meteor storm because the debris it leaves in Earth's orbit does not have a chance to spread out in space due to how closely the comet passes our planet by.

A Leonid meteor storm doesn't exactly match the close approach of Tempel-Tuttle to the year, but usually occurs within a few years. Tempel-Tuttle last reached perihelion in 1998 and will return again in 2031, making this a Halley-type comet. Viewers in 1966 experienced a spectacular Leonid storm following the 1965 perihelion, when thousands of meteors per minute fell through Earth's atmosphere during a 15-minute period. The last Leonid meteor storm took place in 2002, and we can expect the next Leonid storm in the 2030s.

The Leonids

During the peak of the Leonids meteor shower in mid-November each year, you can typically see between 15–50 meteors per hour. This number varies widely from year to year. Additionally, the Leonids occasionally produce a so-called meteor storm in excess of 1000 meteors per hour. Though the Leonids are not always the most active meteor shower of the year, in certain years they surpass the rate of all other meteor showers combined!

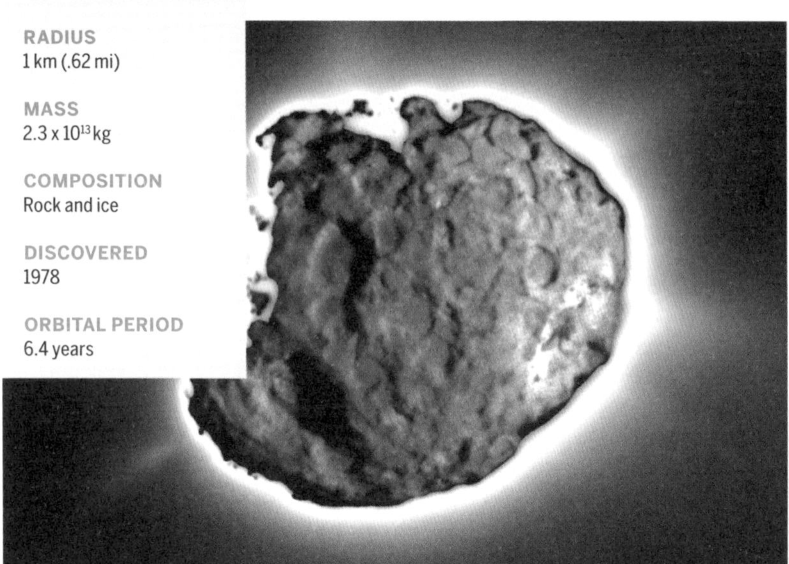

RADIUS
1 km (.62 mi)

MASS
2.3×10^{13} kg

COMPOSITION
Rock and ice

DISCOVERED
1978

ORBITAL PERIOD
6.4 years

Stardust captured this composite image of Wild 2 on its flyby in 2004.

Wild 2

Proof that the orbits of small solar system objects are not static, Comet Wild 2 gained attention when it changed orbit and moved significantly closer to the Sun.

Comet 1P/Wild, called Wild 2 and pronounced 'vilt', is what's known as a fresh periodic comet. A small comet with the shape of a flattened sphere, Wild 2 orbits the Sun between Mars and Jupiter, but it did not always travel the course of its orbit here. Originally, this comet's orbit lay between that of Uranus and Jupiter. In 1974, gravitational interactions between this comet and the planet Jupiter pulled the comet's orbit into a new shape. Astronomer Paul Wild discovered this comet during its first revolution of the Sun on the new orbit, and it was thus named after him by convention. In its relatively new path around the Sun, it takes Wild 2 nearly six and a half years (6.41 to be exact) to orbit the Sun once. Wild 2 last reached perihelion in 2016.

Since Wild 2 is a fresh comet – practically speaking, it has not had as many orbits around the Sun at close quarters – it is an ideal specimen for discovering more about the early solar system. NASA sent the Stardust mission (which also visited Comet Tempel 1) to fly by Wild 2 in 2004 and gather particles. This was the first such gathering of extra-terrestrial materials from a comet. These samples were gathered in an aerogel collector as the spacecraft flew by and were then returned to Earth in an Apollo-like capsule in 2006. In those samples, scientists discovered glycine: a fundamental building block of life. Images taken during the flyby also revealed intricate basins and craters on its rocky surface. And the 'wildest' feature? Violent jets that form when the Sun shines on the comet's surface. Solid ice becomes a gas and escapes into the vacuum of space; the jets blast out at hundreds of kilometres per hour.

An artist's conception of a Voyager probe in the Oort Cloud.

Oort Cloud

Not a comet at all but a possible home to the long-period comets, the Oort Cloud is believed to be a thick bubble of icy debris that surrounds our solar system.

Far beyond the edges of the distant Kuiper Belt, a giant spherical shell of objects may surround everything within our solar system, anchored by the Sun's gravity. It is too far to be seen with current telescopes, so it hasn't been directly observed or discovered, but the Oort Cloud is scientists' best guess about where long-period comets come from. Astronomers have studied several comets believed to have come from this distant region of our solar system, as far as three light-years away from the Sun.

This distant, predicted cloud may extend a third of the way from our Sun to the next star – somewhere between 1,000 and 100,000 AU. In 1950, Dutch astronomer Jan Oort first proposed the idea of this sphere of icy bodies to explain the origins of comets that take thousands of years to orbit the Sun. These long-period comets have likely been seen only once in recorded history.

There may be hundreds of billions, even trillions, of icy bodies in the Oort Cloud. Every now and then, something disturbs or collides with one of these icy worlds and it begins a long fall toward our Sun. Two recent examples are comets C/2012 S1 (ISON) and C/2013 A1 (Siding Spring). ISON was destroyed when it passed too close the Sun, while Siding Spring, which made a very close pass by Mars, will not return to the inner solar system for about 740,000 years.

For context, at its current speed of about 17 km per second (11 mi/s), the Voyager 1 spacecraft – launched in 1977 – probably won't even reach the Oort Cloud for about 300 years. Once it crosses into the Oort Cloud, it will take about 30,000 years to reach the other side. Perhaps by then, we'll have telescopes powerful enough to confirm the Oort Cloud's existence and begin studying the objects there.

EXOPLANETS

The above is the first ever visual image capture taken of an exoplanet, 2M1207b.

Exoplanets at a Glance

Discovering thousands of planets beyond our solar system counts as a 'eureka' moment in human exploration. But the biggest payoff is yet to come: capturing evidence of a distant world hospitable to life.

Since the first confirmation of an exoplanet orbiting a Sun-like star in 1995, and with only a few, narrow slices of our Milky Way Galaxy so far surveyed, we've already struck many rich veins. After the initial discovery the others came – first by the dozens, then by the hundreds. A recent statistical estimate places, on average, at least one planet around every star in the galaxy. That means there's something on the order of a trillion planets in our galaxy alone, many of them in Earth's size range.

While visiting even the closest of these is beyond the realm of current possibility, even for the Sun's nearest stellar neighbours and their orbiting companions, exploring the surprising variety of these has been an incredible boon over the past few decades, both for the trove of scientific knowledge it is unearthing and for its ability to expand our understanding of worlds beyond our own, whether they have multiple stars in their skies or travel alone through space as a rogue planet.

Hardly least of all, it reminds us of how fortunate, unusual and special an outlier our own Earth really is. Though many 'Super-Earths' and almost-Earth twins exist in the findings thus far, together with speedy Mercury analogues and massive 'Hot Jupiters', none can definitely be said to nurture life, though our search will continue.

Top Highlights

Epsilon Eridani

1 This system is our chance to watch a solar system like our own develop in its early stages.

Fomalhaut b

2 Fomalhaut b is unique for several reasons: it sits within a disc of debris around its star, and it's been directly imaged.

Gliese 504 b

3 Mysterious Gliese 504 b orbits 43.5 AU from what would be its star – unless it's a brown dwarf instead of an exoplanet.

Gliese 876 c

4 Like Saturn, Gliese 876 c is adorned with a dramatic ring system.

HAT-P-7b

5 With precious stones in its atmosphere and 500 times larger than Earth, windy HAT-P-7b would be a wonder to see.

HD 69830

6 This system has been found to have its own asteroid belt, filled with 26 times the material of that in the Solar System's equivalent.

HD 149026 b

7 This tidally locked, almost totally non-reflective exoplanet has a dayside that is warmed by its star to three times the temperature of Venus.

HIP 68468

8 Current theory states that this yellow dwarf, otherwise quite like our Sun, swallowed one of its own planets whole.

Kepler-11

9 Kepler-11 has a familiar looking system, which hosts a robust six known planets.

Kepler-70

10 The exoplanets of this star orbit at a dizzyingly fast speed, completing their year in mere hours.

Lich

11 Spooky planets orbit this pulsar, a stellar corpse.

Proxima b

12 Our closest planetary neighbour outside the solar system has a mass similar to Earth.

An artist's vision of what the landscape on Proxima b might look like.

How Do Exoplanet Detection Methods Work, Anyway?

The first exoplanet discovered in 1995 was a hot, star-hugging gas giant believed to be about half the size of Jupiter. It tugged so hard on its parent star as it raced around in a four-day orbit that the star's wobbling was obvious to earthly telescopes – once astronomers knew what to look for. Finding this fast-moving giant, known as 51 Pegasi b, kicked off what might be called the 'classical' period of planet hunting. The early technique of tracking wobbling stars revealed one planet after another, many of them large 'Hot Jupiters' with tight, blistering orbits.

The wobble method measures changes in a star's 'radial velocity'. The wavelengths of starlight are alternately squeezed and stretched as a star moves slightly closer, then slightly farther away from us. Those gyrations are caused by gravitational tugs, this way and that, from orbiting planets. Tracking them allows careful investigators to discover hidden exoplanets, typically of large masses.

Enter NASA's Kepler Space Telescope, launched in 2009 to inaugurate what we could call the 'modern' era of planet hunting. Kepler settled into an Earth-trailing orbit, then fixed its gaze on a small patch of sky. It stared at that patch for four years.

Within that small patch were some 150,000 stars. Kepler was waiting to catch tiny dips in the amount of light coming from individual stars, caused by planets crossing in front of them. The result: more than 2000 confirmed exoplanets were sifted from the data, the bulk of the more than 3300 confirmed so far, with more than 2400 planetary candidates as scientists continue to mine Kepler's observation data.

The Kepler mission faced its own sceptical audience in the 1990s. Four times, NASA rejected the designs proposed by William Borucki of the NASA Ames Research Center. Borucki,

An artist's rendition of the Kepler spacecraft in orbit around Earth.

now retired, finally won approval in 2001. His idea was proven right; Kepler's four years of data are still revealing new planets.

Other instruments, on the ground and in space, continue to round out the tally of exoplanets counted so far. The European CoRoT satellite preceded Kepler, and also used the transit method to find numerous planets during its functional period from 2006 to 2012.

The Hubble Space Telescope not only has discovered a variety of transiting exoplanets, but has characterised the atmospheres of some of them. As a planet makes its transit across the face of its star, a sliver of starlight shines through the planet's atmosphere. Gases and chemicals in the atmosphere absorb different wavelengths of the light as it passes through. By looking for these missing slices of the star's light spectrum, scientists can tell which constituents are present in that alien atmosphere.

Another skywatcher, NASA's Spitzer Space Telescope, observes transiting exoplanets in infrared wavelengths, and has helped to chart and characterise many, including puzzling out details of planetary atmospheres.

Spitzer often works in conjunction with ground-based telescopes, including OGLE's Warsaw Telescope at the Las Campanas Observatory in Chile. In 2015, a collaboration between Spitzer and Italy's 3.6-m Galileo National Telescope in the Canary Islands revealed the closest known rocky planet: HD 219134b, only 21 light years away from Earth. Disappointingly, however, the planet orbits its star too closely to make it suitable for life. Yet more planets are always being found.

The Transiting Exoplanet Survey Satellite (TESS) discovered its first Earth-size world in April 2019 after a 2018 launch. In a planned two-year survey of the solar neighbourhood, TESS will monitor the brightness of stars for periodic drops caused by planet transits. The TESS mission is expected to find planets ranging from small, rocky worlds to giant ones, showcasing the diversity of planets in the galaxy.

Astronomers predict that TESS will discover dozens of Earth-sized planets and up to 500 planets less than twice the size of Earth. In addition to Earth-sized planets, TESS is expected to find some 20,000 additional exoplanets in its two-year prime mission. TESS will likely find upwards of 17,000 planets larger than Neptune.

All but a handful of the thousands of exoplanets observed so far have been detected via indirect methods, such as watching for transits or measuring star wobbles. We've only just begun to enter a new era of planet hunting: direct imaging.

Astronomers say the future of exoplanet exploration is all about direct observation. Missions like the James Webb Space Telescope, now under construction, and the planned WFIRST (Wide-Field Infrared Survey Telescope) will expand and sharpen our ability to capture actual images of distant planets.

New technology under development will boost these capabilities, allowing us to snap portraits of smaller and smaller exoplanets. The WFIRST mission, for example, will use an internal instrument called a coronagraph to selectively block and process incoming starlight to reveal the planets hidden in the glare.

Something similar could be done outside the telescope by a device called the starshade, being developed at NASA's JPL. The starshade would deploy in deep space like a sunflower the size of a baseball diamond. Tens of thousands of miles away, a space telescope would point toward it; the starshade would block unwanted starlight, allowing the space telescope to capture images of the planets around the target star.

In coming decades, as space telescopes grow larger and more refined, perhaps we'll finally capture the iconic image of another Earth – a faraway world of continents, clouds and oceans. The ideal candidate is an Earth-sized, rocky world nestled comfortably within its star's habitable zone.

Even with the expected advances in observing technology in years to come, we're unlikely to know the precise nature of any life we might detect, be they crusts of algae or crawling creatures. Still, the explosion of new exoplanets opens fresh frontiers for exploring the nature of the Universe.

TYC 9486-927-1 DATA

CELESTIAL COORDINATES
Right Ascension 21h 25m 27.4899s
Declination -81° 38' 27.673"

DISTANCE
87 light years

CONSTELLATION
Octans (the Octant)

VISUAL MAGNITUDE
11.821

MASS
0.4 solar masses

RADIUS
Unknown

STAR TYPE
Red dwarf, M1

TEMPERATURE
3490 K

ROTATION PERIOD
Unknown

NUMBER OF PLANETS
1

2MASS J2126 DATA

PLANET TYPE
Gas Giant

MASS
13.3 x Jupiter

RADIUS
Unknown

ORBITAL PERIOD
900,000 years

ORBITAL RADIUS
6900 AU

DETECTION TYPE
Imaging

DISCOVERED
2009, 2016

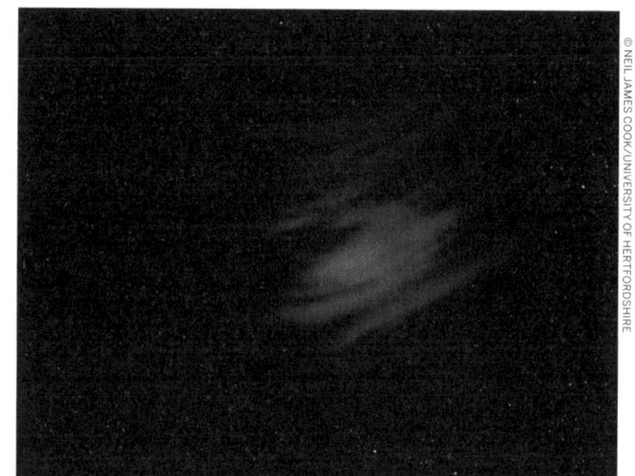

An artist's impression of 2MASS J2126, which orbits almost 7000 AU from its star.

2MASS J2126-8140

One of the loneliest known planets, 2MASS J2126 orbits its parent star from an incredible distance of one-tenth of a light year.

The gas giant 2MASS J2126-8140 is about 13 times the mass of Jupiter, but it's distinctive in other ways too. Almost 7000 AU away from its parent star (i.e. 700 times the distance that Earth orbits our own Sun), it has the widest and longest orbit of any planet found around another star. At such an enormous distance, it takes roughly 900,000 years to complete one orbit, and it's likely not tidally locked either. There is little prospect of any life on an exotic and cold world like this, but any inhabitants would see their 'Sun' as no more than a bright star, and might not even imagine they were connected to it at all.

Upon its initial discovery in 2009 using data from the Two Micron All-Sky Survey – hence 2MASS – the exoplanet was classified by the research team as a brown dwarf due to its size and infrared light emissions. When 2MASS J2126-8140 was first found to be an exoplanet rather than a brown dwarf, in 2016, astronomers assumed it was an orphan, or free-floating planet, having been ejected somehow from its original system and banished to roam the space between the stars. But after some clever detective work, researchers from the University of Hertfordshire, UK, found that 2MASS J2126-8140 is actually tied to a star after all, the prosaically named red dwarf TYC 9486-927-1 (it's also referred to as 2MASS J21252752-8138278). 'Nobody had made the link between the objects before', said lead scientist Dr Niall Deacon. 'The planet is not quite as lonely as we first thought, but it's certainly in a very long-distance relationship.'

A concept illustration of 51 Pegasi b.

51 Pegasi b

The planet 51 Pegasi b is almost astronomical royalty. It was one of the earliest exoplanets ever found, and the first seen to orbit a Sun-like star.

Discovered in 1995 in orbit around the yellow star 51 Pegasi, the planet 51 Pegasi b is a hot, star-hugging gas giant believed to be about half the size of Jupiter. It tugs so hard on its parent star as it races around in a four-day orbit that the star's wobbling was obvious to earthly telescopes – once astronomers knew what to look for. Finding this fast-moving giant kicked off what might be called the 'classical' period of planet hunting. The early technique of tracking wobbling stars revealed one planet after another, many of them large hot Jupiters with tight, blistering orbits.

The wobble method measures changes in a star's 'radial velocity'. The wavelengths of starlight are alternately squeezed and stretched as a star moves slightly closer, then slightly farther away from us. Those gyrations are caused by gravitational tugs, this way and that, from orbiting planets. The European team of Michel Mayor and Didier Queloz, from the University of Geneva, announced their discovery of 51 Pegasi b using this method, and the race was on to find others. It favours the discovery of high-mass exoplanets on a tight orbit.

Recently, the IAU renamed the exoplanet Dimidium, the Latin meaning 'half', in reference to its mass at half the size of Jupiter. It's unclear whether this name will supplant its initial title, however, especially as the moniker Bellepheron, the name of the Greek hero who defeated the Chimera, has also been in use by one of the astronomers who confirmed the exoplanet's existence.

51 PEGASI DATA

CELESTIAL COORDINATES
Right Ascension 22h 57m 27.988s
Declination 20° 46' 7.7912"

DISTANCE
50 light years

CONSTELLATION
Pegasus (the Winged Horse)

VISUAL MAGNITUDE
5.49

MASS
1.11 solar masses

RADIUS
1.237 solar radii

STAR TYPE
Sun-like, G2V

TEMPERATURE
5768 K

ROTATION PERIOD
21.9 days

NUMBER OF PLANETS
1

51 PEGASI B DATA

PLANET TYPE
Hot Jupiter

MASS
0.47 x Jupiter

RADIUS
0.5 x Jupiter

ORBITAL PERIOD
4.23 days

ORBITAL RADIUS
0.0527 AU

DETECTION TYPE
Radial Velocity

DISCOVERED
1995

55 CANCRI (COPERNICUS) DATA

CELESTIAL COORDINATES
Right Ascension 8h 52m 35.81s
Declination 28° 19' 50.96"

DISTANCE
41 light years

CONSTELLATION
Cancer (the Crab)

VISUAL MAGNITUDE
5.5

MASS
0.96 solar masses

RADIUS
0.96 solar radii

STAR TYPE
Yellow dwarf, G8V

TEMPERATURE
5165 K

ROTATION PERIOD
42.2 days

NUMBER OF PLANETS
5

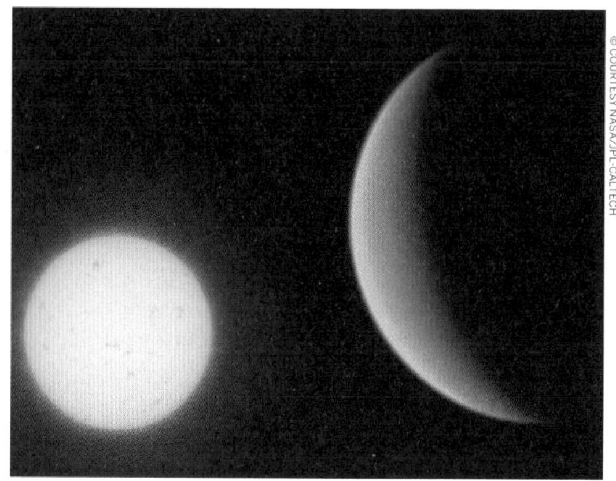

This rendering of 55 Cancri e shows it with a hazy atmosphere.

55 Cancri

This binary star system, 14 light years away in the constellation of Cancer, is a veritable planetary treasure trove. At least five planets are known here – and there could be more.

The larger star of this binary system, 55 Cancri A, is similar to the Sun, while its companion, 55 Cancri B, is smaller and dimmer – a red dwarf. In 1995, astronomers discovered that 55 Cancri A hosts a gas giant planet, dubbed 55 Cancri b. It was one of the first planets found outside the solar system. Since then, subsequent searches have increased the planet count to five. This makes 55 Cancri one of the most heavily populated known planetary systems.

Conventionally, the planets are given lower-case suffixes in order of discovery, beginning with b. So 55 Cancri b was applied to the first planet, while the others are labelled 55 Cancri c to f, regardless of their relative positions around the star. In 2015, the public voted to give the star 55 Cancri A the name 'Copernicus' after one of the fathers of observational astronomy, which was approved by the International Astronomical Union. The planets in the system are also named for famous astronomers of the Renaissance: Galileo, Brahe, Lippershey, Janssen and Harriot for planets b to f respectively.

Explore the 55 Cancri System

55 Cancri b

1 'Galileo', or 55 Cancri b, is a gas giant exoplanet. It has a mass 0.83 times that of Jupiter, takes 14.65 days to complete one orbit, and is 0.115 AU from its star, or just over a tenth of the Earth-Sun distance (the Astronomical Unit, or AU). It was announced to the world in January 1997, two years after its detection. One of the very first exoplanets ever found, it helped to herald the age of exoplanet research and discovery.

55 Cancri c and d

2 The next planets to be discovered were 55 Cancri c ('Brahe') and d ('Lippershey'), in 2002. Brahe is an ice giant exoplanet, somewhat like Uranus or Neptune. It has a mass 0.17 that of Jupiter, takes 44.3 days to complete one orbit, and is 0.24 AU from its star. Lippershey is more substantial – a gas giant some four times the mass of Jupiter – taking a much longer 14.3 years to complete one orbit of its star from a distance of 5.7 AU. This is similar to the distance of Jupiter from the Sun.

55 Cancri e

3 'Janssen', or 55 Cancri e, was discovered in August 2004. It is the innermost planet. Unlike its siblings, this is a so-called 'Super-Earth' – a rocky world eight times the mass of Earth. It orbits very close to its star, whipping around it every 18 hours. Because of the planet's proximity to the star, it is tidally locked by gravity just as our moon is to Earth. That means one side of 55 Cancri e, referred to as the dayside, is always cooking under the intense heat of its star, while the nightside remains in the dark and is much cooler. The hottest side is nearly 2430°C (4406°F), and the coolest is 1130°C (2066°F).

55 Cancri e Conditions

4 In 2016, scientists trained the Spitzer Space Telescope on this world. Interestingly, it likely has an atmosphere whose ingredients could be similar to those of Earth, but thicker. Lava lakes directly exposed to space without an atmosphere would create local hot spots of high temperatures, so they are not the best explanation for the Spitzer observations, scientists said. 'If there is lava on this planet, it would need to cover the entire surface', said Dr Renyu Hu, astronomer at NASA's JPL. 'But the lava would be hidden from our view by the thick atmosphere.'

55 Cancri f

5 Finally, 55 Cancri f, or 'Harriot', is another ice giant, like Brahe. It was discovered in April 2005. The second-most distant in the system, it tips the scales at 0.14 Jupiter masses, and orbits at 0.78 AU – roughly the position of Venus in our solar system. It completes a single orbit in 260 days.

Top Tip

Strange things transpire in the twilight zone, and stranger still is the place where the Sun never rises nor sets, but remains trapped at dusk. The planet Janssen (55 Cancri e) is tidally locked, a two-faced 'Super-Earth' whose dayside is molten from the heat of its star Copernicus while its nightside is plunged permanently into darkness. You might think you'd survive in the twilight or 'terminator' zone, where the day and nightsides meet, but Janssen's year is only 18 hours long. That means the backside of the planet is just cool enough to harden the dayside's boiling hellish world of possible lava flows. Don't get on this planet's bad side – either of them – or you'll be toast.

Getting There & Away

Aboard a Space Shuttle, you can nip across to 55 Cancri in only 1.6 million years. You can get there in under 200,000 years, though, if you travel as quickly as the Parker Solar Probe, the fastest ever probe.

An artist's impression of the surface of Barnard's Star b.

Barnard's Star b

Discovered in late 2018, Barnard's Star b is the second-closest known exoplanet, just six light years from the Earth. But it's like a frozen ball of rock and ice, with a surface temperature of around -170°C (-274°F).

Barnard's Star is the fastest star in the night sky, moving at about 500,000 km/h (310,000mph) in relation to our Sun, or the width of the moon every 180 years as viewed from Earth's surface. That's a full arcminute, or 1/2 degree. While it may not sound like much, for a distant object it's quite a move. (This is also why the zodiac no longer lines up with the positions of its constellations as they once were.) Barnard's Star is a red dwarf – a cool, low-mass star, which only dimly illuminates this newly-discovered world. Light from Barnard's Star provides its planet with only 2% of the energy the Earth receives from the Sun.

The data indicate that the planet, which orbits its host star in roughly 233 days, could be a 'Super-Earth' with a mass at least 3.2 times that of Earth. Called Barnard's Star b, it is a frozen, dimly lit world. To zero in on the planet, astronomers used multiple telescopes to measure its star's 'wobble' – how much the star is tugged to-and-fro by the planet in orbit around it. They made such precise measurements that they could account for a shift in the star's motion at roughly human walking speed. They also had to account for two other kinds of motion: Barnard's Star's rapid motion across the sky, and the fact that the star is moving toward us.

BARNARD'S STAR DATA

CELESTIAL COORDINATES
Right Ascension 17h 57m 48.498s
Declination 4° 41' 36.2072"

DISTANCE
5.958 light years

CONSTELLATION
Ophiuchus (the Serpent Bearer)

VISUAL MAGNITUDE
9.9511

MASS
0.144 solar masses

RADIUS
0.196 solar radii

STAR TYPE
Red dwarf, M4V

TEMPERATURE
3134 K

ROTATION PERIOD
130.4 days

NUMBER OF PLANETS
1 or 2

BARNARD'S STAR B DATA

PLANET TYPE
Super-Earth

MASS
3.23 x Earth

RADIUS
Unknown

ORBITAL PERIOD
232.8 days

ORBITAL RADIUS
0.404 AU

DETECTION TYPE
Radial Velocity

DISCOVERED
2018

Explore Barnard's Star b

Why Barnard?

1 This star is, very simply, named after its discoverer Edward Emerson Barnard (1857-1923), an American astronomer who discovered it in 1916. A prominent observational astronomer, he made groundbreaking efforts in the area of celestial photography (what we now call 'astrophotography', albeit done by very different methods).

Sister Stars?

2 Okay, so Barnard's Star is a red dwarf, not a yellow one like our Sun, but as the fourth nearest star to our Sun (that's including all the stars of the Centauri system), it still counts as family, and like all main sequence stars, it's made of hydrogen and helium. It is also the closest star in the northern hemisphere. Other than that the similarities are few. Barnard Star's mass is 15% of the Sun's, and 150 times Jupiter's mass packed into a package only about twice as large.

Population I or II?

3 Barnard's Star is much older than our own Sun, formed prior to the Universe acquiring large amounts of heavy elements, which require stars to generate them during their life cycle. Its rapid speed and lower metallicity indicate that it is an intermediate Population II star.

A Frozen 'Super-Earth'

4 The designation of Super-Earth relies on size and rocky core, rather than indicating life on an exoplanet of this designation would be at all Earth-like. In this case, Barnard's Star's faint luminosity indicates that its exoplanet would be frozen over. It has an orbital distance of 60 million km (37 million mi).

A New Class?

5 As a red dwarf that is estimated to be 11 or 12 billion years old, Barnard's Star is still relatively young for its stellar type. It could continue to burn as a red dwarf for 40 billion years before moving on to its next evolutionary stage, a cool black dwarf. For now the existence of these stellar remnants is only hypothetical, since the Universe's age of 13.8 billion years hasn't yet allowed any red dwarfs to evolve this far!

Top Tip

Want to have your very own star named after yourself too? Though companies do 'sell' these naming rights, discovering one for yourself is still the best way to achieve your aim. The IAU oversees all naming of astronomical objects, but it can be an anarchic process before that stage (and after).

Getting There & Away

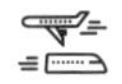

If you want to make an interstellar journey and the Alpha Centauri system is off limits for some reason, Barnard's Star is where you'd head. It helps that this star is moving in our direction anyway. At about 6 light years away, it's still a daydream rather than a feasible destination unless technology advances radically, but it's still extremely close in cosmic terms.

COROT-7 DATA

CELESTIAL COORDINATES
Right Ascension 6h 43m 49.47s
Declination -1° 3' 46.82"

DISTANCE
520 light years

CONSTELLATION
Monoceros (the Unicorn)

VISUAL MAGNITUDE
11.67

MASS
0.91 solar masses

RADIUS
0.82 solar radii

STAR TYPE
Yellow dwarf, G9V

TEMPERATURE
5250 K

ROTATION PERIOD
23 days

NUMBER OF PLANETS
2, maybe 3

COROT-7B DATA

PLANET TYPE
Super-Earth

MASS
2.3–8.5 x Earth

RADIUS
1.6 x Earth

ORBITAL PERIOD
0.85 days

ORBITAL RADIUS
0.017 AU

DETECTION TYPE
Transit

DISCOVERED
2009

CoRoT-7 and its exoplanet are extremely close, as shown in this concept illustration.

CoRoT-7b

CoRoT-7b is one of at least two 'Super-Earths' orbiting a Sun-like star. With dayside temperatures in the thousands of degrees, it's a molten, planetary hell.

When astronomers first discovered CoRoT-7b, they were amazed at its proximity to its star, at just 1/60th of the Earth-Sun separation, as well as its correspondingly high surface temperature – a blistering 1982°C (3560°F). 'The first planets detected outside our solar system turned out to be enormous gas giants in very tight orbits around their stars', said Brian Jackson at NASA's Goddard Space Flight Center in Greenbelt, Maryland. 'Now, we're beginning to see Earth-sized objects in similar orbits. Could there be a connection?'

CoRoT-7b's size (70% larger than Earth) and mass (2.3–8.5 times the Earth) indicate that the world is probably made of rocky materials. 'But with such a high dayside temperature, any rocky surface facing the star must be molten, and the planet cannot retain anything more than a tenuous atmosphere, even one of vaporised rock', Jackson said. CoRoT-7b is so close to its star that it's literally being eaten away by the extreme environment. Jackson estimates that solar heating may have already cooked off several Earth masses of material from CoRoT-7b. Not only extreme in temperature, it also orbits at great speed: more than 750,000 km/h (466,000mph).

Explore CoRoT-7b

CoRoT-7b's Star

1 Discovered in February 2009 by the Convection, Rotation and Planetary Transits (CoRoT) satellite, a mission led by the French Space Agency, CoRoT-7b takes just 20.4 hours to circle its Sun-like star, located 480 light years away in the constellation Monoceros. Astronomers believe the star is about 1.5 billion years old, or about one-third the Sun's age. Its planet is so close that, from there, the parent star 'appears almost 360 times larger than the Sun does in our sky', Jackson said.

Sister Planets

2 CoRoT-7b has at least one sibling, and possibly two. The confirmed planet is CoRoT-7c. Weighing in at around 13 Earth masses, it's very probably an ice giant like Neptune or Uranus, but much hotter. It also orbits its star very closely, around one-tenth of the distance of Mercury from the Sun. Another planet, called CoRoT-7d, is as yet unconfirmed. This one, if it exists, is even more massive and still very close to its star, orbiting once in just nine days.

Tidally Locked

3 Owing to CoRoT-7b's extremely close proximity to its star – just 0.017 AU – it is quite likely tidally locked. That is, it keeps one face towards its star at all times, just as the moon does to the Earth. The probable temperature on its dayside is about that of the tungsten filament of an incandescent light bulb, but a frigid -200°C (-328°F) on its locked, dark nightside. Each hemisphere is expected to have very different geology as a result.

Evaporating Planet

4 'There's a complex interplay between the mass the planet loses and its gravitational pull', Jackson explained. Tidal forces gradually change the planet's orbit, drawing it inward in a process called tidal migration. But closer proximity to the star then increases the mass loss, which in turn slows the rate of orbital change. After accounting for the give-and-take of mass loss and tidal migration, Jackson's team found that CoRot-7b could have weighed in at 100 Earth masses – or about the heft of Saturn – when it first formed.

Remnant Core

5 Jackson suggests that similar processes have likely influenced many other exoplanets that lie close to their stars. In fact, several recent studies suggest that many hot Jupiters have undergone similar mass loss and tidal evolution, perhaps leaving behind remnant cores similar to CoRoT-7b. 'CoRoT-7b may be the first in a new class of planet – evaporated remnant cores'.

Top Tip

The exact mass of planet CoRoT-7b is still being debated. At the time of its discovery, the authors, led by Dr Didier Queloz at the Cavendish Laboratory and Geneva University, put the mass at 4.8 times the Earth. Since then, other researchers have measured values in the range 2.3 to 8.5 times the Earth. At the upper end of this range, the planet would have a density far higher than Earth's, comparable to that of another exoplanet, Kepler-10b. It may also have an entirely solid core of iron, like Mercury, and is thus, like that planet, unable to generate a global magnetic field.

Getting There & Away

With existing rocket technology, a trip to this system will take 20 million years, give or take. As it's a planetary hell, you might want to pick a calmer, closer target.

The aptly named Very Large Telescope in Chile's Atacama Desert.

Artwork of the dark exoplanet CVSO 30c.

CVSO 30b and c

CVSO 30 is a young red star with one, possibly two planets. It is the first star where planets have been found both directly and by using the transiting planet method.

CVSO 30 is a young stellar object called a T Tauri star, 1140 light years away in Orion. The Sun underwent a similar stage in its evolution, when it was still surrounded by dust clouds and gas. Potential exoplanets have been found in this system using two methods, direct imaging and the transit method. Neither exoplanet has yet been confirmed, however. 'In 2012, astronomers found that CVSO 30 hosted one exoplanet (CVSO 30b) using the detection method known as transit photometry', says a press release from the European Southern Observatory (ESO), 'where the light from a star observably dips as a planet travels in front of it.'

This planet's existence remains controversial, or at least unverified, but in 2016, astronomers went back to look at the system using a number of telescopes. 'The study combined observations obtained with the ESO's Very Large Telescope (VLT) in Chile, the W. M. Keck Observatory in Hawaii, and the Calar Alto Observatory facilities in Spain', wrote Tobias Schmidt from the University of Hamburg. 'While the previously detected planet, CVSO 30b, orbits very close to the star, whirling around CVSO 30 in just under 11 hours at an orbital distance of 0.008 AU, CVSO 30c orbits significantly further out, at a distance of 660 AU, taking a staggering 27,000 years to complete a single orbit.' Both planets are gas giants, with CVSO 30b a possible hot Jupiter.

CVSO 30 DATA

CELESTIAL COORDINATES
Right Ascension 5h 25m 7.56s

Declination 1° 34' 24.35"

DISTANCE
1140 light years

CONSTELLATION
Orion (the Hunter)

VISUAL MAGNITUDE
16.26

MASS
0.39 solar masses

RADIUS
1.39 solar radii

STAR TYPE
T Tauri, MV

TEMPERATURE
3740 K

ROTATION PERIOD
Unknown

NUMBER OF PLANETS
2 candidates

EPS ERI DATA

CELESTIAL COORDINATES
Right Ascension 3h 32m 55.84496s
Declination -9° 27' 29.7312"

DISTANCE
10.46 light years

CONSTELLATION
Eridanus (the River)

APPARENT MAGNITUDE
3.74

MASS
0.82 solar masses

RADIUS
0.74 solar radii

STAR TYPE
Orange dwarf, K2V

TEMPERATURE
5084 K

ROTATION PERIOD
11.2 days

NUMBER OF PLANETS
1

Artist's illustration of the Epsilon Eridani system showing Epsilon Eridani b.

Epsilon Eridani

Double the rubble. Two asteroid belts and two possible planets encircle this popular science-fiction locale.

Located 10.5 light years away in the southern hemisphere of the constellation Eridanus, the star Epsilon Eridani (Eps Eri for short) – popularised by the TV show *Babylon 5* – is the closest planetary system around a star similar to the early Sun. It is a prime location to research how planets form around stars like our Sun.

Studies indicate that Eps Eri has two debris discs, which is the name astronomers give to leftover material still orbiting a star after planetary construction has completed. The debris can take the form of gas and dust, as well as small rocky and icy bodies. Debris discs can be broad, continuous discs or concentrated into belts of debris, similar to our solar system's asteroid belt and the Kuiper Belt – the region beyond Neptune where hundreds of thousands of icy-rocky objects reside. Furthermore, careful measurements of the motion of Eps Eri indicate that a planet with nearly the same mass as Jupiter circles the star at a distance comparable to Jupiter's distance from the Sun. Other planets may also be present.

One model for the system indicates that warm material is in two narrow rings of debris, which would correspond respectively to thc positions of the asteroid belt and the orbit of Uranus in our solar system. The other model attributes the warm material to dust originating in the outer Kuiper-Belt-like zone and filling in a disc of debris toward the central star. In this model, the warm material is in a broad disk, not concentrated into asteroid belt-like rings or associated with any planets in the inner region.

Explore the Epsilon Eridani System

The Star

1 Epsilon Eridani is very young. At less than a billion years old, it is around one-fifth of the age of the Sun. It therefore has a much more powerful magnetic field than the Sun, with a more blustery stellar wind 30 times as strong. This is the stream of particles flowing away from the star. Being a main-sequence star of spectral class K2, it is also cooler than the Sun, with a surface temperature of around 5000K, giving it an orange hue.

A Young System

2 'This system probably looks a lot like ours did when life first took root on Earth', said Dr Dana Backman, an astronomer at the SETI Institute, in Mountain View, California. 'The main difference we know of so far is that it has an additional ring of leftover planet-construction material.'

Two Belts and a Comet Ring

3 The asteroid belt orbits at a distance of about three astronomical units from its star – or about the same position as the asteroid belt in our own solar system. An astronomical unit (AU) is the distance between Earth and our Sun. The second asteroid belt lies at about 20 AU from the star, a position comparable to Uranus in our solar system. There is also an outer comet ring, which orbits from 35 to 90 AU from the star; our solar system's analogous Kuiper Belt extends from about 30 to 50 AU from the Sun.

Two Possible Planets

4 One of the two planets, Epsilon Eridani b, is a gas giant about 0.77 times the mass of Jupiter. It takes 6.9 years to complete one orbit and is 3.39 AU from its star. Its discovery was announced in 2000. The other, Epsilon Eridani c, is unconfirmed. It is about 0.1 times the mass of Jupiter and orbits in 280 years at a distance of 40 AU.

A Third Planet?

5 The intermediate belt detected by the Spitzer Space Telescope, launched by NASA in 2003, suggests that a third planet could be responsible for creating and shepherding its material. This planet would orbit at approximately 20 AU and lie between the other two planets. 'Detailed studies of the dust belts in other planetary systems are telling us a great deal about their complex structure', said Michael Werner, co-author of the study and project scientist for Spitzer at JPL. 'It seems that no two planetary systems are alike.'

Top Tip

Because the star is so close and similar to the Sun, it is a popular locale in science fiction. The television series *Star Trek* and *Babylon 5* referenced Epsilon Eridani, and it has been featured in novels by Isaac Asimov and Frank Herbert, among others. The popular star was also one of the first to be searched for signs of advanced alien civilisations using radio telescopes in 1960. At that time, astronomers did not know of the star's young age.

Getting There & Away

At a distance of just over 10 light years, it would take a Space Shuttle some 400,000 years to reach Epsilon Eridani. Best bring your e-reader.

FOMALHAUT DATA

CELESTIAL COORDINATES
Right Ascension 22h 57m 39.05s
Declination -29° 37' 20.05"

DISTANCE
25 light years

CONSTELLATION
Piscis Austrinus (the Southern Fish)

VISUAL MAGNITUDE
1.16

MASS
1.92 solar masses

RADIUS
1.84 solar radii

STAR TYPE
Blue main-sequence, A3V

TEMPERATURE
8590 K

ROTATION PERIOD
Unknown

NUMBER OF PLANETS
1 and debris disc

FOMALHAUT B DATA

PLANET TYPE
Gas Giant

MASS
< 2 x Jupiter

RADIUS
Unknown

ORBITAL PERIOD
1700 years

ORBITAL RADIUS
177 AU

DETECTION TYPE
Direct Imaging

DISCOVERED
2008

A Saturn-like ring may encircle Fomalhaut, with its planet amidst the particles.

Fomalhaut b

Fomalhaut b is a rarity – an exoplanet that astronomers imaged directly, rather than by using the transit or radial velocity methods.

NASA Hubble Space Telescope images of a vast debris disc encircling the nearby star Fomalhaut, and a mysterious planet circling it, may provide forensic evidence of a titanic planetary disruption in the system. The planet Fomalhaut b is almost twice Jupiter's size. It's also one of the few planets with a name chosen by the public – Dagon, a Semitic deity associated with vegetation and fertility. Scientists were able to image the planet directly because it orbits its star at a huge distance – on average, 177 AU. They used the coronagraph of the HST's Advanced Camera for Surveys to block out the star's bright glare so that the dim planet could be seen. Fomalhaut b is one billion times fainter than its star; that's quite a significant amount of interference.

Fomalhaut is a special system because it looks like it may have given scientists a snapshot of what our solar system was doing four billion years ago. Images clearly showed a ring of protoplanetary debris approximately 21.5 billion miles across and having a sharp inner edge. The planetary architecture is being redrawn, the comet belts are evolving, and planets may be gaining and losing their moons. Astronomers will continue monitoring Fomalhaut b for decades to come because they may have a chance to observe a planet entering an icy debris belt that is like the Kuiper Belt at the fringe of our own solar system. The debris in the planetary disc is accreting to make new objects, which only time will reveal.

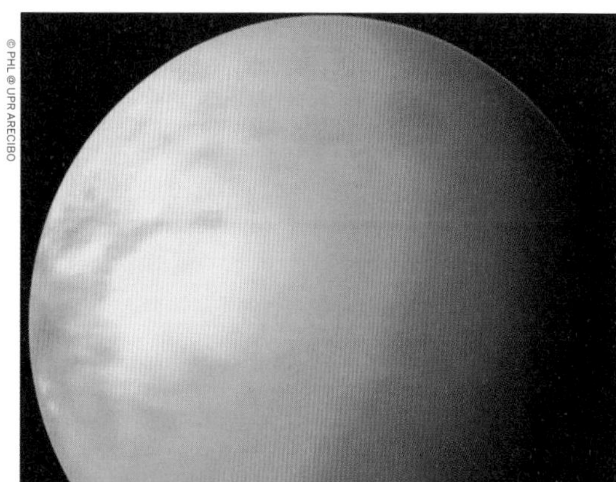

Gliese 163 c, pictured in an illustration, might be able to support microbial life.

Gliese 163 b, c and d

Gliese 163 is a red dwarf star with at least three planets, one of which is a potentially habitable 'Super-Earth'. The system is 49 light years from the Sun.

Astronomers targeted this system in 2012, using the High Accuracy Radial velocity Planet Searcher (HARPS) at the La Silla Observatory in Chile. The search initially turned up two planets, called Gliese 163 b and c, with a third being confirmed a while later, Gliese 163 d. Little is known about them except their masses and orbital characteristics. Still, this has not stopped astronomers from speculating.

The smaller planet, Gliese 163 c, is around 6.8 times the mass of the Earth, but as its radius is unknown, so is its nature. 'We do not know for sure that it is a terrestrial planet', said Xavier Bonfils, of France's Joseph Fourier University-Grenoble, the lead author of the discovery paper. 'Planets of that mass regime can be terrestrial, ocean, or Neptune-like planets.' Meanwhile, Abel Mendez, a Professor of Physics and Astrobiology at the University of Puerto Rico at Arecibo, believes that Gliese 163 c may be a water world, with a global ocean hidden under a dense, cloud-covered sky. If so, it could be habitable, as it orbits in its parent star's habitable zone. Questions over red dwarf stars' suitability for fostering life are also open – as more active stars, their stellar wind might be a risk to planetary atmospheres in their reach.

GLIESE 163 DATA

CELESTIAL COORDINATES
Right Ascension 4h 9m 15.66s
Declination -53° 22' 25.31"

DISTANCE
49 light years

CONSTELLATION
Dorado (the Swordfish)

VISUAL MAGNITUDE
11.8

MASS
0.4 solar masses

RADIUS
Unknown

STAR TYPE
Red dwarf, M3.5V

TEMPERATURE
3500 K

ROTATION PERIOD
61 days

NUMBER OF PLANETS
3, possibly more

GLIESE 176 DATA

CELESTIAL COORDINATES
Right Ascension 4h 42m 55.77s

Declination 18° 57' 29.40"

DISTANCE
30.7 light years

CONSTELLATION
Taurus (the Bull)

VISUAL MAGNITUDE
9.95

MASS
0.5 solar masses

RADIUS
0.45 solar radii

STAR TYPE
Red dwarf, M2V

TEMPERATURE
3679 K

ROTATION PERIOD
40 days

NUMBER OF PLANETS
1

GLIESE 176 B DATA

PLANET TYPE
Super-Earth or Neptune-like

MASS
9.06 x Earth

RADIUS
Unknown

ORBITAL PERIOD
8.78 days

ORBITAL RADIUS
0.066 AU

DETECTION TYPE
Radial Velocity

DISCOVERED
2007

The Hobby-Eberly telescope, pictured here, discovered Gliese 176 b in 2007.

Gliese 176 b

A potential 'Super-Earth' orbits the red dwarf Gliese 176 (or GJ 176). But the planet, Gliese 176 b, has uncertain mass, and may be a 'Neptune-Like' world instead.

Gliese 176 b has gone through a bit of an evolution since its initial discovery in 2007. Astronomers Michael Endl and William D. Cochran, from McDonald Observatory at the University of Texas, initially located the planet while performing a search for planets around red dwarf stars – also called M dwarfs – using the Hobby-Eberly Telescope (HET) located at their observatory. 'The orbital period of the planet is 10.24 days. GJ 176 thus joins the small (but increasing) sample of M dwarfs hosting short-period planets with minimum masses in the Neptune-mass range', said the authors, writing for the *Astrophysical Journal*.

However, it turned out that the orbital period was wrong – the researchers had confused the orbital period with the rotation of the star, at 40 days. Later measurements, using the HARPS instrument at La Silla, Chile, filtered out the spurious signal and found a new orbital period estimate of 8.78 days, and subsequently a smaller mass too. Until astronomers can estimate the planet's radius, however, it will remain uncertain whether this is a 'Super-Earth', or an ice giant like Neptune.

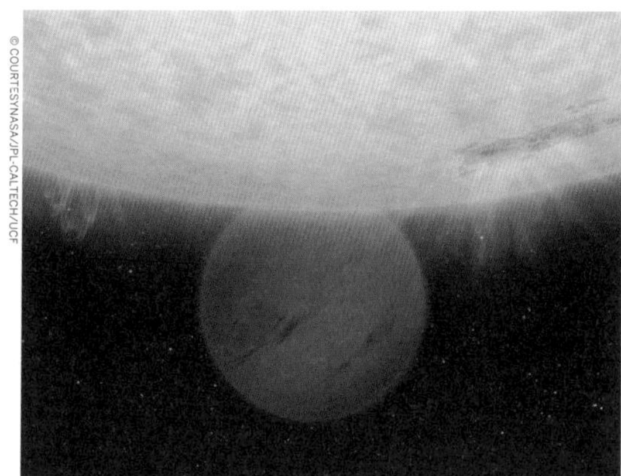

Gliese 436 b gives 'icy hot' a new meaning.

Gliese 436 b

This mysterious world, just under 32 light years away, has been dubbed the 'planet of burning ice'. It's a Neptune-sized world, but very much hotter than that planet.

Gliese 436 b leapt to fame when it was announced in 2004 as the first of its kind – a so-called 'Hot Neptune'. 'The Planet of Burning Ice orbits a red-dwarf star less luminous than our star, called Gliese 436, which can be viewed in the zodiac constellation Leo', says the website astronaut.com. 'The fact that Gliese 436 b is located so close to its star is buttressed by the observation that it completes a whole revolution in only two days and 15.5 hours. So, the surface temperature of this exoplanet is around 439°C. But the boiling point of water is 100°C. So how does this puzzle of 'ignited ice' even exist?'

The answer is that it's ice – but not as we know it. As explained on astronaut.com, 'There are many different states of water, rather than the three forms we are aware of, and the water on the Planet of Burning Ice is subjected to conditions that make it much denser than the familiar ice seen on Earth.' Just as carbon turns to diamond under great pressure and temperature, the water on this enigmatic world has turned into a hot crystalline substance called 'ice VII', which can remain solid even under extreme heat. It's unclear where else this phenomenon might occur in the Universe. It's certainly an impressive party trick.

Two other planets have been detected around the same star, but their status is not currently confirmed.

GLIESE 436 DATA

CELESTIAL COORDINATES
Right Ascension 11h 42m 11.09s

Declination 26° 42' 23.65"

DISTANCE
31.8 light years

CONSTELLATION
Leo (the Lion)

VISUAL MAGNITUDE
10.67

MASS
0.41 solar masses

RADIUS
0.42 solar radii

STAR TYPE
Red dwarf, M2.5V

TEMPERATURE
3318 K

ROTATION PERIOD
39.9 days

NUMBER OF PLANETS
1, possibly 3

GLIESE 436 B DATA

PLANET TYPE
Hot-Neptune

MASS
21.36 x Earth

RADIUS
4.33 x Earth

ORBITAL PERIOD
2.64 days

ORBITAL RADIUS
0.028 AU

DETECTION TYPE
Radial Velocity

DISCOVERED
2004

GLIESE 504 DATA

CELESTIAL COORDINATES
Right Ascension 13h 16m 47.0s
Declination 9° 25' 27"

DISTANCE
57.3 light years

CONSTELLATION
Virgo (the Virgin)

VISUAL MAGNITUDE
5.22

MASS
1.16 solar masses

RADIUS
1.36 solar radii

STAR TYPE
Yellow dwarf, GOV

TEMPERATURE
6205 K

ROTATION PERIOD
3.33 days

NUMBER OF PLANETS
1

GLIESE 504 B DATA

PLANET TYPE
Gas Giant or Brown Dwarf

MASS
Various estimates, 4–30 x Jupiter

RADIUS
0.96 x Jupiter

ORBITAL PERIOD
Various estimates, 155–1557 years

ORBITAL RADIUS
Various estimates, 31–129 AU

DETECTION TYPE
Imaging

DISCOVERED
2013

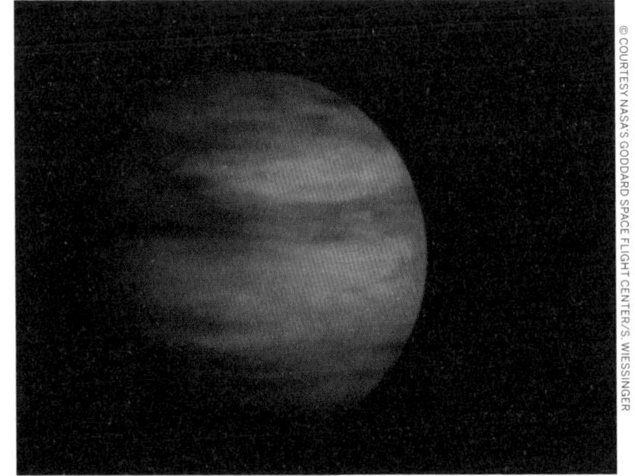

Gliese 504b weighs in with about four times Jupiter's mass.

Gliese 504 b

Orbiting a star in Virgo, Gliese 504 b (also called GJ 504 b) is the lowest-mass planet candidate ever detected around a star like the Sun using direct imaging techniques.

Gliese 504 b orbits its star at nearly nine times the distance Jupiter orbits the Sun. It might be a type of failed star called a brown dwarf, but if it's a planet, this poses a challenge to theoretical ideas of how giant planets form.

According to consensus, Jupiter-like planets get their start in the gas-rich debris disc that surrounds a young star. A core produced by collisions among asteroids and comets provides a seed, and when this core reaches sufficient mass, its gravitational pull rapidly attracts gas from the disc to form the planet. While this works fine for planets out to where Neptune orbits, about 30 times Earth's average distance from the Sun (or 30 AU), it's more problematic for worlds located further from their stars. Gliese 504b lies at a projected distance of 43.5 AU from its star; the actual distance depends on how the system tips to our line of sight, which is not precisely known.

Until it can be firmly placed in the proper category, Gliese 504b's mass is not well known. If it is a planet, it's about four times more massive than Jupiter, but if it's a brown dwarf, it could be much more massive and dense. If we could travel there, we would see a world still glowing from the heat of its formation with a colour reminiscent of a dark cherry blossom, a dull magenta.

Brown Dwarf or Exoplanet?

Searching for space objects with their size as a main way of classifying them into types is quite a useful shorthand, but it can be perilous at the border zone between the enormous gas giant hot Jupiters and small, cool brown dwarfs, which lack nuclear fusion at their core to heat them up as a recognisable star.

Astronomers are hopeful that the powerful infrared capability of NASA's James Webb Space Telescope will resolve a puzzle as fundamental as stargazing itself – what *is* that dim light in the sky? Brown dwarfs muddy a clear distinction between stars and planets, throwing established understanding of those bodies, and theories of their formation, into question.

Several research teams will use Webb to explore the mysterious nature of brown dwarfs, looking for insight into both star formation and exoplanet atmospheres, and the hazy territory in-between where the brown dwarf itself exists. Previous work with Hubble, Spitzer, and ALMA have shown that brown dwarfs can be up to 70 times more massive than gas giants like Jupiter, yet they do not have enough mass for their cores to burn nuclear fuel and radiate starlight. Though brown dwarfs were theorised in thc 1960s and confirmed in 1995, there is not an accepted explanation of how they form: like a star, by the contraction of gas, or like a planet, by the accretion of material in a protoplanetary disc? Adding to the uncertainty, some have a companion relationship with a star, while others drift alone in space.

At the Université de Montréal, Étienne Artigau leads a team that will use Webb to study a specific brown dwarf, labelled SIMP0136. It is a low-mass, young, isolated brown dwarf – one of the closest to our Sun – and it has many features of a planet without being too close to the blinding light of a star. SIMP0136 was the object of a past scientific breakthrough, when evidence was found suggesting it has a cloudy atmosphere. He and his colleagues will use Webb's spectroscopic instruments to learn more about the chemical elements and compounds in those clouds.

These observations could lay groundwork for future exoplanet exploration with Webb, including which worlds could support life: Webb's infrared instruments will be capable of detecting the types of molecules in the atmospheres of exoplanets. The search for low-mass, isolated brown dwarfs is a good start. Brown dwarfs have a lower mass than stars and do not 'shine' but merely emit the dim afterglow of their birth, and so they are best seen in infrared light. Brown dwarf infrared sources are relatively cool compared to the energy emitted from hot stars and other celestial objects.

The relative sizes of brown dwarfs and gas giants can be quite (confusingly) similar.

GLIESE 581 DATA

CELESTIAL COORDINATES
Right Ascension 15h 19m 26.83s

Declination -7° 43' 20.19"

DISTANCE
20.4 light years

CONSTELLATION
Libra (the Scales)

VISUAL MAGNITUDE
10.56

MASS
0.31 solar masses

RADIUS
0.30 solar radii

STAR TYPE
Red dwarf, M3V

TEMPERATURE
3480 K

ROTATION PERIOD
133 days

NUMBER OF PLANETS
3, possibly 6

Artist's impression of the five-Earth mass planet, Gliese 581 c.

Gliese 581 b, c and e

This red dwarf in Libra is the 89th closest star to the Sun – a stone's throw away at just 20 light years. It hosts three confirmed planets, and an equal number awaiting verification.

In 2005, astronomers at the European Southern Observatory (ESO) found the first planet around Gliese 581. Called Gliese 581 b (or GJ 581 b), it is a Neptune-sized world that orbits its host star in 5.4 days. At the time, according to ESO, the astronomers had already seen hints of another planet. 'They therefore obtained a new set of measurements and found a "Super-Earth" (Gliese 581 c), and also clear indications for another one, an eight-Earth-mass planet completing an orbit in 84 days', said an ESO press release. This latter exoplanet, however, Gliese 581 d, is unconfirmed.

The discoveries were made thanks to HARPS (High Accuracy Radial velocity Planetary Searcher), perhaps the most precise spectrograph in the world, located on the ESO 3.6-m telescope at La Silla, Chile. 'HARPS is a unique planet-hunting machine', said Michel Mayor, from Geneva Observatory, and HARPS Principal Investigator. 'Given the incredible precision of HARPS, we have focused our effort on low-mass planets. And we can say without doubt that HARPS has been very successful: out of the 13 known planets with a mass below 20 Earth masses, 11 were discovered with HARPS.'

The Gliese 581 star has about 30% the mass of our Sun, and the outermost planet is closer to its star than we are to the Sun. The fourth planet candidate, Gliese 581 g, would thus be a planet that could possibly sustain life.

Explore the Gliese 581 System

The Star

1 The European Southern Observatory (ESO) notes that Gliese 581 is among the 100 closest stars to us, located only 20.5 light years away in the constellation Libra. It has a mass and radius of only one-third that of the Sun. Such red dwarfs are intrinsically at least 50 times fainter than the Sun and are the most common stars in our galaxy. Of the 100 closest stars to the Sun, 80 belong to this class.

Habitable Zone

2 'Red dwarfs are ideal targets for the search for low-mass planets where water could be liquid', said astronomer Xavier Bonfils. Because Gliese 581 is so dim, though, planets have to orbit it much closer than they do our Sun if they are to be habitable, or else they will be too cold to host liquid water, considered by most scientists as essential for life.

Gliese 581 b

3 This is the most massive of the three confirmed exoplanets around Gliese 581. With a mass of 15.8 times that of the Earth, it is probably an ice giant, somewhat like Neptune or Uranus. At the time of its discovery in 2005, it was the fifth planet known to orbit a red dwarf star. It orbits between planets c (closest) and e (furthest).

Gliese 581 c

4 According to ESO, this rocky exoplanet, found in 2007, is one of the smallest ever discovered up to now, and it completes a full orbit in 13 days. 'We have estimated that the mean temperature of this 'Super-Earth' lies between 0 and 40°C, and water would thus be liquid', explains Stéphane Udry, from Switzerland's Geneva Observatory and lead author of the paper reporting the result. 'Moreover, its radius should be only 1.5 times the Earth's radius, and models predict that the planet should be either rocky – like our Earth – or fully covered with oceans', he adds.

Gliese 581 e

5 After more than four years of observations using HARPS, the most successful low-mass-exoplanet hunter in the world, astronomers discovered what was at the time (2009) the lightest known exoplanet. Gliese 581 e is only about twice the mass of our Earth and circles its host star in 3.15 days. It is the most distant confirmed planet in the system, but still much closer in than Mercury is to the Sun.

Top Tip

Gliese 581 has been the subject of controversy, because three of the six planets discovered there have since been cast into doubt. As well as using the HARPS spectrograph, astronomers sponsored by NASA have also studied the system with the W. M. Keck Observatory in Hawaii, one of the world's largest optical telescopes. The problem is that there are multiple ways to address the data and astronomers are currently uncertain whether this system is best described by a model with three, four, five or even six planets. The three unconfirmed planets are all much further from their star, at distances of 0.13 AU, 0.22 AU, and 0.75 AU respectively.

Getting There & Away

Despite its relative proximity, a trip to the Gliese 581 system will still take a good 790,000 years by Space Shuttle. A visit to Proxima Centauri or Barnard's Star instead will cut that travel time considerably.

GLIESE 625 DATA

CELESTIAL COORDINATES
Right Ascension 18h 25m 24.6233s

Declination 54° 18' 14.7658"

DISTANCE
21.11 light years

CONSTELLATION
Draco (the Dragon)

VISUAL MAGNITUDE
10.17

MASS
0.3 solar masses

RADIUS
0.31 solar radii

STAR TYPE
Red dwarf, M2V

TEMPERATURE
3499 K

ROTATION PERIOD
Unknown

NUMBER OF PLANETS
1

GLIESE 625 B DATA

PLANET TYPE
Super-Earth

MASS
2.82 x Earth

RADIUS
Unknown

ORBITAL PERIOD
14.628 days

ORBITAL RADIUS
0.078 AU

DETECTION TYPE
Radial Velocity

DISCOVERED
2017

Gliese 625 b may share some of Earth's characteristics, but on a larger scale.

Gliese 625 b

Gliese 625 b is a monster-sized rocky exoplanet that orbits a red dwarf star. If there is water there, it could be habitable.

Situated a mere 21 light years away in Draco, Gliese 625 is a red dwarf star roughly one third of the mass and radius of our Sun. In 2017, a team of astronomers led by Alejandro Suarez Mascareño of the Canary Islands Institute of Astrophysics completed an in-depth, three-year study of this star. They used the HARPS-N spectrograph on the island of La Palma, one of Spain's Canary Islands. Once they had analysed the data, they detected a tell-tale variation in the radial velocity pointing to a new exoplanet in orbit, now dubbed Gliese 625 b.

The new discovery is a so-called 'Super-Earth' – rocky but 2.8 times the mass of the Earth. It swings around its host star in 14.6 days, at a close distance of just 0.08 AU. The team found that the planet orbits within its star's habitable zone – the region in which water ought to be liquid. It may be too close to its star, but as a red dwarf, Gliese 625 is smaller and dimmer than our Sun, and therefore less apt to blast objects with intense radiation. With a temperate surface temperature in the region of 77°C (170°F), Gliese 625 b could turn out to be habitable for some forms of life. However, more research is required before jumping to that conclusion, specifically targeting the planet's atmosphere, if there is one, for analysis using spectrometers.

A conceptualisation of rocky planet Gliese 667 Cc.

Gliese 667 Cb and Cc

Astronomers struck gold here: a triple star with a system of at least two planets orbiting the smallest member, a red dwarf called Gliese 667 C. Both are likely rocky and potentially habitable.

The innermost planet in this system around a multiple star is Gliese 667 Cb, discovered in 2009. It's about 5.6 times the mass of the Earth, orbiting in just seven days at 1/20th of the Earth-Sun separation. The outermost planet, Gliese 667 Cc, has a mass of about 3.7 times that of Earth and came to light in 2011, discovered by astronomers combing through data from the European Southern Observatory's 3.6-meter telescope in Chile. It orbits its red dwarf in the habitable zone, though closely enough - with a mere 28-day orbit - to make the planet subject to intense flares that could erupt periodically from the star's surface. Still, its sun is smaller and cooler than ours, and Gliese 667 Cc's orbital distance means it probably receives around 90% of the energy we get from the Sun. That's a point in favour of life, if the planet's atmosphere is something like ours.

In total, six planets have been proposed as existing in this system. However, this may be a spurious conclusion. 'Our analysis shows that the data only provide strong evidence for the presence of two planets', wrote Frank Perez of Carnegie Observatories, Pasadena, in a 2014 paper for *Monthly Notices of the Royal Astronomical Society*. More studies will have to confirm the actual number.

GLIESE 667 C DATA

CELESTIAL COORDINATES
Right Ascension 17h 18m 57.16s

Declination -34° 59' 23.14"

DISTANCE
23.6 light years

CONSTELLATION
Scorpius (the Scorpion)

VISUAL MAGNITUDE
10.2

MASS
0.31 solar masses

RADIUS
0.42 solar radii

STAR TYPE
Red dwarf, M1.5V

TEMPERATURE
3700 K

ROTATION PERIOD
105 days

NUMBER OF PLANETS
2; 4 candidates

GLIESE 832 DATA

CELESTIAL COORDINATES
Right Ascension 2h 33m 33.98s

Declination -49° 0' 32.40"

DISTANCE
16.19 light years

CONSTELLATION
Grus (the Crane)

VISUAL MAGNITUDE
8.66

MASS
0.45 solar masses

RADIUS
0.48 solar radii

STAR TYPE
Red dwarf, M2V

TEMPERATURE
3620 K

ROTATION PERIOD
45.7 days

NUMBER OF PLANETS
2

A visualisation of Gliese 832 c, this system's rocky planet.

Gliese 832 b and c

This red dwarf star in the southern constellation of Grus has a rocky planet in a very tight orbit and a gas giant orbiting much further out.

Gliese 832 is a system of two extremes. On the one hand there is a likely rocky planet, five times the mass of the Earth, swinging around its parent star in a matter of just 35.68 days. This is Gliese 832 c. The short orbital period is due to its proximity to its dim red dwarf star, well inside the orbit of Mercury around our Sun. Its additional planet, Gliese 832 b, is the other extreme - a massive gas giant comparable to Jupiter on a 10-year orbit. It is 21 times more distant in the system than its rocky sibling.

An interesting coincidence, however, is that Gliese 832 c receives just about the same average 'flux' (sunlight) from its parent star as does the Earth. Much remains unknown about this exoplanet's true mass, size and atmosphere, though. If Gliese 832 c has an atmosphere like Earth, it may be a 'Super-Earth' undergoing strong seasons but capable of supporting life. Alternatively, if Gliese 832 c has a thick atmosphere like Venus, it is unlikely to support life as we know it, potentially the victim of a similar runaway greenhouse affect. The 16-light year distance makes the Gliese 832 planetary system currently the fifth nearest to Earth that could theoretically support life if conditions were right.

The rings of Gliese 876 c.

Gliese 876 b, c, d and e

Gliese 876 is a multi-planet system centred on a red dwarf star 15 light years away in Aquarius. There are two gas giants, a rocky planet, and another possibly like Neptune.

Scientists' knowledge of Gliese 876 is constantly evolving. They discovered its first exoplanet, the gas giant Gliese 876 b, in 1998, making it one of the first few alien worlds known. Since then, subsequent studies have found three additional worlds, in 2001, 2005 and most recently in 2010. These are another gas giant (Gliese 876 c), a Super-Earth (Gliese 876 d), and another which may be either a Super-Earth or a Neptune-like world (Gliese 876 e).

In 2002, Gliese 876 b became the first extrasolar world whose mass was confirmed using the astrometry technique. As a planet orbits its star, the star makes a small ellipse on the sky in response. Precise measurements of this motion can yield the planet's mass.

Dr George Benedict, from the University of Austin, Texas, had to observe the star's yo-yo motion for over two years, using a total of 27 orbits' worth of Hubble Space Telescope observations. 'Making these kinds of measurements of a star's movement on the sky is quite difficult', Benedict emphasised. 'We're measuring angles (0.5 milliarcseconds) equivalent to the size of a quarter seen from 3000 miles away.'

Another distinctive feature of this system is the rings believed to encircle gas giant Gliese 876 c, which are thought to be similar to those of Saturn. That means it likely has moons too, which might have liquid water.

GLIESE 876 DATA

CELESTIAL COORDINATES
Right Ascension 22h 53m 16.73s
Declination -14° 15' 49.30"

DISTANCE
15.25 light years

CONSTELLATION
Aquarius (the Water Bearer)

VISUAL MAGNITUDE
10.15

MASS
0.37 solar masses

RADIUS
0.376 solar radii

STAR TYPE
Red dwarf, M4V

TEMPERATURE
3129 K

ROTATION PERIOD
96.9 days

NUMBER OF PLANETS
4

GLIESE 3470 DATA

CELESTIAL COORDINATES
Right Ascension 7h 59m 6.0s
Declination 15° 23' 30"

DISTANCE
100 light years

CONSTELLATION
Cancer (the Crab)

VISUAL MAGNITUDE
12.3

MASS
0.51 solar masses

RADIUS
0.48 solar radii

STAR TYPE
Red dwarf, M1.5

TEMPERATURE
3652 K

ROTATION PERIOD
Unknown

NUMBER OF PLANETS
1

GLIESE 3470 B DATA

PLANET TYPE
Hot Neptune

MASS
13.73 x Earth

RADIUS
3.88 x Earth

ORBITAL PERIOD
3.3366 days

ORBITAL RADIUS
0.03557 AU

DETECTION TYPE
Transit

DISCOVERED
2012

Neptune-like Gliese 3470 b compared to Earth's size.

Gliese 3470 b

Gliese 3470 b is a rarity – a so-called 'Hot Neptune' – similar in mass to Neptune but far hotter. Only a handful are known.

This is a planet that is vanishing at an alarming rate. Gliese 3470 b orbits its parent star so closely - around ten times closer to its star than Mercury is to the Sun - that it is literally being evaporated while astronomers watch.

Hot Jupiters are everywhere - massive gas planets in star-hugging orbits with concomitant blisteringly high surface temperatures. Hundreds are now known. But hot Neptunes are far rarer. The first one ever found was Gliese 436 b in 2007 and at the time of its discovery, in 2012, Gliese 3470 b was only the second, although a few others have been found since then. Its radius is estimated to be about 4 times that of Earth's radius, and its mass approximates the mass of (what else) Neptune.

'This is the smoking gun that planets can lose a significant fraction of their entire mass', said physicist and planetary scientist Dr David Sing of Johns Hopkins University. 'Gliese 3470 b is losing more of its mass than any other planet we have seen so far; in only a few billion years from now, half of the planet may be gone.' Once the closeness of Gliese 3470 causes the atmosphere to fully boil or evaporate away, Gliese 3470 b may be left only as a rocky core. It's a vivid example of why closely orbiting planets are considered to be outside their star's habitable zone.

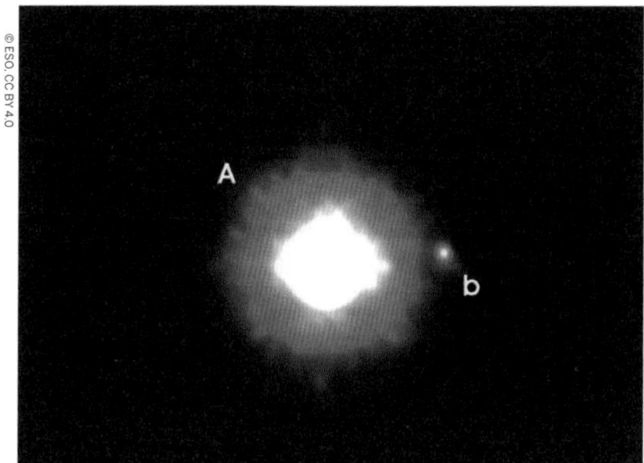

A visual of GQ Lupi taken by the Very Large Telescope (VLT).

GQ Lupi b

GQ Lupi b is somewhat of a mystery. Like Gliese 504 b, astronomers can't quite decide if it's an exoplanet or a much more massive brown dwarf.

Astronomers are finding more exoplanets directly, but this is a difficult task as these are generally hidden in the 'glare' of the host stars. According to the European Southern Observatory, 'To partly overcome this problem, astronomers study very young objects.' This is because planetary mass objects are much hotter and brighter when young and therefore can be more easily detected next to their bright stars. GQ Lupi is one such young star, studied by astronomers from the University of Jena, Germany.

ESO continues: 'The astronomers observed GQ Lupi ... with the adaptive optics instrument NaCo attached to Yepun, the fourth 8.2-m Unit Telescope of the Very Large Telescope located on top of Cerro Paranal (Chile). The instrument's adaptive optics overcomes the distortion induced by atmospheric turbulence, producing extremely sharp near-infrared images.' The data showed the signature of a very cool object with a radius that is twice as large as Jupiter. However, whether GQ Lupi b is a large, cold planet or a small 'failed star' more properly called a brown dwarf remains to be seen.

GQ LUPI DATA

CELESTIAL COORDINATES
Right Ascension 15h 49m 12.14s
Declination -35° 39' 3.95"

DISTANCE
500 light years

CONSTELLATION
Lupus (the Wolf)

VISUAL MAGNITUDE
11.4

MASS
0.7 solar masses

RADIUS
Unknown

STAR TYPE
T Tauri, K7V

TEMPERATURE
5150 K

ROTATION PERIOD
Unknown

NUMBER OF PLANETS
1

GQ LUPI B DATA

PLANET TYPE
Gas Giant or Brown Dwarf

MASS
Various estimates, 1–36 x Jupiter

RADIUS
Various estimates, 3–4.6 x Jupiter

ORBITAL PERIOD
Approx. 1200 years

ORBITAL RADIUS
Approx. 100 AU

DETECTION TYPE
Imaging

DISCOVERED
2005

Serving Up Hot Jupiters

The last decade has seen a bonanza of exoplanet discoveries. Over 4000 exoplanets – planets outside our solar system – have been confirmed so far, and many more candidate exoplanets have been identified. Many of these exotic worlds belong to a class known as hot Jupiters. These are gas giants like Jupiter but much hotter, with orbits that take them feverishly close to their stars.

At first, hot Jupiters were considered oddballs, since we don't have anything like them in our own solar system. But as more were found, in addition to many other smaller planets that orbit very closely to their stars, our solar system started to seem like the real misfit.

'We thought our solar system was normal, but that's not so much the case', said astronomer Greg Laughlin of the University of California, Santa Cruz, co-author of a study from NASA's Spitzer Space Telescope that investigates hot Jupiter formation.

As common as hot Jupiters are now known to be, they are still shrouded in mystery. How did these massive orbs form, and how did they wind up so shockingly close to their stars?

The Spitzer telescope found new clues by observing a hot Jupiter known as HD 80606b, situated 190 light years from Earth. This planet is unusual in that it has a wildly eccentric orbit almost like that of a comet, swinging very close to its star and then back out to much greater distances over and over again every 111 days. One side of the planet is thought to become dramatically hotter than the other during its harrowing close approaches. In fact, when the planet is closest to its host star, the side facing the star quickly heats up to more than 1100°C (2000°F).

HD 80606b is thought to be in the process of migrating from a more distant orbit to a much tighter one typical of hot Jupiters. One of the leading theories of hot-Jupiter formation holds that gas giants in distant orbits become hot Jupiters when the gravitational influences from nearby stars or planets drive them into closer orbits. The planets start out in eccentric orbits, then, over a period of hundreds of millions of years, are thought to gradually settle down into tight, circular orbits.

Spitzer previously studied HD 80606b in 2009. The latest observations are more detailed, thanks to a longer observing time of 85 hours and improvements in Spitzer's sensitivity to exoplanets.

A key question addressed in the new study is: how long is HD 80606b taking to migrate from an eccentric to a circular orbit? One way to assess this is to look at how 'squishy' the planet is. When HD 80606b whips closely by its star, the gravity of the star squeezes it. If the planet is squishier, or more pliable, it can better dissipate this gravitational energy as heat. And the more heat that is dissipated, the faster the planet will transition to a circular orbit, a process known as circularisation.

'If you take a Nerf ball and squeeze it a bunch of times really fast, you'll see that it heats up', said Laughlin. 'That's because the Nerf ball is good at transferring that mechanical energy into heat. It's squishy as a result.'

The Spitzer results show that HD 80606b does not dissipate much heat when it is squeezed by gravity during its close encounters – and thus is not squishy, but rather stiffer as a whole. This suggests the planet is not circularising its orbit as fast as expected, and may take another 10 billion years or more to complete. That's almost the same amount of time that our Universe has existed!

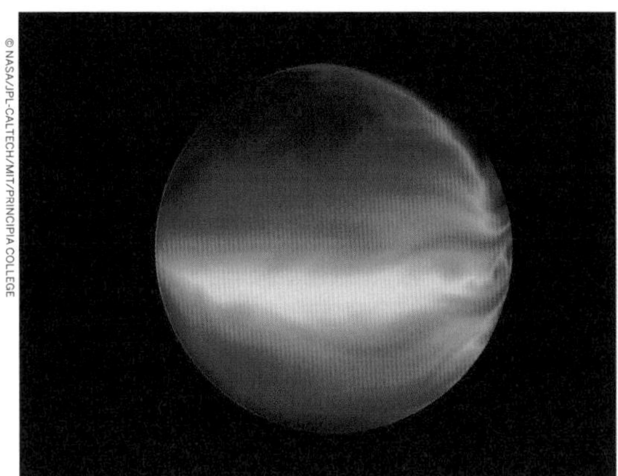

The turbulent atmosphere of a hot Jupiter is shown in this simulation.

HAT-P-7b

This monster planet, mightier than Jupiter, had a surprise for astronomers – ferocious winds in an atmosphere of rubies and sapphires.

First discovered in 2008, HAT-P-7b (or Kepler-2b) is about 1040 light years away from us. It is an exoplanet 40% larger than Jupiter and 500 times more massive than the Earth - and orbits a star 50% more massive, and twice as large, as the Sun. In 2016, astronomers led by Dr David Armstrong from the UK's Warwick University found signs of 'powerful changing winds' there. It is among the first weather systems discovered on a gas giant outside our solar system.

Armstrong's research team discovered that the gas giant HAT-P-7b is affected by large-scale changes in the strong winds moving across the planet, likely leading to catastrophic storms. This shift is caused by an equatorial jet with dramatically variable wind-speeds - at their fastest, pushing vast amounts of cloud across the planet. The clouds themselves would be visually stunning - likely made up of corundum, the mineral which forms rubies and sapphires. 'These results show that strong winds circle the planet, transporting clouds from the nightside to the dayside. The winds change speed dramatically, leading to huge cloud formations building up then dying away.' Thanks to this pioneering research, astrophysicists can now begin to explore how weather systems on planets outside our solar system change over time.

HAT-P-7 DATA

CELESTIAL COORDINATES
Right Ascension 19h 28m 59.3534s
Declination 47° 58' 10.229"

DISTANCE
1040 light years

CONSTELLATION
Cygnus (the Swan)

VISUAL MAGNITUDE
10.46

MASS
1.47 solar masses

RADIUS
1.84 solar radii

STAR TYPE
Yellow-white dwarf, F8V

TEMPERATURE
6441 K

ROTATION PERIOD
Unknown

NUMBER OF PLANETS
1

HAT-P-7B DATA

PLANET TYPE
Hot Jupiter

MASS
1.806 x Jupiter

RADIUS
1.363 x Jupiter

ORBITAL PERIOD
2.204 days

ORBITAL RADIUS
0.0381 AU

DETECTION TYPE
Transit

DISCOVERED
2008

HAT-P-11 DATA

CELESTIAL COORDINATES
Right Ascension 19h 50m 50.25s
Declination 48° 4' 51.10"

DISTANCE
123 light years

CONSTELLATION
Cygnus (the Swan)

VISUAL MAGNITUDE
9.473

MASS
0.81 solar masses

RADIUS
0.683 solar radii

STAR TYPE
Orange dwarf, K4

TEMPERATURE
4780 K

ROTATION PERIOD
Unknown

NUMBER OF PLANETS
1

HAT-P-11B DATA

PLANET TYPE
Hot Neptune

MASS
23.4 x Earth

RADIUS
4.36 x Earth

ORBITAL PERIOD
4.89 days

ORBITAL RADIUS
0.052 AU

DETECTION TYPE
Radial Velocity

DISCOVERED
2009

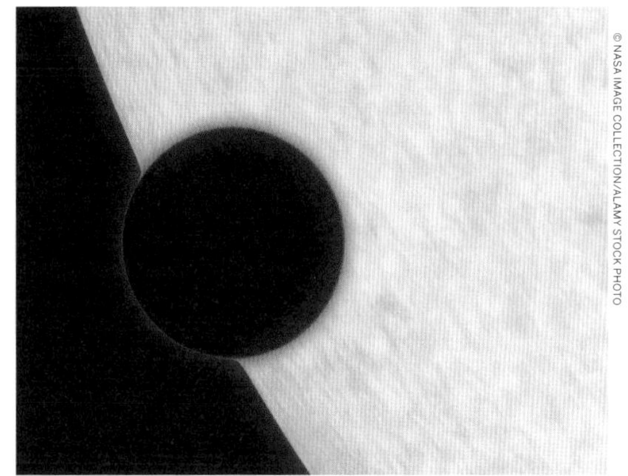

HAT-P-11b orbits much closer in than does its Neptune analogue in our solar system.

HAT-P-11b

Using data from three of NASA's space telescopes, astronomers discovered clear skies and steamy water vapour on this gaseous Neptune-sized planet.

The planet HAT-P-11b is categorised as an exo-Neptune – a Neptune-sized planet that orbits the star HAT-P-11. It is located about 120 light years away in the constellation Cygnus. This planet orbits much closer to its star than does our Neptune, making one lap roughly every five days. It is a warm world thought to have a rocky core and gaseous atmosphere. As a gas giant rather than an ice giant, apart from the size, this exoplanet might resemble Saturn and Jupiter more than Neptune in terms of its make-up. After its discovery in 2009, not much else was known about the composition of the planet, or other exo-Neptunes like it, until 2014, when a team led by Jonathan Fraine of the University of Maryland was able to analyse its atmosphere.

Using Hubble's Wide Field Camera 3, Fraine's group employed a technique called transmission spectroscopy, in which a planet is observed as it crosses in front of its parent star. Starlight filters through the rim of the planet's atmosphere and if molecules like water vapour are present they absorb some of the starlight, leaving distinct signatures in the light that reaches our telescopes. Using this strategy, Hubble was able to detect water vapour in HAT-P-11b, indicating the possibilities of using this technique on exoplanet targets to learn how our solar system's planets compare.

Explore HAT-P-11b

Parent Star

1 The parent star, HAT-P-11, is a metal-rich orange dwarf, about 81% the mass and 68% the radius of the Sun and somewhat cooler at 4780K. The term 'metal-rich' is one scientists use to describe stars that have a higher than usual abundance of elements heavier than helium. Astronomers believe HAT-P-11 has starspots, cooler 'freckles' on the face of stars similar to sunspots. At magnitude 9.5, you'll need a small amateur telescope if you want to view it.

Detecting Water?

2 Clouds in a planet's atmosphere can block the view to underlying molecules that reveal information about the planet's composition and history. Finding clear skies on a Neptune-size planet is a good sign that smaller planets might have similarly good visibility. But before the team could celebrate clear skies on the exo-Neptune, they had to show that its starspots were not the real sources of water vapour. Cool starspots on the parent star can contain water vapour that might erroneously appear to be from the planet.

Further Observations

3 The team turned to Kepler and Spitzer. Kepler had been observing one patch of sky for years, and HAT-P-11b happens to lie in that field. Those visible-light data were combined with targeted Spitzer observations taken at infrared wavelengths. By comparing these observations, the astronomers figured out that the starspots were too hot to have any steam. It was at that point the team could celebrate detecting water vapour on a world unlike any in our solar system. This discovery indicates the planet did not have clouds blocking the view, a hopeful sign that more cloudless planets can be located and analysed in the future.

Atmospheric Composition

4 The observational results demonstrate that HAT-P-11b is blanketed not only in water vapour, but also hydrogen gas and likely other yet-to-be-identified molecules. Theorists will be drawing up new models to explain the planet's makeup and origins. 'We think that exo-Neptunes may have diverse compositions, which reflect their formation histories', said study co-author Heather Knutson of the California Institute of Technology. 'Now, with data like these, we can begin to piece together a narrative for the origin of these distant worlds.'

Future Applications

5 The astronomers plan to examine more exo-Neptunes in the future, and hope to apply the same method to 'Super-Earths' – massive, rocky cousins to our home world with up to 10 times the mass. NASA's James Webb Space Telescope will search 'Super-Earths' for signs of water vapour and other molecules; however, finding signs of oceans and potentially habitable worlds is likely some way off.

Top Tip

Lightning cracks against the maelstrom of a Neptune-like sky, illuminating the hideous figure of Frankenstein's monster – just like the day of the monster's unnatural birth, when the harnessed power of lightning brought his scavenged parts to life. But don't judge him by his looks; Frankenstein's monster (we like to call him Frankenstein Jr.) is a gentle giant seeking to be understood. Scorned by humankind as a science experiment gone horrifically awry, we've found a home planet for Frank Jr on HAT-P-11b, 123 light years away. Water in the atmosphere and a weak radio signal once led astronomers to believe this safe haven for monster misfits had enormous lightning storms, many times stronger than those on Jupiter or Earth.

Getting There & Away

With today's typical technology, the journey to HAT-P-11b will require in the region of five million years. Let's hope the spacecraft offers cryogenic suspension.

HD 40307 DATA

CELESTIAL COORDINATES
Right Ascension 5h 54m 4.2409s
Declination -60° 1' 24.498"

DISTANCE
41.8 light years

CONSTELLATION
Pictor (the Painter)

VISUAL MAGNITUDE
7.17

MASS
0.75 solar masses

RADIUS
0.716 solar radii

STAR TYPE
Orange dwarf, K2.5V

TEMPERATURE
4977 K

ROTATION PERIOD
31.8 days

NUMBER OF PLANETS
6

HD 40307 G DATA

PLANET TYPE
Super-Earth or Neptune-Like

MASS
7.09 x Earth

RADIUS
2.39 x Earth

ORBITAL PERIOD
197.8 days

ORBITAL RADIUS
0.600 AU

DETECTION TYPE
Radial Velocity

DISCOVERED
2012

Modestly-sized gas giant or habitable terrestrial world?

HD 40307 g

A string of six planets orbits HD 40307, and one of them, HD 40307 g, may just be a habitable terrestrial world. But it could equally be a small gas giant.

More than double the volume of the Earth, HD 40307 g straddles the line between 'Super-Earth' and 'Neptune-like', and scientists aren't sure if it has a rocky surface or one that's buried beneath thick layers of gas and ice. One thing is certain though: at seven times the Earth's mass, its gravitational pull is much, much stronger.

HD 40307 g is one of a set of six planets in orbit about the orange star HD 40307, nearly 43 light years away in Pictor. The others are all more massive than Earth as well, with HD 40307 g being the second heaviest. What makes this planet more interesting than its siblings is that it's situated much further from the star and therefore within its habitable zone. The others are too close and therefore too hot for liquid water. But for the planet to be habitable it must have a solid surface, like Earth, and this remains uncertain. When asked whether he thought the planet was a 'Super-Earth', Mikko Tuomi, one of the researchers who discovered it in 2012, said, 'If I had to guess, I would say 50-50 ... But the truth at the moment is that we simply do not know whether the planet is a large Earth or a small, warm Neptune without a solid surface.' In classifying space objects, an abundance of caution often proves useful.

This star seems to not only have exoplanets in orbit, but also an asteroid belt.

HD 69830 b, c and d

The star HD 69830 made headlines in 2005 when astronomers discovered clear signs of an extrasolar asteroid belt. The following year, three planets were also confirmed.

Asteroid belts are the junkyards of planetary systems. They are littered with the rocky scraps of failed planets, which occasionally crash into each other, kicking up plumes of dust. In our own solar system, asteroids have collided with Earth, the moon and other planets. The belt in HD 69830 is thicker than our own asteroid belt, with 25 times as much material. If our solar system had a belt this dense, its dust would light up the night skies as a brilliant band. The alien belt is also much closer to its star. Our asteroid belt lies between the orbits of Mars and Jupiter, whereas this one is located inside an orbit equivalent to that of Venus.

According to a press release by the European Southern Observatory, where the planets were discovered, 'The three planets have minimum masses between 10 and 18 times the mass of the Earth. Extensive theoretical simulations favour an essentially rocky composition for the inner planet, and a rocky/gas structure for the middle one. The outer planet has probably accreted some ice during its formation, and is likely to be made of a rocky/icy core surrounded by a quite massive envelope. Further calculations have also shown that the system is in a dynamically stable configuration.'

HD 69830 DATA

CELESTIAL COORDINATES
Right Ascension 8h 18m 23.947s
Declination -12° 57' 5.8116"

DISTANCE
40.7 light years

CONSTELLATION
Puppis (the Poop Deck)

VISUAL MAGNITUDE
5.98

MASS
0.863 solar masses

RADIUS
0.905 solar radii

STAR TYPE
Sun-like, G2V

TEMPERATURE
5394 K

ROTATION PERIOD
35.1 days

NUMBER OF PLANETS
3

HD 149026 DATA

CELESTIAL COORDINATES
Right Ascension 19h 30m 29.6185s
Declination 38° 20' 50.308"

DISTANCE
250 light years

CONSTELLATION
Hercules

VISUAL MAGNITUDE
8.15

MASS
1.345 solar masses

RADIUS
1.541 solar radii

STAR TYPE
Yellow subgiant, GOIV

TEMPERATURE
6147 K

ROTATION PERIOD
Unknown

NUMBER OF PLANETS
1

HD 149026 B DATA

PLANET TYPE
Hot Jupiter

MASS
0.36 x Jupiter

RADIUS
0.725 x Jupiter

ORBITAL PERIOD
2.876 days

ORBITAL RADIUS
0.042 AU

DETECTION TYPE
Radial Velocity

DISCOVERED
2005

HD 149026 b broils on its light side.

HD 149026 b

This two-faced gas giant is blazing hot on its sunlit side, and much cooler on its dark side. It emits almost no light.

Some 250 light years away is a planet which is so hot that astronomers believe it is absorbing almost all of the heat from its star, and reflecting very little to no light. With this almost total lack of reflectivity, HD 149026 b is one of the blackest known planets in the universe – as well as one of the hottest. The planet is a sweltering 2038°C (3700°F), about three times hotter than the rocky surface of Venus, itself the hottest planet in our solar system.

The temperature of this dark and balmy planet was taken with the Spitzer Space Telescope. While the planet reflects no visible light, its heat causes it to radiate a little visible and a lot of infrared light. Spitzer, an infrared observatory, was able to measure this infrared light through a technique called secondary eclipse. HD 149026 b (dubbed Smertrios by the IAU) is what is known as a transiting planet, which means that it crosses in front of and passes behind its star – the secondary eclipse – when viewed from Earth. By determining the drop in total infrared light that occurs when the planet disappears, astronomers can figure out how much infrared light is coming from the planet alone, in effect taking its temperature. The Spitzer observations of HD 149026b also suggest a hot spot in the middle of the side of the planet that always faces its star. Even though the planet is black, the spot would glow like a black lump of charcoal. HD 149026b is thought to be tidally locked, such that one side of the planet perpetually bakes under the heat of its sun.

This artist's concept illustrates the hottest planet yet observed in the Universe, seen from its nightside.

HD 189733 DATA

CELESTIAL COORDINATES
Right Ascension 20h 00m 43.71s
Declination 22° 42' 39.1"

DISTANCE
63.4 light years

CONSTELLATION
Vulpecula (the Little Fox)

VISUAL MAGNITUDE
7.66

MASS
0.85 solar masses

RADIUS
0.81 solar radii

STAR TYPE
Orange dwarf, K1.5V

TEMPERATURE
4875 K

ROTATION PERIOD
11.95 days

NUMBER OF PLANETS
1

HD 189733 B DATA

PLANET TYPE
Hot Jupiter

MASS
1.16 x Jupiter

RADIUS
1.14 x Jupiter

ORBITAL PERIOD
2.22 days

ORBITAL RADIUS
0.031 AU

DETECTION TYPE
Transit

DISCOVERED
2005

Artist's impression of the deep blue planet HD 189733 b.

HD 189733 b

To the human eye, this far-off planet looks bright blue. But any space traveller confusing it with the friendly skies of Earth would be fatally mistaken.

Some 63 light years away in the constellation Vulpecula lies the enigmatic planet called HD 189733 b. This so-called hot Jupiter was discovered in 2005 as it passed in front of its orange star, betraying its presence as the star's light dipped by some 3%. It was the first exoplanet whose atmosphere was confirmed to contain water vapour. But don't let that fool you – this is a far from hospitable place, not with temperatures of 1000°C (1832°F).

The nightmare world of HD 189733 b is the killer you never see coming. The weather on this world is deadly. Its winds blow up to 8700 km/h (4971 mph), at seven times the speed of sound, whipping all would-be travellers in a sickening spiral around the planet. And getting caught in the rain on this planet is more than an inconvenience; it's death by a thousand cuts. This scorching alien world possibly rains glass – sideways – in its howling winds. That's because the cobalt blue colour comes not from the reflection of a tropical ocean, as on Earth, but rather a hazy, blow-torched atmosphere containing high clouds laced with silicate particles.

At a distance of 63 light years from us, this turbulent alien world is one of the nearest exoplanets to Earth that can be seen crossing the face of its star. By observing this planet before, during and after it disappeared behind its host star during orbit, astronomers were able to deduce that HD 189733b is a deep, azure blue – reminiscent of Earth's colour as seen from space.

Explore HD 189733 b

Star System

1 The parent star, HD 189733, is about 80% of the radius and mass of the Sun, but is cooler and therefore more orange. It actually belongs to a binary star system. Its stellar partner is a much smaller star, a red dwarf. However, the two stars are far apart, about 210 times the Earth-Sun distance, and take 3200 years to orbit each other. The planet orbits the larger orange star, not both of them.

Planet Vital Statistics

2 HD 189733 b is a so-called hot Jupiter. It's similar in mass and radius to Jupiter in our own solar system, but orbits its star in a much tighter gravitational bind. It is around 30 times closer to its star than Earth is to the Sun, and whips around it in a mere 2.2 days. According to this planet's calendar, a 70-year-old Earthling would have a rip-old-age of 11,613!

X-ray Transit

3 Nowadays, planets are routinely discovered as they transit their stars optically. But in 2013, astronomers trained NASA's Chandra X-ray Observatory and the European Space Agency's XMM Newton Observatory on HD 189733 and for the first time observed a planet transiting in the X-ray region of the electromagnetic spectrum. 'Finally being able to study [an exoplanet atmosphere] in X-rays is important because it reveals new information about the properties of an exoplanet', said Katja Poppenhaeger of the Harvard-Smithsonian Center for Astrophysics (CfA).

Atmosphere

4 The authors of the Chandra work, mentioned above, estimate that the percentage decrease in X-ray light during the transits is about three times greater than the corresponding decrease in optical light. This tells them that the region blocking X-rays from the star is substantially larger than the region blocking optical light from the star, helping to determine the size of the planet's atmosphere.

Evaporating World

5 Astronomers have known for more than a decade that ultraviolet and X-ray radiation from the main star in HD 189733 are evaporating the atmosphere of HD 189733 b over time. Scientists estimate it is losing 100 million to 600 million kilograms of mass per second. HD 189733 b's atmosphere appears to be thinning 25–65% faster than it would be if the planet's atmosphere were smaller.

Top Tip

HD 189733 b is much stranger than our own planet. On this turbulent alien world, the daytime temperature is nearly 1093°C (2000°F). A vast range of metals will readily puddle under those conditions, including brass, aluminium, bronze, copper, magnesium – possibly even iron. Compared to HD 189733 b, Venus is like a balmy afternoon in the Maldives. 'No amount of Sun lotion or ice cream would make this a nice place to visit', warned NASA in a press release. Our advice? Find somewhere closer to home for a vacation.

Getting There & Away

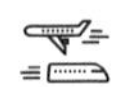

With existing rocket technology, a trip to this fiery planet will take 2.5 million years, give or take. Plenty of time to prepare yourself for the conditions you'll face.

HD 209458 DATA

CELESTIAL COORDINATES
Right Ascension 22h 3m 10.77s
Declination 18° 53' 3.55"

DISTANCE
159 light years

CONSTELLATION
Pegasus (the Winged Horse)

VISUAL MAGNITUDE
7.65

MASS
1.13 solar masses

RADIUS
1.14 solar radii

STAR TYPE
Sun-like, GOV

TEMPERATURE
6071 K

ROTATION PERIOD
Unknown

NUMBER OF PLANETS
1

HD 209458 B DATA

PLANET TYPE
Hot Jupiter

MASS
0.71 x Jupiter

RADIUS
1.35 x Jupiter

ORBITAL PERIOD
3.52 days

ORBITAL RADIUS
0.045 AU

DETECTION TYPE
Transit

DISCOVERED
1999

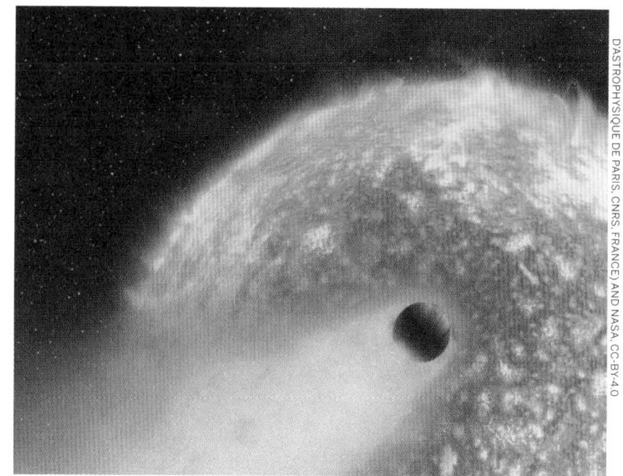

A dramatic close-up of the scorched extrasolar planet HD 209458 b.

HD 209458 b

HD 209458 b is a scorched gas giant skimming its Sun-like star from a distance of just four million miles. Its atmosphere is being blasted away at an alarming rate.

Located 150 light years away in the constellation Pegasus, a planet is metaphorically having its flesh stripped from its bones. HD 209458 b, nicknamed 'Osiris' after an Egyptian god, is a gas giant being destroyed by its voracious star at a rate of more than 35,000 km/h (21,747mph). The parts that once made up the doomed world's atmosphere are smeared through space like an ominous tail. Soon, the planet (and any skeleton armies on it), will be destroyed by gravity.

HD 209458 b, among the earliest exoplanets found, in 1999, is another hot Jupiter – a vast gas world in a tight orbit about its star, heated to extremes of around 1093°C (2000°F). It is an extrasolar planet with an astounding list of firsts: the first discovered transiting its Sun; the first with an atmosphere; the first observed to have an evaporating hydrogen atmosphere (in 2003 by the same team of scientists); and, most recently, the first to have an atmosphere containing oxygen and carbon. Furthermore, the atmospheric 'blow-off' effect was actually observed by the same team during 2003 using the Hubble Space Telescope. The research was led by Alfred Vidal-Madjar of L'Institut d'Astrophysique de Paris, CNRS, France.

Explore HD 209458 b

The Star

1 HD 209458 is a Sun-like star of spectral type G0 (the Sun is G2). Its mass and radius are about 20–25% greater than the Sun's, it spins twice as fast on its axis, and it is around 65°C (149°F) warmer. It isn't visible to the naked eye but, at magnitude 7.65, is still bright enough to be picked up with good binoculars or a telescope.

Evaporating Planet

2 Using the Hubble Space Telescope, scientists observed HD 209458 b passing in front of its parent star, and found oxygen and carbon surrounding the planet in an extended ellipsoidal envelope about the shape of a rugby-ball. These atoms are swept up from the lower atmosphere with the flow of the escaping atmospheric atomic hydrogen, like dust in a supersonic whirlwind. Astronomers estimate the amount of hydrogen gas escaping HD 209458 b to be at least 10,000 tons per second, but possibly much more. The planet may therefore already have lost quite a lot of its mass.

Shedding Light on the Early Solar System

3 According to the scientists, the discovery of the fierce evaporation process is 'highly unusual', but may indirectly confirm theories of our own Earth's childhood. 'This is a unique case in which such a hydrodynamic escape is directly observed. It has been speculated that Venus, Earth and Mars may have lost their entire original atmospheres during the early part of their lives. Their present atmospheres have their origins in asteroid and cometary impacts and outgassing from the planet interiors', said Vidal-Madjar.

Oxygen

4 Oxygen is one of the possible indicators of life that is often looked for in experiments searching for extraterrestrial life (such as those onboard the Viking probes and the Spirit and Opportunity rovers), but according to Vidal-Madjar: 'Naturally this sounds exciting – the possibility of life on Osiris – but it is not a big surprise as oxygen is also present in the giant planets of our solar system, like Jupiter and Saturn.'

Osiris

5 This extraordinary extrasolar planet has provisionally been dubbed 'Osiris'. This is the Egyptian god of the afterlife, the underworld and rebirth, who lost part of his body – like HD 209458 b – after his brother killed and cut him into pieces to prevent his return to life. Charming.

Top Tip

The whole evaporation mechanism of HD 209458 is so distinctive that there is reason to propose the existence of a new class of extrasolar planets – the chthonian planets, a reference to the Greek god Khtôn, used for Greek deities from the hot infernal underworld. The chthonian planets are thought to be the solid remnant cores of 'evaporated gas giants', orbiting even closer to their parent star than Osiris.

Getting There & Away

With existing rocket technology, a trip to this planet will only take 6.1 million years or so. You can get there in about one-tenth of that time if you're pressed – and if you can hijack the solar probes Helios or Parker.

HIP 68468 DATA

CELESTIAL COORDINATES
Right Ascension 14h 1m 3.69s
Declination -32° 45' 24"

DISTANCE
286 light years

CONSTELLATION
Centaurus (the Centaur)

VISUAL MAGNITUDE
9.39

MASS
1.05 solar masses

RADIUS
1.19 solar radii

STAR TYPE
Yellow dwarf, G3V

TEMPERATURE
5857 K

ROTATION PERIOD
Unknown

NUMBER OF PLANETS
2

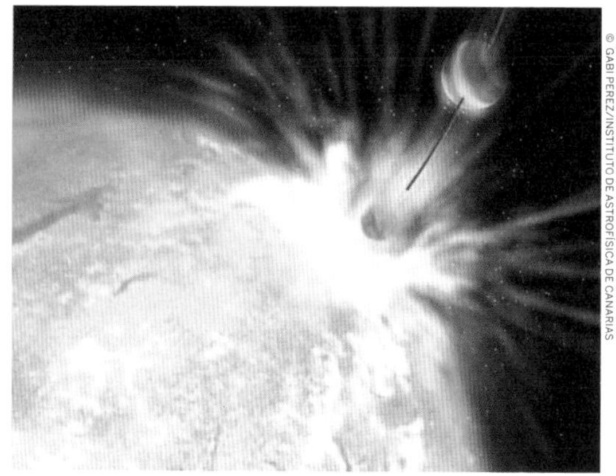

A dramatic visualisation of this 'cannibal' yellow star consuming a planet.

HIP 68468 b and c

Three hundred light years away lurks a seemingly sedate yellow star. But don't let it fool you – this ogre has fed on its planets, and only two remain.

HIP 68468 seems like an ordinary star, so much like the Sun it's uncanny. It's almost the same size, mass, colour, age and temperature as our own local star. In 2016, astronomers headed by Jorge Meléndez used the HARPS spectrograph at La Silla Observatory, Chile, and found two very different worlds here. The outermost is a Neptune-like giant called HIP 68468 c. It is 26 times the mass of the Earth and orbits 0.66AU from the star with a period of 194 days. Its sibling is much closer in, at 0.03AU. This is HIP 68468 b, a massive rocky planet called a Super-Earth.

But there is more to this system than meets the eye. The star's atmosphere is choked with a metal called lithium. Lithium is quickly destroyed in the atmosphere of stars, so there is only one place this surplus could have come from - cannibalized planets. Two may remain now, but the over-abundance of lithium, the authors argue, suggests that the larger planet migrated inwards in the past. They wrote, in the *Astronomy and Astrophysics* journal, that this inward migration 'could also have driven other planets towards the host star, enhancing thus the abundance of lithium ... in HIP 68468.' They conclude: 'The intriguing evidence of planet accretion warrants further observations to verify the existence of the planets ... and to better constrain the nature of the planetary system around this unique star.' In other words? Yellow star HIP 68468 may have already taken its first planetary victim.

An artist's representation of Kapteyn b.

Kapteyn b and c

Speeding across the sky faster than almost any other star, Kapteyn's Star is accompanied by at least one 'Super-Earth' planet, and possibly two.

Kapteyn's Star is a K-type red subdwarf, so called because it is dimmer than a normal red dwarf of comparable size. It is known to have a very rapid proper motion – its speed across the sky as it orbits the galaxy. It moves across the celestial sphere a distance equal to about the diameter of the moon once every 225 years. Just 12.8 light years from Earth, until the discovery of Proxima Centauri b at around one-third the distance these were thought to be the closest known exoplanets to our solar system.

In 2014, astronomers using data from the HARPS instrument at the European Southern Observatory reported the discovery of two rocky planets, weighing five and seven times the mass of the Earth. The innermost planet, Kapteyn b, was touted as being the oldest known potentially habitable exoplanet, with more than double the age of the Earth. However, only the outermost planet, Neptune-like Kapteyn c, is confirmed. Questions have arisen as to whether the original Kapteyn b signal may have come from stellar activity. More research is needed before Kapteyn b can be promoted to a status of confirmed. As the system itself is believed to be 11 billion years old, over twice the age of our own solar system, whether it has one or two exoplanets, it's still a remarkable find.

KAPTEYN'S STAR DATA

CELESTIAL COORDINATES
Right Ascension 5h 11m 50s
Declination 45° 2' 30"

DISTANCE
12.76 light years

CONSTELLATION
Pictor (the Painter)

VISUAL MAGNITUDE
8.853

MASS
0.274 solar masses

RADIUS
0.291 solar radii

STAR TYPE
Subdwarf, M1

TEMPERATURE
3550 K

ROTATION PERIOD
Unknown

NUMBER OF PLANETS
1 or 2

KELT-9 DATA

CELESTIAL COORDINATES
Right Ascension 20h 31m 26.4s
Declination 39° 56' 20"

DISTANCE
620 light years

CONSTELLATION
Cygnus (the Swan)

VISUAL MAGNITUDE
7.56

MASS
2.8x Jupiter

RADIUS
~2 solar radii

STAR TYPE
Blue, A0

TEMPERATURE
10,170 K

ROTATION PERIOD
Unknown

NUMBER OF PLANETS
1

KELT-9B DATA

PLANET TYPE
Hot Jupiter

MASS
2.8 x Jupiter

RADIUS
1.888 x Jupiter

ORBITAL PERIOD
1.481 days

ORBITAL RADIUS
0.03462 AU

DETECTION TYPE
Transit

DISCOVERED
2017

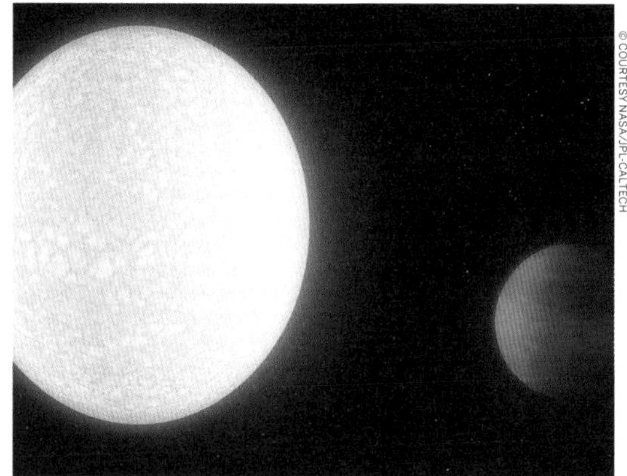

Gas giant KELT-9b is smaller than its star, but not by all that much.

KELT-9b

KELT-9b, also known as HD 195689 b, is a Jupiter-like world that is so hot, it's being vaporised by its own star. Guilty as charged in an act of astronomical cannibalism.

With a dayside temperature of more than 4315°C (7799°F), KELT-9b is a planet that is hotter than most stars. But its blue A-type star, called KELT-9, is even hotter; in fact, it is probably unravelling the planet through evaporation.

'This is the hottest gas giant planet that has ever been discovered', said Scott Gaudi, astronomy professor at Ohio State University in Columbus, who led a study on the topic. Since then, scientists have found Kepler-70b, which is ever hotter. Both of them have higher temperatures than most stars.

KELT-9b is 2.8 times more massive than Jupiter, but only half as dense. Scientists would expect the planet to have a smaller radius, but the extreme radiation from its host star has caused the planet's atmosphere to puff up like a balloon. Because the planet is tidally locked to its star – as the moon is to Earth – one side of the planet is always facing toward the star, and one side is in perpetual darkness. Molecules such as water, carbon dioxide and methane can't form on the dayside because it is bombarded by too much ultraviolet radiation. The properties of the nightside are still mysterious – molecules may be able to form there, but probably only temporarily.

Explore KELT-9b

The Star

1 KELT-9 is only 300 million years old, which is young in star time. It is more than twice as large and nearly twice as hot as our Sun. Given that the planet's atmosphere is constantly blasted with high levels of ultraviolet radiation, it may even be shedding a tail of evaporated planetary material like a comet. 'KELT-9 radiates so much ultraviolet radiation that it may completely evaporate the planet', said Keivan Stassun, a professor of physics and astronomy at Vanderbilt University, Nashville, Tennessee, who directed the study with Gaudi.

Discovery

2 The KELT-9b planet was found using one of the two telescopes called KELT, or Kilodegree Extremely Little Telescope. In late May and early June 2016, astronomers using the KELT-North telescope at Winer Observatory in Arizona noticed a tiny drop in the star's brightness – only about 0.5% – which indicated that a planet may have passed in front of the star. The brightness dipped once every 1.5 days, which is how long it takes for the planet to complete a 'yearly' circuit around its star.

Doomed Planet

3 KELT-9b is nowhere close to habitable, but there's a good reason to study worlds that are unliveable in the extreme.'KELT-9 will swell to become a red giant star in a few hundred million years', said Stassun. 'The long-term prospects for life, or real estate for that matter, on KELT-9b are not looking good.'

Strange Orbit

4 The planet is unusual in that it orbits perpendicular to the spin axis of the star. That would be analogous to the planet orbiting perpendicular to the plane of our solar system.

Future Word

5 'Thanks to this planet's star-like heat, it is an exceptional target to observe at all wavelengths, from ultraviolet to infrared, in both transit and eclipse. Such observations will allow us to get as complete a view of its atmosphere as is possible for a planet outside our solar system', said Knicole Colon, who was based at NASA Ames Research Center in California's Silicon Valley during the time of this study.

Top Tip

Astronomers hope to take a closer look at KELT-9b with other telescopes – including the Spitzer and Hubble space telescopes, and eventually the James Webb Space Telescope now scheduled to launch in 2021 (delayed from 2018). Observations with Hubble would enable them to see if the planet really does have a cometary tail, and allow them to determine how much longer that planet will survive its current hellish condition.

Getting There & Away

You'll need to book 23 million years off work if you plan to visit this system aboard a shuttle. Hopefully, you have understanding bosses.

KEPLER-10 DATA

CELESTIAL COORDINATES
Right Ascension 19h 2m 43.0612s
Declination 50° 14' 28.701"

DISTANCE
608 light years

CONSTELLATION
Draco (the Dragon)

VISUAL MAGNITUDE
10.96

MASS
0.910 solar masses

RADIUS
1.065 solar radii

STAR TYPE
Yellow dwarf, G

TEMPERATURE
5643 K

ROTATION PERIOD
Unknown

NUMBER OF PLANETS
2 or 3

Kepler-10 star system, located about 560 light years away near Cygnus and Lyra.

Kepler-10b and c

The star Kepler-10 is home to two confirmed rocky planets, both of them massive and exceedingly hot.

In January 2011, astronomers using the Kepler spacecraft announced the first of two confirmed worlds around the Sun-like star Kepler-10. Called Kepler-10b, it was the first rocky planet that Kepler had found outside the solar system. This planet, which has a radius of 1.4 times that of Earth, whips around its star every 0.8 days. In May that same year, the Kepler team announced another member of the Kepler-10 family, called Kepler-10c. It's bigger than Kepler-10b, with a radius of 2.35 times that of Earth, and it orbits the star every 45 days. Both planets would be blistering hot worlds, with temperatures of 310°C (590°F) on Kepler-10c and 1560°C (2840°F) on Kepler-10b. Additionally, there is possibly a third world, also rocky, slighter farther from the star, but it awaits confirmation.

The existence of Kepler-10c was later validated using a combination of a computer simulation technique called 'Blender', and NASA's Spitzer Space Telescope. Both of these methods are powerful ways to validate the Kepler planets that are too small and faraway for ground-based telescopes to confirm using the radial velocity technique. The Kepler team says that a large fraction of their discoveries will be validated with both of these methods, flipping candidate planets to confirmed ones. Before Kepler's full attention was focused on this star, it was known as KOI 72, standing for 'Kepler Object of Interest'.

Sun-like star Kepler-11 has six exoplanets in orbit around it.

Kepler-11b to g

At the time of its discovery in 2011, Kepler-11 was a marvel of astronomical research – a star system with no fewer than six orbiting planets.

Six worlds orbit Kepler-11, a Sun-like star about 2000 light years distant in the constellation Cygnus. The find was based on data from NASA's planet-hunting Kepler spacecraft. Compared to our solar system, five of Kepler-11's planets orbit closer to their parent star than the Mercury-Sun distance, with orbital periods ranging from 10 to 47 days. The innermost planet, Kepler-11b, is ten times closer to its star than Earth is to the Sun. The outermost planet, Kepler-11g, is twice as close to its star than Earth is to the Sun. If placed in our solar system, Kepler-11g would orbit between Mercury and Venus, and the other five planets would orbit between Mercury and our Sun. The orbits of the five inner planets in the Kepler-11 planetary system are much closer together than any of the planets in our solar system. The inner five exoplanets have orbital periods between 10 and 47 days around the dwarf star, while Kepler-11g has a period of 118 days.

Their presence, sizes, and masses have been determined by carefully watching the planets dim the light of Kepler-11 while transiting or crossing in front of the star itself. In fact, in August 2010, Kepler's telescope and camera recorded a simultaneous transit of three of the planets in the system. All of the planets orbiting yellow-dwarf Kepler-11 are larger than Earth, with the largest ones being comparable in size to Uranus and Neptune.

KEPLER-11 DATA

CELESTIAL COORDINATES
Right Ascension 19h 48m 27.6228s
Declination 41° 54' 32.903"

DISTANCE
2150 light years

CONSTELLATION
Cygnus (the Swan)

VISUAL MAGNITUDE
14.2

MASS
0.961 solar masses

RADIUS
1.065 solar radii

STAR TYPE
Yellow dwarf, G6V

TEMPERATURE
5663 K

ROTATION PERIOD
Unknown

NUMBER OF PLANETS
6

KEPLER-16 AB BINARY DATA

CELESTIAL COORDINATES
Right Ascension 19h 16m 18.1759s
Declination 51° 45' 26.778"

DISTANCE
254 light years

CONSTELLATION
Cygnus (the Swan)

VISUAL MAGNITUDE
Unknown

MASSES
0.6897 (A) and 0.20255 (B) solar masses

RADII
0.6489 (A) and 0.20255 (B) solar radii

STAR TYPES
K and M

TEMPERATURE
4450 K (A) and 3311 K (B)

ROTATION PERIOD
35.1 days (A)

NUMBER OF PLANETS
1

KEPLER-16 (AB)-B DATA

PLANET TYPE
Gas Giant

MASS
3.33 x Earth

RADIUS
Unknown

ORBITAL PERIOD
228.776 days

ORBITAL RADIUS
0.7048 AU

DETECTION TYPE
Transit

DISCOVERED
2011

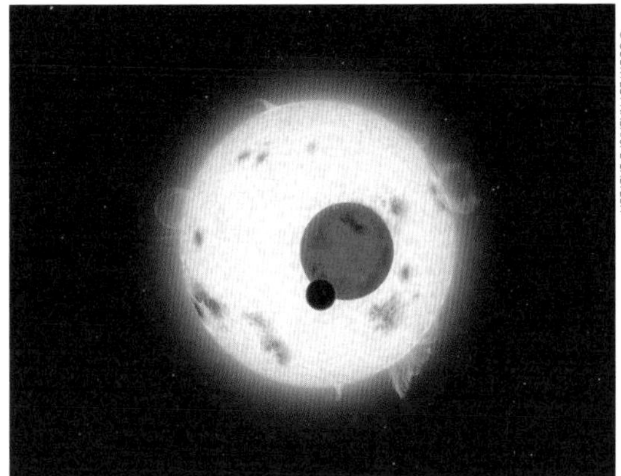

The Kepler-16 binary system treats its exoplanet to two sunsets a day.

Kepler-16 (AB)-b

The existence of a world with a double sunset, as portrayed in the film Star Wars *more than 30 years ago, is now scientific fact. Meet Kepler-16b.*

Like Luke Skywalker's home world of Tatooine, Kepler-16 (AB)-b (or, more informally, Kepler-16b) is a circumbinary planet – one that orbits not a single star, but a pair. The binary star is called Kepler-16 AB, whose component stars are both smaller and cooler than our Sun. The two stars are separated by only one-fifth of the Earth-Sun distance.

Unlike *Star Wars*' Tatooine, the planet Kepler-16b is cold, gaseous and not thought to harbour life, but its discovery demonstrates the diversity of planets in our galaxy. Previous research has hinted at the existence of circumbinary planets, but clear confirmation proved elusive. Kepler detected such a planet by observing transits, where the brightness of a parent star dims from the planet crossing in front of it.

'This discovery confirms a new class of planetary systems that could harbour life', Kepler principal investigator William Borucki said. 'Given that most stars in our galaxy are part of a binary system, this means the opportunities for life are much broader than if planets form only around single stars. This milestone discovery confirms a theory that scientists have had for decades but could not prove until now.'

As for why it's the search for life that underlies most excitement over findings? The Universe as a whole may not be heliocentric, but it certainly is anthropocentric as we see it.

A visualisation of Kepler-22b if it were covered by a liquid ocean.

Kepler-22b

Kepler-22b is a 'Super-Earth' that could be covered in a vast ocean. The jury is still out on the planet's true nature; at 2.4 times Earth's radius, it might even be gaseous.

Back in 2011, the celebrated Kepler planet-hunting spacecraft chalked up its first confirmed planet orbiting within a star's habitable zone, the region where liquid water could exist on a planet's surface. The crowd went wild. At the time, the planet Kepler-22b was the smallest yet found to orbit in the middle of the habitable zone of a star similar to our Sun. Scientists don't yet know if Kepler-22b has a predominantly rocky, gaseous or liquid composition – they don't even know its mass – but its discovery was hailed as a step closer to finding Earth-like planets.

Kepler-22b is located 600 light years away in the Cygnus constellation. While the planet is larger than Earth, its orbit of 290 days around a Sun-like star resembles that of our world. The planet's host star also belongs to the same class as our Sun, called G-type, although it is slightly smaller and cooler.

'This is a major milestone on the road to finding Earth's twin', said Douglas Hudgins, Kepler program scientist at NASA headquarters in Washington, D.C. By now, the idea that there will only be a singular twin has been disproven, as there have been many close matches discovered, though it will be a long time before we know the actual composition of any of these worlds.

KEPLER-22 DATA

CELESTIAL COORDINATES
Right Ascension 19h 16m 52.1904s
Declination 47° 53' 3.948"

DISTANCE
638 light years

CONSTELLATION
Cygnus (the Swan)

VISUAL MAGNITUDE
11.664

MASS
0.970 solar masses

RADIUS
0.979 solar radii

STAR TYPE
Yellow dwarf, G5V

TEMPERATURE
5518 K

ROTATION PERIOD
Unknown

NUMBER OF PLANETS
1

KEPLER-22B DATA

PLANET TYPE
Super-Earth or Neptune-like

MASS
Unknown, max 52.8 x Earth

RADIUS
2.4 x Earth

ORBITAL PERIOD
289.862 days

ORBITAL RADIUS
0.849 AU

DETECTION TYPE
Transit

DISCOVERED
2011

KEPLER-62 DATA

CELESTIAL COORDINATES
Right Ascension 18h 52m 51.0519s
Declination 45° 20' 59.4"

DISTANCE
1200 light years

CONSTELLATION
Lyra (the Lyre)

VISUAL MAGNITUDE
13.75

MASS
0.99 solar masses

RADIUS
0.94 solar radii

STAR TYPE
Orange dwarf, K2V

TEMPERATURE
4925 K

ROTATION PERIOD
39.3 days

NUMBER OF PLANETS
5

Look familiar? This artist's illustration depicts habitable planet Kepler-62f.

Kepler-62b to f

Like Kepler-11, TRAPPIST-1 and a few others, Kepler-62 is a multi-planet system, boasting no fewer than five planets, all of them rocky or terrestrial like Earth.

The five planets of the Kepler-62 system orbit a star classified as a K2 dwarf, measuring just two-thirds the size of the Sun and only one-fifth as bright. At seven billion years old, the star is somewhat older than the Sun. It is about 1200 light years from Earth in the constellation Lyra. Three of its planets are rocky, two of which lie within the habitable zone: Kepler-62e and Kepler-62f, pictured above. The discoveries were announced in April 2013, just as our ability to detect these types of planets was becoming obvious.

Per John Grunsfeld, who was associate administrator of the Science Mission Directorate at NASA, as he described the Kepler-62 system with two of its five planets sitting squarely in the habitable zone, 'The discovery of these rocky planets in the habitable zone brings us a bit closer to finding a place like home. It is only a matter of time before we know if the galaxy is home to a multitude of planets like Earth, or if we are a rarity.'

Of the five known planets, four are more massive and larger than the Earth, and one is similar in size and mass to Mars. But the system is much more compact than our solar system, with all of the planets closer to their sun than Venus is to ours. Since the star Kepler-62 is two-thirds the size of our Sun, the smaller scale of the system itself makes sense.

Kepler-70's radiated energy assaults its orbiting exoplanets with a fury.

Kepler-70b and c

Kepler-70b and Kepler-70c are two of the fieriest worlds ever found, whizzing around their hot blue parent star in mere hours.

Two of the more intriguing Kepler finds are Kepler-70b and Kepler-70c, a pair of planets orbiting a subdwarf B star. On the continuous band that makes up main sequence (dwarf) stars, B-type stars are between twice and sixteen times the Sun's size; they are very hot, luminous, and blue. True to form, Kepler-70 is a super-hot blue star some 4200 light years away in Cygnus. What's so special about its orbiting worlds is their temperature. At about 6800°C (12272°F), Kepler-70b holds the dubious honour of the hottest planet discovered so far, and Kepler-70c is not much cooler. Indeed, both planets are hotter than the Sun – hotter, in fact, than the vast majority of stars, as they absorb the immense energies of Kepler-70.

Kepler-70b and c, announced in 2011, used to be Jupiter-size giants. However, their star became a red giant and engulfed them. As a result, the planets spiralled inwards. It's a trip that would end most worlds, but it left these two as Freddy Krueger-like burnt-out survivors, smaller than Earth. The inner planet is so close to its parent star – 160 times closer than the Earth is to the Sun – that a year there only takes five hours. But a trip to its surface would be much shorter. A spaceship wouldn't even have time to melt in its extreme heat; rather it would simply vaporise.

KEPLER-70 DATA

CELESTIAL COORDINATES
Right Ascension 19h 45m 25.48s
Declination 41° 5' 33.88"

DISTANCE
4200 light years

CONSTELLATION
Cygnus (the Swan)

VISUAL MAGNITUDE
14.87

MASS
0.496 solar masses

RADIUS
0.203 solar radii

STAR TYPE
Subdwarf B

TEMPERATURE
27,730 K

ROTATION PERIOD
Unknown

NUMBER OF PLANETS
2

KEPLER-78 DATA

CELESTIAL COORDINATES
Right Ascension 19h 34m 58.0143s
Declination 44° 26' 53.961"

DISTANCE
410 light years

CONSTELLATION
Cygnus (the Swan)

VISUAL MAGNITUDE
11.72

MASS
0.81 solar masses

RADIUS
0.74 solar radii

STAR TYPE
Yellow dwarf, late G

TEMPERATURE
5089 K

ROTATION PERIOD
Unknown

NUMBER OF PLANETS
1

KEPLER-78B DATA

PLANET TYPE
Super-Earth

MASS
1.86 x Earth

RADIUS
1.173 x Earth

ORBITAL PERIOD
0.355 days

ORBITAL RADIUS
0.089 AU

DETECTION TYPE
Transit

DISCOVERED
2013

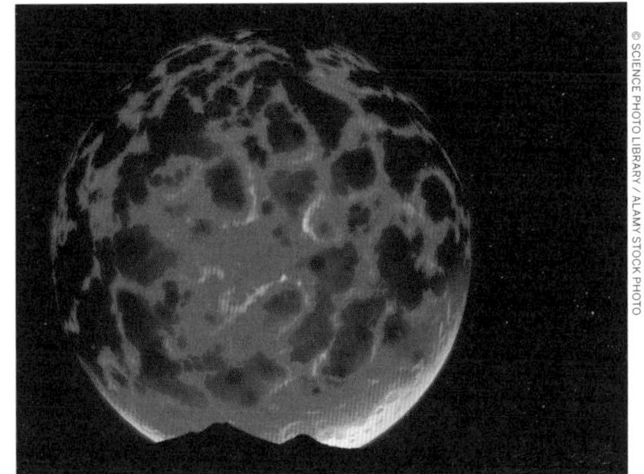

Artist's conception of the molten exoplanet Kepler-78b and its star.

Kepler-78b

The discovery of Kepler-78b in 2013 caused much excitement. It was the first Earth-size exoplanet known to have a rocky composition like that of Earth.

A handful of planets the size or mass of Earth were known before Kepler-78b, but this was the first world to have both a measured mass and size. With both quantities known, scientists can calculate a density and determine what the planet is made of. Earth-size does not mean Earth-like, though. Kepler-78b whizzes around its host star every 8.5 hours, making it a blazing inferno and not suitable for life as we know it, rocky or not.

Two independent research teams used ground-based telescopes to confirm and characterise Kepler-78b. To determine the planet's mass, the teams employed the radial velocity method to measure how much the gravitational tug of an orbiting planet causes its star to wobble. Kepler, on the other hand, determines the size or radius of a planet by the amount of starlight blocked when it passes in front of its host star (during a transit).

Kepler-78b is 1.2 times the size of Earth and 1.7 times more massive, resulting in a density that is the same as Earth's. This suggests that Kepler-78b is also made primarily of rock and iron. Its star is slightly smaller and less massive than the Sun and is located about 400 light years from Earth in the constellation Cygnus (the Swan).

Comparative planet sizes between our solar system and the Kepler-90 system.

Kepler-90b

Do other stars have planetary systems like our own? Yes – one such system is Kepler-90.

Our solar system now is tied for the most number of planets around a single star, with the recent discovery of an eighth planet circling Kepler-90, a Sun-like star 2545 light years from Earth. The planet was discovered in data from NASA's Kepler Space Telescope. The newly discovered eighth planet is Kepler-90i – a sizzling hot, rocky planet that orbits its star once every 14.4 days. It was found using machine learning from Google. Machine learning is an approach to artificial intelligence in which computers 'learn'. In this case, computers learned to identify planets by finding instances in Kepler data where the telescope recorded changes in starlight caused by planets beyond our solar system.

Similarities between Kepler-90 and our system include a G-type star comparable to our Sun, rocky planets comparable to our Earth, and large planets comparable in size to Jupiter and Saturn. Differences include that all of the known Kepler-90 planets orbit relatively close in – closer than Earth's orbit around the Sun – making them possibly too hot to harbour life. However, observations over longer time periods may discover cooler planets further out.

KEPLER-90 DATA

CELESTIAL COORDINATES
Right Ascension 18h 57m 44.0384s
Declination 49° 18' 18.4958"

DISTANCE
2500 light years

CONSTELLATION
Draco (the Dragon)

VISUAL MAGNITUDE
14.0

MASS
1.2 solar masses

RADIUS
1.2 solar radii

STAR TYPE
Sun-like, G0V

TEMPERATURE
6080 K

ROTATION PERIOD
Unknown

NUMBER OF PLANETS
8

KEPLER-186 DATA

CELESTIAL COORDINATES
Right Ascension 19h 54m 36.6536s
Declination 43° 57' 18.0259"

DISTANCE
582 light years

CONSTELLATION
Cygnus (the Swan)

VISUAL MAGNITUDE
15.29

MASS
0.544 solar masses

RADIUS
0.523 solar radii

STAR TYPE
Red dwarf, M1V

TEMPERATURE
3755 K

ROTATION PERIOD
34.404 days

NUMBER OF PLANETS
5

An artist's impression of Earth-like exoplanet Kepler-186f.

Kepler-186b to f

Kepler-186 hosts five rocky planets. One of them, Kepler-186f, was the first validated Earth-size planet to orbit a distant star in the habitable zone.

Kepler-186f resides in the Kepler-186 system about 500 light years from Earth in the constellation Cygnus. It orbits its star once every 130 days and receives one-third the energy that Earth does from the Sun, placing it near the outer edge of the habitable zone. If you could stand on the surface of Kepler-186f, the brightness of its star at high noon would appear similar to that of our Sun about an hour before sunset on Earth. Kepler-186f is known to be less than 10% larger than Earth, but its mass, composition and density are not known. Previous research suggests that a planet the size of Kepler-186f is likely to be rocky. Prior to this discovery, the record for the most 'Earth-like' planet went to Kepler-62f, which is 40% larger than the size of Earth and also orbits in its star's welcoming habitable zone.

The system is also home to four inner planets, orbiting closer to the star. These four companion planets are significantly smaller in size to the Earth. Kepler-186b, Kepler-186c, Kepler-186d and Kepler-186, orbit every 4, 7, 13 and 22 days, respectively, making them very hot and inhospitable for life as we know it – Mercury twins rather than Earth ones.

Explore the Kepler-186 System

Record Holder No Longer

1 While Kepler-186f was the smallest exoplanet that had been found upon its discovery, that's no longer the case. Kepler 37b, discovered in 2013, is on the scale not of Earth but of our very own moon.

Confirmation Party

2 Before a planet is confirmed as an exoplanet, it is (usually) registered as an exoplanet candidate. Initial findings have to be double-checked and verified, as the measurements involved can be quite minute, though some confirmed planets never were in the candidate category. Either way, verification is cause for celebration.

Crowded Inner Circle

3 Whereas our solar system is composed of a majority of gas and ice giants, Earth-like Kepler-186f is the largest of its five sibling planets. The inner four companion terrestrial planets are less than half the size of our own planet.

Kepler Kudos

4 The five exoplanets of Kepler-186 are only a tiny percentage of the 2662 confirmed exoplanets that Kepler discovered in its nine-year lifetime. The space telescope observed 530,506 stars before ending its mission in 2018, when the spacecraft ran out of fuel. It lasted four and half years longer than its planned duration, and returned a trove of new information about the range of exoplanets to be found beyond our solar system.

Looking Up?

5 It's not known whether Kepler-186f has its own atmosphere, or what it consists of. Future analysis of spectrographic data from the system might reveal more about its chemical composition. The sun would certainly be different in the sky either way, as Kepler-186, a much luminous M dwarf, is much cooler and smaller than our Sun.

Top Tip

Other exoplanets have seemed as if they're in the right range of their star, and composed of the right materials, to potentially harbour life. Chief on the list is Kepler-452b, somewhat older and larger than Earth but positioned in the right spot to thrive, as far as we know. After a spate of hot Jupiter discoveries, there have been more terrestrial planets in the habitable zone than astronomers hoped to be able to find, with more sure to come.

Getting There & Away

Kepler-186 is some 582 light years away from Earth. A family reunion between the cousin or sibling-like planets isn't likely anytime soon!

KEPLER-444 DATA

CELESTIAL COORDINATES
Right Ascension 19h 19m 1.0s
Declination 41° 38' 05"

DISTANCE
116 light years

CONSTELLATION
Lyra (the Lyre)

VISUAL MAGNITUDE
8.86

MASS
0.758 solar masses

RADIUS
0.752 solar radii

STAR TYPE
Orange dwarf, KOV

TEMPERATURE
5040 K

ROTATION PERIOD
49.4 days

NUMBER OF PLANETS
5

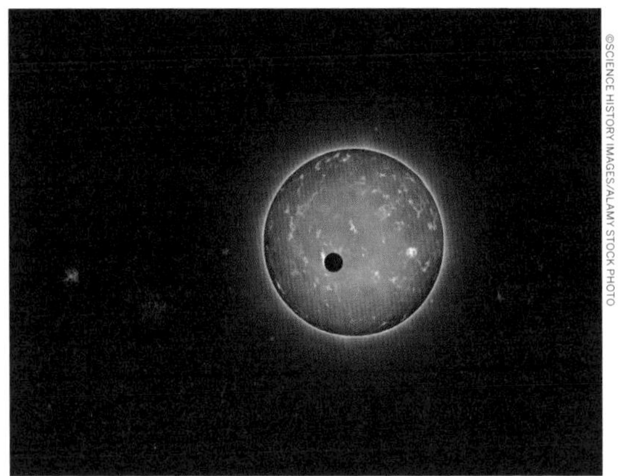

Kepler-444 is home to five known planets within tight orbits.

Kepler-444b to f

Using data from the Kepler mission, astronomers discovered this system of five small planets dating back to when the Milky Way Galaxy was a youthful two billion years old.

The Kepler-444 home star is approximately 117 light years away toward the constellation Lyra. The tightly packed system is host to five rocky planets that range in size between Mercury and Venus. That's a very low-mass family of worlds, as exoplanets go. All five planets orbit their Sun-like star in less than ten days, which makes their orbits much closer in than Mercury's already sweltering 88-day orbit around the Sun. As such, these planets are all well outside the temperate Goldilocks zone.

Kepler-444 formed 11.2 billion years ago, when the universe was less than 20% its current age. To determine the age of the star, and thus its planets, scientists measured the very small change in brightness of the host star caused by pressure waves within the star. The boiling motion beneath the surface of the star generates these pressure waves, affecting the star's temperature and luminosity. These fluctuations lead to miniscule changes or variations in a star's brightness. This study of the interior of stars is called asteroseismology and allows the researchers to measure the diameter, mass and age of a star.

'While this star formed a long time ago, in fact before most of the stars in the Milky Way, we have no indication that any of these planets have now or ever had life on them', said Steve Howell, Kepler/K2 project scientist at NASA's Ames Research Center.

An artist's rendering of what could be the first moon found outside our solar system.

Kepler-1625b and Exomoon

Using the Hubble and Kepler space telescopes, astronomers have uncovered tantalising evidence of what could be the first discovery of a moon orbiting a planet outside our solar system.

This moon candidate, which is 8000 light years from Earth in the Cygnus constellation, orbits a gas-giant planet that in turn orbits a star called Kepler-1625. Researchers caution that the moon hypothesis is tentative and must be confirmed by follow-up observations. The planetary wobble could be caused instead by the gravitational pull of a hypothetical second planet in the system, rather than a moon. While Kepler has not detected a second planet in the system, it could be that the planet is there, but not detectable using Kepler's techniques.

Since moons outside our solar system – known as exomoons – cannot be imaged directly, their presence is inferred when they pass in front of a star, momentarily dimming its light. Such an event is called a transit, and has been used to detect many of the exoplanets catalogued to date. However, exomoons are harder to detect because they are smaller than their companion planet, so their transit signal is weaker when plotted on a light curve that measures the duration of the planet crossing and the amount of momentary dimming. Exomoons also shift position with each transit because the moon is orbiting its planet. The moon candidate is estimated to be only 1.5% the mass of its planet, and the planet is estimated to be several times the mass of Jupiter, a similar mass-ratio to the one between Earth and the moon.

KEPLER-1625 BINARY DATA

CELESTIAL COORDINATES
Right Ascension 19h 41m 43.0402s
Declination 39° 53' 11.4990"

DISTANCE
8000 light years

CONSTELLATION
Cygnus (the Swan)

VISUAL MAGNITUDE
13.916

MASS
1.079 solar masses

RADIUS
1.793 solar radii

STAR TYPE
Unknown

TEMPERATURE
5548 K

ROTATION PERIOD
Unknown

NUMBER OF PLANETS
1

KEPLER-1625B DATA

PLANET TYPE
Gas Giant

MASS
3 x Jupiter

RADIUS
0.6 x Jupiter

ORBITAL PERIOD
287.37 days

ORBITAL RADIUS
0.811-0.8748 AU

DETECTION TYPE
Transit

DISCOVERED
2016

KEPLER-1647 AB BINARY DATA

CELESTIAL COORDINATES
Right Ascension 19h 52m 36.02s
Declination 40° 39' 22.2"

DISTANCE
3700 light years

CONSTELLATION
Cygnus (the Swan)

VISUAL MAGNITUDE
13.78

MASS
1.22 (A) and 0.97 (B) solar masses

RADIUS
1.79 (A) and 0.966 (B) solar radii

STAR TYPES
F and G

TEMPERATURE
6210 (A) and 5770 (B)

ROTATION PERIOD
Unknown

NUMBER OF PLANETS
1

KEPLER-1647 (AB)-B DATA

PLANET TYPE
Gas Giant

MASS
1.52 x Jupiter

RADIUS
1.06 x Jupiter

ORBITAL PERIOD
3.03 years

ORBITAL RADIUS
2.72 AU

DETECTION TYPE
Transit

DISCOVERED
2016

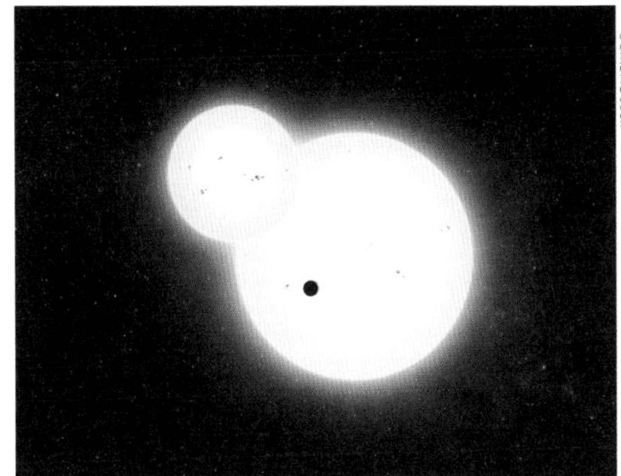

Kepler-1647b shown in syzygy, or stellar eclipse and simultaneous planetary transit.

Kepler-1647 (AB)-b

Like Kepler-16, Kepler-1647 is a binary star, with a single planet orbiting the two stars as if they are one – like the planet Tatooine in Star Wars.

The binary star Kepler-1647 is 3700 light years away and approximately 4.4 billion years old, roughly the same age as Earth. The two stars are similar to the Sun, with one slightly larger than our home star and the other slightly smaller. The planet, Kepler-1647 (AB)-b (or, more informally, Kepler-1647b), has a mass and radius nearly identical to that of Jupiter, making it the largest transiting circumbinary planet ever found. A circumbinary planet is one that orbits two stars as if they were one.

Circumbinary planets are sometimes called 'Tatooine' planets, after Luke Skywalker's home world in *Star Wars*, which was depicted with a double sunset. Using Kepler data, astronomers search for slight dips in brightness that hint a planet might be passing or transiting in front of a star, blocking a tiny amount of the star's light. 'But finding circumbinary planets is much harder than finding planets around single stars', said astronomer William Welsh, one of the co-authors of the research. 'The transits are not regularly spaced in time and they can vary in duration and even depth.' So far, Kepler-1647b is the largest planet orbiting a binary pair as one, with the longest orbit. This may be due to the greater difficulty of finding circumbinary planets from their transits at greater orbits.

The planet sits in the habitable zone of its stars, and while as a likely gas giant it isn't expected to harbour life itself, it's a strong candidate for having many captured rocky moons which might play host to the elements needed for life.

Explore Kepler-1647 (AB)-b

Better Late Than Never

1 The research was conducted by astronomers from NASA's Goddard Space Flight Center in Greenbelt, Maryland, and San Diego State University (SDSU) in California, who used NASA's Kepler Space Telescope. 'It's a bit curious that this biggest planet took so long to confirm, since it is easier to find big planets than small ones', said SDSU astronomer Jerome Orosz, a co-author on the study. 'But it is because its orbital period is so long.'

A Planet Is Confirmed

2 Once a candidate planet is found, researchers employ advanced computer programs to determine if it really is a planet. It can be a gruelling process. Laurance Doyle, a co-author on the paper and astronomer at the SETI Institute, noticed a transit back in 2011, but more data and several years of analysis were needed to confirm the transit was indeed caused by a circumbinary planet.

Vital Statistics

3 The planet takes 1107 days – just over three years – to orbit its host stars, the longest period of any confirmed transiting exoplanet found so far. The planet is also much further away from its stars than any other circumbinary planet, breaking with the tendency for such planets to have close-in orbits. Interestingly, its orbit puts the planet with in the so-called habitable zone – the range of distances from a star where liquid water might pool on the surface of an orbiting planet.

Habitability

4 Like Jupiter, however, Kepler-1647b is a gas giant, making the planet unlikely to host life. Yet if the planet has large moons, they could potentially be suitable for life. 'Habitability aside, Kepler-1647b is important because it is the tip of the iceberg of a theoretically predicted population of large, long-period circumbinary planets', said Welsh.

Amateurs Step-Up

5 A network of amateur astronomers in the Kilodegree Extremely Little Telescope (KELT) 'Follow-Up Network' provided additional observations that helped the researchers estimate the planet's mass. Pro-amateur collaborations like this are particularly fruitful in astronomy, where the volume of data (and sky) to analyse is so large.

Top Tip

Kepler-1647 is in good company. It joins the likes of Kepler-16, HW Virginis, Kepler-453 and a handful of others in a growing list of circumbinary planetary systems. Indeed, one of the more famous ones was also one of the very first exoplanetary systems ever found, back in 1993 when this area of research was in its infancy. That system was PSR B1620-26 – a white dwarf orbiting a neutron star – and the planet is dubbed, unofficially, Methuselah, owing to its extreme age (see page 418 for more).

Getting There & Away

If you want to have your own Luke Skywalker experience, you'll need to wait to get there – a good 140 million years, plus or minus.

PSR B1257+12 (LICH) DATA

CELESTIAL COORDINATES
Right Ascension 13h 0m 01s
Declination 12° 40' 57"

DISTANCE
2300 light years

CONSTELLATION
Virgo (the Virgin)

VISUAL MAGNITUDE
12.2

MASS
1.4 solar masses

RADIUS
10 km (6 mi)

STAR TYPE
Pulsar

TEMPERATURE
28,856 K

ROTATION PERIOD
0.006219s

NUMBER OF PLANETS
3

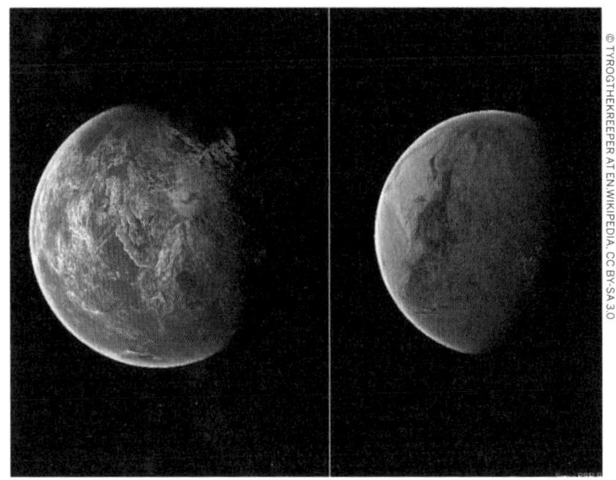

Artist's impression of the planets of pulsar PSR B1257+12.

Lich System (PSR B1257+12)

This entire system is a graveyard, the remnants of what used to be a normal, functional planetary system before the star blew apart in a giant cosmic explosion.

The very first planetary bodies outside our solar system to be clearly identified orbit a bizarre object called PSR B1257+12. This is a dead star – a pulsar – the extremely dense, rapidly spinning core of a star that has died a spectacular death; blowing itself apart in a supernova explosion. Pulsars shoot intense beams of radio waves in opposite directions as they spin, like rapidly rotating lighthouse beacons. This characteristic came in handy during early attempts to locate planets circling other stars.

By measuring changes in the pulsing beat from just such a spinning, stellar corpse, Dr Alexander Wolszczan of Pennsylvania State University found three 'pulsar planets' orbiting this exotic star. The planets' gravitational tugs altered the rhythm of the pulsar, revealing their existence by a kind of interstellar Morse code. Wolszczan announced the discovery of two in 1992, confirming the third two years later. The massive shockwave from the supernova stripped away any atmosphere or living creatures that might have once lived on these planets, leaving behind ghostly, rocky shells; dead planets orbiting the corpse of an extinct star.

Explore the Lich System

The Star

1 Like all pulsars, PSR B1257+12 – also called Lich – is tiny. At only 10 km (6 mi) or so in radius, it is no larger than a small city. Literally spinning in its grave, Lich makes a full rotation every 6.22 milliseconds and emits an intense beam of radiation that can be detected from Earth. The star's unfortunate planets are thus bathed in deadly radiation on a regular basis, ensuring that this system remains a cosmic no-man's land.

Planet Discovery

2 Discovered in 1992 and 1994, using the unusual method of pulsar timing where the gravitational pulls of the planets create tiny changes in the ticking of the pulsar's otherwise highly accurate clock, Lich's three planets were the first ever found orbiting a star other than the Sun. Also unusually for planets, these have been given names. The innermost (PSR 1257+12b) is called Draugr. Planets c and d are called Poltergeist and Phobetor, respectively.

Draugr

3 Planet Draugr (PSR B1257+12b) is named after a Norse undead creature. Weighing in at a mere 0.02 Earth masses – about twice our moon's heft – it is not only the lightest-known exoplanet, it is also less massive than any of the planets in our solar system. Its radius remains unknown but with such little mass it must be a rocky world.

Poltergeist

4 The spookily named Poltergeist is PSR B1257+12c, the second innermost world in the system. More than four times the mass of the Earth, it orbits the pulsar at a distance of 0.36 astronomical units (AU, the Earth-Sun distance), similar to Mercury's location relative to the Sun. It is probably a rocky 'Super-Earth'.

Phobetor

5 The final planet, PSR B1257+12d, has been dubbed Phobetor, named after the Greek deity of nightmares. At 3.9 Earth masses, it is another 'Super-Earth', slightly lighter than Poltergeist, and probably about 50 percent larger than our planet. It orbits Lich at roughly half the Earth-Sun distance and has a temperature estimated to be about 269°C (516°F).

Top Tip

Lich is 2300 light years away, so a trip to this system will take a great deal of determination and patience. If a T. Rex had left Earth 65 million years ago aboard a Space Shuttle, it would still have another 22 million years to wait until arrival.

Getting There & Away

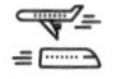

If you can't stand the stillness of the grave or the rotting flesh of zombies, then the pulsar planets are off-limits. Nothing can live in this most inhospitable corner of the galaxy. An astronaut who flew to the planet Poltergeist (PSR B1257+12c) would find herself in the midst of three dead planet cores shambling through the twisted magnetic fields of their corpse star. Lich has twin beams of radiation spinning faster than you can blink, which can instantly incinerate any spaceship. The radiation constantly rains down on Poltergeist and its neighbouring worlds creating silent nights, possibly lit with sickly irradiated auroras.

BINARY STAR DATA

CELESTIAL COORDINATES
Right Ascension 18h 23m 38.2218s
Declination -26° 31' 53.769"

DISTANCE
12,400 light years

CONSTELLATION
Scorpius (the Scorpion)

APPARENT MAGNITUDE
21.3

STAR TYPE
Pulsar/White Dwarf Binary

NUMBER OF PLANETS
1

PLANET DATA

PLANET TYPE
Gas Giant

MASS
2.5 x Jupiter

RADIUS
Unknown

ORBITAL PERIOD
100 years

ORBITAL RADIUS
23 AU

DETECTION TYPE
Radial Velocity

DISCOVERED
1994

This illustration shows the binary stars of Methusalah's Planet.

Methuselah's Planet

Meet the great-grandfather of all exoplanets, Methuselah. At a ripe-old-age of 13 billion years, it was created when the Universe itself was just getting out of bed.

Methuselah is an important figure in the Bible. The grandfather of Noah, he's said to have lived to an age of 969 years, finally passing in the year of the Great Flood. So it seems fitting that the oldest known planet, at an estimated 13 billion years of age, should share his name. It is also referred to as the Genesis planet, or with less pizzazz, PSR B12620-26 b. Whatever the name, it has a mighty history.

Long before our Sun and Earth ever existed, a Jupiter-sized planet formed around a Sun-like star. Now, 13 billion years later, NASA's Hubble Space Telescope has precisely measured the mass of this farthest and oldest known planet. The ancient planet has had a remarkable history because it has wound up in an unlikely, rough neighbourhood. It orbits a peculiar pair of burned-out stars, one of them a rapidly spinning neutron star or pulsar, the other a white dwarf - the dead core of the planet's original parent star - in the crowded core of a globular star cluster. The planet is 2.5 times the mass of Jupiter. Its very existence provides tantalizing evidence that the first planets were formed rapidly, within a billion years of the Big Bang, leading astronomers to conclude that planets may be much more abundant in the Universe than it was first believed.

Explore the Methuselah System

Parent Stars

1 The story of this planet's discovery began in 1988, when a pulsar, called PSR B1620-26, was discovered in the globular cluster M4, located 12,400 light years away in the summer constellation Scorpius. It is a neutron star spinning just under 100 times per second and emitting regular radio pulses like a lighthouse beam. Its white dwarf partner was quickly found through its effect on the clock-like pulsar, as the two stars orbited each other twice per year.

Planet or Brown Dwarf?

2 Eventually, astronomers monitoring this pulsar noticed further irregularities that implied that a third object was orbiting the others. At an estimated age of 13 billion years, it's about three times older than the Earth. Until Hubble's measurement, astronomers had debated the identity of this object. Was it a planet or a brown dwarf? Hubble's analysis shows that the object is 2.5 times the mass of Jupiter, confirming that it is a planet.

Habitability

3 It is likely that Methuselah's planet is a gas giant, without a solid surface like the Earth. Because it was formed so early in the life of the universe, it probably doesn't have abundant quantities of elements such as carbon and oxygen. For these reasons, it is very improbable the planet would host life. Even if life arose on, for example, a solid moon orbiting the planet, it is unlikely to have survived the intense environments the planet has witnessed.

Unlikely Home

4 The Methuselah planet came as a surprise. Globular clusters are deficient in heavier elements because they formed so long ago that heavier elements had not been cooked up in abundance in the nuclear furnaces of stars. Some astronomers had therefore argued previously that globular clusters cannot contain planets.

Early Planet Formation

5 'Our Hubble measurement offers tantalising evidence that planet formation processes are quite robust and efficient at making use of a small amount of heavier elements. This implies that planet formation happened very early in the universe', says Steinn Sigurdsson of Pennsylvania State University, State College.

Top Tip

Methuselah's planet has had a rough road over the last 13 billion years. When it was born, it probably orbited its then youthful yellow star at approximately the same distance Jupiter is from our Sun. The planet survived blistering ultraviolet radiation, supernova radiation and shockwaves, which must have ravaged the young globular cluster in a furious firestorm of star birth in its early days. Around the time multicelled life appeared on Earth, the planet and star were plunging into the core of the M4 globular cluster.

Getting There & Away

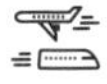

Methuselah's planet is exceptionally distant, at 12,400 light years. Travelling there by Space Shuttle will require a good 480 million years of your time. Even one of the fastest ever probes, Helios B, travelling at 70 km/s (43mps), would take 54 million years to traverse this expanse.

PI MENSAE DATA

CELESTIAL COORDINATES
Right Ascension 5h 37m 9.89s
Declination -80° 28' 8.8"

DISTANCE
59.62 light years

CONSTELLATION
Mensa (the Table)

VISUAL MAGNITUDE
5.65

MASS
1.11 solar masses

RADIUS
1.15 solar radii

STAR TYPE
Sun-like, G0V

TEMPERATURE
6013 K

ROTATION PERIOD
Unknown

NUMBER OF PLANETS
2

Artist's illustration showing the view of Pi Mensae c from Pi Mensae b's surface.

Pi Mensae b and c

Pi Mensae is a system of two worlds – a whopping supermassive gas giant, and a 'Super-Earth', the first ever exoplanet discovered with the new TESS satellite.

The bright star Pi Mensae is similar to the Sun in mass and size. Astronomers discovered its first known exoplanet, Pi Mensae b, in 2001, using the Anglo-Australian Telescope in Australia. With a mass around ten times that of Jupiter, it was heralded at the time as the most massive known exoplanet, although that record now goes to HR 2562 b, which beats it by a factor of three. In fact, Pi Mensae b may not be a planet at all. It could be a sort of failed star called a brown dwarf. Its distance from Pi Mensae is an average of 3 AUs, giving it an orbit of 2083 Earth days.

The second planet in orbit around this star, Pi Mensae c, is a much more recent find, from 2018, and marks the first ever discovery by TESS (the Transiting Exoplanet Survey Satellite), launched in April that year. Called Pi Mensae c, it is about twice Earth's size, with an orbital period of six days. It appears to be terrestrial in nature, with an iron core, though it orbits too close to its star to have liquid water. 'This star was already known to host a planet, called Pi Mensae b, which ... follows a long and very eccentric orbit', said Chelsea Huang, a Juan Carlos Torres Fellow at the Massachusetts Institute of Technology's Kavli Institute for Astrophysics and Space Research (MKI). 'In contrast, the new planet, called Pi Mensae c, has a circular orbit close to the star, and these orbital differences will prove key to understanding how this unusual system formed.'

A visualisation of Pollux b, which orbits a star that has been known since antiquity.

Pollux b

Pollux b is a rarity – a bona fide exoplanet whose parent star is clearly visible to the naked eye, 33 light years from Earth in Gemini.

If you've ever looked up on a clear winter night, you've probably seen it: Pollux, also known as Beta Geminorum, one of the brightest and most familiar stars in the night sky. This orange giant is one of the famous twins in Gemini, the other one being Castor. And now, scientists have discovered that Pollux harbours a secret: a hidden planet, about three times the size of Jupiter. The discovery of the new planet, dubbed Pollux b, was confirmed in 2006 by two separate planet-finding teams, one led by Sabine Reffert and the other by Artie Hatzes, both of Germany. Research by the latter team was supported, in part, by a grant from NASA. Both teams found the planet using the Doppler (or radial velocity) method, which infers the existence of a planetary companion from its gravitational tug on the host star.

As well as being interesting in orbiting a readily visible star, Pollux b is also one of the few exoplanets to have received a proper name. In 2014, the International Astronomical Union (IAU) invited the public to name Pollux b and other planets. The winning name was Thestias, from Leda in Greek mythology. She was Thestius' daughter, hence the formulation, necessary because 'Leda' is already the name of one of Jupiter's moons as well as an asteroid. It remains to be seen how widely adopted the new name will become, however.

POLLUX DATA

CELESTIAL COORDINATES
Right Ascension 7h 45m 18.94987s
Declination 28o 1' 34.316"

DISTANCE
33.78 light years

CONSTELLATION
Gemini (the Twins)

VISUAL MAGNITUDE
1.14

MASS
1.91 solar masses

RADIUS
8.8 solar radii

STAR TYPE
Orange giant, KOIII

TEMPERATURE
4666 K

ROTATION PERIOD
558 days

NUMBER OF PLANETS
1

POLLUX B (THESTIAS) DATA

PLANET TYPE
Gas Giant

MASS
2.3 x Jupiter

RADIUS
Unknown

ORBITAL PERIOD
1.61 years

ORBITAL RADIUS
1.64 AU

DETECTION TYPE
Radial Velocity

DISCOVERED
2006

PROXIMA DATA

CELESTIAL COORDINATES
Right Ascension 14h 29m 43.94853s
Declination -62° 40' 46.1631"

DISTANCE
4.244 light years

CONSTELLATION
Centaurus (the Centaur)

APPARENT MAGNITUDE
11.13

MASS
0.122 solar masses

RADIUS
0.154 solar radii

STAR TYPE
Red dwarf, M5.5V

TEMPERATURE
3042 K

ROTATION PERIOD
82.6 days

NUMBER OF PLANETS
1

PROXIMA B DATA

PLANET TYPE
Terrestrial or Super-Earth

MASS
1.3 x Earth

RADIUS
0.8–1.5 x Earth

ORBITAL PERIOD
11.18 days

ORBITAL RADIUS
0.049 AU

DETECTION TYPE
Radial Velocity

DISCOVERED
2016

Proxima Centauri circled in red, below Alpha Centauri (left) and Beta Centauri (right).

Proxima b

Proxima Centauri is the closest star outside the solar system. It caused much excitement when astronomers announced that it has at least one planet – and it might even be habitable.

At just over four light years away, Proxima Centauri is the closest star to the Sun – the clue is in the name. This cool star in the constellation of Centaurus is too faint to be seen with the unaided eye and lies near to the much brighter pair of stars known as Alpha Centauri AB.

In 2016, astronomers using the European Southern Observatory's 3.6-meter telescope in Chile discovered that this star has a planet orbiting around it. 'This is really a game-changer in our field', said Olivier Guyon, a planet-hunting affiliate at NASA's Jet Propulsion Laboratory, California, and associate professor at the University of Arizona, Tucson. 'The closest star to us has a possible rocky planet in the habitable zone. That's a huge deal. It also boosts the already existing, mounting body of evidence that such planets are near, and that several of them are probably sitting quite close to us.'

They determined that the new planet, dubbed Proxima b, is at least 1.3 times the mass of Earth. It orbits its low mass red dwarf star far more closely than Mercury orbits our Sun, taking only 11 days to complete a single orbit. With that near an orbital distance, this likely rocky planet is surely well outside of the habitable zone, especially since Proxima Centuari is a flare star that emits periodic X-ray blasts, but it's still nice to have a neighbour. We should have it for quite a while yet: Proxima b's star is expected to last as a main sequence star for another four trillion years.

Explore the Proxima System

Habitability

1 While Proxima b lies within its star's 'habitable zone' – a distance at which temperatures are right for liquid water – scientists do not yet know if the planet has an atmosphere. It also orbits a red-dwarf star, far smaller and cooler than our Sun. The planet likely presents only one face to its star, as the moon does to Earth, instead of rotating through our familiar days and nights. And Proxima b could be subject to potentially life-extinguishing stellar flares.

Stability

2 Just because Proxima b's orbit is in the habitable zone doesn't mean it's habitable. It doesn't take into account, for example, whether water actually exists on the planet, or whether an atmosphere could survive at that orbit. Atmospheres are also essential for life as we know it: having the right atmosphere allows for climate regulation, maintenance of a water-friendly surface pressure, shielding from hazardous space weather, and housing life's chemical building blocks.

Climate

3 If Proxima b does have an atmosphere, liquid water may be present, but probably only on the surface of the planet in the sunniest regions. Proxima b's rotation, the strong radiation from its star and the formation history of the planet all conspire to make its climate quite different from that of the Earth, and it is unlikely that Proxima b has seasons.

Triple Star

4 Proxima Centauri is actually part of a triple star system. Two other, bright stars, called Alpha Centauri A and B, form a close binary system, visible in the southern hemisphere; they are separated by only 23 times the Earth-Sun distance. This is slightly greater than the distance between Uranus and the Sun. Proxima b orbits the AB binary once every half a million years.

Potential Exploration

5 Proxima b could potentially reignite the admittedly far-off goal of sending a probe to another planetary system. Bill Borucki, an exoplanet pioneer, said the new discovery might inspire more interstellar research, especially if Proxima b proves to have an atmosphere. Coming generations of space- and ground-based telescopes, including large ground telescopes now under construction, could yield more information about the planet, perhaps inspiring ideas on how to pay it a visit.

Top Tip

In Proxima Centauri's habitable zone, Proxima b encounters bouts of extreme ultraviolet radiation hundreds of times greater than Earth does from the Sun. That radiation generates enough energy to strip away not just the lightest molecules such as hydrogen, but also, over time, heavier elements such as oxygen and nitrogen. Theoretical modelling shows that Proxima Centauri's powerful radiation would drain any Earth-like atmosphere as much as 10,000 times faster than what happens at Earth. 'This simple calculation', said Katherine Garcia-Sage, at NASA's Goddard Space Flight Center in Maryland, 'doesn't consider variations like extreme heating in the star's atmosphere or violent stellar disturbances to the exoplanet's magnetic field.'

Getting There & Away

Don't let Proxima's closeness lull you into false hope. It's nearby in space terms, but 4.24 light years is still 40 trillion km (almost 25 million mi). The trip by Space Shuttle will take a sedate 163,000 years.

PSO J318.5-22 DATA

PLANET TYPE
Gas Giant

MASS
6.5 x Jupiter

RADIUS
1.53 x Jupiter

ORBITAL PERIOD
n/a

ORBITAL RADIUS
n/a

DETECTION TYPE
Direct Imaging

DISCOVERED
2013

Faint, cool rogue planet PSO J318.5-22, unattached to any star, is easy to miss.

PSO J318.5-22

This planet is an oddity – it has no stellar host. It is one of a class of planets doomed to roam the space between the stars.

Discovered in October 2013 using direct imaging (wide-field photographs of the deep sky), PSO J318.5-22 belongs to a special class of planets called rogue, or free-floating, planets. Wandering alone in the galaxy, they do not orbit a parent star. Not much is known about how these planets come to exist, but scientists theorise that they may be either failed stars or planets ejected from very young systems after an encounter with another planet. These rogue planets glow faintly from the heat of their formation. Once they cool down, they will be dancing in the dark.

Astronomers unearthed this planetary oddball using images taken by the Panoramic Survey Telescope And Rapid Response System (Pan-STARRS) telescope. The lead astronomer involved in the research, Michael Liu of the Institute for Astronomy at the University of Hawaii, stated: 'We have never before seen an object free-floating in space that looks like this. It has all the characteristics of young planets found around other stars, but it is drifting out there all alone.' Indeed, it was discovered by mere happenstance in a search for brown dwarfs, which are also cool, low-mass objects.

PSO J318.5-22 is associated with a group of young stars called Beta Pictoris. Despite not having a nearby star to draw energy from, it appears quite similar to other discovered exoplanets in orbit around a star. Cold and faint in appearance, it mostly emits in the infrared wavelength; its visual magnitude is 100,000 less than that of the planet Venus.

Artist's impression of temperate planet Ross 128 b orbiting its red dwarf parent star.

Ross 128 b

Ross 128 b is a nearby rocky planet with a very temperate climate. It is the fourth closest known exoplanet system, only 11 light years away.

A temperate Earth-sized planet has been discovered only 11 light years from our solar system by a team using ESO's unique planet-hunting HARPS instrument. Ross 128 b is expected to be temperate, with a surface temperature that may also be close to that of the Earth. Many red dwarf stars, including the celebrated Proxima Centauri, are subject to flares that occasionally bathe their orbiting planets in deadly ultraviolet and X-ray radiation. However, it seems that Ross 128 is a much quieter star, and so its planets may be the closest known comfortable abode for possible life.

With the data from HARPS, the team found that Ross 128 b orbits 20 times closer than the Earth orbits the Sun. Its orbital period is only 9.9 days. Despite this proximity, Ross 128 b receives only 1.38 times more irradiation than the Earth. As a result, Ross 128 b's equilibrium temperature is estimated to lie between -60 and 20°C (-76°F to 68°F), thanks to the cool and faint nature of its small red dwarf host star, which has just over half the surface temperature of the Sun.

Although it is currently 11 light years from Earth, Ross 128 is moving towards us and is expected to become our nearest stellar neighbour in just 79,000 years – a blink of the eye in cosmic terms. Ross 128 b will then take the crown from Proxima b and become the closest exoplanet to Earth!

ROSS 128 DATA

CELESTIAL COORDINATES
Right Ascension 11h 47m 44.3974s
Declination 0° 48' 16.395"

DISTANCE
11.03 light years

CONSTELLATION
Virgo (the Virgin)

VISUAL MAGNITUDE
11.13

MASS
0.168 solar masses

RADIUS
0.1967 solar radii

STAR TYPE
Red dwarf, M4V

TEMPERATURE
3192 K

ROTATION PERIOD
Unknown

NUMBER OF PLANETS
1

ROSS 128 B DATA

PLANET TYPE
Super-Earth

MASS
1.35 x Earth

RADIUS
Unknown

ORBITAL PERIOD
9.8596 days

ORBITAL RADIUS
0.0493 AU

DETECTION TYPE
Radial Velocity

DISCOVERED
2017

TRAPPIST-1 DATA

CELESTIAL COORDINATES
Right Ascension 23h 6m 29.283s
Declination -5° 2' 28.59"

DISTANCE
39.6 light years

CONSTELLATION
Aquarius (the Water Carrier)

VISUAL MAGNITUDE
18.789

MASS
0.089 solar masses

RADIUS
0.121 solar radii

STAR TYPE
Red dwarf, M8V

TEMPERATURE
2511 K

ROTATION PERIOD
3.295 days

NUMBER OF PLANETS
7

This illustration shows the seven TRAPPIST-1 planets as they might look from Earth.

TRAPPIST-1

TRAPPIST-1 shook the world with its discovery: a system of seven rocky planets, three of which lie within the star's habitable zone.

The star we now call TRAPPIST-1 was first discovered in 1999 by astronomer John Gizis and colleagues. At that time, the ultra-cool dwarf star got the unwieldy name 2MASS J23062928-0502285, because it was spotted with the Two Micron All-Sky Survey (2MASS). Then, in May 2016, scientists announced they had found three planets around this star using the Transiting Planets and Planetesimals Small Telescope (TRAPPIST) in Chile. In honour of this telescope, scientists began referring to the star as TRAPPIST-1.

In February 2017, astronomers using NASA's Spitzer Space Telescope and ground-based telescopes announced a further surprise. The system actually has seven planets. Three of them are in the theoretical 'habitable zone', the area around a star where rocky planets are most likely to hold liquid water. The discovery sets a new record for the greatest number of habitable-zone planets found around a single star outside our solar system. All of these seven planets could have liquid water – key to life as we know it – under the right atmospheric conditions, but the chances are highest with the three in the habitable zone.

Though we are now accustomed to routine discoveries of new exoplanet systems, even ones with Earth-like candidates, peering into space to find planets is still a quite recent, and revolutionary, ability of astronomy. TRAPPIST-1 was part of a revolution in our conception of and ability to probe the larger expanses of space.

Explore the TRAPPIST-1 System

A Mini Solar System

1 In contrast to our Sun, the TRAPPIST-1 star – classified as an ultra-cool dwarf – is so cool that liquid water could survive on planets orbiting very close to it, closer than is possible on planets in our solar system. All seven of the TRAPPIST-1 planetary orbits are closer to their host star than Mercury is to our Sun. The planets also are very close to each other. If a person was standing on one of the planets' surfaces, they could gaze up and potentially see geological features or clouds on neighbouring worlds, which would sometimes appear larger than the moon in Earth's sky.

Seven Rocky Worlds

2 Throughout 2017, scientists worked on creating sophisticated computer models to simulate the planets based on available information. They used additional data from Spitzer, Kepler and ground-based telescopes to come up with the best estimates for the planets' densities. The results are consistent with all of the TRAPPIST-1 planets being mostly made of rock. This result was published in February 2018.

Atmospheric Studies

3 NASA's Hubble Space Telescope was used to find that TRAPPIST-1 b and c were unlikely to have hydrogen-dominated atmospheres like those we see in gas giants. This strengthens the case that these planets could be rocky and possibly hold onto water. As of February 2018, continued observations with Hubble showed that TRAPPIST-1d, e and f are unlikely to have puffy, hydrogen-dominated atmospheres. Scientists will need more data to determine how much hydrogen TRAPPIST-1g has.

Planet of Ice?

4 Researchers using NASA's Kepler space telescope determined that the furthest planet from the star, TRAPPIST-1h, orbits its star every 19 days. This is still much shorter than the orbit of Mercury, which goes around the Sun every 88 days. But because TRAPPIST-1 is so faint – it outputs only 0.05% of the amount of energy of the Sun – planet h receives a lot less heat than Mercury, and may be covered in ice.

Determining the Age

5 The age of a star is important for understanding whether planets around it could host life. Scientists wrote in an August 2017 study that TRAPPIST-1 is between 5.4 and 9.8 billion years old. This is up to twice as old as our own solar system, which formed some 4.5 billion years ago. Shorter-lived stars are less able to host planets that are viable for life.

Top Tip

The night comes alive as the clouds part to reveal the full moon, and once-human lips release an unearthly howl. Werewolves could roam free on TRAPPIST-1 b, the innermost of seven tidally-locked planets. Only one side of the planet faces its red star; the other side is plunged in eternal night. On the nightside, it's always dark enough to see the other six planets, which reflect the light of their red star like moons. With six worlds looming large in the sky, the chances of a full 'moon' every night are high. Likely enough to keep a savage night creature in wolfish form forever.

Getting There & Away

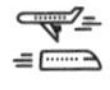

TRAPPIST-1 is an exciting system, but you'll need determination to get there. Despite its relative proximity, a trip by shuttle will take 1.5 million years.

TRES-2 A DATA

CELESTIAL COORDINATES
Right Ascension 19h 7m 14.04s
Declination 49° 18' 59.09"

DISTANCE
707 light years

CONSTELLATION
Draco (the Dragon)

APPARENT MAGNITUDE
11.41

MASS
1.05 solar masses

RADIUS
1.00 solar radii

STAR TYPE
Yellow, G05V, with an unknown binary companion

TEMPERATURE
5850 K

ROTATION PERIOD
Unknown

NUMBER OF PLANETS
1

TRES-2B DATA

PLANET TYPE
Hot Jupiter

MASS
1.20 x Jupiter

RADIUS
1.27 x Jupiter

ORBITAL PERIOD
2.47 days

ORBITAL RADIUS
0.0356 AU

DETECTION TYPE
Transit

DISCOVERED
2006

TrES-2b

TrES-2b reflects less than 1% of the sunlight falling on it, making it blacker than coal or any planet or moon in our solar system.

Are you afraid of the dark? Do the hairs on the back of your neck lift when the lights are out, waiting for the touch of spectral fingers? Then don't visit the world of TrES-2b. Welcome to the planet of eternal night. The darkest planet ever discovered orbiting a star (TrES-2 A), this alien world, found by NASA's Kepler space telescope, is less reflective than coal. Inside its atmosphere, you'd be flying blind in the dark. But fear not, traveller – it's not pitch black. Some scientists think an eerie deep red glow would emanate from its burning atmosphere. The air of this planet is the same temperature as the hottest lava, like an infernal nightlight to guide your way.

'By combining the impressive precision from Kepler with observations of over 50 orbits, we detected the smallest-ever change in brightness from an exoplanet: just six parts per million', said astronomer David Kipping of the Harvard-Smithsonian Center for Astrophysics (CfA) in 2006. 'In other words, Kepler was able to directly detect visible light coming from the planet itself.' Reasons for TrES-2b's darkness remain unknown and are an active topic of research. The gas giant planet is roughly the size of Jupiter, and goes by the nickname of the Coal Planet. It may be that a lack of reflective clouds in its atmosphere is responsible for its exceptionally low albedo. While the interior composition of the gas giants may be roughly the same, no Great Red Spot is likely to be in evidence at this hot Jupiter.

TrES-2b orbits its star, one of a binary pair, in the same direction of the stellar rotation at a slight tilt off of the equator. One of Kepler's first finds, this exoplanet was an exceptionally flashy discovery for its extremely low light-absorption rates.

The distant exoplanet TrES-2b, shown here in an artist's conception, is darker than the blackest coal.

WASP-12 DATA

CELESTIAL COORDINATES
Right Ascension 6h 30m 32.79s
Declination 29° 40' 20.29"

DISTANCE
1300 light years

CONSTELLATION
Auriga (the Charioteer)

VISUAL MAGNITUDE
11.69

MASS
1.35 solar masses

RADIUS
1.57 solar radii

STAR TYPE
Sun-like, G0

TEMPERATURE
6300 K

ROTATION PERIOD
Unknown

NUMBER OF PLANETS
1

WASP-12B DATA

PLANET TYPE
Hot Jupiter

MASS
1.39 x Jupiter

RADIUS
1.9 x Jupiter

ORBITAL PERIOD
1.09 days

ORBITAL RADIUS
0.023 AU

DETECTION TYPE
Transit

DISCOVERED
2008

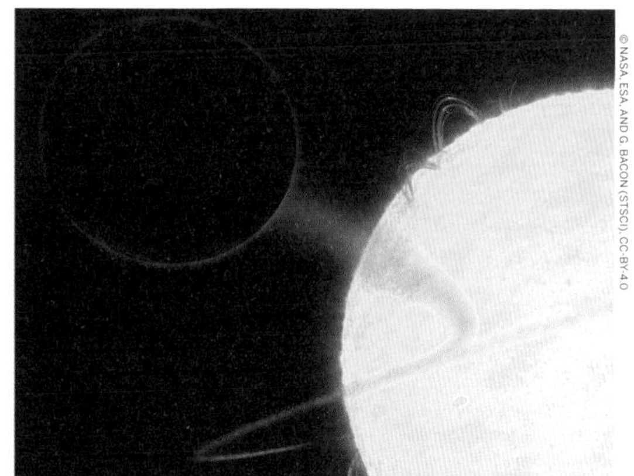

This artist's impression shows the exoplanet WASP-12b orbiting a star like our Sun.

WASP-12b

WASP-12b is a doomed planet. This dark world is being pulled into an egg shape by its parent star, and will likely be entirely devoured in an astronomical wink of an eye.

The Hubble Space Telescope's Cosmic Origins Spectrograph (COS) has observed a planet outside our solar system that looks as black as fresh asphalt because it eats light rather than reflecting it back into space. This light-eating prowess is due to the planet's unique capability to trap at least 94% of the visible starlight falling into its atmosphere.

The oddball exoplanet, called WASP-12b, is yet one more of the class of hot Jupiters, gigantic, gaseous planets that orbit very close to their host star and are heated to extreme temperatures. The planet's atmosphere is so hot that most molecules are unable to survive on its blistering dayside, where the temperature is 2538°C (4600°F). That's comparable to the temperature of red dwarf stars. Therefore, clouds probably cannot form on the dayside to reflect light back into space. Instead, incoming light penetrates deep into the planet's atmosphere, where it is absorbed by hydrogen atoms and converted to heat energy. The researchers determined the planet's light-eating capabilities by searching in mostly visible light for a tiny dip in starlight as the planet passed directly behind the star. The amount of dimming tells astronomers how much reflected light is given off by the planet. The observations did not detect reflected light, meaning that the daytime side of the planet is absorbing almost all the starlight falling onto it.

Explore Wasp-12b

The Star

1 WASP-12 is a yellow dwarf star located approximately 1300 light years away in the winter constellation Auriga. Its exoplanet was discovered by the United Kingdom's Wide Area Search for Planets (WASP) in 2008, hence the name. The automated survey looks for the periodic dimming of stars from planets passing in front of them, an effect called transiting. The hot planet is so close to the star that it completes an orbit in a quick 1.1 days.

Cannibalized

2 WASP-12b is among the hottest known planets in the Milky Way Galaxy, and may also be its shortest-lived world. The doomed planet is being stretched into an egg shape and eaten by its parent star, according to observations made by a new instrument on NASA's Hubble Space Telescope, the Cosmic Origins Spectrograph (COS). The planet may only have another 10 million years left before it is completely devoured. The atmosphere has ballooned to nearly three times Jupiter's radius and is spilling material onto the star.

Mass Transfer

3 This effect of matter exchange between two stellar objects is commonly seen in close binary star systems, but this is the first time it has been seen so clearly for a planet. 'We see a huge cloud of material around the planet, which is escaping and will be captured by the star. We have identified chemical elements never before seen on planets outside our own solar system', said team leader Carole Haswell of the Open University in Great Britain.

Carbon Planet

4 NASA's Spitzer Space Telescope discovered that WASP-12b has more carbon than oxygen, making it the first carbon-rich planet ever observed. Our planet Earth has relatively little amounts of carbon – it is made largely of oxygen and silicon. Other gas planets in our solar system, for example Jupiter, are expected to have less carbon than oxygen, but this is not known. Unlike WASP-12b, these planets harbour water, the main oxygen carrier, deep in their atmospheres, where it is difficult to measure from Earth.

Atmospheric Study

5 WASP-12b has a fixed dayside and nightside because it orbits so close to the star that it is tidally locked. The nightside is more than 1093°C cooler (2000°F), which allows water vapour and clouds to form. Previous Hubble observations of the day/night boundary detected evidence of water vapour and possibly clouds and hazes in the atmosphere.

Top Tip

Not a fan of medical experiments? Watch out for the hulking monster of a star stealing pieces of its nearby planet to assemble itself into the ultimate Frankenstein creation. The extreme force of this star's gravity is stretching WASP-12b into the shape of an egg, all the while slowly cannibalising pieces of the planet and sucking them into its scorching surface. Relatively soon (10 million years – a fleeting moment in space time) this planet will be completely devoured by its hungry star. If you enjoy watching your world fall to pieces, then feel free to touch down on the doomed planet.

Getting There & Away

WASP-12b is pretty far as exoplanets go, at 1300 light years. Expect to be aboard your interstellar shuttle for at least 50 million years.

WASP-121 DATA

CELESTIAL COORDINATES
Right Ascension 7h 10m 25.0595s
Declination -39° 5' 50.682"

DISTANCE
850 light years

CONSTELLATION
Puppis (the Poop Deck)

VISUAL MAGNITUDE
10.4

MASS
1.353 solar masses

RADIUS
1.458 solar radii

STAR TYPE
Yellow-white dwarf, F6V

TEMPERATURE
6460 K

ROTATION PERIOD
Unknown

NUMBER OF PLANETS
1

WASP-121 B DATA

PLANET TYPE
Hot Jupiter

MASS
1.184 x Jupiter

RADIUS
1.81 x Jupiter

ORBITAL PERIOD
1.275 days

ORBITAL RADIUS
0.0254 AU

DETECTION TYPE
Transit

DISCOVERED
2015

A visualisation of WASP-121 b as its star pulls on its atmosphere.

WASP-121 b

WASP-121 b is unique – an egg-shaped planet, distorted by the gravity of its star, with a seething hot stratosphere of 'glowing water'.

In WASP 121 b, scientists have discovered the strongest evidence to date for a stratosphere on an exoplanet. A stratosphere is a layer of atmosphere in which temperature increases with higher altitudes.

Reporting in the journal *Nature*, scientists used data from NASA's Hubble Space Telescope to study WASP-121 b, another example of the type of exoplanet called a hot Jupiter. Its mass is 1.2 times that of Jupiter, and its radius is about 1.9 times Jupiter's - making it puffier. But while Jupiter revolves around our Sun once every 12 years, WASP-121 b has an orbital period of just 1.3 days. This exoplanet is so close to its star that if it got any closer, the star's gravity would start ripping it apart. It also means that the top of the planet's atmosphere is heated to a blazing 2538°C (4530°F), hot enough to boil some metals.

'This result is exciting because it shows that a common trait of most of the atmospheres in our solar system - a warm stratosphere - also can be found in exoplanet atmospheres', said Mark Marley, study co-author based at NASA's Ames Research Center in California. 'We can now compare processes in exoplanet atmospheres with the same processes that happen under different sets of conditions in our own solar system.'

Explore WASP-121 b

A New Class of Planet?

1 'Theoretical models have suggested stratospheres may define a distinct class of ultra-hot planets, with important implications for their atmospheric physics and chemistry', said Tom Evans, lead author and research fellow at the University of Exeter, United Kingdom. 'Our observations support this picture.'

Stratospheric Study

2 To study the stratosphere of WASP 121 b, scientists analysed how water molecules in the atmosphere react to particular wavelengths of light, using Hubble's capabilities for spectroscopy. Starlight is able to penetrate deep into a planet's atmosphere, where it raises the temperature of the gas there. This gas then radiates its heat into space as infrared light. However, if there is cooler water vapour at the top of the atmosphere, the water molecules will prevent certain wavelengths of this light from escaping to space. But if the water molecules at the top of the atmosphere have a higher temperature, they will glow at the same wavelengths.

Fireworks

3 The phenomenon is similar to what happens with fireworks, which get their colours from chemicals emitting light. When metallic substances are heated and vaporised, their atoms move into higher energy states. Depending on the material, these atoms will emit light at specific wavelengths as they lose energy: sodium produces orange-yellow and strontium produces red in this process, for example. The water molecules in the atmosphere of WASP-121 b similarly give off radiation as they lose energy, but in the form of infrared light, which the human eye is unable to detect.

Benchmark

4 'This super-hot exoplanet, WASP-121 b, is going to be a benchmark for our atmospheric models, and it will be a great observational target moving into the Webb era', said Hannah Wakeford, study co-author who worked on this research while at NASA's Goddard Space Flight Center, Greenbelt, Maryland. She refers to the James Webb telescope, due for launch in 2021.

Other Exoplanet Stratospheres

5 WASP-121 b is not the first exoplanet with a well-defined stratosphere. Previous research found possible signs of a stratosphere on the exoplanet WASP-33 b as well as some other hot Jupiters. But the WASP-121 b study presents the best evidence yet because of the signature of hot water molecules that researchers observed for the first time.

Top Tip

In solar system planets, the change in temperature within a stratosphere is typically around 38°C (100°F). On WASP-121 b, the temperature in the stratosphere rises by 538°C (1000°F). Scientists do not yet know what chemicals are causing the temperature increase in WASP-121 b's atmosphere. Vanadium oxide and titanium oxide are candidates, as they are commonly seen in brown dwarfs, 'failed stars' that have some commonalities with exoplanets. Such compounds are expected to be present only on the hottest of hot Jupiters, as high temperatures are needed to keep them in a gaseous state.

Getting There & Away

A trip to this system aboard a shuttle will take a cool 33 million years. Make sure you don't leave the oven on.

WOLF 1061 DATA

CELESTIAL COORDINATES
Right Ascension 16h 30m 18.06s
Declination -12° 39' 45.33"

DISTANCE
14.04 light years

CONSTELLATION
Ophiuchus (the Serpent Bearer)

VISUAL MAGNITUDE
10.07

MASS
0.294 solar masses

RADIUS
0.307 solar radii

STAR TYPE
Red dwarf, M3.5V

TEMPERATURE
3342 K

ROTATION PERIOD
94 days

NUMBER OF PLANETS
3

Wolf 1061 c is shown orbiting its star.

Wolf 1061 b, c and d

With at least three planets, including one in the so-called Goldilocks zone, Wolf 1061's system is prime real estate waiting to be explored.

In 2015, astronomers from the University of New South Wales (UNSW), Australia, discovered three planets orbiting the red dwarf called Wolf 1061. The UNSW team made the discovery using observations of Wolf 1061 collected by the HARPS spectrograph on the European Southern Observatory's 3.6 meter telescope in La Silla in Chile. According to Duncan Wright, the study's lead author, 'It is a particularly exciting find because all three planets are of low enough mass to be potentially rocky and have a solid surface, and the middle planet, Wolf 1061 c, sits within the "Goldilocks" zone where it might be possible for liquid water – and maybe even life – to exist.' While a few other planets have been found that orbit stars closer to us than Wolf 1061, many are not considered to be remotely habitable.

The three newly detected planets orbit the small, relatively cool and stable star about every five, 18 and 217 days. Their masses are at least 1.9, 4.3 and 7.7 times that of Earth, respectively. The larger outer planet falls just outside the outer boundary of the habitable zone and is also likely to be rocky, while the smaller inner planet is too close to the star to be habitable. Small rocky planets like our own are now known to be abundant in our galaxy, and multi-planet systems also appear to be common. However most of the rocky exoplanets discovered so far are hundreds or thousands of light years away.

Explore the Wolf 1061 System

The Star

1 It was the German astronomer Max Wolf who first catalogued this star, among others, in 1919. They now share his name. Like all red dwarfs, Wolf 1061 is an M-class star. It is the 36th closest star system to the Sun, located 14 light years away in the constellation of Ophiuchus, from which it takes its alternate name, V2306 Ophiuchi. Another moniker is HIP 80824, named for ESA's Hipparcos satellite. The star is only 30% of the mass and radius of the Sun, and much cooler at just 3342 K. Its total, or bolometric, luminosity is just one percent of our Sun's.

Wolf 1061 b

2 Wolf 1061 b, the innermost planet so far discovered in this system, is also the most lightweight, at about 1.91 times the mass of the Earth. It is probably slightly larger than our home world, maybe by around 20%. Its orbital period is incredibly short, at just 4.9 days, owing to its proximity to its parent star of just 0.0375 AU – just one-tenth that of Mercury's distance from the Sun. This means it's likely to be too hot for liquid water to exist on its surface.

Wolf 1061c

3 The most exciting planet around Wolf 1061 is the second-most distant, Wolf 1061 c, and that's because it's located on the inner edge of the planet's so-called habitable zone – the region within which conditions are favourable for the existence of liquid surface water. Wolf 1061 c is a rocky exoplanet at least 4.3 times the mass of the Earth. It takes just 17.9 days to complete one orbit of its star, from a close distance of only 0.089 AU. That's well inside the orbit of Mercury. It is likely around 1.5 times the radius of the Earth, with about 1.6 times its gravity. A human there would weigh about 50% more their usual weight.

Wolf 1061d

4 The most remote of Wolf 1061's three known planets, Wolf 1061 d, is also its most massive. It is called a 'Super-Earth'. At an orbital period of 217 days, it must reside no closer than 0.47 AU from its star – comparable to the orbit of Mercury. With a mass of around eight Earths, Wolf 1061 d is close to the upper limit of 'Super-Earths', so it may not be rocky. It could be like Neptune or Uranus. With an average temperature of -157°C (-250°F), it is one of the coldest known 'Super-Earths'.

More Planets?

5 While Wolf 1061 has at least three planets to explore, there may well be more, which we cannot currently detect with our technology. In all likelihood, some of the planets will also have satellites, or exomoons. Wolf 1061 is a very compact system. Even its most distant planet, Wolf 1061 d, is not much further from its star than Mercury is from the Sun.

Top Tip

'Our team has developed a new technique that improves the analysis of the data from this precise, purpose-built, planet-hunting instrument (the HARPS Spectrograph), and we have studied more than a decade's worth of observations of Wolf 1061', says Professor Chris Tinney, head of the Exoplanetary Science at UNSW group. 'These three planets right next door to us join the small but growing ranks of potentially habitable rocky worlds orbiting nearby stars cooler than our Sun.'

Getting There & Away

Despite its relative proximity of 14 light years, Wolf 1061 is still 29,000 times further than the most distant solar system planet, Neptune. Be prepared for a trip lasting nearly 540,000 years with present rocket technology.

Danger Zones

Violent outbursts of seething gas from young red dwarf stars may make conditions uninhabitable on fledgling planets. In this artist's rendering (right), an active, young, red dwarf (top) is stripping the atmosphere from an orbiting planet (bottom). Scientists found that flares from the youngest red dwarfs they surveyed – approximately 40 million years old – are 100 to 1000 times more energetic than when the stars are older. They also detected one of the most intense stellar flares ever observed in ultraviolet light – more energetic than the most powerful flare ever recorded from our Sun.

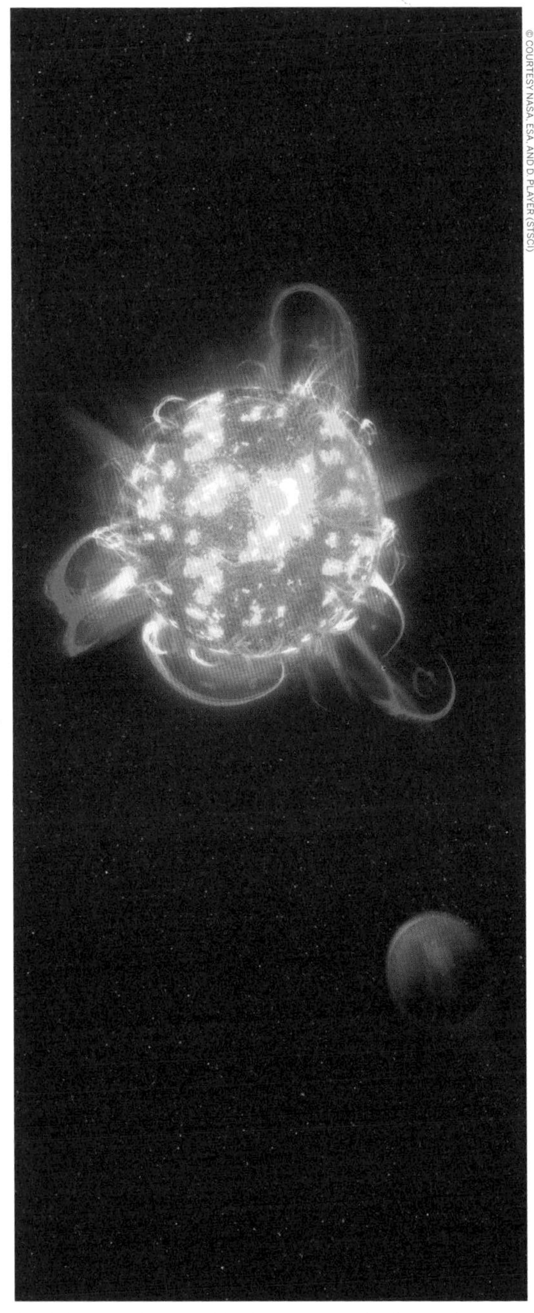

Artist's illustration of a young red dwarf stripping away a planet's atmosphere.

YZ Ceti b, c and d

YZ Ceti is the nearest multi-planet system found orbiting a red dwarf star or M dwarf – three rocky planets reside there.

The YZ Ceti system is a recent find. The paper announcing its discovery was published in 2017, after research using the High Accuracy Radial velocity Planet Searcher (HARPS) at the European Southern Observatory's La Silla Observatory in Chile. It was a first in two ways. Not only is it the closest known multi-planet system, at just over 12 light years away, but it also harbours the lowest-mass planets ever found using the radial velocity method - where the gravitational tugs on the star by its planets causes the star to 'wobble' subtly back and forth.

Like most red dwarf planetary systems, this one is very compact. The innermost planet, YZ Ceti b, is about 0.016 AU away from YZ Ceti, YZ Ceti c is 0.021 AU and the outermost, YZ Ceti d, is 0.028 AU. Their orbits take just a few days. Because of this proximity, the planets are likely too close to the star to be within its so-called habitable zone - water would simply evaporate. But as YZ Ceti is a flare star - a type of variable star that exhibits regular eruptions - it is not a ripe environment for life anyway.

Stellar flares from red dwarfs are particularly bright in ultraviolet wavelengths, compared with Sun-like stars. It's believed the flares are powered by intense magnetic fields that get tangled by the roiling motions of the stellar atmosphere. When the tangling gets too intense, the fields break and reconnect, unleashing tremendous amounts of energy. Super-flares of such frequency and intensity could potentially bathe young planets in so much ultraviolet radiation that they forever doom chances of habitability. Because approximately three-quarters of the stars in our galaxy are red dwarfs, most of the galaxy's 'habitable-zone' planets - planets orbiting their stars at a distance where temperatures are moderate enough for liquid water to exist on their surface - likely orbit red dwarfs. However, young red dwarfs are active stars, producing ultraviolet flares that blast out so much energy they could influence atmospheric chemistry and possibly strip off the atmospheres of these fledgling planets. Hubble's HAZMAT mission, short for Habitable Zones and M dwarf Activity across Time, looks at red dwarfs of all ages to glean data on how their activity may impact exoplanets. This research will tell scientists more about the possibility of continuous habitability on the planets orbiting variable red dwarfs such as YZ Ceti.

YZ CETI DATA

CELESTIAL COORDINATES
Right Ascension 1h 12m 30.64s
Declination -16° 59' 56.36"

DISTANCE
12.11 light years

CONSTELLATION
Cetus (the Whale)

VISUAL MAGNITUDE
12.03–12.18

MASS
0.130 solar masses

RADIUS
0.168 solar radii

STAR TYPE
Red dwarf, M4V

TEMPERATURE
3056 K

ROTATION PERIOD
68-83 days

NUMBER OF PLANETS
3, possibly 4

STELLAR OBJECTS

Star clusters merge in the Doradus Nebula.

Stellar Objects

Beyond our solar system, the Universe gets really interesting. If we consider that our knowledge of objects in the Universe is strongest the closer an object is to home, it's astounding how much we've learned about various objects and structures, and how diverse these distant sources of light can be.

What we think of simply as 'stars' encompasses an entire range of life-cycles and stages of stellar activity, from early star cradles forming new objects to supernova and black holes on the other end of the spectrum. Stellar objects come in a variety of shapes and sizes – or more accurately, a variety of wavelengths and masses. These objects are full of surprises, from newer discoveries (such as black holes) to long-known bright variable stars.

Included are stars from the beginning to the end of the stellar life cycle, and everywhere in between, as well as a variety of unusual objects which only roughly fit our known models to explain them – or which have redefined the theories and principles entirely upon their discovery. Also explored are so-called deep sky objects, meaning anything that's not properly classified as a star: nebulae and star clusters are prominent examples of these, and are beloved targets of amateur and professional astronomers alike. From the closest celestial neighbours to the furthest reaches of the known Universe, these stars, nebula and clusters give a sense of the incredible diversity and wonder of the objects that fill the night sky above.

Top Highlights

Betelgeuse

1 The 'cannibal' runaway star Betelgeuse has been a striking feature of the night sky since humans started watching it.

Cat's Eye Nebula

2 The cat's eye is a classic planetary nebula; the mesmerising shape comes from gas ejected by this once red giant.

Epsilon Aurigae

3 The mystery behind this dimming star was recently revealed to be an eclipsing companion.

Heavy Metal Subdwarfs

4 HE 1256-2738 and HE 2359-2844 are rich with lead and rare heavy metals, hence their star type's name.

HLX-1

5 HLX stands for 'hyper-luminous X-ray source'. These X-rays were how this black hole, 20,000 solar masses dense, was found.

Horsehead Nebula

6 This dark molecular cloud is a striking presence in the heavens.

Kepler's Supernova

7 The most recent supernova witnessed within the Milky Way, Kepler himself observed this 1604 supernova, now a remnant.

MY Camelopardalis

8 Intertwined massive stars dance together in MY Camelopardalis, potentially headed for a merger.

Pleiades

9 The 'Seven Sisters' are famous in myth, but these visible stars represent only the brightest members of an immense cluster.

Rigel

10 Though it appears as one bright star in the sky, Rigel is actually a multiple star system.

Tabby's Star

11 Tabby's Star has such unusual brightness changes that it's a head-scratching mystery.

UY Scuti

12 Classed as a hypergiant, this massive star nearly 2000 times the Sun's size is the largest yet discovered.

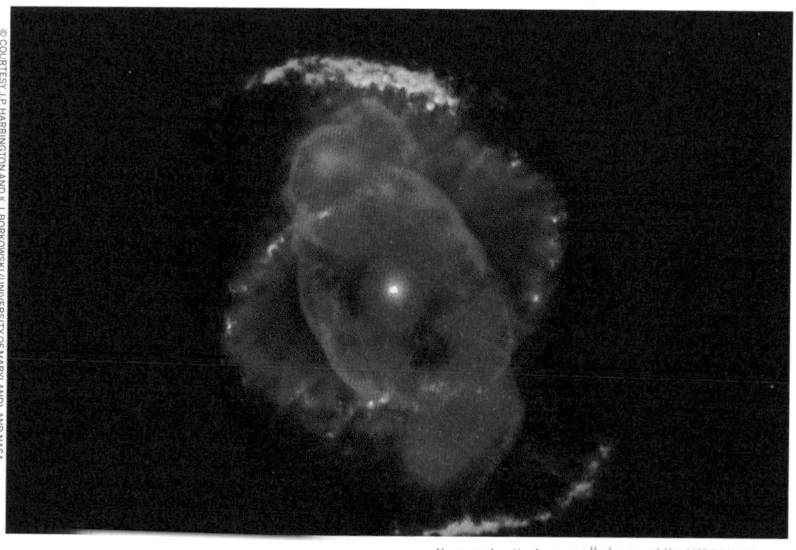

X-ray and optical composite image of the Cat's Eye Nebula.

Types of Stellar Objects

Hubble captured this image of a star forming region within the constellation Cygnus.

Star Formation: Nebula and Protostars

A nebula is a giant cloud of dust and gas in space. Nebulae exist in the space between the stars – also known as interstellar space. They come in a variety of types. Some nebulae come from the gas and dust thrown out by the explosion of a dying star, such as a supernova. Other nebulae are regions where new stars are beginning to form. For this reason, some nebulae are called 'star nurseries'.

Planetary Nebula

The term planetary nebula is a misnomer since they don't harbour planets but were rather named because of their visual similarity to planets when first observed. This nebula type doesn't form stars; instead, the ring of the planetary nebula is formed during the death of a red giant star, as ejected material is transmitted to the interstellar medium through this ionised gas cloud. Many notable nebulae are this type, such as the Cat's Eye Nebula.

Emission Nebula

These hot star-forming regions are clouds of high temperature gas which emit their own light across a variety of wavelengths It is often ionised by a nearby hot star's high-energy photons.

Reflection Nebula

As their name indicates, these clouds of dust reflect the light from nearby stars, giving them a typically blue appearance. They may also be sites of star formation.

Dark Nebula

Also called absorption nebula, rather than reflecting nearby light, dark nebula absorb it.

Protostar

Protostars are the hot core at the heart of the collapsing cloud that will one day become a star, in the stage of mass accretion. Turbulence deep within nebulae gives rise to knots with sufficient mass that the gas and dust can begin to collapse under its own gravitational attraction. As the cloud collapses, the material at the centre begins to heat up.

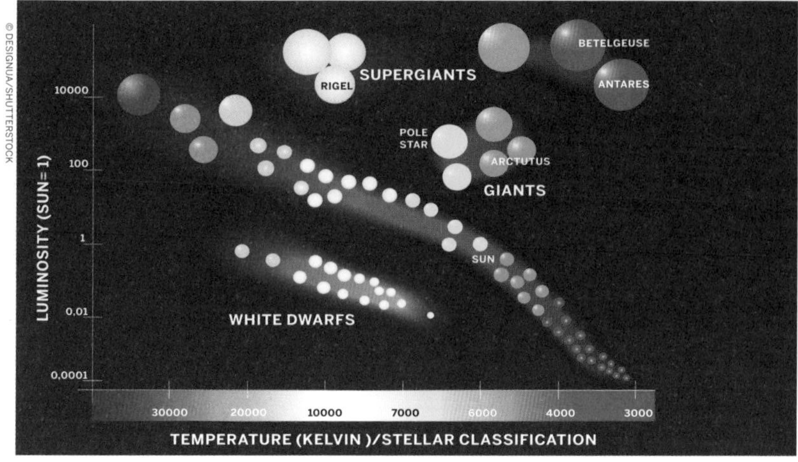

'Main sequence' or dwarf stars form the central band on this stellar classification chart.

Main Sequence Star Types

Once a forming star contracts enough that its central core can burn hydrogen to helium, it becomes a 'main sequence' star, which fuse hydrogen atoms together to make helium atoms in their cores. The more massive a main sequence star, the brighter and bluer it is. Another term for this type of star is a dwarf star, as opposed to giant stars on the larger end of the spectrum. There are several types of main sequence or dwarf stars, which are one of the most common types of stellar objects in the Universe.

Yellow Dwarf

Surely the most familiar star type to Earthlings is the yellow dwarf, also known as a G dwarf star. It's the type of star that our own Sun is classified as. In fact, in mass and colour, the Sun is a representative example of this type of main sequence star.

Orange Dwarf

The orange dwarf or K-type main sequence star is intermediate in size between red and yellow type stars; Alpha Centauri B is one.

Red Dwarf

Also called an M dwarf, red dwarfs are a class of small, relatively cool stars, emitting dim, red light. The most common type of star, they comprise about 75 percent of all stars in the galaxy.

Brown Dwarf

Brown dwarfs are objects which have a size between that of a giant planet like Jupiter and that of a small star. In fact, most astronomers would classify any object with between 15 and 75 times the mass of Jupiter to be a brown dwarf. Given that range of masses, the object would not have been able to sustain the fusion of hydrogen like a regular star; thus, many scientists have dubbed brown dwarfs as 'failed stars', and there can be overlap in classificatiom between brown dwarfs and exoplanets.

Heavy Metal Subdwarf

One of several types of smaller 'subdwarf' type stars, heavy metal subdwarfs have large quantities of germanium, strontium, yttrium, zirconium, and lead, among other heavy metals.

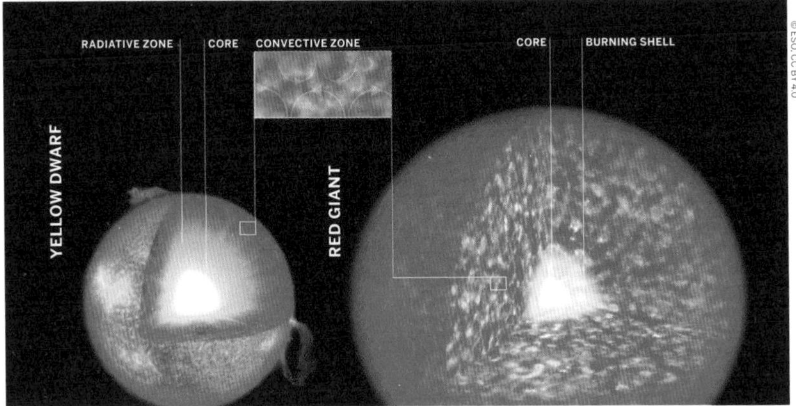

A red giant has consumed all of the hydrogen in its core and begins to burn the hydrogen in its expanding shell.

Giant Star Types

Giant stars have masses from eight times to as much as 100 times that of the mass of our Sun. These massive stars have hotter and denser cores than dwarf stars (stars that are smaller than five times the mass of our Sun). Therefore, giant stars have a greater rate of the nuclear reactions that light up stars. Massive stars also use up the hydrogen fuel in their core faster, despite starting out with much more of it, meaning they live much shorter lives than dwarf stars. A giant star may end its life in a spectacular fashion, via a supernova explosion, leaving behind a strange object such as a neutron star or an even more bizarre black hole.

Red Giant

When a yellow dwarf star has fused all the hydrogen in its core, nuclear reactions cease. Deprived of the energy production needed to support it, the core begins to collapse into itself and becomes much hotter. Hydrogen fusion continues in a shell surrounding the core; increasingly hot, the core pushes the outer layers of the star outward, causing them to expand and cool, transforming the star into a red giant. This is the Sun's own fate.

Yellow Giant

Yellow giants form from stars with higher masses than those that become red giants, and last for less time. They are typically variable.

Supergiant

As their name indicates, these are some of the most luminous and massive stars that exist, with a visual magnitude between -3 and -8 and temperatures of over 20,000 K. They are typically eight to twelve times the size of our Sun.

Hypergiant

The most massive stars, known as hypergiants, may be 100 or more times more massive than the Sun, and have surface temperatures of more than 30,000 K. Hypergiants emit hundreds of thousands of times more energy than the Sun, but have lifetimes of only a few million years. Although extreme stars such as these are believed to have been common in the early Universe, today they are extremely rare - the entire Milky Way Galaxy contains only a handful of hypergiants.

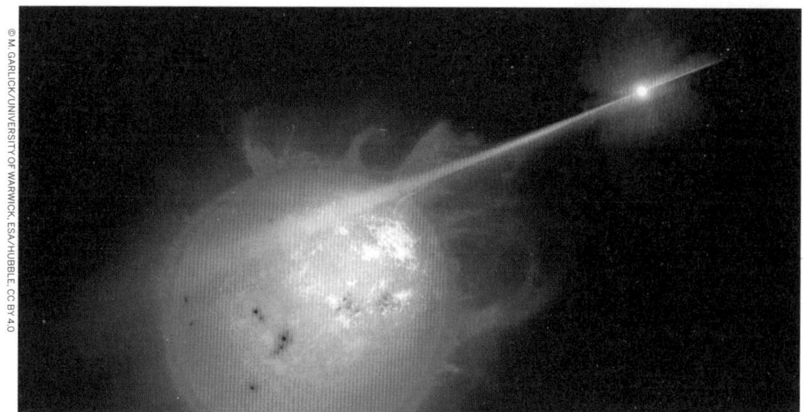

Artist's visualisation of the binary pair AR Scorpii, consisting of a rapidly spinning white dwarf and red dwarf star.

Binaries and Clusters

Many stars, like the Sun, are single field or loner stars. But many are gravitationally bound, either to another star or to a collection of hundreds or thousands of them. These clusters are considered deep-sky objects (as are nebulae and galaxies).

Binary Stars

William Herschel created the term binary star after discovering that stars close together in the sky could be bound pairs travelling together in tandem.
The two stars in a binary system exert a gravitational pull on each other, allowing astronomers to determine exactly what their respective masses are. Sometimes the stars will be in a stable relationship, but other times, they slowly pull together on a collision course. Frequently, the interaction of their masses leads to eccentric orbits. Binary star systems can even have their own exoplanet systems, orbiting either in very near or very far orbit. It is believed that most are formed together from the nebula stage, though instances of stellar capture are expected in globular clusters. Binaries can exist in tight or wide proximity, and either spiral closer together or slowly diverge depending on their orbital trajectories.
In fact, some theorise that the Sun once had its own binary. Stars can also come in multiple star groupings of more than two.

Open Cluster

Smaller and less dense than globular clusters, these are groupings of stars that are gravitationally linked; the Pleiades is one. It contains over 3000 stars though only seven are typically visible to the naked eye. A more common number of stars for an open cluster is a few hundred. The member stars are usually young, and may disassociate with time.

Globular Cluster

Globular star clusters are snow-globe-shaped islands of several hundred thousand ancient stars, comprising the oldest known in the Universe. They usually share a similar age, though they may contain populations of stars of different ages. Often, these were first classed by observers as nebula; the term used to mean any hazy light source which could not be otherwise identified. To our knowledge there are about 200 roaming within the Milky Way alone. They may even be small galaxy remnants.

An artist's impression a distant quasar powered by a supermassive black hole.

End of Life

After millions to billions of years, depending on their initial masses, stars run out of their main fuel - hydrogen. Once the ready supply of hydrogen in the core is gone, nuclear processes occurring there cease. Without the outward pressure generated from these reactions to counteract the force of gravity, the outer layers of the star begin to collapse inward toward the core.

Neutron Star

Neutron stars are formed at the atomic level when a massive, or giant, star runs out of fuel and collapses. The very central region of the star - the core - collapses, crushing together every proton and electron into a neutron. If the core of the collapsing star is between about 1 and 3 solar masses, these newly-created neutrons can stop the collapse, leaving behind a neutron star. (Stars with higher masses will continue to collapse into stellar-mass black holes.)

This collapse leaves behind an object with the mass of a star squished down to the size of a city, to be found scattered throughout the galaxy in the same places where we find stars. Many neutron stars are likely undetectable because they simply do not emit enough radiation. However, under certain conditions, they can be easily observed. A handful of neutron stars have been found sitting at the centres of supernova remnants quietly emitting X-rays. In binary systems, some neutron stars can be found accreting materials from their companions, emitting electromagnetic radiation powered by the gravitational energy of the accreting material. There are two main sub-types of neutron stars we have observed, magnetars and pulsars.

Magnetar

In a typical neutron star, the magnetic field is trillions of times that of the Earth's magnetic field; however, in a magnetar, the magnetic field is another 1000 times stronger. In all neutron stars, the crust of the star is locked together with the magnetic field so that any change in one affects the other. In a magnetar, with its huge magnetic field, movements in the crust cause the neutron star to release a vast amount of energy in the form of electromagnetic radiation.

Pulsar

Most neutron stars are observed as pulsars. Pulsars are rotating neutron stars observed to have pulses of radiation at very regular intervals that typically range from milliseconds to seconds. Pulsars have very strong magnetic fields which funnel jets of particles out along the two magnetic poles. Similarly to a lighthouse, the jet can only be seen when it is directed to an observer.

White Dwarf

While the name might seem to indicate that these stars are still part of the main sequence like the other dwarfs, actually white dwarf stars are dim, dense, compact stars at the end of their life; the remnant core that remains after intermediate-mass stars (similar to the Sun) exhaust their nuclear fuel and blow off their outer layers. They are dominated by oxygen and carbon, but often have thin layers of hydrogen and helium. Usually only the size of a planet, these stellar core remnants can no longer generate energy. Often they are surrounded by planetary nebula from ejected material. This is the least dramatic, but likely most common, way for a low mass star (one that's less than 10 times the mass of the Sun) to end its life.

Black Dwarf

Considered the final stage of stellar evolution for those stars that move through the dwarf phases, a black dwarf is a white dwarf star which has cooled so much that it no longer emits significant heat or light, and is thus invisible. No black dwarf stars have been discovered, but they are considered to be a theoretical possibility. These stellar remnants are one possible endpoint of stellar evolution.

Blue Dwarf

Currently hypothetical, a blue dwarf might be the remnant outcome of an exhausted red dwarf star. The Universe isn't yet old enough for any to have formed!

Nova or Supernova

A supernova happens where there is a change in the core of a star. The first type of supernova happens in binary star systems. One of the stars, a carbon-oxygen white dwarf, steals matter from its companion star. Eventually, the white dwarf accumulates too much matter. Having too much matter causes the star to explode, resulting in a supernova.

The second type occurs at the end of a single star's lifetime. As the star runs out of nuclear fuel, some of its mass flows into its core. Eventually, the core is so heavy that it cannot withstand its own gravitational force. The core collapses, which results in the giant explosion of a supernova.

Quasar

Many astronomers believe that quasars are the most distant objects yet detected in the Universe. The word quasar is short for 'quasi-stellar radio source'. This name, which means star-like emitters of radio waves, was given in the 1960s when quasars were first detected. The name is retained today, even though astronomers now know most quasars are faint radio emitters. In addition to radio waves and visible light, quasars also emit ultraviolet rays, infrared waves, X-rays, and gamma-rays. Most quasars are larger than our solar system.

Quasars give off enormous amounts of energy - they can be a trillion times brighter than the Sun! Quasars are believed to produce their energy from massive black holes in the centre of the galaxies in which the quasars are located. Because quasars are so bright, they drown out the light from all the other stars in the same galaxy. Despite their brightness, due to their great distance from Earth, no quasars can be seen with an unaided eye. Energy from quasars takes billions of years to reach the Earth's atmosphere. For this reason, the study of quasars can provide astronomers with information about the early stages of the Universe.

Black Hole

A black hole is anything but empty space. Rather, it is a great amount of matter packed into a very small area - think of a star ten times more massive than the Sun squeezed into a sphere approximately the diameter of New York City. The result is a gravitational field so strong that nothing, not even light, can escape. Astronomers believe that supermassive black holes lie at the centre of virtually all large galaxies, even the Milky Way. They can be detected by watching for their effects on nearby stars and gases. One possible mechanism for the formation of supermassive black holes involves the merger of intermediate-mass black holes, which combine to form a supermassive black hole from their combined masses.

The Missing Middle

Although the basic formation process for black holes is understood, one perennial mystery in the science of black holes is that they appear to exist on two radically different size scales. On the one end, there are the countless smaller black holes that are the remnants of individual massive stars. On the other end of the size spectrum are the giants known as 'supermassive' black holes, which are millions, if not billions, of times as massive as the Sun. Historically, astronomers have long believed that no mid-sized black holes exist in between these scales. However, recent evidence from Chandra, XMM-Newton and Hubble strengthens the case that mid-size black holes *do* exist.

The Theory of Black Holes

Black holes were predicted by Einstein's theory of general relativity, which showed that when a massive star dies, it leaves behind a small, dense remnant core. If the core's mass is more than about three times the mass of the Sun, the equations showed, the force of gravity overwhelms all other forces and produces a black hole. Scientists can't directly observe black holes. We can, however, infer the presence of black holes and study them by detecting their effect on other matter nearby. If a black hole passes a star or a cloud of interstellar matter, for example, it will draw matter inward in a process known as accretion. As the attracted matter accelerates and heats up, it emits radiating X-rays.

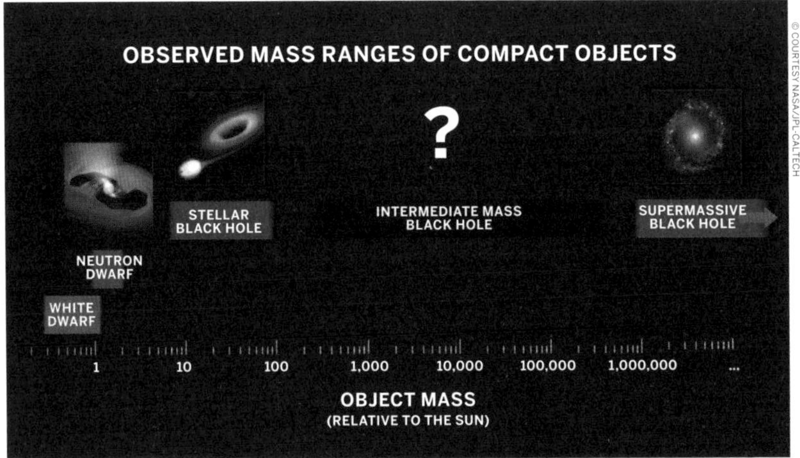

There's a large gap under investigation between the masses of stellar black holes and supermassive black holes.

©COURTESY NASA/CXC

Chandra X-ray Observatory captured this image of Cygnus X-1, a cosmic X-ray source.

The Life Cycle of Stars

For all types of stars, a star's life cycle is determined by its mass. The larger its mass, the shorter its life cycle. A star's mass is determined by the amount of matter that is available in its nebula, the giant cloud of gas and dust from which it was born. Over time, the hydrogen gas in the nebula is pulled together by gravity and it begins to spin. As the gas spins faster, it heats up and becomes a protostar. Eventually, the temperature reaches 15,000,000 degrees and nuclear fusion occurs in the cloud's core. The cloud begins to glow brightly, contracts a little and becomes stable. It is now a main sequence star and will remain in this stage, shining for millions to billions of years to come. This is the stage our Sun is at right now.

For average stars like the Sun, the process of ejecting its outer layers continues until the stellar core is exposed. This dead, but still ferociously hot stellar cinder is called a white dwarf. White dwarfs, which are roughly the size of our Earth despite containing the mass of a star, once puzzled astronomers - why didn't they collapse further? What force supported the mass of the core? Quantum mechanics provided the explanation. Pressure from fast-moving electrons keeps these stars from collapsing. The more massive the core, the denser the white dwarf that is formed. Thus, the smaller a white dwarf is in diameter, the larger it is in mass! These paradoxical stars are very common – our own Sun will become a white dwarf star billions of years from now. White dwarfs are intrinsically very faint because they are so small and, lacking a source of energy production, they fade into oblivion as they gradually cool down to become black dwarfs.

This fate awaits only those singular stars with a mass up to about 1.4 times the mass of our Sun. Above that mass, electron pressure cannot support the core against further collapse. Such stars suffer a different fate; stars in binary or multiple star systems also experience a different set of stages in their life cycle.

If a white dwarf forms in a binary or multiple star system, it may experience a more eventful demise as a nova. Nova is Latin for 'new' – novae were once thought to be new stars. Today, we understand that they are in fact, very old stars – white dwarfs. If a white dwarf is close enough to a companion star, its gravity may drag matter, mostly hydrogen, from the outer layers of that star onto itself, building up its surface layer. When enough hydrogen has accumulated on the surface, a burst of nuclear fusion occurs, causing the white dwarf to brighten substantially and expel the remaining material. Within a few days, the glow subsides and the cycle starts again. Sometimes, particularly massive white dwarfs (those near the 1.4 solar mass limit mentioned above) may accrete so much mass in the manner that they collapse and explode completely, becoming what is known as a supernova.

If the collapsing stellar core at the centre of a supernova contains between about 1.4 and 3 solar masses, the collapse continues until electrons and protons combine to form neutrons, producing a neutron star. Neutron stars are incredibly dense – similar to the density of an atomic nucleus. Because it contains so much mass packed into such a small volume, the gravitation at the surface of a neutron star is immense. Like the white dwarf stars above, if a neutron star forms in a multiple star system it can accrete gas by stripping it off any nearby companions.

Neutron stars also have powerful magnetic fields which can accelerate atomic particles around its magnetic poles producing powerful beams of radiation. Those beams sweep around like massive searchlight beams as the star rotates. If such a beam is oriented so that it periodically points towards the Earth, we observe it as regular pulses of radiation that occur whenever the magnetic pole sweeps past the line of sight. In this

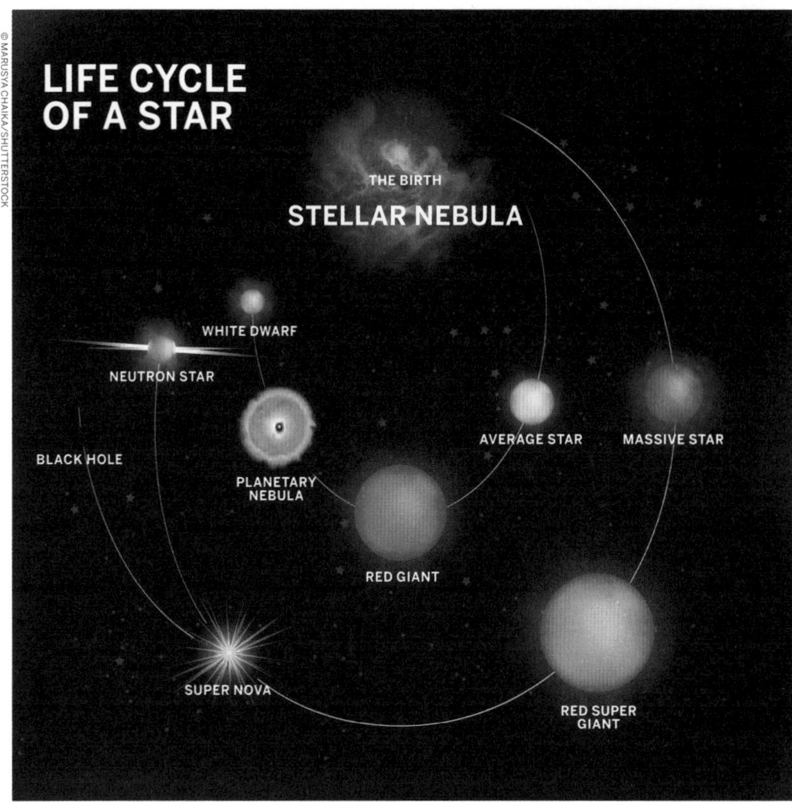

Each star type follows its own evolutionary path over time.

case, the neutron star is known as a pulsar.

A neutron star can also be called a magnetar. In a typical neutron star, the magnetic field is trillions of times that of the Earth's magnetic field; however, in a magnetar, the magnetic field is another 1000 times stronger. Since the crust and magnetic field are tied together, any stellar explosion ripples through the magnetic field. In a magnetar, with its huge magnetic field, any movements in the crust of the star cause the neutron star to release a vast amount of energy in the form of electromagnetic radiation.

If the collapsed stellar core is larger than three solar masses, it collapses completely to form a black hole: an infinitely dense object whose gravity is so strong that nothing can escape its immediate proximity, not even light. Since photons are what our instruments are designed to see, black holes can only be detected indirectly. Indirect observations are possible because the gravitational field of a black hole is so powerful that any nearby material - often the outer layers of a companion star - is caught up and dragged in. As matter spirals into a black hole, it forms a disc that is heated to enormous temperatures, emitting copious quantities of X-rays and Gamma-rays that indicate the presence of the underlying hidden companion.

Main sequence stars over eight solar masses are destined to die in a titanic explosion called a supernova. A supernova is not merely a bigger nova. In a nova, only the star's surface explodes. In a supernova, the star's core collapses and then explodes. In massive stars, a complex series of nuclear reactions leads to the production of iron in the core. Having achieved iron, the star has wrung all the energy it can out of nuclear fusion – fusion reactions that form elements heavier than iron actually consume energy rather than produce it. The star no longer has any way to support its own mass, and the iron core collapses. In just a matter of seconds, the core shrinks from roughly 5000 miles across to just a dozen, and the temperature spikes 100 billion degrees kelvin or more. The outer layers of the star initially begin to collapse along with the core, but rebound with the enormous release of energy and are thrown violently outward.

Supernovae release an almost unimaginable amount of energy. For a period of days to weeks, a supernova may outshine an entire galaxy. Likewise, all the naturally occurring elements and a rich array of subatomic particles are produced in these explosions. On average, a supernova explosion occurs about once every hundred years in the typical galaxy. About 25 to 50 supernovae are discovered each year in other galaxies, but most are too far away to be seen without a telescope.

The dust and debris left behind by novae and supernovae eventually blend with the surrounding interstellar gas and dust, enriching it with the heavy elements and chemical compounds produced during stellar death. Eventually, those materials are recycled, providing the building blocks for a new generation of stars and accompanying planetary systems.

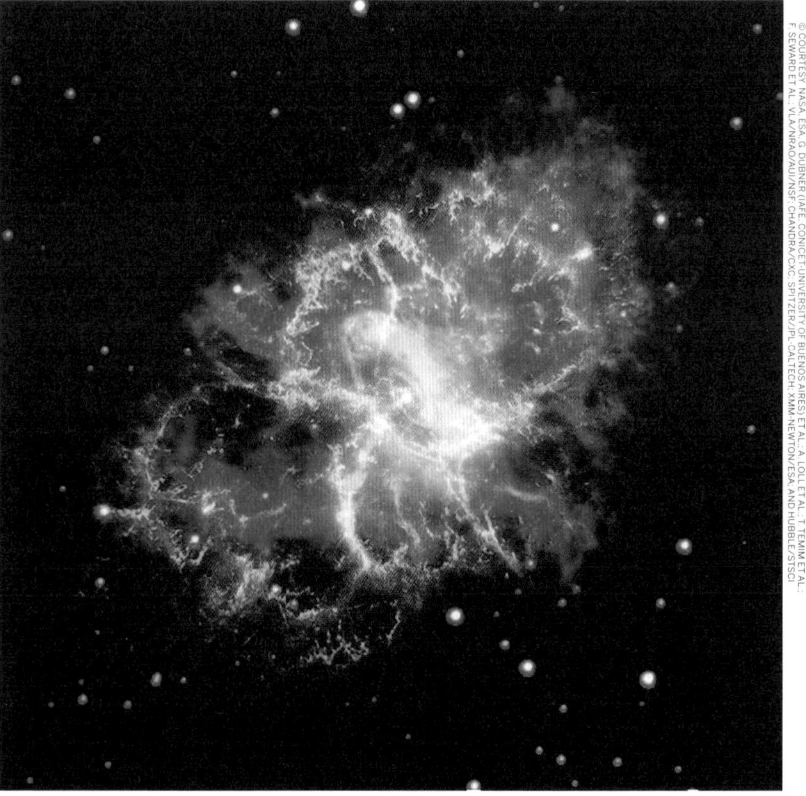

The vibrant Crab Nebula's reflective gases, a supernova remnant.

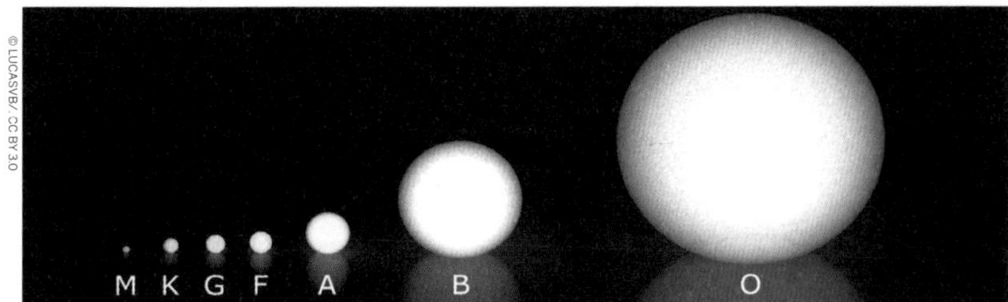

Stellar classifications indicate the mass, luminosity, temperature and size.

Using Spectra to Classify Stars

When light from the Sun or stars is displayed according to wavelength, the result is said to be a spectrum. More than one spectrum are called spectra (not spectrums).

Astronomers get a lot of information about the Sun and stars from what is called spectra, which form the basis of the current spectral classification system. Main sequence stars are currently classified by this Morgan-Keenan (MK) system under the letters O, B, A, F, G, K, and M. M type is the coolest, with O type the hottest, and all other star types arranged in between. The Sun is a G type, for instance. O, B and A type stars are referred to as early spectral types, and cool stars like G, K and M are late type stars (though the belief that led to this coinage has since been proved false). Stars can also be referred to by their luminosity class as well as by the temperature range. There are also C type carbon stars, S type stars (often variable) T (T Tauri) young stars and W (Wolf-Rayet) class stars.

Astronomers rely upon atomic physics in order to dig information out of solar and stellar spectra. Atomic physics is the branch of science that deals with atoms and ions, and the light that comes from them. Each different type of atom or ion emits light waves at a combination of wavelengths that are special to that particular type of atom or ion, and different from the wavelengths of light waves that are sent out by any other kind of atom or ion. These light waves have become known as emission lines because light at these particular wavelengths looked like many straight lines in a spectrum when astronomers first obtained them.

Each different type of atom or ion has its own special, unique set of emission lines. Astronomers use these emission lines to identify the atoms or ions that send out light from the Sun and stars. This is similar to the way a detective uses fingerprints to determine whose hands have touched an object. Once astronomers have determined what ions are present on the Sun and stars, they know immediately what elements are there. (Remember that ions are just atoms of a given element that have lost one or more of their electrons.) Furthermore, astronomers know how hot the Sun and stars are because each different type of ion is found only in a certain temperature range. Hydrogen, helium, iron and calcium are all important elements for understanding what type of star is being observed.

In addition to providing information about the composition and temperatures of the Sun and stars, emission lines also provide information. For example, astronomers can sometimes measure densities by comparing how bright some emission lines are relative to others. Astronomers can also measure motions on the Sun and stars by measuring changes in the wavelengths of emission lines, or by the shapes of emission lines in the spectra.

STELLAR TYPE
Magnetar

CONSTELLATION
Cassiopeia

MASS
Unknown

TEMPERATURE
Unknown

RADIUS
Unknown

DETECTION TYPE
X-ray & Gamma ray

DISCOVERED
1981

DISTANCE FROM SUN
10,000 light years

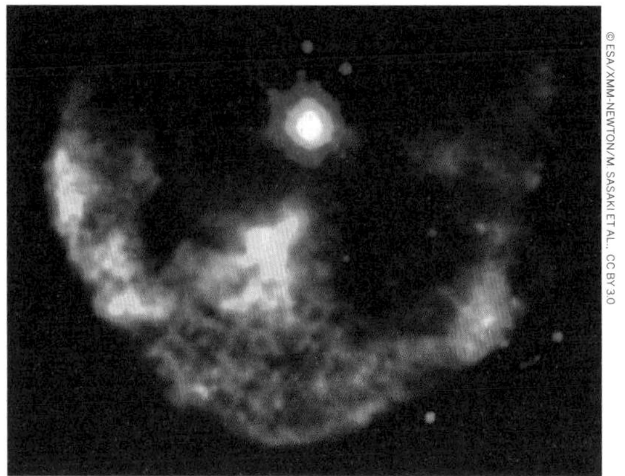

Magnetar 1E 2259+586 shines a brilliant blue-white in this false-colour X-ray image.

1E 2259+586

One of only two dozen known magnetars, 1E 2259+586 fascinates researchers. However, when they observed a sudden decrease in rotational speed, this star became one of the more puzzling objects in the Universe.

In the whole Universe of unknown wonders to discover, surprises are actually quite uncommon. Most objects in the Universe behave in well-understood and predictable ways, in line with modern principles of physics. 1E 2259+586 surprised researchers not long after its discovery when it behaved very oddly.

The star 1E 2259+586 is located about 10,000 light years away toward the constellation Cassiopeia. As a neutron star, it is the crushed core of a massive star that ran out of fuel, collapsed under its own weight and exploded as a supernova. Matter within a neutron star is so dense a teaspoonful would weigh about a billion tons on Earth. But 1E 2259+586 is special in that it is also known as a magnetar, a class of stars which have very powerful magnetic fields and occasionally produce high-energy explosions or pulses.

Researchers observing 1E 2259+586 have discovered that its rotation is actually slowing down, as indicated by observations of X-ray pulses from 1E 2259+586 from July 2011 through mid-April 2012. Typically, neutron stars increase their rotational speed – a phenomenon called a glitch – so any decrease is certainly unusual; researchers at NASA's Goddard Space Flight Center called it an 'anti-glitch'. They continue to study 1E 2259+586 to see what it will do next.

3C 273's light has taken some 2.5 billion years to reach Earth.

3C 273

Quasar 3C 273 may be over two billion light years from Earth, but it still holds a special place in the hearts of astronomers: it the most distant celestial object amateur astronomers are likely to see through their telescopes.

Quasar 3C 273 is an extremely bright high-energy source and was the first quasar ever to be identified. It was discovered in the early 1960s by astronomer Allan Sandage. Despite being 2.4 billion light years away, it is still one of the closest quasars to our home – but that great distance is fortunate because having a quasar nearby is not good thing for a planet like Earth!

The term quasar is an abbreviation of the phrase 'quasi-stellar radio source', as they appear to be star-like in the sky. In fact, quasars are the intensely powerful centres of distant, active galaxies, powered by a huge disc of particles surrounding a supermassive black hole. Some quasars, including 3C 273, have been observed to fire off super-fast jets into the surrounding space. In addition to radio waves and visible light, quasars also emit ultraviolet rays, infrared waves, X-rays, and gamma-rays. Most quasars are larger than our solar system. A quasar is approximately 1 kiloparsec in width.

Quasars are capable of emitting hundreds or even thousands of times the entire energy output of our galaxy, making them some of the most luminous and energetic objects in the entire Universe. Of these very bright objects, 3C 273 is the brightest in our skies. If 3C 273 was located just 30 light years from our own planet, it would still appear as bright as the Sun in the sky.

STELLAR TYPE
Quasar

CONSTELLATION
Virgo (the Virgin)

MASS
Unknown

TEMPERATURE
Unknown

RADIUS
~1 kiloparsec

DETECTION TYPE
Radio

DISCOVERED
1963

DISTANCE FROM SUN
2.4 billion light years

STELLAR TYPE
Binary star

CONSTELLATION
Eridanus (the River)

MASS
6.7 solar masses

TEMPERATURE
15,000 K

RADIUS
7.3–11.4 solar radii

DETECTION TYPE
Visual

DISCOVERED
Known to antiquity

DISTANCE FROM SUN
139 light years

Bright Achernar is actually two stars, not just one.

Achernar

As children, we draw stars as pointed objects or spherical dots. Achernar proves our artistic talents inaccurate, as the least 'round' star studied in the Milky Way.

Achernar, the 10th brightest star in the night sky, is actually a binary system. It comprises Alpha Eridani A ('Achernar'), the primary, and Alpha Eridani B ('Achernar B'), a secondary star that orbits the primary from roughly 12 astronomical units away. Achernar is a B-type star, so it appears both bright and blue in the night sky. In fact, it is only seven times larger than the Sun but over 3000 times more luminous. It lies at the southern tip of the constellation Eridanus, which represents the river Po in Italy. This large constellation is only visible from the southern hemisphere between October and December, and was outlined by the Greek astronomer Ptolemy.

Achernar is one of the least spherical stars we have studied in our galaxy, flattened by its massive rotational speed. It is difficult to measure some properties of Achernar because it spins at a rate of nearly 250 km/s (155mps) and is significantly wider than it is tall, as Achernar's equatorial diameter is 56% greater than the star's polar diameter. Its unusual properties have made it especially fascinating to astronomers, including those at the European Space Agency who used the Very Large Telescope (VLT) in Chile to study try and study Achernar - and to prove that the telescopes at the VLT were up to the challenge of measuring an unusually shaped, fast-spinning star.

Aldebaran dominates the Taurus constellation.

Aldebaran

It's hard to miss the fiery red glow of Aldebaran, the eye of Taurus the Bull. This red giant star has fascinated amateur and professional astronomers for millennia with its distinctive hue.

STELLAR TYPE
Red giant

CONSTELLATION
Taurus (the Bull)

MASS
1.16 solar masses

TEMPERATURE
3910 K

RADIUS
44.13 solar radii

DETECTION TYPE
Visual

DISCOVERED
Known to antiquity

DISTANCE FROM SUN
68 light years

Aldebaran is one of the oldest mythologised stars in the night sky, as well as one of the brightest. Ancient astronomers in the Middle East, India, Greece, Mexico, and Australia all had stories to explain Aldebaran's reddish glow, which is actually a product of its large size and relatively cool surface temperature. This red glow made it a natural for the eye of the constellation Taurus. In the life cycle of stars, giant Aldebaran is likely an older star, having moved out of the main sequence after spending eons as a yellow dwarf star like the Sun. This makes it brighter in our sky on Earth, and the 14th brightest star we can see with the naked eye.

Aldebaran, also known as Alpha Tauri, is home to Aldebaran b, an exoplanet 6.5 times larger than Jupiter. Aldebaran b was initially detected in 1993, but it took until 2015 for its existence to be confirmed. Unfortunately, Aldebaran b is an unlikely candidate for carbon-based life, as its surface temperature is roughly 1500 K and it receives large amounts of radiation from its expanding host star, just as Earth will eventually.

In 2003, the Pioneer 10 spacecraft sent its last weak message back to Earth. After exploring the solar system, Pioneer 10 set out on a trajectory toward Aldebaran, which it should reach in roughly two million years. At this point, Pioneer 10 is relying solely on its forward momentum to take it towards the giant star.

STELLAR TYPE
Multiple star

CONSTELLATION
Perseus

MASS
3.17 solar masses (Aa1),
0.7 solar masses (Aa2),
1.76 solar masses (Ab)

TEMPERATURE
13,000 K (Aa1),
4500 K (Aa2),
7500 K (Ab)

RADIUS
2.73 solar radii (Aa1),
3.48 solar radii (Aa2),
1.73 solar radii (Ab)

DETECTION TYPE
Visual; optical telescope

DISCOVERED
Known to antiquity; 1889 as a multiple star system

DISTANCE FROM SUN
90 light years

Algol is part of a multiple star system, seen in the distance in this illustration.

Algol

Nicknamed the 'Demon Star' for its location in the constellation Perseus, luminous Algol is actually a family of stars engaged in a unique celestial dance.

Algol, or Beta Persei, is a multiple star system comprising three stars: Beta Persei Aa1, Aa2, and Ab (sometimes called Algol A, B, and C). What makes Algol most fascinating is its consistent dimming, which occurs as hot, luminous Beta Persei Aa1 is partially eclipsed by large, cool Beta Persei Aa2. This makes Beta Persei Aa2 a so-called occulting star. For roughly 10 hours every 2.86 days, you can observe Algol changing luminosity, dimming during the partial eclipse as it is obscured from view and then returning to full brightness. This makes it an extrinsic variable star rather than an intrinsic one; Algol doesn't dim due to reasons of its internal composition, but only because its partner star is coming between Earth and the star's light source.

Explaining Algol's unusual dimming behaviour took centuries of observation, conjecture, and confirmation. While Italian astronomer Geminiano Montanari noted the variability in Algol's brightness in 1667, an explanation of this dimming wasn't proposed until 1783. British amateur astronomer John Goodricke suggested that Algol was either rotating a dark face toward Earth or that it was being eclipsed. This was verified by astronomers in 1881 when Harvard astronomer Edward Charles Pickering proposed that Algol was an eclipsing binary star, and was finally confirmed in 1889. It took four more decades to discover the third star in the family, Beta Persei Ab.

Alpha Centauri is the bright star to the left, Beta Centauri is to the right.

Alpha Centauri A

Part of a system that comprises our nearest cosmic neighbours, Alpha Centauri A is akin to our own Sun – and may have similarly life-sustaining planets like Earth.

STELLAR TYPE
Yellow dwarf (class G)

CONSTELLATION
Centaurus (the Centaur)

MASS
1.1 solar masses

TEMPERATURE
5790 K

RADIUS
1.2 solar radii

DETECTION TYPE
Visual; optical telescope

DISCOVERED
Known to antiquity; 1689 as a binary system

DISTANCE FROM SUN
4.37 light years

The closest star system to the Earth is the famous Alpha Centauri group. Located in the constellation of Centaurus (the Centaur), at a distance of 4.3 light years, this system is made up of the binary formed by the Sun-like stars Alpha Centauri A and Alpha Centauri B plus the faint red dwarf Alpha Centauri C, also known as Proxima Centauri (seen in the red circle above).

Alpha Centauri A, officially named Rigil Kentaurus, is a near twin of our Sun in almost every way, including age and stellar type, but it is slightly bigger. With Alpha Centauri B, the two stars orbit a common centre of gravity once every 80 years, with a minimum distance of about 11 times the distance between Earth and the Sun.

Because Alpha Centauri A, together with its siblings Alpha Centauri B and Proxima Centauri, are the closest to Earth, they are among the best studied by astronomers. They are also among the prime targets in the hunt for habitable exoplanets, though Proxima b is the main candidate thus far. A recent study involving long-term monitoring of Alpha Centauri by NASA's Chandra X-ray Observatory indicates that any planets orbiting the two brightest stars are likely not being pummeled by large amounts of X-ray radiation from their host stars. This is important for the viability of life in the nearest star system outside the solar system. JPL has tentatively discussed a plan to send a mission to the system in 2069!

STELLAR TYPE
K dwarf (class K)

CONSTELLATION
Centaurus (the Centaur)

MASS
0.9 solar masses

TEMPERATURE
5260 K

RADIUS
0.86 solar radii

DETECTION TYPE
Visual

DISCOVERED
1689 as a binary system

DISTANCE FROM SUN
4.37 light years

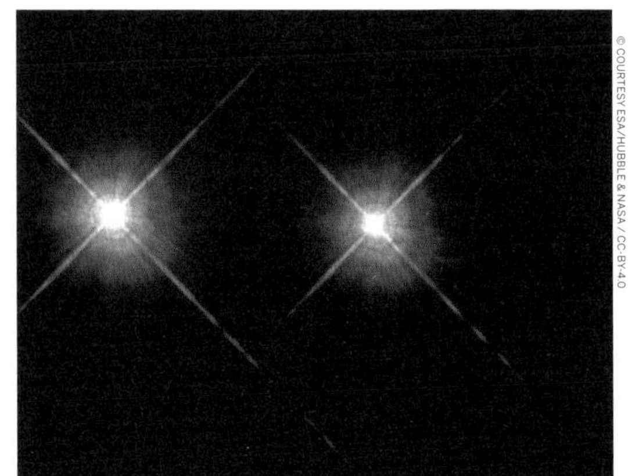

Hubble took this bright image of Alpha Centauri A (left) and Alpha Centauri B (right).

Alpha Centauri B

Scientifically overshadowed by its sister Alpha Centauri A, Alpha Centauri B is among the best and closest candidates for life-sustaining planets in our part of the cosmos.

Alpha Centauri B, officially named Toliman, and its sister star Alpha Centauri A (preceding page) are textbook examples of a binary star system, just 0.21 light years apart. They are also part of the closest star system to Earth, the Alpha Centauri group (along with their sibling Proxima Centauri, a red dwarf).

The Sun-like star Alpha Centauri B is a K1-type star, slightly smaller than our Sun – but it is among our Sun's closest neighbours, barely a stone's throw away in celestial terms. As such, Alpha Centauri B has been extensively studied. In particular, researchers have looked to it as a prime candidate for having Earth-like planets in its orbit, and it's a constant target of sci-fi inspired plans for space exploration. Exoplanets of Alpha Centauri B have been posited in both 2013 and 2015, and research continues to confirm whether these, or any Alpha Centauri A exoplanets, exist to keep Proxima Centauri's known exoplanet company.

Research also continues to determine how life-sustaining any of these planets might be. In 2018, research from scientists at the University of Colorado Boulder revealed that Alpha Centauri B likely produces 5 to 6 times more X-ray radiation, while Alpha Centauri A produces less X-ray activity than the Sun overall. That might make them good candidates for possibly sustaining life on any exoplanets which might exist in their habitable zones.

Altair as seen from Mt Wilson Observatory.

Altair

As a star in the highly visible 'Summer Triangle', Altair has the opposite of a stellar beach body: its rapid rotation makes it wider than it is tall.

STELLAR TYPE
Main sequence star

CONSTELLATION
Aquila (the Eagle)

MASS
1.79 solar masses

TEMPERATURE
6900–8500 K

RADIUS
1.63 to 2.03 solar radii

DETECTION TYPE
Visual

DISCOVERED
Known to antiquity

DISTANCE FROM SUN
16.7 light years

Altair, also called Alpha Aquilae, is a well-known member of the Summer Triangle, clearly visible in the summer night sky across the northern hemisphere. Altair is the brightest star in the constellation Aquila, or the Eagle, comprising the bird's head. The name Altair even comes from the Arabic word for eagle. In Japan, Altair (named Hikobooshi) is linked to the holiday of Tanabata. It is located in the G-Cloud, an interstellar cloud (also home to Alpha Centauri) that sits next to our own Local Interstellar Cloud. At just 16.7 light years from Earth, it is one of the closest objects you can view in the night sky.

'Altair is the twelfth brightest star in the sky - you'd think that everything there is to know about this star would have been discovered already', said Dr David Ciardi of the University of Florida, Gainesville, after a team he was working on studied Altair and discovered that it is not perfectly round. Like Achernar, Altair rotates at over 200km/s (124mps), and has formed a 'midriff bulge' due to the gravitational pressure of its rapid spin, giving it the shape of an oblate spheroid; its equatorial diameter is 14% larger than its polar diameter. It rotates on its axis once every ten hours! Because of its rapid spinning, Altair is a variable star, though taking note of its nine different levels of brightness requires special attention and sensitive equipment. While Altair is only about twice as large as our Sun, it shines at over 11 times the luminosity, yet it's not much hotter than the Sun, at 7550 K.

STELLAR TYPE
Binary star (variable)

CONSTELLATION
Scorpius (the Scorpion)

MASS
12 solar masses

TEMPERATURE
3570 K

RADIUS
680–800 solar radii

DETECTION TYPE
Visual observing

DISCOVERED
Known to antiquity; 1844 as a binary system

DISTANCE FROM SUN
550 light years

Cassini observed Antares through the rings of Saturn.

Antares

Located at the centre of the constellation Scorpius, Antares is the 15th brightest star in the night sky. It is also visible almost year round, and since ancient times, astronomers around the world have wondered at its reddish hue.

In a class of stars called red supergiants, Antares, or Alpha Scorpii, is huge. It is about 700 times the diameter of our own Sun, roughly 15 times more massive, and 10,000 times brighter. Cool superstars like this are expected to end their lives in a supernova, an event which may occur in the next ten thousand years. Antares is the brightest star in the constellation of Scorpius and one of the brighter stars in all the night sky, but it is actually a binary star system, consisting of bright red supergiant Alpha Scorpii A and less luminous Alpha Scorpii B. Antares is surrounded by a nebula of gas it has expelled, making it difficult to study Alpha Scorpii B. Occasionally it's occulted (obscured from view by an object passing in front of it) by the moon or solar system planets.

The name Antares comes from its reddish hue; ancient Greek astronomers named it 'ant-Ares' or 'rival to Ares', the Greek name for the planet we now call Mars. But Antares' bright and distinctive presence in the constellation Scorpius has been noted since ancient times, and records of Australian Aboriginal astronomy note that Antares, under the name Waiyungari, has been part of oral traditions for centuries in tales that specifically relate to its variability. Other red stars including Betelgeuse and Aldebaran have been found as part of the same oral traditions and stories.

The Arcturus star in Boötes constellation.

Arcturus

One of the bright stars that dominate the night sky, Arcturus is a great anchor for stargazing its surrounding modern constellations – or navigating the seas in ancient times.

Arcturus is a red giant in the northern hemisphere's diamond-shaped constellation Boötes, aka the Herdsman. It is the fourth-brightest star in the night sky. The name comes from the Greek 'arktus', meaning bear watcher, as it trails Ursa Major, the great bear, across the spring sky. With Regulus and Spica, it forms the Spring Triangle as the northern hemisphere tips back toward the Sun each year. Arcturus is an aging red giant star, estimated to be roughly seven billion years old. It shines roughly 110 times brighter than the Sun, and can be found by following the arc of the Big Dipper's handle. During the fall when it is low to the horizon, it may appear orangeish due to the light refracting through the atmosphere.

Arcturus has been known to astronomers in various cultures since ancient times. It is likely that it played a significant part in helping Polynesian navigators make their way across the Pacific Ocean to Hawai'i; on a return trip, they would use bright Sirius to navigate at night. Centuries later, the star's light was used to open the 1933 Chicago World Fair. Today, Arcturus is currently nearing its closest point to the Sun in the movement of objects in the Universe. In roughly 4000 years Arcturus will reach its closest proximity to us – though this will only be a few hundredths of a light year nearer, and is unlikely to be noticeable here on Earth. Later in its life, the star will become a white dwarf as the outer layers of the star create a surrounding planetary nebula.

STELLAR TYPE
Red giant

CONSTELLATION
Boötes (the Herdsman)

MASS
1.08 solar masses

TEMPERATURE
4286 K

RADIUS
25.4 solar radii

DETECTION TYPE
Visual observing

DISCOVERED
Known to antiquity

DISTANCE FROM SUN
36.7 light years

STELLAR TYPE
Red dwarf

CONSTELLATION
Ophiucus (the Serpent-Bearer)

MASS
0.144 solar masses

TEMPERATURE
3134 K

RADIUS
0.196 solar radii

DETECTION TYPE
Visible to infrared

DISCOVERED
1888 or 1916

DISTANCE FROM SUN
6 light years

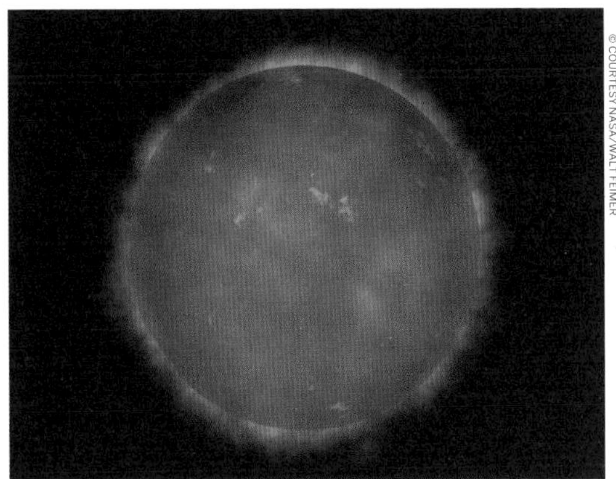

An artist's conception of a low-mass red dwarf star with a temperature of 3200 K.

Barnard's Star

Named for the astronomer who charted its motion through the Universe, Barnard's Star is a small, cool star with a super-Earth-like exoplanet that has astronomers lined up to observe this celestial neighbour.

At just six light years from Earth, Barnard's Star is a close neighbour – but you'll be hard pressed to find it in the night sky. Barnard's Star is a low-mass red dwarf star, invisible to the unaided eye. With telescopes, especially infrared telescopes, Barnard's Star appears brightly, and is the closest star in the celestial northern hemisphere. It also moves at a rapid pace across the night sky over time, giving it the nickname 'runaway star' due to its rate of movement at 10.3 arcseconds a year. That may not sound like much, as each degree in the sky is divided into 60 arcminutes, but it outpaces all other stars in the sky. Because of its proximity and astrophysical characteristics, researchers have studied Barnard's Star for decades, especially curious about whether this small, cool star might have exoplanets capable of sustaining life.

In late 2018, after decades of speculation and observation, the existence of a super-Earth-like exoplanet was discovered orbiting Barnard's Star. Named Barnard's Star b, this super-Earth is 1.3 times the size of Earth and over 3.2 times our planet's mass. It is hypothesised that Barnard's Star b has a rocky mantle covered with ice or snow (Ribas, Tuomi, Reiners, et al., *Nature*, 2018). Barnard's Star b is also relatively close to Barnard's Star, making it a great candidate for future observation.

Orion's Betelgeuse, an easily located star in our skies.

Betelgeuse

Bright, orange-hued Betelgeuse is easily spotted in the sky as part of the constellation Orion. Don't worry, you can say the name of this star three times while stargazing with no ill effects.

STELLAR TYPE
Red supergiant

CONSTELLATION
Orion (the Hunter)

MASS
11.6 solar masses

TEMPERATURE
3590 K

RADIUS
887–955 solar radii

DETECTION TYPE
Visual observing

DISCOVERED
Known to antiquity

DISTANCE FROM SUN
640 light years

Red supergiant Betelgeuse is also known as Alpha Orionis, and informally, as a 'cannibal' star. It is one of the stars in the familiar winter constellation of Orion. While Betelgeuse is the ninth brightest star in the night sky, it is only the second brightest in Orion; the brightest is the multiple star Rigel. As Betelgeuse is a variable star, its luminosity changes, sometimes dropping below other bright stars in the sky and sometimes becoming one of the five brightest stars you can see. Its unpredictable behaviour, including a very fast spin, may be explained by Betelgeuse having previously absorbed a stellar companion. Asteroseismology data suggests this might account for its present instability. Otherwise, a star at Betelgeuse's stage of evolution would spin at a rate 150 times more slow.

Betelgeuse is cooler than the Sun, at 3600 K, but it is more massive and over 1000 times larger. If placed at the centre of our Solar System, it would extend past the orbit of Jupiter. Betelgeuse is one of the largest stars you can see with the unaided eye.

Betelgeuse is actually considered a runaway star, moving at a speed of 30 km/s (18mps) through space and creating a bow shock over four light years wide. From Earth's perspective, however, Betelgeuse won't move significantly in the approximately million years before it inevitably goes supernova.

STELLAR TYPE
Emission nebula

CONSTELLATION
Perseus

MASS
Unknown

TEMPERATURE
Variable

RADIUS
50 light years

DETECTION TYPE
Photographic plate

DISCOVERED
1884

DISTANCE FROM SUN
1500 light years

California's shape is (roughly) represented in the gas clouds of its namesake nebula.

California Nebula

Named for its distinctive state-like shape when viewed from our perspective, the hydrogen-rich California Nebula gives us a look inside another arm of our Milky Way Galaxy.

What's California doing in space? Drifting through the Orion Arm of the spiral Milky Way Galaxy, a cosmic cloud echoes the outline of California on the west coast of the United States. Our own Sun also lies within the Milky Way's Orion Arm, only about 1500 light years from the California Nebula. Also known as NGC 1499, this classic emission nebula is around 100 light years long.

A regular target for astrophotographers due to its striking elongated appearance, the California Nebula can be spotted with a wide-field telescope under a dark sky toward the constellation of Perseus, not far from the Pleiades. Unfortunately, it's hard to spot the California Nebula with the unaided eye because despite its large size (it stretches nearly 2.5° in the night sky), it has a low surface brightness and is difficult to discern from the blackness of space. When viewed, it has a reddish colour imbued by its high hydrogen content. Ultraviolet light radiating from nearby blue giant star Xi Persei, also known as Menkib, ionises the gas in the nebula. Xi Persei has about 40 times the mass of the sun and gives off 330,000 times the amount of light.

The California Nebula was discovered by E. E. Barnard in 1884, who also discovered Barnard's Star and the first non-Galilean moon of Jupiter (its fifth moon, Amalthea) while working at Lick Observatory in California.

A picture of Canopus taken from the ISS.

Canopus

Canopus, the second brightest star in our sky, has been one of the most important stars for navigation throughout history – that is, for societies that were far enough south to see it. This southern hemisphere star shines bright.

Also known as Alpha Carinae (in other words, the first star of the Carina constellation, of which Canopus is the brightest member), Canopus is a rare aging yellow-white supergiant only approximately 300 light years from Earth. These two facts - its proximity and its luminosity - combine to make Canopus the second brightest star in the sky after Sirius. Canopus is over 60 times wider and 15,000 times more luminous than the Sun and is large enough to stretch three-quarters of the way across Mercury's orbit. Now past its red giant stage, the hydrogen in Canopus is exhausted and the star has now moved on to burning the helium within its core.

As an especially bright star, Canopus has been well known since ancient times. Since Canopus is visible only below certain latitudes, Greek and Roman astronomers did not write about its distinctive presence in the sky. Instead, Indian, Egyptian, Bedouin, and Navajo cultures all had beliefs about Canopus and its placement in the sky. Most of these mythologies credit Canopus with religious or royal properties, and in Egypt its rising opposite sunset indicated the start of an annual festival. In modern times, it's easy to view Canopus with the unaided eye, and for those viewing from the southern hemisphere (or even in the southern US during winter), you can also easily spot Sirius, the Small Magellanic Cloud, and Large Magellanic Cloud on a night of stargazing.

STELLAR TYPE
Supergiant

CONSTELLATION
Carina (the Keel)

MASS
8.0 solar masses

TEMPERATURE
10,700 K

RADIUS
65–71 solar radii

DETECTION TYPE
Visual observing

DISCOVERED
Known to antiquity

DISTANCE FROM SUN
310 light years

STELLAR TYPE
Multiple star

CONSTELLATION
Auriga (the Charioteer)

MASS
2.57 solar masses (Aa),
2.48 solar masses (Ab)

TEMPERATURE
4970 K (Aa),
5730 K (Ab)

RADIUS
11.98 solar radii (Aa),
8.83 solar radii (Ab)

DETECTION TYPE
Photographic plates as a binary system

DISCOVERED
1899 as a multiple star system

DISTANCE FROM SUN
42.9 light years

Capella, in the Auriga constellation, looks like one star to us but is actually four.

Capella

Luminous Capella dominates the winter sky, but calling it a single star obscures an important fact: two binary pairs combine in luminosity to help Capella shine so brightly.

Capella, the bright star also known as Alpha Aurigae, is the dominant feature in the constellation Auriga, the charioteer. In fact, Capella comprises four stars: two binary pairs known as Capella Aa and Capella Ab, and two red dwarfs orbiting much further out, Capella H and Capella L. Interestingly, each of these binary pairs is made of similar stars. Capella Aa and Ab are both luminous yellow giant stars, while Capella H and L are both red dwarfs. Divided by only .76 AU, the two yellow giants orbit around each other once every 104 days. The red dwarf pair is separated by a far greater distance. Together, these four stars make Capella the sixth brightest 'star' in the night sky, though Capella Aa and Ab have been studied far more than the other binary pair of fainter stars.

Capella forms part of an asterism known as the Winter Hexagon, similar to other seasonal asterisms like the Spring Triangle and Summer Triangle. These are groupings or patterns of stars in the sky that are smaller than a constellation. The Hexagon is formed by Capella together with Aldebaran in Taurus, Rigel in Orion, Sirius in Canis Major, Procyon in Canis Minor, and the twins Castor and Pollux in the constellation Gemini. All of these bright stars and their respective constellations dominate the night sky during the winter months in the northern hemisphere.

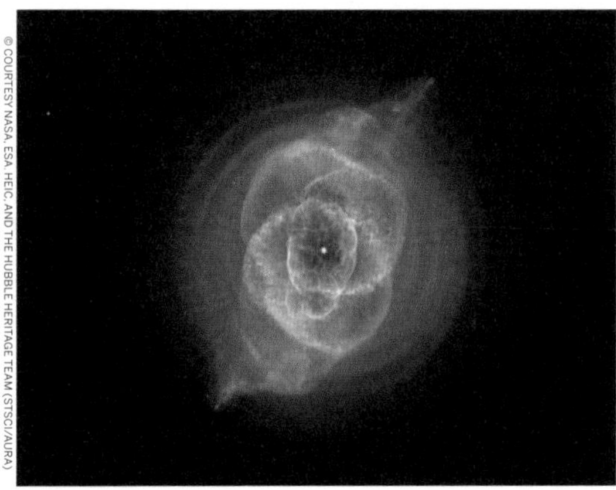

© COURTESY NASA, ESA, HEIC, AND THE HUBBLE HERITAGE TEAM (STSCI/AURA)

The Cat's Eye Nebula is a classic example of a planetary nebula.

Cat's Eye Nebula

Staring across interstellar space, the alluring Cat's Eye Nebula lies thousands of light years from Earth, giving us a peek at a possible future form of our own Sun in its stellar life cycle.

The Cat's Eye Nebula, also classified as NGC 6543, is home to a dying central star throwing off shells of glowing gas. That star may have produced the simple, outer pattern of dusty concentric shells by ejecting stellar gas and material in a series of regular convulsions every 1500 years; some photos make its outer layers appear like a sliced onion. Inner layers near the core appear similar to the pupil of a cat's eye, from Earth's perspective. At its core, the Cat's Eye Nebula is home to a former red giant star, or possibly a binary star system. This deep-sky object can be found within the region of the Draco constellation, or the dragon.

While the Cat's Eye Nebula is formally termed a 'planetary nebula', this term is misleading. Although some celestial objects in the nebula may appear round and planet-like in small telescopes, high-resolution images reveal them to be stars surrounded by cocoons of gas blown off in the late stages of stellar evolution.

In the case of the Cat's Eye, material shed by the star or stars is ejected away from the nebula core at a high rate to form a complex network of bubbles, arcs, jets and knots. These are buffeted by a fast stellar wind that drives a mass loss of twenty two trillion tons per second from the system. The star itself is expected to collapse to become a white dwarf star in a few million years.

STELLAR TYPE
Planetary nebula and possible binary star

CONSTELLATION
Draco (the Dragon)

MASS
1 solar mass at its core (per Bianchi, Cerrato, and Grewing, Harvard, 1986)

TEMPERATURE
7000–9000 K at its core; the stellar dust is as low as 85 K

RADIUS
0.2 light years in the core

DETECTION TYPE
Radio to X-ray

DISCOVERED
1786

DISTANCE FROM SUN
3000 light years

STELLAR TYPE
Supernova remnant nebula

CONSTELLATION
Taurus (the Bull)

MASS
Unknown

TEMPERATURE
Variable

RADIUS
5.5 light years

DETECTION TYPE
Visual; telescope; radio

DISCOVERED
1054 as a supernova; 1731 as a nebula; 1968 as a pulsar

DISTANCE FROM SUN
6500 light years

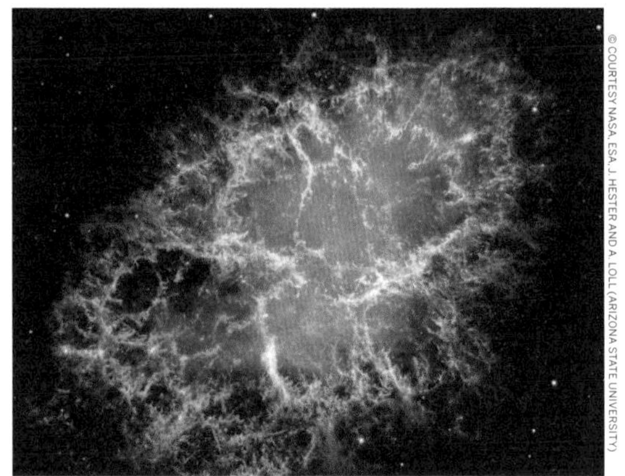

This mosaic image of the Crab Nebula is one of the largest ever taken by Hubble.

Crab Nebula

Few stellar objects can be traced back to their astronomical birth, but the Crab Nebula is one of them. Over the past thousand years, this supernova remnant has captivated observers.

When a star explodes into a supernova, it tends to leave quite an interesting mess behind. The Crab Nebula is no exception. The remnant of a massive star that ended its life in the supernova explosion known as SN 1054, it's a stunning deep-sky object. Nearly a thousand years old, the supernova was noted by Chinese astronomers in the year 1054.

Centuries later, astronomers began to document the new nebula that had formed as a result. That makes the Crab Nebula unique in that astronomers have more comprehensive documentation of its origin and development than many other nebulae. It is the first astronomical object that was able to be linked to a supernova on the historical record. The Crab Nebula is catalogued as M1, the first object on Charles Messier's famous list of things which are not comets. Astronomers today continue to study the Crab Nebula to unpack its components, which include a rapidly spinning neutron star at the middle and an interior pulsar wind nebula. The neutron star (PSR B0531+21, or the Crab Pulsar) is the most persistent source of X-rays in the sky, emitting radiation as it spins at a rate of 30.2 times per second.

The Crab Nebula is not visible to the unaided eye, but with a good pair of astronomical binoculars, you can pick out the cloudy form of this nebula in the constellation of Taurus. It's a common target for astrophotography.

An image of Cygnus X-1 taken by Chandra X-ray Observatory.

Cygnus X-1

A powerful source of X-ray radiation, Cygnus X-1 first puzzled astrophysicists looking for evidence of Einstein's theory of black holes. Today, it continues to interest researchers as the first stellar object to be given the title.

STELLAR TYPE
Black hole

CONSTELLATION
Cygnus (the Swan)

MASS
14–16 solar masses

TEMPERATURE
31,000 K

RADIUS
20–22 solar radii

DETECTION TYPE
X-ray

DISCOVERED
1964

DISTANCE FROM SUN
6100 light years

Is Cygnus X-1 a black hole or not? This question was the basis of a bet between well-known astrophysicists Stephen Hawking and Kip Thorne in the 1970s. Since its discovery 45 years ago, Cygnus X-1 has been one of the most intensively studied cosmic X-ray sources. About a decade after its discovery, Cygnus X-1 secured a place in the history of astronomy when a combination of X-ray and optical observations led to the conclusion that it was a black hole, the first such identification.

The Cygnus X-1 system consists of a black hole with a mass of about 10 times that of the Sun, in a close orbit with a blue supergiant star with a mass of about 20 Suns. Gas flowing away from the supergiant in a fast stellar wind is focused by the black hole, and some of this gas forms a disc that spirals into the black hole. The gravitational energy release by this infalling gas powers the X-ray emission from Cygnus X-1. It may have formed in an unusual type of supernova that somehow prevented the newly formed black hole from acquiring as much spin as other stellar black holes.

Although more than a thousand scientific articles have been published on Cygnus X-1, its status as a bright and nearby black hole continues to attract the interest of scientists seeking to understand the nature of black holes and how they affect their environment.

STELLAR TYPE
Supergiant

CONSTELLATION
Cygnus (the Swan)

MASS
19 solar masses

TEMPERATURE
8525 K

RADIUS
203 solar radii

DETECTION TYPE
Visual

DISCOVERED
Known to antiquity

DISTANCE FROM SUN
2615 light years

Deneb in the Cygnus constellation.

Deneb

Deneb, a hot bright star in the constellation Cygnus, is an eye-catching object in the sky year-round. If you have trouble spotting it, look for the bright star in the popular Northern Cross asterism.

A bright blue-white supergiant also known as Alpha Cygni, Deneb may be the 'tail' of the Swan constellation Cygnus, but it is also its brightest star. Deneb also happens to be between 55,000 and 196,000 times as luminous as the Sun. This makes it among the most luminous stars in the sky, rivalling Rigel for the most inherently bright object in we can observe here on Earth. Both are first-magnitude stars, which have apparent magnitudes lower than +1.50. The scale is logarithmically defined, meaning that a 6.00 magnitude star is 100 times dimmer than a 1.00 magnitude star. Objects on the negative scale are rare, and easy to find with the naked eye; the Sun is -26.7 mag, and a full moon is -12.7 mag.

With its distinctive white hue, Deneb is also part of two asterisms: the Northern Cross which forms a major part of Cygnus, and the Summer Triangle, with bright Vega and Altair making up the other vertices. As Earth follows its orbit around the Sun, the stars it faces change over the seasons, making these asterisms come and go.

While both Altair and Vega are relatively close to the Sun (17 and 25 light years away, respectively), Deneb is over ten times further away. Its luminosity at that distance shows just how much brighter it is than the other two stars that form the Summer Triangle.

The Hubble Space Telescope took this picture of the Dumbbell Nebula's gases.

Dumbbell Nebula

The first of its kind to be discovered, this planetary nebula is a popular sight for amateur astronomers, and can easily be viewed with binoculars or an amateur telescope.

The Dumbbell Nebula goes by many alternate names: Apple Core Nebula, M 27, or NGC 6853. It was discovered Charles Messier, who included it as the 27th member of his famous catalogue of nebulous objects. Though he did not know it at the time, this was the first in a class of objects now known as 'planetary nebulae' to make it into the catalogue. The nebula represents the ejected gases from a red giant's discarded outer layers as it contracts and evolves.

At its core, the Dumbbell Nebula is home to a white dwarf star and many knots of gas and dust. These dense knots seem to be a natural part of the evolution of planetary nebulae. Similar knots have been discovered in other nearby planetary nebulae that are all part of the same evolutionary scheme. They form when the stellar winds are not powerful enough to blow away a larger clump of matter but are able to blow away smaller particles, creating a trail behind the clump. The shapes of these knots change as the nebula expands.

By viewing the Dumbbell Nebula, astronomers glimpse into the future of our own solar system: after expanding its current bounds to become a red giant, our own Sun will eventually become a white dwarf like the one at the core of the Dumbbell Nebula.

STELLAR TYPE
Planetary nebula

CONSTELLATION
Vulpecula (the Little Fox)

MASS
Unknown

TEMPERATURE
Unknown

RADIUS
1.44 light years

DETECTION TYPE
Refracting telescope

DISCOVERED
1764

DISTANCE FROM SUN
1360 light years

STELLAR TYPE
Eclipsing binary star

CONSTELLATION
Auriga (the Charioteer)

MASS
2.2–15 solar masses

TEMPERATURE
7750 K

RADIUS
143–358 solar radii

DETECTION TYPE
Visual

DISCOVERED
1821 as a variable system

DISTANCE FROM SUN
1350 light years

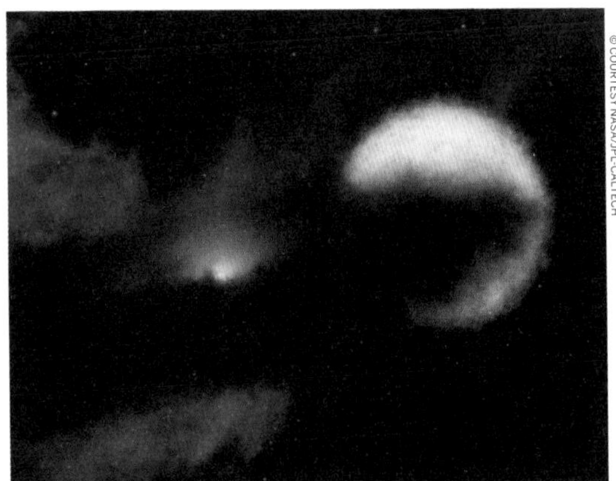

An artist's concept of the dusty disc and companion star of Epsilon Aurigae.

Epsilon Aurigae

Epsilon Aurigae's consistent dips in luminosity have led astronomers to use modern technology to solve a centuries-old celestial puzzle.

For centuries, humans have looked up at a bright star called Epsilon Aurigae and watched with their own eyes as it seemed to disappear into the night sky, slowly fading every 27 years for a two-year period until it is 2.5 times dimmer before coming back to life again. Within the last decade, researchers began to comprehend the puzzling nature of Epsilon Aurigae as an 'eclipsing binary system'. Known as Almaaz or the he-goat in Arabic, its mysteries have been slow to unfurl.

Originally, scientists believed that Epsilon Aurigae might consist of three stars: a bright supergiant being eclipsed by a companion binary star system, which was itself surrounded by a dusty disc. When these binary stars and their dusty surroundings passed in front of the supergiant from our perspective, we observed the relative dimming of the star system.

New observations using NASA's Spitzer Space Telescope suggest that Epsilon Aurigae's large primary star is not a supergiant as a favoured theory had proposed, but is instead a dying star with a lot less mass, eclipsed by a single companion B star surrounded by dust. In its uneclipsed state, Epsilon Aurigae can be seen at night from the northern hemisphere with the naked eye, even in some urban areas. By studying star systems like Epsilon Aurigae, astronomers can better understand the evolution of stars (and encounter the limits of our knowledge).

The South Pillar region of the Carina Nebula in the infrared, seen by Spitzer.

Eta Carinae

When a star dims, disappears, then reappears, and later doubles in luminosity, you know something interesting is happening. In the case of Eta Carinae, two stars are slowly dancing toward an explosive stellar conclusion.

Eta Carinae, the most luminous and massive stellar system within 10,000 light years of Earth, is known for its surprising behaviour, erupting twice in the 19th century for reasons scientists still don't understand. Located in the southern constellation of Carina, Eta Carinae comprises two massive stars (Eta Carinae A and Eta Carinae B), whose eccentric orbits bring them unusually close every 5.5 years. At closest approach, the stars are 225 million kilometres apart, or about the average distance between Mars and the Sun.

Their celestial tango will likely produce an explosive end: both of the massive stars of Eta Carinae may one day end their lives as a supernova. For stars, mass is destiny, and what will determine their ultimate fate is how much matter they can lose – through stellar winds or as-yet-inexplicable eruptions – before they run out of fuel and collapse under their own weight. For this reason, the Eta Carinae star system expected to have at least onc supernova explosion in the future (though in astronomical timescales the 'near future' could still be a million years away). Eta Carinae's great eruption in the 1840s already created the billowing Homunculus Nebula. The stars are contained within the larger Carina Nebula, part of the Trumpler 16 open cluster in the Milky Way's Carina-Sagittarius Arm.

STELLAR TYPE
Binary star

CONSTELLATION
Carina (the Keel)

MASS
120–200 solar masses (A),
30–80 solar masses(B)

TEMPERATURE
9400–35,200 K (A),
37,200 K (B)

RADIUS
60–881 solar radii (A),
14.3–23.6 solar radii (B)

DETECTION TYPE
Optical telescope

DISCOVERED
1677

DISTANCE FROM SUN
7500 light years

STELLAR TYPE
Planetary nebula

CONSTELLATION
Hydra (the Snake)

MASS
Unknown

TEMPERATURE
Unknown

RADIUS
Unknown

DETECTION TYPE
Optical telescope

DISCOVERED
1785

DISTANCE FROM SUN
1400 light years

An infrared view of Ghost of Jupiter from Spitzer.

Ghost of Jupiter

A misnomer on many accounts, the Ghost of Jupiter planetary nebula offers insight into the future of our Sun when it becomes a white dwarf billions of years from now.

Named for its faint but similar appearance to our solar system's ruling gas giant planet, the Ghost of Jupiter, also classified NGC 3242, is much farther away than the measly 40 light-minutes distance from Earth to the gas giant.

After a star like the Sun completes fusion in its core, it throws off its outer layers in a brief, beautiful cosmic display called a planetary nebula. The Ghost of Jupiter is just such a phenomenon, with the stellar remnant white dwarf star visible at the centre. The nebula even has a massive outer halo, which appears as wispy sphere around the system to astronomers. The precise origin and composition of this halo is not known for certain. It is most likely material ejected during the star's red-giant phase before the white dwarf was exposed. However, it may be possible that the extended material is simply interstellar gas that, by coincidence, is located close enough to the white dwarf to be energised by it and induced to glow with ultraviolet light. It has FLIER regions at its ends, or Fast Low-Ionization Emission Regions; these are younger tails of fast moving gas.

Stretching two light years from end to end, the Ghost of Jupiter is a common target for amateur astronomers. In smaller telescopes, the nebula appears as a bluish-green object; larger telescopes allow you to see the outer halo too. It was discovered by William Herschel.

The region that is home to GRS 1915+105.

GRS 1915+105

Awkwardly named GRS 1915+105 is a powerful X-ray binary system, comprising a star and a black hole. It's also one of the fascinating stellar objects that researchers discovered has a 'heartbeat' at times.

STELLAR TYPE
Binary stellar black hole system

CONSTELLATION
Aquila (the Eagle)

MASS
14 solar masses

TEMPERATURE
Unknown

RADIUS
Unknown

DETECTION TYPE
X-ray

DISCOVERED
1992

DISTANCE FROM SUN
35,000 light years

GRS 1915+105 is a two-star system which contains a black hole about 14 times the mass of the Sun that is feeding off material from a nearby companion star. Gas from the 'normal' star spills towards the black hole, lured by gravity. The gas spirals into the black hole like water down a drain: it doesn't just fall in all at once but first swirls around the black hole, forming a reservoir of matter called an accretion disc. GRS 1915+105 is one of the heaviest known stellar black hole systems discovered.

This system shows remarkably unpredictable and complicated variability in its emissions of particles and radiation, ranging from timescales of seconds to months. So far, researchers have discovered 14 different patterns of variation. In one of the most interesting patterns, GRS 1915+105 gives off a short, bright pulse of X-ray light approximately every 50 seconds. This type of rhythmic cycle closely resembles an electrocardiogram of a human heart, though at a slower pace. The jet in GRS 1915 may be periodically choked off when a hot wind, seen in X-rays, is driven off the accretion disc around the black hole. The wind is believed to shut down the jet by depriving it of matter that would have otherwise fueled it. Conversely, once the wind dies down, the jet can re-emerge. These results suggest that these black holes have a mechanism for regulating the rate at which they grow

STELLAR TYPE
Heavy metal subdwarf

CONSTELLATION
Hydra (the Snake)

MASS
Unknown

TEMPERATURE
38,000 K

RADIUS
Unknown

DETECTION TYPE
ESO's Very Large Telescope

DISCOVERED
2003

DISTANCE FROM SUN
1000 light years

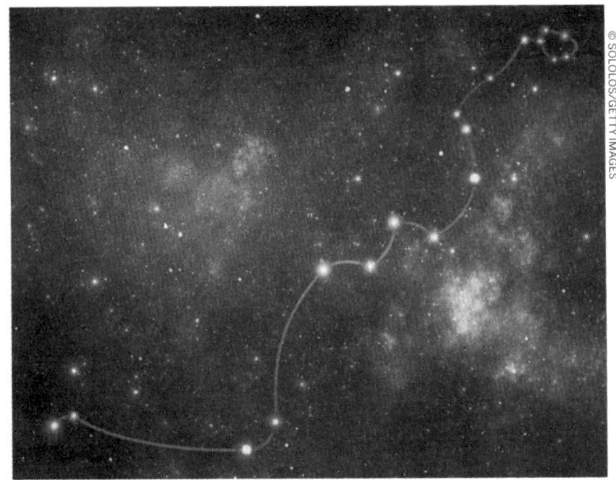

HE 1256-2738 can be found in the constellation Hydra, pictured above.

HE 1256-2738

You won't find heavy metal subdwarf star HE 1256-2738 rocking out at a concert. Instead, in discovering this new class of stars with high concentrations of heavy metals, researchers found a missing link in the evolution of stars.

Located a relatively close 1000 light years from Earth in the vicinity of the Hydra constellation, it might seem surprising that researchers only discovered HE 1256-2738 in the past few decades. As we've studied the sky, we've become surprisingly adept at discovering all of the stars currently visible across the electromagnetic spectrum. However, HE 1256-2738 and a similar star, HE 2359-2844, are so unique that their discovery prompted researchers to create a new class of stars: heavy metal subdwarfs. Subdwarf stars are a class of stars between main sequence stars like our Sun and white dwarf stars like Vega or the star at the heart of the Dumbbell Nebula.

Passing on the chance for astronomy to be more oriented to pop culture, heavy metal subdwarfs are not named for their love of a specific music genre. Instead, this special class of subdwarfs is so named for their exceptionally high concentrations of heavy metals, including germanium, strontium, yttrium, zirconium and lead. In the case of HE 1256-2738, the primary heavy metal is lead, which is strongly concentrated in the star's atmosphere. Its surface temperature has been measured as 38,000°C (68,400°F). This is so hot that the iron atoms three electrons removed. A team from Taiwan and the UK made the finding.

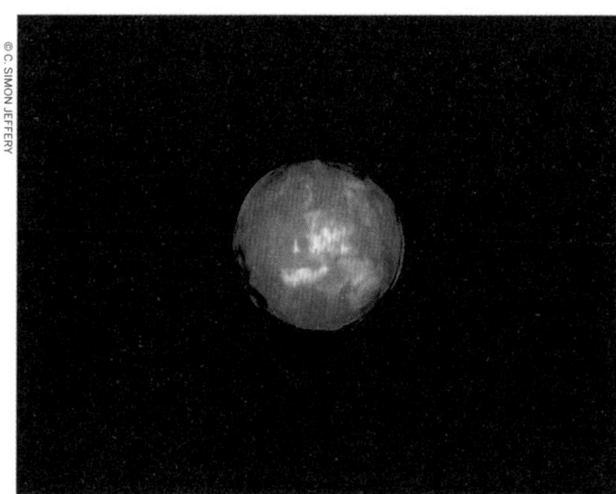

Dr Simon Jeffery created this visualisation of the new heavy metal subdwarf class.

HE 2359-2844

HE 2359-2844, a heavy metal subdwarf, is home to high concentrations of lead, yttrium, and zirconium. When researchers discovered HE 2359-2844, it helped increase our knowledge of the life cycle of stars.

STELLAR TYPE
Heavy metal subdwarf

CONSTELLATION
Sculptor

MASS
Unknown

TEMPERATURE
38,000 K

RADIUS
Unknown

DETECTION TYPE
ESO's Very Large Telescope

DISCOVERED
2003

DISTANCE FROM SUN
800 light years

Discovered at the same time as HE 1256-2738, HE 2359-2844 helped establish a new class of star known as heavy metal subdwarfs. These stars have extremely high concentrations of heavy metals, including germanium, strontium, yttrium, zirconium and lead; in the case of HE 2359-2844, lead is the predominant heavy metal of the star, 10,000 times more concentrated than it is in the Sun. Researchers also found HE 2359-2844 has ten thousand times more zirconium and yttrium (a rare-earth element) than our Sun.

In discovering this new class of stars, researchers gained a better understanding of the spectrum of stars. Heavy metal subdwarfs are a class of stars that exist between main sequence stars like our own Sun and white dwarf stars. Heavy metal subdwarfs like HE 2359-2844 are hot subdwarfs, but unlike more common hot subdwarf stars which are primarily helium and hydrogen, they contain heavy metals in much greater proportions. In addition, those discovered so far have very elliptical and varied orbits, though the reason why is not yet understood; it might indicate that the objects are quite old. Unlike most main sequence stars, their interiors also don't feature convection. Heavy metal subdwarf stars give researchers greater insight into the wide variety of stars that exist in our Universe.

STELLAR TYPE
Planetary nebula

CONSTELLATION
Aquarius (the Water Bearer)

MASS
Unknown

TEMPERATURE
Unknown

RADIUS
2.87 light years

DETECTION TYPE
Optical telescope

DISCOVERED
Karl Ludwig Harding, 1824

DISTANCE FROM SUN
650 light years

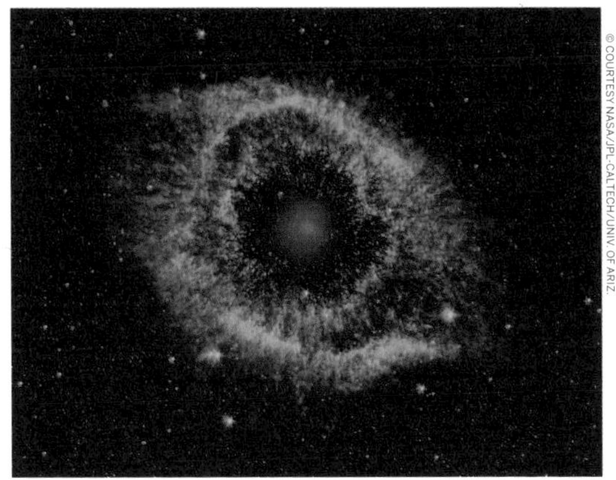

The Helix Nebula in an infrared image from Spitzer Space Telescope.

Helix Nebula

Curious about what happens when a star dies? Look to the Helix Nebula, which shows the death throes of a star much like our Sun.

The Helix Nebula is not going quietly into the night; instead, this dying star is throwing a cosmic tantrum as the star's dusty outer layers are unravelling into space, glowing from the intense ultraviolet radiation being pumped out by the hot stellar core.

Also known by the catalogue number NGC 7293, the Helix Nebula is a typical example of a planetary nebula. Discovered in the 18th century, these cosmic works of art were erroneously named for their resemblance to gas-giant planets. In fact, they are the remains of stars that once looked a lot like our Sun, but which have run out of hydrogen and exploded into a supernova. The Helix Nebula is one of the closest planetary nebulae to our Sun; its supernova likely put on an amazing stellar show that would have been visible from Earth had any humans been around to see it.

Before the star that preceded the Helix Nebula died, its comets, and possibly planets, would have orbited in an orderly fashion, much like our own solar system. When the star ran out of hydrogen to burn and blew off its outer layers, the icy bodies and outer planets would have been tossed about and into each other, kicking up an ongoing cosmic dust storm. Any inner planets in the system would have burned up or been swallowed as their dying star expanded into the beautiful nebula visible to observers today.

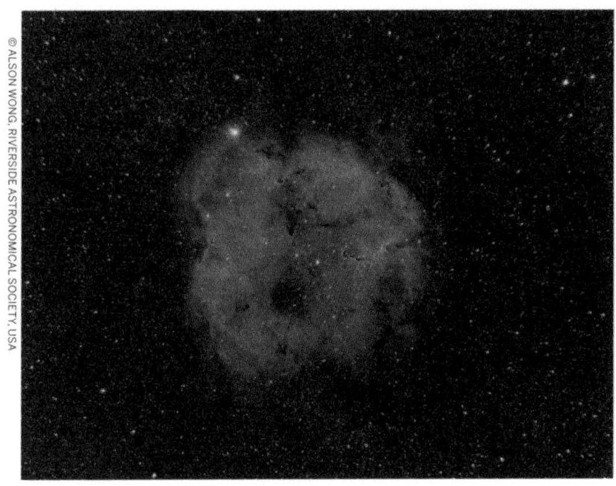

The orange star in the upper left of the image is Herschel's Garnet Star (Mu Cephei).

Herschel's Garnet Star

Visible as a dark reddish star in the northern sky, Herschel's Garnet Star is one of the biggest and brightest objects we can observe.

STELLAR TYPE
Red supergiant

CONSTELLATION
Cepheus

MASS
19.2 solar masses

TEMPERATURE
3750 K

RADIUS
1260–1650 solar radii

DETECTION TYPE
Orbital telescope

DISCOVERED
1848 as a variable star

DISTANCE FROM SUN
2840 light years

Also known as Mu Cephei and, less commonly, Erakis, Herschel's Garnet Star takes its name from the colour, as it was described by astronomer William Herschel in the late 18th century. A red supergiant or hypergiant, Herschel's Garnet Star is also a variable star, and astronomers have been continuously studying its variability since 1881. At 4.08 magnitude it can be seen through binoculars in the constellation Cepheus, or the King.

Herschel's Garnet Star is considered a runaway star, moving through the cosmos at a rate of roughly 80 km/s (50mps). It is nearly 100,000 times more luminous than our own Sun, and about 1000 times larger. In fact, it's large enough that if it were our own solar system's star, it would extend to between the orbits of Jupiter and Saturn. This makes it among the largest stars you can see with your unaided eye. Since 1943, its spectrum has been used as a standard for comparing and classifying newly discovered stars of similar types. By observing Herschel's Garnet Star and changes in its radiation emissions, astronomers believe the star is dying. Herschel's Garnet Star has begun to fuse helium into carbon; 'healthy' main sequence stars fuse helium into hydrogen, like our own Sun does. After Herschel's Garnet Star runs out of fuel and explodes into a supernova, it is likely it will leave a black hole where this massive star once existed.

STELLAR TYPE
Intermediate black hole

CONSTELLATION
Phoenix

MASS
20,000 solar masses

TEMPERATURE
Unknown

RADIUS
Unknown

DETECTION TYPE
ESA's XMM-Newton

DISCOVERED
2009

DISTANCE FROM SUN
290 million light years

This spectacular edge-on galaxy, ESO 243-49, is home to HLX-1.

HLX-1

Just as astronomers study the life cycle of stars and galaxies, they also seek to understand how black holes form, evolve, and grow. HLX-1 offers an important insight into this process.

Astronomers know how massive stars collapse to form black holes but it is not clear how supermassive black holes – which can weigh billions of times the mass of our Sun – form in the cores of galaxies. One idea is that they may build up through the merger of smaller black holes. A discovery by an astronomer at the Sydney Institute for Astronomy in Australia in 2009 helped researchers gain more evidence for this theory.

Known as HLX-1 (Hyper-Luminous X-ray source 1), the black hole has an estimated weight of 'only' about 20,000 solar masses, making it small on the cosmic scale. 'Before this latest discovery, we suspected that intermediate-mass black holes could exist, but now we understand where they may have come from', said Sean Farrell, the astronomer who discovered HLX-1, which is located in the galaxy ESO 243-49, about 290 million light years from Earth. 'The fact that there seems to be a very young cluster of stars indicates that the intermediate-mass black hole may have originated as the central black hole in a very-low-mass dwarf galaxy. The dwarf galaxy might then have been swallowed by the more massive galaxy, just as happens in our Milky Way'.

The discovery and study of HLX-1 helps astronomers gain insight into how black holes evolve, an important piece of the life cycle of Sagittarius A*, the supermassive black hole at the centre of our own galaxy.

A composite colour image of the Horsehead Nebula and its immediate surroundings.

Horsehead Nebula

By fortuitous chance, the Horsehead Nebula has a familiar shape to us on Earth. One of the most recognisable nebulae in the sky, it is easily spotted and photographed by amateur and professional astronomers alike.

The Horsehead Nebula in Orion is part of a large, dark, molecular cloud. Also known as Barnard 33, the unusual shape was first discovered on a photographic plate in the late 1800s. The dark molecular cloud, roughly 1400 light years distant, is visible only because its obscuring dust is silhouetted against another, brighter nebula. The prominent horse head portion of the nebula is really just part of a larger cloud of dust. Inside this dust, stars are in the process of forming; in photos of the Horsehead Nebula, you can see them as lighter areas in the dusty horse's head shape. After many thousands of years, the internal motions of the cloud and nearby radiation will surely alter its appearance.

The Horsehead Nebula is located just south of Alnitak, the easternmost star in Orion's Belt. It is a favourite target for amateur and professional astronomers, with its shadowy shape easily visible through a telescope. Inside the dust and gas that make up its distinctive appearance, a cold molecular cloud contains the raw materials of possible future stars and planetary systems, while other regions of the nebula hold already-formed stars radiating hot plasma. The Horsehead's seeming darkness is caused by thick clouds of dust which block light from passing through.

STELLAR TYPE
Dark nebula

CONSTELLATION
Orion (the Hunter)

MASS
Unknown

TEMPERATURE
Unknown

RADIUS
3.5 light years

DETECTION TYPE
Photographic plate, Harvard College Observatory

DISCOVERED
Williamina Fleming, 1888

DISTANCE FROM SUN
1500 light years

STELLAR TYPE
Variable star

CONSTELLATION
Tucana (the Toucan) / Small Magellanic Cloud

MASS
Unknown

TEMPERATURE
3450 K

RADIUS
916 solar radii

DETECTION TYPE
Photographic plate, Harvard College Observatory

DISCOVERED
Henrietta Swan Leavitt, 1908

DISTANCE FROM SUN
Unknown

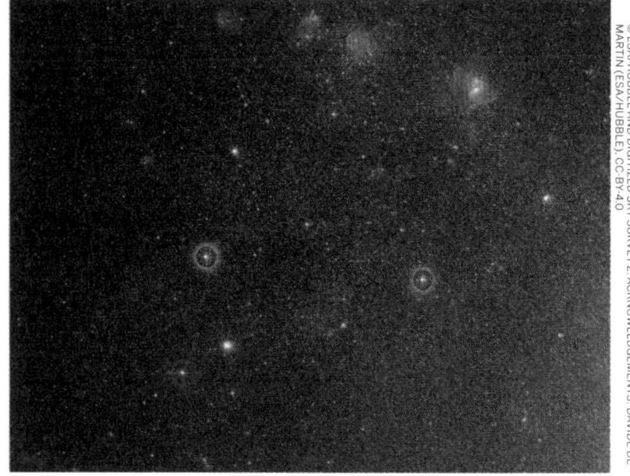

A portion of the Small Magellanic Cloud around HV 2112.

HV 2112

Little is understood about HV 2112, a star in the direction of – or perhaps within – the Small Magellanic Cloud, further proof that there's a long way to go in our understanding of the Universe.

HV 2112, so called because it was identified by astronomer Henrietta Leavitt in the Harvard catalogue under that same name, is an unusual star with properties and behaviours which are not fully understood. HV 2112 is located in the Small Magellanic Cloud, but astronomers aren't even sure if it is within the Cloud, or appears in front of it from our perspective here on Earth.

For over a century, astronomers have known that HV 2112 is a variable star, but they aren't even sure exactly which kind of star it is. Two hypotheses have been debated over the years. The less likely is that HV 2112 is a Thorne-Żytkow object, a red giant or supergiant star which has a neutron star at its core after the two collided in space. This existence of this theoretical class of star was first proposed in 1977 by noted astrophysicists Kip Thorne and Anna Żytkow. More likely is that HV 2112 is an asymptotic giant branch (AGB) red giant, a cool luminous star which is at least 1000 times brighter than our Sun. If this is the case, HV 2112 likely has a companion star that causes its variability.

Leavitt's work examining and cataloguing the stars on photographic plates according to their relative brightness led her to discover the rules driving the activity of Cepheid variables, which pulsate and alter their brightness over regular periods. HV 2112 is a notable case of a much less predictable variable star.

A visualisation of the solar wind emitted by this black hole's accretion disc.

IGR J17091-3624

As black holes go, IGR J17091-3624 is the runt of the litter, but it creates some of the fastest solar wind ever measured and has helped researchers better understand how the radiation 'heartbeat' of black holes works.

STELLAR TYPE
Black hole

CONSTELLATION
Scorpius (the Scorpion)

MASS
3–10 solar masses

TEMPERATURE
Unknown

RADIUS
Unknown

DETECTION TYPE
ESA's INTEGRAL satellite

DISCOVERED
2003

DISTANCE FROM SUN
28,000 light years

Discovered in 2003 and named IGR J17091-3624 after the astronomical coordinates of its sky position, the binary system combines a normal star with a black hole that may weigh less than three times the Sun's mass. That is near the theoretical mass boundary where black holes become possible, making it one of the smallest black holes ever discovered. It is found in the bulge of the Milky Way Galaxy, and in the sky as seen from Earth it is to be found in the constellation Scorpius.

Like black hole GRS 1915+105, IGR J17091-3624 has a heartbeat that suggests it emits a particle jet from its core. Researchers say that this system's heartbeat emission of radiation can be 20 times fainter than GRS 1915+105 and can cycle some eight times faster, in as little as five seconds. Using the heartbeat data from these two black holes has helped researchers understand how emissions change as black holes gain mass. Like the difference in heart rate between a human and an elephant, IGR J17091-3624 has a faster heartbeat than the significantly larger GRS 1915+105. Winds exit the IGR J17091-3624 black hole accretion disc at ten times the rate of the next fastest yet observed: 3% of light speed, or 32 million km/h (20 million mph). Based on its attributes, the black hole may have a low or retrograde spin.

STELLAR TYPE
Reflection nebula

CONSTELLATION
Cepheus

MASS
Unknown

TEMPERATURE
Unknown

RADIUS
3 light years

DETECTION TYPE
Telescope

DISCOVERED
William Herschel, 1794

DISTANCE FROM SUN
1300 light years

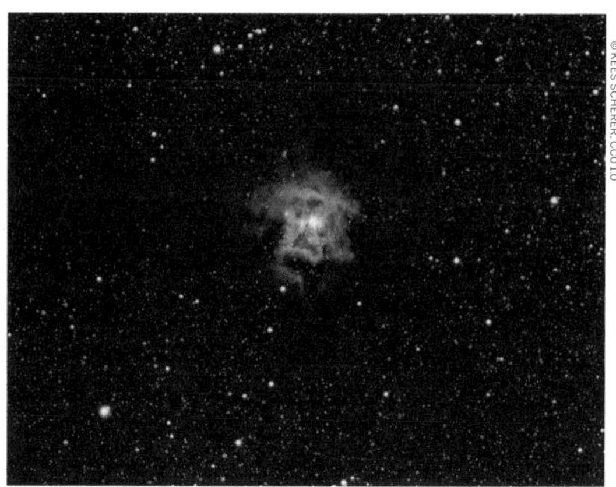

As a reflection nebula, the Iris Nebula's blue light is enhanced by surrounding dust.

Iris Nebula

One of many flowers in the universal garden, the Iris Nebula is a choice option for amateur and professional astronomers alike thanks to its distinctive blue hue.

Like delicate cosmic petals, these clouds of interstellar dust and gas resembling a garden flower have blossomed in fertile starfields. Sometimes called the Iris Nebula and dutifully catalogued as NGC 7023 or Caldwell 4, this is not the only nebula to evoke the imagery of flowers; the Rosette Nebula is another well-known example. Within the Iris itself, dusty nebular material surrounds a hot, young star in its formative years. Dark clouds of surrounding dust make the bright star SAO 19158 at the core of the Iris Nebula even more eye-catching. The brown tint of the pervasive dust comes partly from photoluminescence – dust converting ultraviolet radiation to red light. The dominant colour of the brighter reflection nebula is blue, characteristic of dust grains reflecting starlight. This is the same effect that makes Earth's sky blue. The bright blue portion of the Iris Nebula spans about six light years.

Astronomers frequently study the Iris Nebula because of the unusual prevalence there of Polycyclic Aromatic Hydrocarbons (PAHs), complex molecules that are also released on Earth during the incomplete combustion of wood fires. Amateur astronomers can spot the distinctive blue hue of the Iris Nebula with a small telescope, making it a popular target for observing on a clear night.

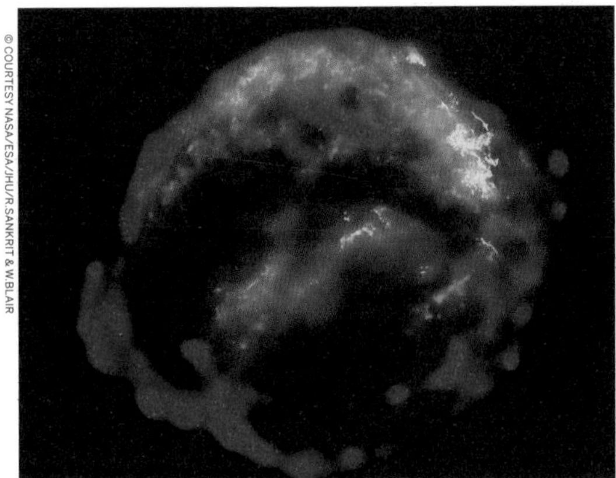

Kepler's Supernova in a false colour composite image.

Kepler's Supernova

Over the centuries, Kepler's Supernova has allowed astronomers a rare opportunity to watch a supernova evolve and to hypothesise about how it happened in the first place.

STELLAR TYPE
Supernova remnant nebula

CONSTELLATION
Ophiuchus (the Serpent Bearer)

MASS
Unknown

TEMPERATURE
Unknown

RADIUS
Unknown

DETECTION TYPE
Visual

DISCOVERED
1604

DISTANCE FROM SUN
20,000 light years

Light from the stellar explosion that created this energised cosmic cloud known as Kepler's Supernova Remnant was first seen in October 1604 by the astronomer of that same name. During the three months the supernova was visible, it was brighter than any star in the sky. Today Kepler's Supernova is still the most recent stellar explosion observed within our Milky Way Galaxy.

As the supernova diminished, it produced a bright new star in early 17th century skies. Without the benefit of a telescope, Kepler and his contemporaries had no way to explain the heavenly apparition. We now know it was a Type 1a supernova, the thermonuclear explosion of a white dwarf star. These supernovas are important cosmic distance markers for tracking the accelerated expansion of the Universe.

Armed with a modern understanding of stellar evolution, early 21st century astronomers continue to explore the expanding debris cloud of Kepler's Supernova Remnant across the electromagnetic spectrum. Recent research suggests that the supernova was caused by the transfer of material between two smaller dwarf stars. The added material brings the total mass of one of the stars beyond the critical threshold for supernova collapse; this amount, named the Chandrasekhar limit, is equivalent to 1.4 solar masses. Going past this amount of transferred mass is too destabilising.

STELLAR TYPE
Supernova remnant and pulsar

CONSTELLATION
Aquila (the Eagle)

MASS
Unknown

TEMPERATURE
Unknown

RADIUS
Unknown

DETECTION TYPE
Chandra X-ray Observatory

DISCOVERED
2000

DISTANCE FROM SUN
19,000 light years

This composite Chandra X-ray Observatory images shows Kes 75 expanding.

Kes 75

Despite being too dim to be seen from Earth, the supernova remnant Kes 75 is home to the youngest known pulsar and an object of fascination to astronomers.

When some massive stars collapse and explode as supernovas, they leave behind dense stellar nuggets called 'neutron stars'. A supernova remnant known as Kes 75 is one such stellar object and contains the youngest known pulsar in the Milky Way Galaxy, PSR J1846.

Kes 75 exploded about five centuries ago. Unlike other supernova remnants from this era, such as the Kepler Supernova, there is no historical evidence that the explosion that created Kes 75 was observed. Why wasn't Kes 75 seen from Earth? Observations indicate that the interstellar dust and gas that fill our galaxy are very dense in the direction of the doomed star. This would have rendered it too dim to be seen from Earth several centuries ago, when visual observing by the naked eye was the only way to collect information on the heavens; the first telescope was only invented in 1608.

The pulsar PSR J1846 is a fascinating subject of astronomic study because it is young, and studying it gives astronomers the opportunity to understand more about the early stages in the life cycle of pulsars. Astronomers know PSR J1846 is very young for several reasons. First, it resides inside the Kes 75 supernova remnant, an indicator that it hasn't had time to wander from its birthplace. Second, based on how rapidly its spin rate is slowing down, astronomers calculate that it can be no older than 884 years – an infant on the cosmic timescale.

Hydrogen molecules in the Little Dumbbell Nebula glow green and red in this image.

Little Dumbbell Nebula

In a cosmic coincidence, two nebulae in different parts of the sky appear similar in shape. Difficult for amateurs to spot, the Little Dumbbell Nebula still made the cut for Messier's catalogue of stellar objects.

'Nebula at the right foot of Andromeda...' begins the description for the 76th object in Charles Messier's 18th century *Catalogue of Nebulae and Star Clusters.* In fact, M76 is one of the fainter objects on the Messier list and is also known by the popular name of the 'Little Dumbbell Nebula'. It is so faint that it is difficult to spot within the Perseus constellation by the untrained or amateur astronomer, though professional astronomers still study it to determine its true size, shape, and distance from Earth, currently thought to be about 2500 light years.

Like its brighter namesake M27 (the Dumbbell Nebula), the Little Dumbbell Nebula is recognised as a planetary nebula - a gaseous shroud cast off by a dying Sun-like star. The nebula itself is thought to be shaped more like a doughnut, while the box-like appearance of its brighter central region is due to our nearly edge-on view. Gas expanding more rapidly away from the doughnut hole produces the fainter loops of far flung material, giving the Little Dumbbell Nebula its distinctive shape. It is also catalogued as NGC 650 and referred to as the Barbell or Cork Nebula. This 'little' object still spans a distance of 1.23 light years across!

STELLAR TYPE
Planetary nebula

CONSTELLATION
Perseus

MASS
Unknown

TEMPERATURE
Unknown

RADIUS
0.61 light years

DETECTION TYPE
Telescope

DISCOVERED
1780

DISTANCE FROM SUN
2500 light years

STELLAR TYPE
Binary system

CONSTELLATION
Cetus (the Whale)

MASS
1.18 solar masses (A)

TEMPERATURE
2900 K (A)

RADIUS
332–402 solar radii (A)

DETECTION TYPE
Visual

DISCOVERED
1596 as a variable star

DISTANCE FROM SUN
300 light years

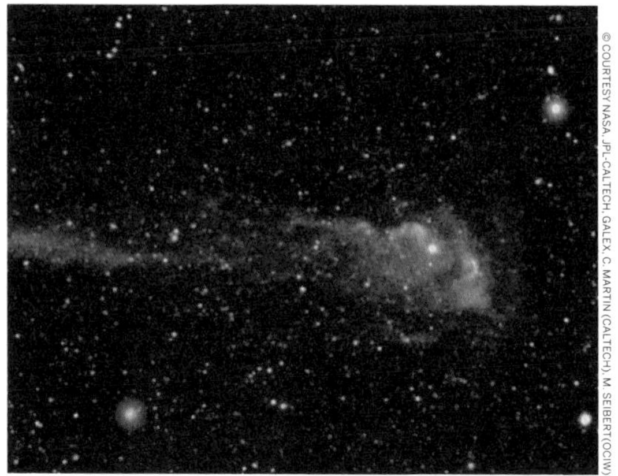

The material in Mira's tail is estimated at 3000 times the mass of Earth.

Mira

Affectionately known as a 'wonderful star' by historic astronomers, binary system Mira continues to enthral researchers with its unusual behaviour and shape.

To seventeenth century astronomers, Omicron Ceti, or Mira, was astonishing – a star whose brightness could change dramatically. Modern astronomers now recognise an entire class of long-period Mira-type variables as cool, pulsating, red giant stars, 700 or so times the diameter of the Sun.

Mira was discovered to be the first variable star about 400 years ago, though it was known to visual observers long before. Details of Mira's variability are still being researched, but the reason for its pulsations are thought related to periodic changes in the thickness of parts of Mira's atmosphere. Recent high resolution images show that Mira is not even round. Instead, it is a binary system consisting of red giant Mira A and a small white dwarf companion star (Mira B) with which it co-orbits.

The red giant star Mira A is undergoing dramatic pulsations, causing it to become more than 100 times brighter over the course of a year. Mira A also has its own, almost comet-like tail, produced by cast off material from the red giant that is flowing through the interstellar medium. Mira B, by comparison, is surrounded by a disc of material drawn from the pulsating giant Mira A. In such a double star system, the white dwarf companion star's hot accretion disc is expected to produce some X-ray emissions.

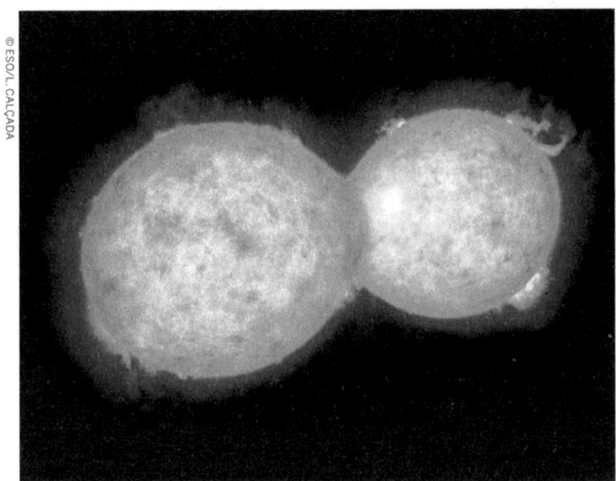

MY Camelopardalis might be thought of as a pair of 'kissing' stars.

MY Camelopardalis

What happens when stars collide? That depends on the stars involved! Astronomers studying the binary system MY Camelopardalis are keen to see if it will become one massive star when its two companion stars collide.

Multiple star systems are fascinating to astronomers, because they behave in interesting and often unique ways depending on how each star formed, evolves, and interacts with the others. MY Camelopardalis is one of the most massive binary star systems astronomers have discovered, comprising two huge blue O-type stars that are slowly spiralling into a cosmic collision.

The stars which comprise MY Camelopardalis are both estimated to be over 30 times the mass of our own Sun, and are so close now that as they orbit and eclipse one another, they actually share the same gaseous envelope – in stellar terms, these two stars are touching, though they are not yet interchanging mass. Their proximity means they now orbit each other once every 1.2 days. While astronomers aren't exactly sure what will happen when the two stars finally collide, they don't expect them to explode in a blaze of supernova as happens when white dwarf stars collide. Instead, astronomers predict MY Camelopardalis will eventually become a jumbo star over 60 times the mass of our own Sun. Another possibility is that the two stars continue their mutually intertwined, but largely separate, existences largely unchanged.

STELLAR TYPE
Binary star

CONSTELLATION
Camelopardalis (the Giraffe)

MASS
37.7 solar masses (A), 31.6 solar masses (B)

TEMPERATURE
42,000 K (A), 39,000 K (B)

RADIUS
7.60 solar radii (A), 7.01 solar radii (B)

DETECTION TYPE
Northern Sky Variability Survey (NSVS)

DISCOVERED
2004 as a binary system

DISTANCE FROM SUN
13,000 light years

STELLAR TYPE
Emission nebula

CONSTELLATION
Cygnus (the Swan)

MASS
Unknown

TEMPERATURE
Unknown

RADIUS
Unknown

DETECTION TYPE
Telescope

DISCOVERED
1786

DISTANCE FROM SUN
1600 light years

The North America Nebula in Cygnus.

North America Nebula

Like so many deep sky objects, the North America Nebula is named for its shape. Within the borders of this massive nebula, especially the area that resembles Central America, stars are currently being born.

Like watching clouds in the sky, astronomers often find familiar shapes in unfamiliar locations while exploring the cosmos. (Okay, and sometimes they're grasping for a connection.) The William Herschel-discovered emission nebula NGC 7000 is famous partly because it resembles our fair planet's continent of North America, which is how it earned its colloquial name as the North America Nebula.

The North America Nebula in the sky can do what North Americans on Earth cannot – form stars. Specifically, in analogy to the Earth-confined continent, the bright part that appears as Central America and Mexico is actually a hotbed of gas, dust, and newly formed stars known as the Cygnus Wall (so named because it is within the constellation of Cygnus, the Swan).

While the North America Nebula is large in the sky, it is dim and cannot easily be seen without binoculars or a telescope. Even with equipment, this massive nebula shows as a foggy patch in the sky; you'll need an astronomical filter to remove some light wavelengths in order to begin seeing the distinctive continental shape of this star nursery.

ESO's La Silla Observatory took this image of globular cluster Omega Centauri.

Omega Centauri

The brightest star cluster of its kind that we can observe from Earth, Omega Centauri is so old that it's home to stars with billions of years difference in age.

STELLAR TYPE
Globular cluster

CONSTELLATION
Centaurus (the Centaur)

MASS
4.05 x 106 solar masses

TEMPERATURE
Unknown

RADIUS
86 light years

DETECTION TYPE
Visual

DISCOVERED
1603 as a cluster

DISTANCE FROM SUN
15.8 light years

Long before humankind evolved, before dinosaurs roamed, and even before our Earth existed, ancient globs of stars condensed and orbited a young Milky Way Galaxy. Of the 200 or so globular clusters that survive today, roaming our galaxy, Omega Centauri is by far the largest, containing over ten million stars.

Globular clusters like Omega Centauri were once thought to be assemblages of stars that share the same birth date. Evidence suggests, however, that Omega Centauri has at least two populations of stars with different ages; the stars in Omega Centauri are between 10 billion and 12 billion years old - those two billion years make a difference in the life-cycle of stars. Some astronomers think that the Omega Centauri cluster may be the remnant of a small galaxy that was gravitationally disrupted long ago by the Milky Way, losing stars and gas.

Omega Centauri is also the brightest globular cluster, visible to southern observers with the unaided eye; it was known to Ptolemy, though not identified as a star cluster until centuries later. If you're trying to spot Omega Centauri, it resembles a small cloud in the southern sky and might easily be mistaken for a comet. Should you find it, take a moment to imagine how many eons these stars have existed, and how many generations of humans have looked up at them over history.

STELLAR TYPE
Reflection / emission nebula

CONSTELLATION
Orion

MASS
2000 solar masses

TEMPERATURE
Up to 10,000 K

RADIUS
20 light years

DETECTION TYPE
Visual

DISCOVERED
Known to antiquity

DISTANCE FROM SUN
1344 light years

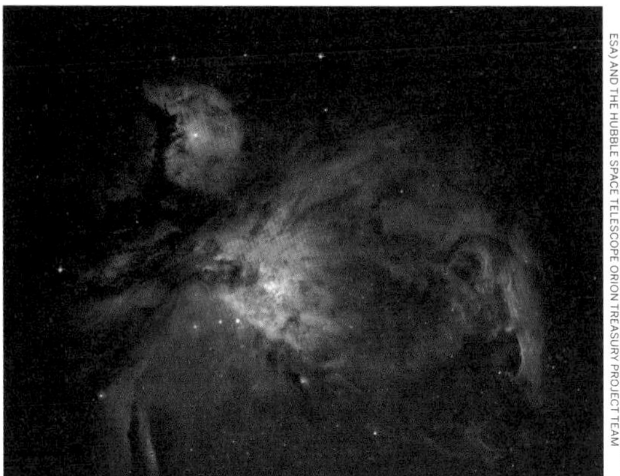

The Orion Nebula seen through three different filters, for sulfur, hydrogen and oxygen.

Orion Nebula

One of the most beautiful and easily spotted objects in the sky, the Orion Nebula has been attributed the powers of creation through both myth and science: it is the stellar nursery of more than a thousand stars.

Few astronomical sights excite the imagination like the nearby stellar nursery that is the Orion Nebula, also known as M42. The Nebula's glowing gas surrounds hot young stars at the edge of an immense interstellar molecular cloud. The Orion Nebula is located in the same spiral arm of our galaxy as the Sun.

Believed to be the cosmic fire of creation by the Maya of Mesoamerica and known to many different cultures throughout human history, the Orion Nebula blazes brightly in the constellation Orion. The nebula is the closest large star-forming region to Earth, and astronomers believe it contains more than 1000 young stars.

Because of its brightness and prominent location just below Orion's belt, the Orion Nebula can be spotted with the naked eye, while offering an excellent peek at stellar birth for those with telescopes. The Orion Nebula is therefore among the best observed and most photographed objects in the night sky. However, this beautiful sight won't last forever: the whole Orion Nebula cloud complex, which includes the Horsehead Nebula, will slowly disperse over the next 100,000 years, once their star-forming period has ended. Sounds like a good reason to go stargazing!

The shimmering Owl Nebula.

Owl Nebula

Like an owl with its face turned toward Earth, the Owl Nebula offers astronomers of all levels the opportunity to examine the future of yellow dwarf stars like our Sun.

The Owl Nebula is perched in the sky toward the bottom of the Big Dipper's bowl. Also catalogued as M97, the 97th object in Messier's well-known list, its round shape along with the placement of two large, dark 'eyes' does suggest the face of a staring owl.

One of the fainter objects in Messier's catalogue, the Owl Nebula is a planetary nebula (one of four that he included in his list), the glowing gaseous envelope shed by a dying Sun-like star as it runs out of nuclear fuel. As such, the Owl Nebula offers yet another example of the fate of our Sun as it runs out of fuel in another five billion years. The vast, diffuse nebula spans roughly 2000 times the diameter of Neptune's orbit; at the very centre, the contracted white dwarf star can still be seen.

The Owl Nebula comprises three concentric shells of gas, expelled as the core white dwarf was born from the red giant that once existed. Astronomers estimate that the Owl Nebula is roughly 8000 years old, based on the size of the outermost shell, which has been expanding ever since.

STELLAR TYPE
Planetary nebula

CONSTELLATION
Ursa Major (the Great Bear)

MASS
Unknown

TEMPERATURE
123,000 K

RADIUS
~1 light year

DETECTION TYPE
Telescope

DISCOVERED
Pierre Méchain, 1781

DISTANCE FROM SUN
2030–2800 light years

STELLAR TYPE
Open star cluster

CONSTELLATION
Taurus (the Bull)

MASS
Unknown

TEMPERATURE
Unknown

RADIUS
Unknown

DETECTION TYPE
Visual

DISCOVERED
Known to antiquity

DISTANCE FROM SUN
444 light years

The Pleiades, or Seven Sisters, is a star cluster containing thousands of stars.

Pleiades

One of the most recognisable star clusters in the entire sky, closer examination unveils thousands of stars that far outnumber the 'Seven Sisters' for which the Pleiades cluster is named.

Most people know the Pleiades open star cluster by its colloquial name: the Seven Sisters. Also classified as M45, the Pleiades contains over three thousand stars that are loosely bound by gravity, but it is visually dominated by a handful of its brightest members. A common legend with a modern twist is that one of the brighter stars has faded since the cluster was named, leaving only six stars visible to the unaided eye. The actual number of visible Pleiades stars, however, may be more or less than seven, depending on the darkness of the surrounding sky and the clarity of the observer's eyesight.

The Pleiades is primarily home to young, hot, and luminous blue stars, which are the primary stars we see when observing the cluster. Astronomers have discovered low-mass brown dwarf stars in the cluster. The Pleiades cluster has been observed since ancient times, so it has no known discoverer. However, Galileo Galilei, the Italian scientist best known for discovering the largest moons of Jupiter and championing a heliocentric model of the solar system, was the first to observe the Pleiades through a telescope. The cluster is visible from across the globe, and featured prominently in a number of mythologies, from Aboriginal oral tales to Egyptian and Celtic legends. The rising of the Pleiades even signalled the beginning of the Aztec New Year.

Multiple star Polaris is a navigational beacon for its brightness as well as its position.

Polaris

A fixed point of light in the northern sky, Polaris actually consists of three stars. Changes in the Earth's spin axis will mean that Polaris too will eventually spin around a different 'North Star'. Turns out Polaris is more changeable than originally believed.

Imagine an Earth where Polaris, the 'North Star' or Alpha Ursae Minoris, is not a fixed point of light in the sky around which all other stars seem to spin. Several thousand years ago - not long in the cosmic timeline - Vega was that star. As the Earth's spin axis changed, Polaris eventually became the star that we can focus on for navigation and orientation in the sky. One day in the future, a new star will replace it in turn. For now, the surface of Polaris, once thought to be a single star, slowly pulsates in its usual home. This causes the star to change its brightness by a few percent over the course of a few days. Given its prominence and interesting behaviour, astronomers have studied Polaris extensively over the centuries.

In 2006, astronomers using the Hubble Telescope made a compelling discovery. While Polaris was thought to be a steady, solitary point of light that guided sailors for ages, there is more to this star than meets the eye. Polaris is actually a triple star system. While one companion (Alpha Ursae Minoris B) is easily viewed with small telescopes, the other (Ab) hugs Polaris (Aa) so tightly that it had never been observed, until astronomers used the high-resolution cameras on Hubble to study it for an extended amount of time and discovered its hidden layers.

STELLAR TYPE
Multiple star system

CONSTELLATION
Ursa Minor (the Small Bear)

MASS
5.4 solar masses (Aa), 1.26 solar masses (Ab), 1.39 solar masses (B)

TEMPERATURE
6015 K (Aa), unknown K (Ab), 6900 K (B)

RADIUS
37.5 solar radii (Aa), 1.04 solar radii (Ab), 1.39 solar radii (B)

DETECTION TYPE
Visual

DISCOVERED
Known to antiquity; 1779 as a multiple-star system

DISTANCE FROM SUN
323–433 light years

STELLAR TYPE
Binary system

CONSTELLATION
Canis Minor (the Dog)

MASS
1.499 solar masses (A),
0.602 solar masses (B)

TEMPERATURE
6530 K (A),
7740 K (B)

RADIUS
2.048 solar radii (A),
0.01 solar radii (B)

DETECTION TYPE
Visual

DISCOVERED
Known to antiquity

DISTANCE FROM SUN
11.46 light years

A view of the sky encompassing Canis Minor, Procyon's constellation.

Procyon

Located relatively near Earth on a cosmic scale, binary star Procyon gives astronomers a peek at the next phase of evolution in stars like our Sun.

Formally called Alpha Canis Minoris, Procyon is a star in the constellation Canis Minor, the Dog, and the eighth-brightest star in the sky. The name Procyon actually translates as 'before the dog' from Ancient Greek, as Procyon precedes Sirius in the night sky. The two stars are surprisingly similar.

Procyon is actually a binary system, comprising a main sequence star (Procyon A) slightly larger than our own Sun, and a faint white dwarf companion star (Procyon B). Procyon A is especially interesting to astronomers, as research suggests it is nearing the end of its life as a main sequence star and will evolve into a red giant within the next 10 to 100 million years; practically no time at all in cosmic terms. According to Carl Sagan's conception of a cosmic year, that's between one to three cosmic days.

Procyon, like Deneb, is part of two asterisms, both seasonal: the Winter Triangle with Sirius and Betelgeuse, and the Winter Hexagon. The Winter Hexagon involves some of the brightest stars visible, together forming a large and easy to find pattern in the winter sky of Earth's northern hemisphere. It is formed by Procyon, Sirius, Rigel, Aldebaran, Capella, and the twins Castor and Pollux. At apparent magnitude 0.4, Procyon is the brightest star in Canis Minor, and the eighth brightest in the sky overall; you might say that it's close to being top dog.

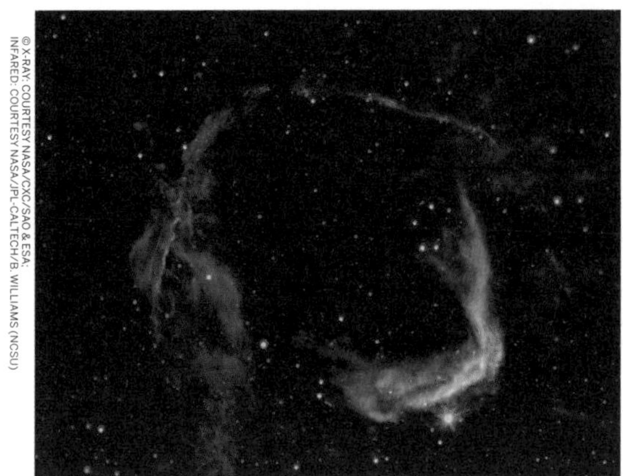

This multi-wavelength view of RCW 86 combines data from four space telescopes.

RCW 86

The remnants of a dazzling supernova that was visible from Earth for eight months, the puzzle of RCW 86's massive solar shell has finally been solved.

An astronomical mystery began nearly 2,000 years ago in 185 AD, when Chinese astronomers noted a 'guest star' that mysteriously appeared in the sky – and remained visible to the naked eye for eight more months. By the 1960s, scientists had determined that the mysterious object was the first documented supernova. Later, they pinpointed RCW 86 as the supernova remnant of this cosmic explosion, but a puzzle persisted. The star's spherical remains in Circinus (the Compass) are larger than expected, 85 light years in diameter. If you could observe RCW 86 in the sky today in infrared light, it would take up more space than our full moon.

Within the last decade, astronomers have studied RCW 86 (sometimes also called SN 185, for 'supernova 185', the year it was observed) to solve the mystery. Their findings show that the stellar explosion of SN 185 took place in a hollowed-out cavity of matter within space. This allowed material expelled by the star to travel much faster and further than it would have otherwise. This event is known as a 'Type Ia' supernova. These explosions are known to occur in binary systems in which one of the members is a white dwarf, a dense star created by the relatively peaceful death of a yellow dwarf like our Sun. The white dwarf later exploded in a supernova after siphoning matter, or fuel, from the nearby binary star. The remnants are what we see today when studying RCW 86

STELLAR TYPE
Supernova remnant nebula

CONSTELLATION
Circinus (the Compass) and Centaurus (the Centaur)

MASS
Unknown

TEMPERATURE
Unknown

RADIUS
42.5 solar radii

DETECTION TYPE
Visual on discovery; Infrared for remnant

DISCOVERED
185 AD as supernova

DISTANCE FROM SUN
9100 light years

STELLAR TYPE
Multiple star system

CONSTELLATION
Leo (the Lion)

MASS
3.8 solar masses (A),
0.8 solar masses (B),
0.3 solar masses (C)

TEMPERATURE
12,460 K (A),
4885 K (B),
unknown K (C)

RADIUS
3.092 solar radii (A),
unknown (B),
unknown (C)

DETECTION TYPE
Visual

DISCOVERED
Known to antiquity

DISTANCE FROM SUN
75 light years

Regulus shines brightly as part of the Leo constellation.

Regulus

Comprising several stars, Regulus offers astronomers a condensed look at main sequence stars like our Sun with a variety of different characteristics.

Regulus, also named Alpha Leonis, is the brightest star in the constellation Leo, the Lion. As the lion was often used as a symbol of royalty in human history, its stars have always had a 'royal' association. In fact, the Latin name 'Regulus' translates as 'little king' or 'prince', lending more weight to the colloquial association of Regulus as a royal star.

Regulus actually consists of four stars, in two binary pairs. The primary star we see is Regulus A, a blue-white main sequence star with a companion that is likely a white dwarf but has actually never been observed. It is hypothesised based on the behaviour and shape of Regulus A. Regulus B and Regulus C are both dim main sequence stars, smaller and less luminous than our Sun. Regulus D is a distant star that appears as part of the star system, but is likely an unrelated background object, merely lending its light to Regulus from our perspective rather than being gravitationally bound to the other stars in their orbits. Parsing which objects merely appear to be in proximity to each other versus which exist in the same region of space can be an ongoing challenge for astronomers, though the increasingly sophisticated detection methods and statistical models available has reduced the confusion.

© SSRO STAR SHADOWS REMOTE OBSERVATORY

Like Regulus, Rigel adds its sparkle to the sky thanks to its multiple stars.

Rigel

Located at Orion's knee, Rigel shines brightly as part of the Winter Hexagon. This single point of light is deceiving: at least three to five stars contribute to Rigel's luminosity.

STELLAR TYPE
Multiple star system

CONSTELLATION
Orion (the Hunter)

MASS
21 solar masses (A), 3.84 solar masses (Ba), 2.94 solar masses (Bb), 3.84 solar masses (C)

TEMPERATURE
12,100 K (A)

RADIUS
78.9 solar radii (A)

DETECTION TYPE
Visual; Telescope (binary classification)

DISCOVERED
Known to antiquity; 1781 as at least a binary

DISTANCE FROM SUN
860 light years

The constellation Orion is among the most recognisable in the night sky, thanks to bright stars like Betelgeuse, Salph, Bellatrix, and Rigel. Rigel is among the 10 brightest stars in the sky, and rivals the distinctively orange Betelgeuse as the most eye-catching star in Orion. Rigel is the most luminous star within 1000 light years of the Sun. Yet despite being the brightest star in Orion, which would conventionally give Rigel the additional name of Alpha Orionis, showy Betelgeuse takes the title, and leaves Rigel to be identified as Beta Orionis.

Rigel – like so many bright stars we can see – is not just one star, but is actually a multiple star system, comprising three to five stars. Two of the stars in Rigel are widely known and studied. Rigel A is a blue-white supergiant, over 60,000 times more luminous than our Sun (and 21 times more mass). Companion Rigel B is 500 times fainter than Rigel A, and is visible only with a telescope. Don't let its faintness fool you: astronomers have determined that Rigel B is itself a multiple star system, likely composed of two main sequence stars, and it has its own companion, Rigel C. While these stars are faint and not extensively studied, Ba, Bb, and C are likely similar in size and stage of stellar evolution. (Confused yet?) Rigel D is a faint, distant star that is likely unrelated to the Rigel star system, but is aligned by coincidence from our perspective.

STELLAR TYPE
White dwarf and planetary nebula

CONSTELLATION
Lyra (the Lyre)

MASS
.62 solar masses

TEMPERATURE
125,000 K

RADIUS
1.3 light years

DETECTION TYPE
Telescope

DISCOVERED
Antoine Darquier de Pellepoix, 1779

DISTANCE FROM SUN
2000 light years

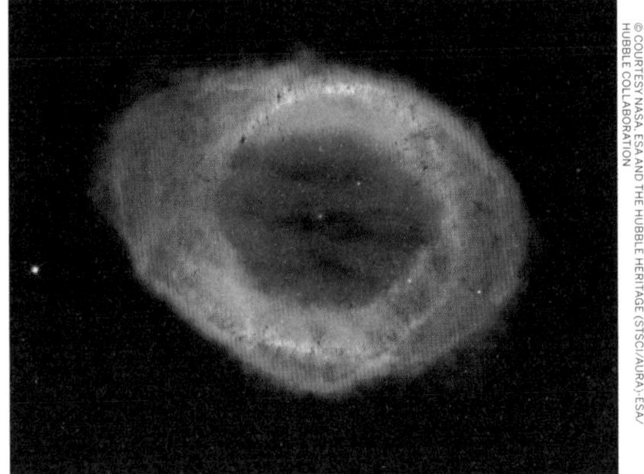

The dramatic Ring Nebula bears some resemblance to the Eye of Sauron.

Ring Nebula

Iconic and easy to spot, the Ring Nebula is a beloved favourite of observers. Don't wait to take a look though: in about 10,000 years, this nebula will fully dissipate.

The Ring Nebula, officially known as M57 in the Messier catalogue, is a planetary nebula, the glowing remains of a Sun-like star much further advanced in its evolutionary path. The tiny white dot in the centre of the nebula is the star's hot core, a small, dense white dwarf. First discovered by the French astronomer Antoine Darquier de Pellepoix in 1779, the Ring Nebula can be spotted with moderately sized telescopes.

M57 is tilted toward Earth so that astronomers see the ring face-on. This initially led researchers to underestimate the complexity of its structure. Though it looks empty, the blue gas in the nebula's centre is actually a football-shaped structure seen end-on that pierces the red, doughnut-shaped material. 'The nebula is not like a bagel, but rather, it's like a jelly doughnut, because it's filled with material in the middle', said C. Robert O'Dell of Vanderbilt University in Nashville, Tennessee. He leads a research team that used the Hubble telescope and several ground-based telescopes to obtain the best view yet of the iconic nebula back in 2013.

The Ring Nebula is about 2000 light years from Earth and measures roughly 1 light year across. Located in the constellation Lyra, the striking nebula is a popular target for amateur astronomers and astrophotographers.

The Rosette Nebula surrounds the NGC 2244 star cluster.

Rosette Nebula

With cosmic petals stretching light years in every direction, the Rosette Nebula gives amateur and professional astronomers plenty to stop and 'smell' – admire – when viewing the night sky.

STELLAR TYPE
Emission nebula

CONSTELLATION
Monoceros (the Unicorn)

MASS
10,000 solar masses

TEMPERATURE
1 to 10 million K

RADIUS
65 light years

DETECTION TYPE
Telescope

DISCOVERED
Observed piecemeal

DISTANCE FROM SUN
5000 light years

Would the Rosette Nebula by any other name look as sweet? The Rosette Nebula, also designated Caldwell 49, is not the only cosmic cloud of gas and dust to evoke the imagery of flowers - but it is the most famous. (The Iris Nebula is another popular one.)

At the edge of a large molecular cloud in the constellation Monoceros, the Unicorn, the petals of the Rosette Nebula are actually a stellar nursery whose lovely, symmetric shape is sculpted by the winds and radiation from its central cluster of hot young stars. The stars in the energetic cluster are only a few million years old, and the solar winds they cause have also opened a hole in the centre of the Rosette, allowing us a peek into the heart of this cosmic bloom. It's so broad that it contains multiple NGC designations for its different segments.

While professional astronomers examine the Rosette Nebula to understand the early stages in the stellar lifecycle, amateur astronomers can enjoy viewing it too. The nebula can be seen first-hand with a small telescope from most dark sky locations.

STELLAR TYPE
Supermassive black hole

CONSTELLATION
Sagittarius (the Archer)

MASS
4.31×10^6 solar masses

TEMPERATURE
10^{-14} K

RADIUS
120 AU

DETECTION TYPE
NRAO interferometer

DISCOVERED
1974

DISTANCE FROM SUN
27,000 light years

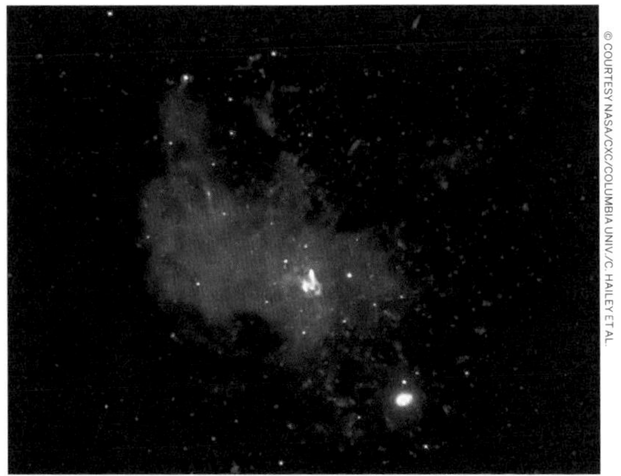

An image of X-rays emitted from Sagittarius A* and nearby stellar-mass black holes.

Sagittarius A*

Black holes generally get a bad reputation for consuming and obliterating all light and matter in their vicinity. Luckily for us, the one at the centre of our Milky Way Galaxy, Sagittarius A, is better behaved than most.*

Even if you know nothing about Sagittarius A*, you've felt its effects every day of your life – it helped shape our entire galaxy. At the centre of the Milky Way Galaxy lies a black hole with four million times the mass of the Sun, fondly known as Sagittarius A* (pronounced 'A-star'). Astronomers have long assumed that supermassive black holes like Sagittarius A* lie at the centre of most spiral and elliptical galaxies like our own, and have come to certain conclusions about how quickly black holes consume the matter and light around them.

Fortunately, Sagittarius A* is mild-mannered compared to the central black holes in distant active galaxies, consuming material around it much more calmly. While Sagittarius A* does have flare ups of emission activity, astronomers still lack a detailed understanding of why Sagittarius A* varies so much in its behaviour compared with other black holes we're observing. It has also collected a large population of stellar black holes around itself, as its gravitational mass causes them to drift inwards to within three light years. By studying emissions and stars in the vicinity of Sagittarius A*, astronomers continue to gain insight into black holes in general – as well as the one propelling our galaxy.

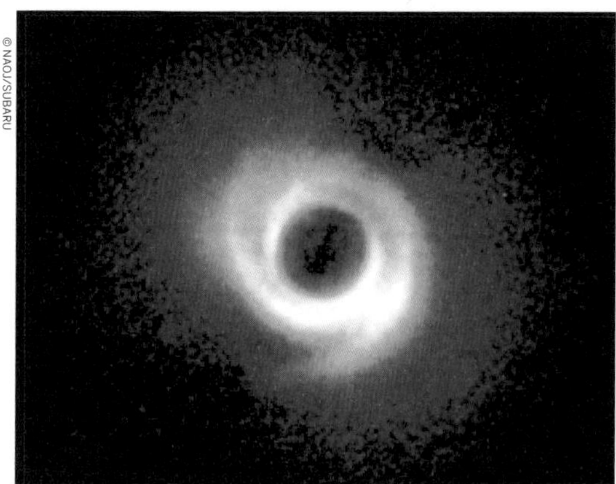

SAO 206462 and its prominent disc.

SAO 206462

Surrounded by an unusually shaped disc of gas and dust, young star SAO 206462 offers astronomers a clear candidate for planet hunting and studying a young solar system.

It's always exciting to find something 'young' in the Universe – on the galactic timescale, that is. SAO 206462 is a relatively young star system at an estimated nine million years old. While this might seem quite old from our human perspective, SAO 206462 is still an adolescent as stars go, and is still figuring out its shape and structure.

In 2011, astronomers discovered that SAO 206462 is surrounded by a disc of gas and dust that has very distinctive shapes: spiral-arm-like structures, similar to the galactic arms we see in the Milky Way. The disc itself is some 14 billion miles across, or about twice the size of Pluto's orbit in our own solar system, and these arms may provide clues to the presence of embedded but as-yet-unseen planets.

The two arms of the disc surrounding SAO 206462 are not a matched pair – they are not symmetrical – suggesting to astronomers the presence of two unseen worlds, one for each arm. While astronomers have not yet confirmed that these hypothetical exoplanets exists, we'll likely continue watching SAO 206462 mature for decades to come. According to computer simulations, planets within the star's disc could create perturbations similar to those witnessed here. The planets might still be in the process of forming, much as our solar system's planets accreted out of a cloud of debris four billion years ago.

STELLAR TYPE
Main sequence star

CONSTELLATION
Lupus (the Wolf)

MASS
Unknown

TEMPERATURE
Unknown

RADIUS
Unknown

DETECTION TYPE
Combined space and earth-based telescopes

DISCOVERED
2011

DISTANCE FROM SUN
460 light years

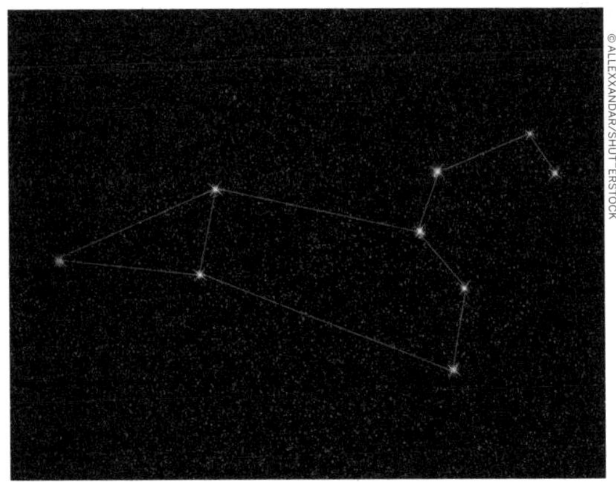

Invisible to observers, this quasar can be found in Leo constellation.

SDSSJ0927+2943

If a black hole typically sits at the centre of a galaxy, why is SDSSJ0927+2943 wandering around the cosmos? Look to gravitational waves for an explanation.

STELLAR TYPE
Recoiling black hole or quasar

CONSTELLATION
Leo (the Lion)

MASS
600 million solar masses

TEMPERATURE
Unknown

RADIUS
4.5 AU

DETECTION TYPE
Sloan Digital Sky Survey

DISCOVERED
2008

DISTANCE FROM SUN
6.85 × 109 light years

With a mouthful of a name, one might think of SDSSJ0927+2943 as some forgotten item in an obscure catalogue of stellar objects. (The initials actually come from the multi-spectral imaging project that found the object, the Sloan Digital Sky Survey.) Instead SDSSJ0927+2943 is as unique as its name: an unusual quasar with unusual emission patterns that appears to be the first evidence of a 'recoiling black hole.'

In the case of SDSSJ0927+2943, it appears that the supermassive black hole was ejected from the centre of its galactic system. Astronomers are using the principles of physics to hypothesise what happened, but the educated idea is that when two black holes combined in SDSSJ0927+2943, the gravitational wave they created actually pushed the newly formed supermassive black hole out of the system. The merging of these immensely massive bodies happens on such a huge scale that they can result in quite dramatic reactions.

At a distance of 6.85 billion light years from Earth, researchers have a limited view into how SDSSJ0927+2943 is behaving today and where its escape trajectory ultimately leads. In any case, they will continue to observe and understand how black holes, gravitational waves, and galaxies can interact – and the fleeing SDSSJ0927+2943 is a great candidate for that observation.

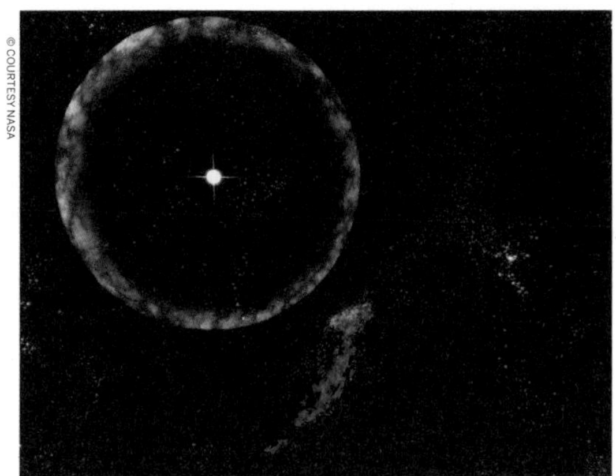

An artist's conception of the gamma ray flare at SGR 1806-20.

SGR 1806-20

One of only four known stars of its type, SGR 1806-20 is also the record-holder for causing the brightest burst of gamma ray light ever recorded in the Milky Way Galaxy.

Scientists in 2004 detected a flash of light from across the galaxy so powerful that it bounced off the moon and lit up the Earth's upper atmosphere. The flash was brighter than anything ever detected from beyond our solar system and lasted over a tenth of a second. What caused this phenomenon? A magnetar near our galactic centre, the source of Soft Gamma Repeater (SGR) 1806-20, had unleashed its largest flare on record, and it finally reached us 50,000 years later. To our knowledge, nothing so bright has ever been seen on Earth from outside of the solar system. The event that caused it is called a starquake, the term for a neutron star that experiences a rapid adjustment of its crust.

A neutron star is the core remains of a star once several times more massive than our Sun. After a supernova, the remaining core is dense, fast-spinning, and highly magnetic. Within the class of neutron star objects, astronomers have discovered a dozen ultra-high-magnetic neutron stars, called magnetars. Four of these magnetars are also called soft gamma repeaters, or SGRs, because they flare up randomly and release gamma rays. SGR 1806-20 is one of these unique stellar objects, as indicated by its name. By studying SGR 1806-20, researchers are learning more about gamma rays, what causes them, and how they impact technology and life on Earth. If a flare of this size were to occur within 10 light years of our planet, it is believed that it would destroy the ozone layer.

STELLAR TYPE
Soft gamma repeater magnetar

CONSTELLATION
Sagittarius (the Archer)

MASS
Unknown

TEMPERATURE
136,000 K

RADIUS
10 km (6 mi)

DETECTION TYPE
Gamma rays

DISCOVERED
1979

DISTANCE FROM SUN
50,000 light years

STELLAR TYPE
Binary star – main sequence and white dwarf

CONSTELLATION
Canis Major (the Greater Dog)

MASS
2.06 solar masses (A), 1.02 solar masses (B)

TEMPERATURE
9940 K

RADIUS
1.71 solar radii (A), 0.008 solar radii (B)

DETECTION TYPE
Visual

DISCOVERED
Known to antiquity; 1862 as a binary system

DISTANCE FROM SUN
8.6 light years

An artists's impression of Sirius A, in the foreground, and Sirius B in the distance.

Sirius

On a dark night, the 'Dog Star' Sirius is a brilliantly alluring object for stargazing. This nearby star is much like our own Sun, but with a surprising celestial companion.

It's nigh impossible to go stargazing and miss Sirius, the brightest star in the night sky. Officially named Alpha Canis Majoris, a title that acknowledges its position as the brightest star in Canis Major, Sirius glows brightly – it is over 20 times brighter than our Sun and over twice as massive. The star's apparent magnitude clocks in at -1.46, where the negative indicates an especially high luminosity.

What appears as a single star is actually two: a luminous main sequence star like our Sun and a small white dwarf. In 1862, Sirius was discovered to be a binary star system with a companion star, Sirius B, which is 10,000 times dimmer than the bright primary, Sirius A. Sirius B was the first white dwarf star discovered, having moved through the life-cycle of a star faster than its sibling Sirius A. The two stars orbit each other every 50 years. Their close proximity has frustrated researchers eager to learn more about the white dwarf part of a star's life-cycle, as Sirius A's luminosity easily obscures Sirius B.

Over the next 60,000 years, Sirius will slowly move closer to Earth, becoming even brighter. It will remain the brightest star in the night sky for the next 210,000 years, by which time perhaps humans will have finally set out on an interstellar journey to visit this celestial neighbour. If dogs are a human's best friend on Earth, Sirius may be our best friend in the sky.

Spica, upper left, helps point the way to the constellation Corvus on the lower right.

Spica

When does one star become two? When astronomers find differences in their light spectra. In the case of nearby Spica, two stars are so close they're pulling each other out of shape

Spica shines constantly as one of the brightest blue stars in the night sky. It is the brightest star in the constellation Virgo, earning its name as Alpha Virginis, and is among the 20 brightest stars in the sky. Pronounced 'spy-kah', the blue-hued star has been visible throughout human history and the sounds that identify it today date back to ancient times. Spica, along with Regulus and Arcturus, is part of the Spring Triangle asterism in the northern hemisphere. With an apparent magnitude of .98, easily visible Spica can help point observers toward the nearby constellations Corvus the Crow, Hydra the Snake and Crater the Cup.

Spica is actually a binary star system, with two stars – a blue giant and a variable star – orbiting so near each other that they pull each other's outlines into an egg shape rather than spheres. These stars are so close that they can't be separated visually through a telescope; instead astronomers must look at the light spectrum for each star to differentiate them. Because the two stars are so different, they each have distinct signatures. Spica is also the closest binary system to our Sun, which makes it the subject of extensive study by astronomers.

STELLAR TYPE
Binary star system

CONSTELLATION
Virgo (the Virgin)

MASS
11.43 solar masses (Primary), 7.21 solar masses (Secondary)

TEMPERATURE
25,300 K (Primary), 20,900 K (Secondary)

RADIUS
7.47 solar radii (Primary), 3.74 solar radii (Secondary)

DETECTION TYPE
Visual

DISCOVERED
Known to antiquity

DISTANCE FROM SUN
250 light years

STELLAR TYPE
Main sequence star

CONSTELLATION
Cygnus (the Swan)

MASS
1.43 solar masses

TEMPERATURE
6750 K

RADIUS
1.58 solar radii

DETECTION TYPE
NASA Kepler

DISCOVERED
2016 for unusual dimming

DISTANCE FROM SUN
1470 light years

This illustration depicts a hypothetical uneven ring of dust orbiting Tabby's Star.

Tabby's Star

Citizen scientists in the Kepler Planet Hunter program discovered something fascinating about Tabby's Star, a star much like our Sun. Unusual changes in brightness have led scientists and stargazers to come up with a myriad of hypotheses to explain this behaviour.

We still have a lot to learn about the Universe, but our understanding of star behaviour is solid. So when citizen scientists hunting for planets discover a star that is behaving differently than any others observed before, it can cause a lot of excitement and interest. Tabby's Star is also known as Boyajian's Star, both after Tabetha S Boyajian, the lead author on an initial study of the star. It also has the official name KIC 8462852; by any name, the intriguing data from this star caused a lot of excitement when astronomers discovered it was dimming at an unusual rate. Was this dimming caused by orbiting planets or some unknown object – perhaps by a massive alien ship orbiting a Sun much like our own?

Initial observations by the Kepler planet-hunting telescope suggested that the unusual light fluctuations in Tabby's Star were likely the result of a swarm of comets; Spitzer Space Telescope studied Tabby's Star too, finding more evidence for the scenario involving a swarm of comets and a dusty cloud. Scientists still aren't totally sure this hypothesis explains all of the changes observed in Tabby's Star, and advocate continued observation to help us better understand this interesting and even mystifying stellar object.

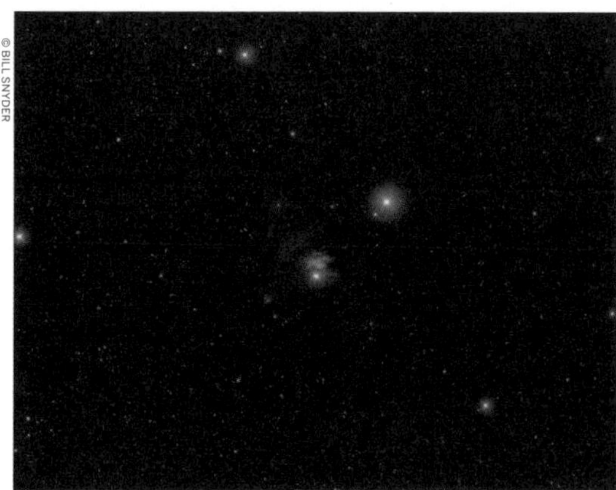
T Tauri is the star at the centre of this image in a disc of dust and gas.

T Tauri

Setting the standard for stars of its type, variable star T Tauri helps astronomers understand the early eons of a star – and possibly grants a peek at younger siblings in earlier stages of formation.

What does a star look like when it is forming? What did the Sun look like before there were planets? The prototypical example is the variable star T Tauri, which is developing in Hind's Variable Nebula. T Tauri is likely part of a multiple star system whose other object(s) are still forming in the cosmic cloud; the primary star is called T Tauri N, and the as-yet-unseen companions are named T Tauri Sa and T Tauri Sb.

Stars like T Tauri are now generally recognised as young - less than a few million years old - Sun-like stars still in the early stages of formation. In young systems like T Tauri, gravity causes a gas cloud to condense, as they have in Hind's Variable Nebula. Some of the infalling gas is heated so much by collisions that it is immediately expelled as an outgoing wind. In a few million years, the central condensate will likely become hot enough to ignite nuclear fusion, by which time much of the surrounding circumstellar material will either have fallen in or have been driven off by the stellar wind. At that time, a new star will shine, and the T Tauri system will begin the next phase of its life as a main sequence star, possibly with neighbouring stars in earlier phases of their own stellar life-cycle. It's a fascinating snapshot of an evolutionary moment in time.

STELLAR TYPE
Variable star

CONSTELLATION
Taurus (the Bull)

MASS
2.12 solar masses(N),
0.53 solar masses(S)

TEMPERATURE
Unknown

RADIUS
Unknown

DETECTION TYPE
Telescope

DISCOVERED
1852

DISTANCE FROM SUN
600 light years

STELLAR TYPE
Quasar

CONSTELLATION
Leo (the Lion)

MASS
2+1.5×109 solar masses
-0.7

TEMPERATURE
Unknown

RADIUS
Unknown

DETECTION TYPE
Infrared

DISCOVERED
2011

DISTANCE FROM SUN
12.9 x 10^9 light years

This impression shows ULAS J1120+0641, a distant black hole-powered quasar.

ULAS J1120+0641

Evidence of a supermassive black hole forming early in the universal timescale, ULAS J1120+0641 is among the most distant objects we've ever observed.

Named for the telescope survey that discovered it (the UKIDSS Large Area Survey) and its place in our sky, ULAS J1120+0641 was the most distant quasar ever found at the time of its discovery in 2011. (It is now the second most distant after the discovery of ULAS J1342+0928 in 2017.)

Observations showed that the mass of the black hole at the centre of ULAS J1120+0641 is about two billion times that of the Sun – a supermassive black hole indeed. This very high mass is hard to explain so early on after the Big Bang. Current theories for the growth of supermassive black holes predict a slow build-up in mass as the compact object pulls in matter from its surroundings, so astronomers are a bit stumped as to how ULAS J1120+0641 grew so big in such a short time on the universal timeline. The quasar it powers is accordingly quite luminous, an estimated 6.3×10^{13} solar luminosities, with a large fraction (10 to 15%) of neutral non-ionised hydrogen. This may indicate it's at the early stages of galaxy formation.

The light observed from ULAS J1120+0641 at the time of its discovery was generated only 770 million years after the Big Bang – but took 12.9 billion light years to reach us on Earth. The supermassive black hole at the core of ULAS J1120+0641 is likely quite different today than we'll ever know. Peering at this object is very much like looking back in time to an earlier Universe.

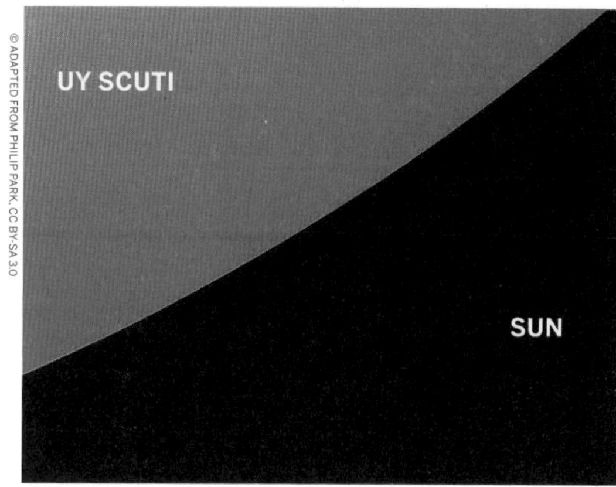

Our yellow dwarf Sun is 1700 times less the estimated radius of hypergiant UY Scuti.

UY Scuti

Located in a cluster of smaller stars, UY Scuti is the biggest star on the block – in fact it's the largest star we've yet observed, though there's plenty of debate and competition.

What does it take to earn the title of 'largest known star in the Universe?' With a radius nearly 2000 times that of our own Sun, red hypergiant UY Scuti currently holds the top spot. If it were in the centre of our Solar System, it would expand past the orbit or Jupiter!

UY Scuti was first discovered by German astronomers in the mid-19th century; they soon realised it was also a pulsating variable star, changing magnitude in a repeating pattern every 740 days. Researchers have since proposed that UY Scuti is shrouded in a cloud of dust that could explain its brightening and dimming behaviours.

Because of this unusual behaviour, UY Scuti is constantly challenged for the throne as the largest star yet found. It's possible that UY Scuti is smaller than current estimates, because at a distance of over 5100 light years, critical information about its composition and nearby structures is hard to confirm. Several other large stars, including VY Canis Majoris, could possibly outrank UY Scuti, but for now it retains its title. Bear in mind that while this hypergiant has a radii far beyond that of our own Sun, its mass is only 30 times greater. Because of the nature of these expanded giants, they are much less dense than a main sequence star.

STELLAR TYPE
Red hypergiant

CONSTELLATION
Scutum (the Shield)

MASS
7–10 solar masses

TEMPERATURE
3365 K

RADIUS
1708 solar radii

DETECTION TYPE
Telescope

DISCOVERED
1860

DISTANCE FROM SUN
5100 light years

STELLAR TYPE
Main sequence star

CONSTELLATION
Lyra (the Lyre)

MASS
2.135 solar masses

TEMPERATURE
9602 K

RADIUS
2.362 solar radii

DETECTION TYPE
Visual

DISCOVERED
Known to antiquity

DISTANCE FROM SUN
25.04 light years

This astrograph depicts Vega (the brightest dot shown) within the constellation Lyra.

Vega

Thanks to Earth's unstable wobble through space, bright Vega is one of the most studied – and most significant – stars in human history, and will be again in the future.

Vega is the fifth brightest star in the night sky and the brightest star in the constellation Lyra, the Harp. Additionally, it is the brightest star in the asterim the Summer Triangle, composed of three bright stars from three different constellations. It has a diameter almost three times that of our Sun – and has an astonishing record in human astronomical history. Some 14,000 years ago, Vega – not Polaris – was the pole star on Earth, the star around which all stars seem to rotate.

Vega went by the name 'Ma'at' 4000 years ago, an example of the trove of ancient human astronomical knowledge and language accumulated by our forebears despite their lack of telescopes. In another 12,000 years, Vega will again become the pole star. Today, however, the name Vega derives from Arabic origins, and means 'stone eagle'. It's also one of the stars representing ill-fated lovers in China's summer Qixi festival.

Astronomers have discovered what appears to be a large asteroid belt around the star Vega, similar to the asteroid and Kuiper belts in our own solar system. They also hypothesise that life-bearing planets, rich in liquid water, could possibly exist around Vega. Astronomers study Vega to understand its composition, as well as that of its solar system. At roughly 25 light years from Earth, Vega is a compelling subject for astronomical study in the future, as well as a link to myths of past and present.

© COURTESY NASA, ESA, AND THE HUBBLE HERITAGE (STSCI/AURA)-ESA/HUBBLE COLLABORATION; ACKNOWLEDGMENT: J. HESTER (ARIZONA STATE UNIVERSITY)

The wide Veil Nebula is one of the largest supernova remnants in the sky.

Veil Nebula

Remnants of a massive explosion thousands of years ago, the Veil Nebula shows all that's left of one massive galactic neighbour 20 times the size of our Sun.

Part of the Cygnus Loop supernova remnant, the Veil Nebula is one of the largest and most spectacular supernova remnants in the sky. It derives its name from its delicate, draped filamentary structures – which are actually the blast wave from an ancient stellar explosion. About 8000 years ago, the Veil Nebula came into being when a massive star in our Milky Way exploded into a supernova. Wisps of gas are all that remain of what was once a star 20 times more massive than our sun. The fast-moving blast wave from the ancient explosion is plowing into a wall of cool, denser interstellar gas, emitting light. The nebula lies along the edge of a large bubble of low-density gas that was blown into space by the dying star prior to its self-detonation.

At the time, the expanding cloud was likely as bright as a crescent moon, remaining visible for weeks to people living at the dawn of recorded history. Today, the resulting supernova remnant has faded and requires at least a small telescope to view it. The remaining Veil Nebula is physically huge, however, and covers over five times the size of the full moon when viewed from Earth. Its extent in space is believed to stretch for a radius of 38.5 light years! It's so wide that several parts of the Veil Nebula have been classified on their own, including the Witch's Broom Nebula and the Filamentary Nebula.

STELLAR TYPE
Supernova remnant

CONSTELLATION
Cygnus (the Swan)

MASS
n/a

TEMPERATURE
n/a

RADIUS
38.5 light years

DETECTION TYPE
Optical telescope

DISCOVERED
William Herschel, 1784

DISTANCE FROM SUN
~1470 light years

STELLAR TYPE
Red hypergiant

CONSTELLATION
Canis Major (the Greater Dog)

MASS
17 solar masses

TEMPERATURE
3490 K

RADIUS
1420 solar radii

DETECTION TYPE
Telescope

DISCOVERED
1801

DISTANCE FROM SUN
3820 light years

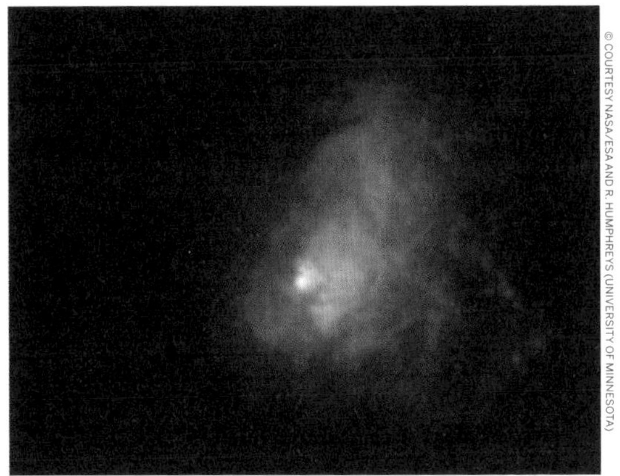

VY Canis Majoris as seen in the visible light spectrum.

VY Canis Majoris

Among the largest stars in the known Universe, VY Canis Majoris allows astronomers to study a class of stars with a shorter lifespan than our Sun.

A massive red hypergiant, VY Canis Majoris is among the largest stars ever discovered. Similar in size to UY Scuti, this huge star is nearly 1500 times larger than our Sun and is one of the most luminous stars in the Milky Way, at 100,000 times the brightness of the Sun. During its life, it emits gas and light and has formed a reflection nebula that obscures it from view, leading VY Canis Majoris to be classified as a variable star. It generates outflows of material flowing into space, analogous to solar flares but on a much, much more powerful and unpredictable scale.

Being such a large star, VY Canis Majoris emits a lot of energy, and will live a shorter stellar life than stars like our Sun, which can live billions of years. Once it explodes into a supernova, the gas and matter in its surrounding nebula may help the system form new stars and possibly planets. Don't hold your breath: VY Canis Majoris will shine for a few million years yet before any of this happens. When it dies, the resulting explosion will release more energy than 100 supernovae and emit enormous quantities of gamma rays.

These gamma rays can actually pose a threat to nearby stars and planets. They are strong enough to destroy any life that may reside there if the planets are unfortunate enough to be in range. Fortunately for us, if this is in fact VY Canis Majoris' fate, it is so far away that Earth will not be affected.

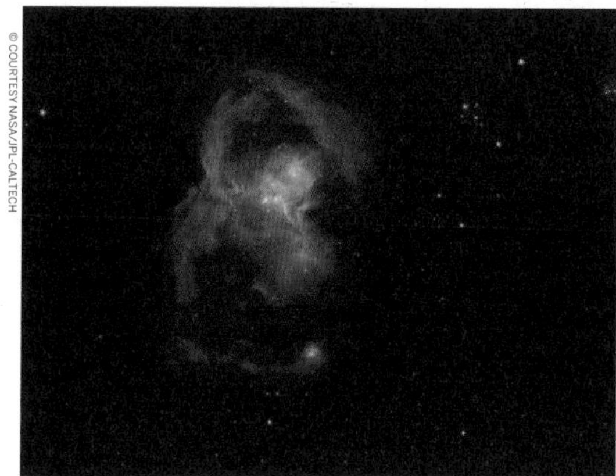

The Spitzer Space Telescope made this image of the W40 nebula.

W40

This nursery for hundreds of baby stars looks somewhat like a red butterfly in space.

Officially named Westerhout 40 (hence W40), this nebula is a giant cloud of gas and dust in space where new stars may form. The butterfly's two 'wings' are giant bubbles of hot, interstellar gas blowing from the hottest, most massive stars in this region. The nebula's cloudy filaments have dense cores of gas, many of which will gravitationally collapse to form stars. Besides being beautiful, W40 exemplifies how the formation of stars results in the destruction of the very clouds that helped create them. Inside giant clouds of gas and dust in space, the force of gravity pulls material together into dense clumps which may reach a critical density that allows stars to form at their cores. Radiation and winds coming from the most massive stars in those clouds - combined with the material spewed into space when those stars eventually explode - sometimes form bubbles in the nebula. These processes also disperse gas and dust, breaking up dense clumps and reducing or halting new star formation.

The material that forms W40's wings was ejected from a dense cluster of stars that lies between the wings. The hottest, most massive of these stars, W40 IRS 1a, lies near the centre of the star cluster. W40 is about 1400 light years from the Sun, about the same distance as the well-known Orion nebula, although the two are almost 180 degrees apart in the sky. They are two of the nearest regions in which massive stars - with masses upwards of 10 times that of the Sun - have been observed to be forming. The cluster's heart is young in astronomical terms at less than a few million years old.

STELLAR TYPE
Diffuse nebula

CONSTELLATION
Serpens Cauda (the Serpent's Tail)

MASS
10^4 solar masses

TEMPERATURE
~250,000 K

SIZE
8 arcminutes

DETECTION TYPE
Herschel Space Observatory

DISCOVERED
2009

DISTANCE FROM SUN
1400 light years

GALAXIES

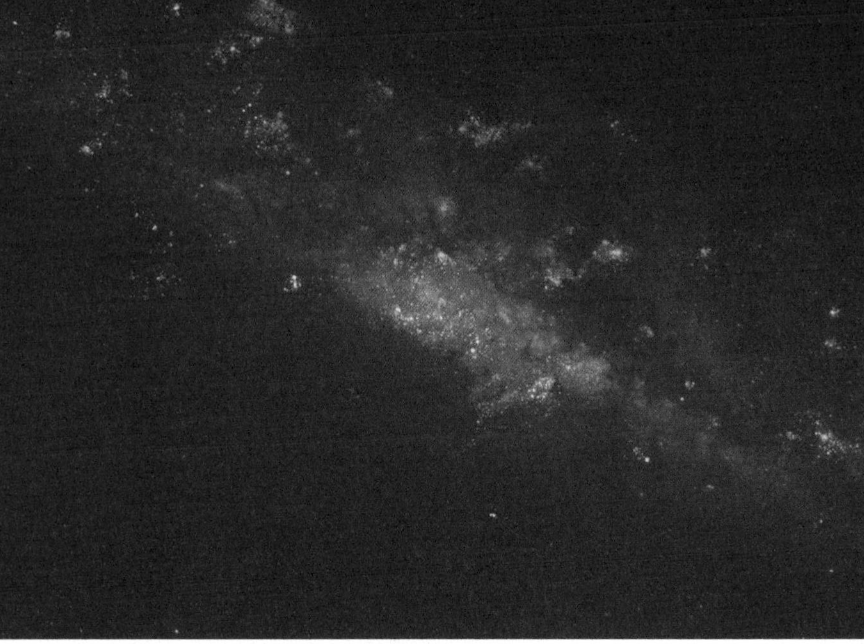

This artist's illustration portrays the gravitationally lensed galaxy cluster SDSS J1110+6459.

Introduction to Galaxies

Beyond solar systems and star clusters, our Universe is arranged into an immense number of galaxies of different ages and masses.

Our galaxy, the Milky Way, has hundreds of billions of stars, enough gas and dust to make billions more stars, and at least ten times as much dark matter as all the stars and gas put together, all held together by gravity. Containing our own solar system, our closest stellar neighbours and a mass of 1.9 trillion suns, our galactic home is only one of two trillion galaxies estimated to exist in the observed Universe.

Like more than two-thirds of the known galaxies, the Milky Way has a spiral shape. At the centre of the spiral, a lot of energy and, occasionally, vivid flares are being generated. Based on the immense gravity that must explain the movement of stars, plus the energy expelled, astronomers conclude that at the centre of nearly all galaxies is a supermassive black hole. Other galaxies have elliptical shapes, and relatively fewer have unusual shapes like toothpicks or rings. When the Hubble Deep Field (HDF) observed a tiny patch of sky for ten days, it found approximately 3000 galaxies of all sizes, shapes, and colours. Not only that: galaxies also form their own massive clusters and even superclusters, groupings of 50 to 50,000 galaxies linked by gravity.

Top Highlights

Andromeda Galaxy

1 The nearest large spiral galaxy to the Milky Way holds a trove of insights for how galaxies evolve.

Black Eye Galaxy

2 This beautiful, dark galaxy filled with obscuring dust has an outer region that rotates in the opposite direction of its core.

Supernova 1993J

3 Bode's Galaxy hosts the remnant of this supernova with a binary companion.

Canis Major Dwarf

4 The closest known galaxy to our own, this dwarf galaxy is part of the Monoceros Ring.

Cigar Galaxy

5 So many stars are forming in this spiral galaxy that it is considered 'starbursting'.

Condor Galaxy

6 At over 522,000 light years across, this galaxy is five times the Milky Way's size.

Hoag's Object

7 This ring galaxy is a rare one, and how it formed its populations of old, red stars at centre and young blue stars in the outer ring isn't yet known.

Magellanic Clouds

8 Not clouds, but galaxies, the Large and Small Magellanic Clouds are the Milky Way's largest satellites.

M87

9 Home to a supermassive black hole that is the first to ever be imaged, M87's vast centre is full of secrets scientists are seeking to unlock.

Pinwheel Galaxy

10 This beautiful pinwheel-shaped galaxy abounds in supernova.

Antennae Galaxies

11 These colliding, once-spiral galaxies are locked in a deadly embrace.

Virgo Cluster

12 If the solar system's scale seems vast, and the Milky Way and Local Group awe, the Virgo Cluster is another level entirely, with almost 50,000 galaxies in its grasp.

This composite NASA image shows the Pinwheel Galaxy, or M101, a spiral galaxy 70% larger than the Milky Way.

OBJECT TYPE
Barred spiral galaxy

WIDTH
250,000 light years

CONSTELLATION
Andromeda

APPARENT MAGNITUDE
3.4

DISTANCE FROM THE SOLAR SYSTEM
2.5 million light years

DISCOVERED
First known description is in 964 by Persian astronomer Abd Al-Rahman Al-Sufi

Andromeda bears the telltale signs of a spiral galaxy.

Andromeda Galaxy

The Andromeda Galaxy, or M31, is our Milky Way's nearest large neighbour in space.

Comprising as many as one trillion stars, the majestic spiral is comparable in size to our home galaxy, and, at just 2.5 million light years away, Andromeda is so close that it can be seen with the unaided eye on a dark night. High in the autumn sky, this galaxy-next-door appears as a cigar-shaped smudge of light in the northern constellation Andromeda (from where it takes its name).

Also referred to by astronomers as M31, Andromeda has been a fixture in the night sky for millennia; in fact, it is impossible to say who discovered it. However, Persian astronomer Abd al-Rahman al-Sufi's *The Book of Fixed Stars* from the year 964 contains the first known report of the galaxy, which for centuries was thought to reside within the Milky Way.

Eventually, that erroneous idea will become true: the Milky Way and Andromeda Galaxies are currently on a collision course, and will merge in about 4.5 billion years and form a large elliptical galaxy.

Top Tip

If you desperately want to visit Andromeda and are running low on gas, time is on your side: In several billion years, our spiral neighbour will literally be on the Milky Way's doorstep, looming large in our skies as it heads for a direct collision with our home galaxy. If you need a quicker fix, Andromeda figures prominently in well-known fictional accounts of space travel and alien worlds, including *Star Trek*, *Doctor Who*, *Superman*, and the Marvel universe.

Getting There & Away

Andromeda might be the nearest large galaxy to our own, but it's still 2.5 million light years away, meaning that flying there in a Boeing 747 (if that were even possible) would take more time than the galaxy has even existed for.

Andromeda Highlights

Impending Collision

1 The Andromeda Galaxy is currently hurtling through space at 250,000 miles per hour – and it's headed right for the Milky Way. Since the early 1900s, scientists have known that our two galaxies are destined to collide, but various teams are still working out the details about how and when that crash will play out. For now, it appears as though Andromeda will strike a glancing blow to the Milky Way in about 4.5 billion years. At that point, the two galaxies will be about 420,000 light years apart – too far away for their sparkling discs to interact but close enough for their dark matter haloes to snag. Then, the galaxies will U-turn, smash into one another, and keep on going. After that, they'll continue whipping around and passing through one another for another billion or so years, until they've merged to form a new, giant elliptical galaxy.

That might sound catastrophic for stars and planets, but space is big and stars are far apart, and there will be very few actual stellar collisions. If Earth is still around at that point, our skies will be filled with a maelstrom of colour as hot gases are mashed together and compressed, igniting bursts of star formation.

Seeing the Galaxy

2 With an apparent magnitude of 3.4, Andromeda is bright enough to see with the unaided eye on a moonless night, even in a moderately light-polluted sky. It's most easily visible in autumn, when it rises high in the northern sky. If you could see the entire galaxy – instead of just a vague smudge indicating its position – it would look like a spindle that's six times longer than the full moon is wide. Binoculars or a small backyard telescope will enhance your view, and the galaxy will likely fill the telescope's eyepiece.

How Many Black Holes?

3 Astronomers have detected as many as 35 known black holes in Andromeda – the largest number of possible black holes found in any galaxy outside of our own. In 2013, astronomers using the Chandra X-ray Observatory announced they'd spotted 26 of these black holes in an observing campaign spanning 13 years of Chandra observations. Unlike the supermassive bruisers churning away at galactic centres, these black holes are stellar-mass, meaning they are about 5 to 10 times as massive as the sun, and were created when a massive star collapsed and died. The real number of black holes is likely to be many more than the estimated 100 million in the Milky Way.

The PHAT Survey: Galactic Cartography

4 Over three years, scientists used 7398 Hubble Space Telescope images to painstakingly map roughly one-third of the stars in Andromeda's disc. Called the Panchromatic Hubble Andromeda Treasury (or PHAT), the result is the sharpest large composite image ever taken of our galactic next-door neighbour. Impressively, Hubble was able to resolve individual stars, which NASA likens to photographing a beach and resolving individual grains of sand.

A panorama, the image stretches over 61,000 light years of Andromeda's pancake-shaped disc and contains more than 100 million stars, some of them in thousands of embedded star clusters. It captures densely packed stars in the galaxy's central bulge, then moves outwards, sweeping across lanes of stars and dust to the sparser outer disc. There, large groups of young blue stars indicate the locations of star clusters and star-forming regions, with dark silhouettes tracing complex dust structures.

A 'Nebula'

5 Until the early 1900s, astronomers thought our galaxy, the Milky Way, spanned the entire universe – and that smudges on the sky, such as Andromeda, were nebulae within our galaxy. But in 1912, Vesto Slipher measured Andromeda's motion in the sky and discovered that it was hurtling towards us at immense speed – so

The bands making up the Andromeda Galaxy's striking blue-white rings are neighbourhoods harbouring hot, young, massive stars.

fast, in fact, that it couldn't be contained within the Milky Way. In 1917, Heber Curtis observed a nova in Andromeda and, based on the explosion's brightness, calculated that the 'nebula' must be about 500,000 light years away. The final nail in the nebula's coffin came in 1925, when Edwin Hubble used variable stars called Cepheids to definitively measure Andromeda's distance, which revealed the fuzzy cluster of stars to be a galaxy in its own right, and tremendously far away.

Immense Gassy Halo

6 Andromeda is big, but it's surrounded by an even bigger, gassy halo that stretches about a million light years from its host galaxy, roughly halfway to our own Milky Way. If you could see the huge bubble of hot, diffuse plasma, it would appear 100 times larger than the full moon. But it's nearly invisible. Astronomers using Hubble identified the gigantic gas blob by studying how it filtered the light of distant bright background objects called quasars – kind of like seeing the glow of a flashlight shining through fog.

The team estimates that the gargantuan halo contains half the mass of the stars in the galaxy, and suspect that it formed at the same time as Andromeda, 10 billion years ago. But it also bears the chemical fingerprints of exploding stars. Called supernova, such explosions in Andromeda's disc have violently blown heavy elements into space and enriched the halo.

M64 is easily identified by the band of absorbing dust obscuring its bright nucleus.

OBJECT TYPE
Spiral galaxy

WIDTH
54,000 – 70,000 light years

CONSTELLATION
Coma Berenices

APPARENT MAGNITUDE
9.8

DISTANCE FROM THE SOLAR SYSTEM
~17 million light years

DISCOVERED
Edward Pigott, 1779

Black Eye Galaxy

Easily identified by the spectacular band of dark, absorbing dust that partially obscures its bright nucleus, the Black Eye Galaxy is home to roughly 100 billion stars. It's also known as M64, NGC 4826, the Evil Eye Galaxy or the Sleeping Beauty Galaxy.

The roiling dust lanes obscuring this heavily lidded beauty hide an appropriately sinister secret: evidence of a relatively recent collision with a small galaxy, which the larger galaxy then devoured. As a result of that collision, the Black Eye Galaxy's gases have been bizarrely stirred up. Gas in the galaxy's outer regions, which extends to 40,000 light years from its core, rotates in the opposite direction from the gas and stars in its inner region, forming an unusual concentric counter-rotating system. At the boundary between those oppositely rotating streams, new stars are bursting to life as compressed, colliding gases ignite.

First discovered by English astronomer Edward Pigott in 1779, the Black Eye Galaxy is located in the northern constellation of Coma Berenices. The name is Latin for 'Berenice's Hair', after Queen Berenice II of Egypt, who ruled during the Ptolemaic dynasty. According to legend, a lock of hair that Queen Berenice II made as an offering to the goddess Aphrodite disappeared from the temple and reappeared in the heavens. For those keen on galaxy-gazing, it's best seen during the month of May and can be spotted with a moderately sized telescope. At the galaxy's core are hot, blue stars that have just formed, along with pink clouds of glowing hydrogen gas that fluoresce when exposed to ultraviolet light from newly formed stars.

OBJECT TYPE
Spiral galaxy

WIDTH
90,000 light years

CONSTELLATION
Ursa Major

APPARENT MAGNITUDE
6.94

DISTANCE FROM THE SOLAR SYSTEM
11.8 million light years

DISCOVERED
Johann Elert Bode, 1774

Bode's Galaxy has a core of older yellow stars and spiral arms of younger bluish stars.

Bode's Galaxy

Bode's Galaxy, also called M81, is a spiral galaxy about 12 million light years away that is both relatively large in the sky and quite bright, making it a frequent target for both amateur and professional astronomers.

Discovered by the German astronomer Johann Elert Bode in 1774, M81 is one of the brightest galaxies in the night sky, with an apparent magnitude of 6.9. Along with several sinuous dust lanes, M81's spiral arms wind all the way into its nucleus. They are made up of young, bluish, hot stars that formed in the past few million years. Those spiralling arms also host a population of stars that were born during a burst of star formation starting about 600 million years ago.

The galaxy's central bulge contains much older, redder stars than the spiral arms and is significantly larger than the Milky Way's bulge. Previous research by Hubble showed that the size of the black hole in a galaxy's nucleus is proportional to the mass of the galaxy's bulge. Bode's Galaxy is a good example of this: a black hole with 70 million solar masses resides at the centre of M81, about 15 times the mass of the Milky Way's central black hole.

M81 is in the constellation Ursa Major and is best observed during April. Through a pair of binoculars, the galaxy appears as a faint patch of light in the same field of view as M82 (the Cigar Galaxy), and a small telescope will resolve its core.

Bode's Galaxy Highlights

Supernova 1993J

1 In 1993, astronomers detected a supernova in M81 – a stellar explosion marking the death of a massive star. After monitoring the light from the supernova, scientists classified it as a rare, unusual Type IIb explosion and called it supernova 1993J. It was the nearest known example of this class of explosion, which is thought to arise from a dying star that has a companion.

For two decades, astronomers closely monitored supernova 1993J's fading light, looking for a surviving companion star that might be hiding in the explosion's residual glow. In 2014, they found that star. It is ultra-hot, radiates ultraviolet light, and confirms that this peculiar type of supernova arises when one star, locked in a deadly dance with a stellar companion, ignites and hurls its guts into space.

Blue Blobs

2 A wispy bridge of gas is strung between M81 and the two galaxies it's colliding with, called NGC 3077 and M82 – and mysterious 'blue blobs' are sprinkled throughout that bridge. About a decade ago, the Hubble Space Telescope stared at the blobs and determined that they are brilliant blue clusters of stars weighing tens of thousands of solar masses. Such clusters are rarely seen in intergalactic locations where the ingredients needed to make new stars are so sparsely distributed.

When scientists measured an age for the clusters, they found that most of their stars are less than 200 million years old – with some being even younger than 10 million years. Not coincidentally, M81 is suspected to have collided with those two galaxies about 200 million years ago, pulling streamers of gas from its neighbours like galactic taffy. Within those streamers, the conditions are hot and turbulent enough to birth young stars.

Pulsing Stars

3 For a long time, M81's distance was uncertain, with estimates placing it between 4.5 and 18 million light years away. But in the early 1990s, astronomers made a targeted search within the galaxy's disc for pulsating stars called Cepheids. Becoming alternately brighter and fainter with periods ranging from 10 to 50 days, Cepheids pulse in a way that allows astronomers to determine their intrinsic brightness, which in turns enables an accurate distance measurement. Now, with more than 30 Cepheids detected in M81, astronomers can precisely pinpoint the galaxy's distance as 11.8 million light years.

Top Tip

A visit to M81 could offer three galaxies for the price of one: The glittering spiral is interacting with its two nearest neighbours and is the largest of 34 galaxies in the M81 group, a cluster of galaxies that is just next door to our own Local Group. M81 is gravitationally perturbing another galaxy called NGC 3077 as well as M82 (the Cigar Galaxy), where the galactic interactions are triggering an intense burst of star formation. Eventually the three will merge.

Getting There & Away

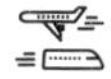

Though M81 is considered relatively close by, a distance of 11.8 million light years, that still means that the light which reaches us from the galaxy has been travelling 60 times longer than Homo sapiens have been known to exist.

OBJECT TYPE
Dwarf Irregular galaxy

WIDTH
Unknown

CONSTELLATION
Canis Major

DISTANCE FROM THE SOLAR SYSTEM
25,000 light years from the Sun, and 42,000 light years from the galactic centre

DISCOVERED
2002

Canis Major Dwarf

The Canis Major Dwarf Galaxy is the nearest galaxy to our own – and it hid in plain sight until 2002. For years, the Large and Small Magellanic Clouds were the closest known galaxies; then, in 1994, astronomers spotted the Sagittarius Dwarf Galaxy. Eight years later, while studying the galactic plane in dust-piercing infrared light, scientists accidentally uncovered the Canis Major Dwarf, at just one-quarter of the distance to the Large Magellanic Cloud.

The Canis Major Dwarf, which contains a paltry one billion stars, is roughly elliptical in shape and is mostly obscured by dust in the Milky Way's plane. It's full of old, red giant stars and has a core that is completely degraded, making it more of a galactic corpse than anything else.

Perhaps surprisingly, the Canis Major Dwarf is in the process of colliding with the Milky Way, whose gravity will completely shred the dwarf and leave nothing but a glimmering stream of stars. Already, it is leaking starry debris: a long filament of stars, gas, and dust spanning 200,000 light years is evidence of the ongoing collision. Called the Monoceros Ring, it actually wraps itself around the Milky Way – three times. (In fact, the Monoceros Ring pointed to the way to this galaxy.) Estimates suggest that the Canis Major Dwarf will merge with the Milky Way in the next billion or so years.

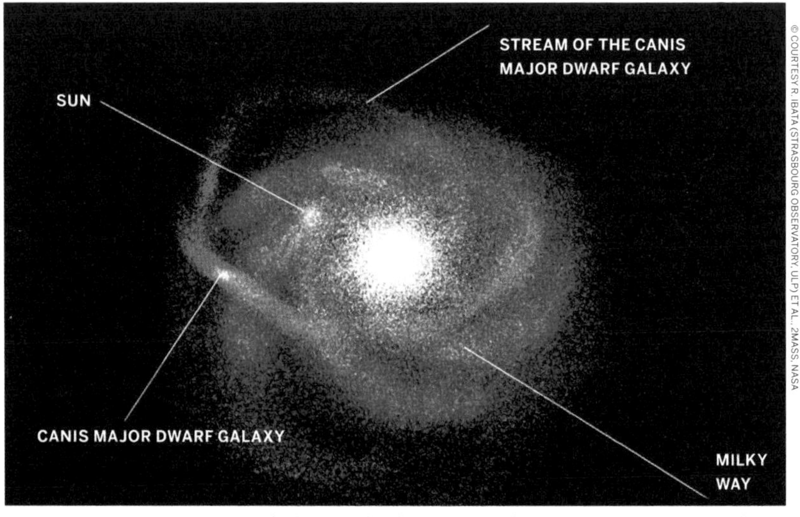

An illustration of the Canis Major Dwarf Galaxy and stream in relation to the Milky Way.

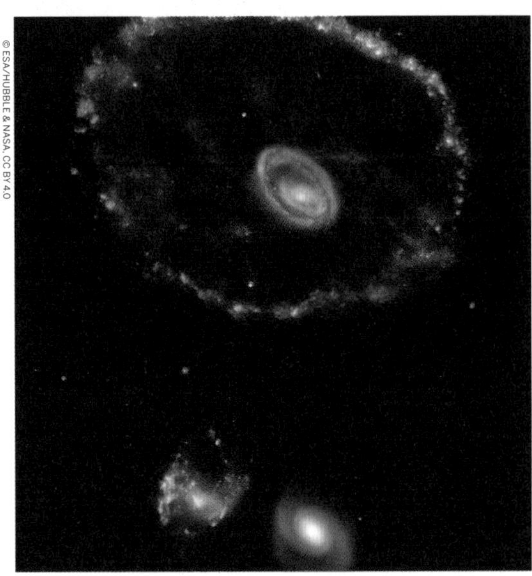

The cartwheel shape seen here is the result of a violent galactic collision.

OBJECT TYPE
Ring galaxy

WIDTH
150,000 light years in diameter

CONSTELLATION
Sculptor

APPARENT MAGNITUDE
15.2

DISTANCE FROM THE SOLAR SYSTEM
~500 million light years

DISCOVERED
Fritz Zwicky, 1941

Cartwheel Galaxy

Located almost 500 million light years away in the constellation Sculptor, the peculiar Cartwheel Galaxy looks like an exploding wagon wheel. The galaxy's nucleus is the bright, swirly object at its centre; a brilliant outer ring of young stars encircles the core, connected by wispy, spoke-like structures of material.

Scientists suspect a nearly head-on collision with a smaller galaxy crafted the Cartwheel's unusual configuration. About 200 million years ago, that galaxy plunged through the Cartwheel and triggered ripples of sudden star formation that expanded outward, much like waves spreading after a stone is dropped into a pond. Now that wave front of concentric rings is the bright ring of massive stars circling the galaxy's core. Before the collision, the Cartwheel Galaxy was likely a spiral galaxy similar in shape to the Milky Way. The energy of that impact is part of what has driven the intense star creation here; the galaxy is thought to have at least several billion young stars, with a total estimated mass of 3 billion solar masses.

Today, the Cartwheel is one of the brightest objects in the ultraviolet sky. It also has a dozen bright X-ray sources, which are usually associated with gas falling onto a black hole from a companion star. That makes sense, because black holes thrive in areas where massive stars are forming quickly and dying fast – like they are seen to do in the wake of a galactic smash-up. Together with the other galaxies in the above image, it belongs to a group of galaxies that are at least 400 million light years distant from our solar system.

OBJECT TYPE
Lenticular/Elliptical (active) galaxy

WIDTH
>60,000 light years

CONSTELLATION
Centaurus

APPARENT MAGNITUDE
6.8

DISTANCE FROM THE SOLAR SYSTEM
~11 million light years

DISCOVERED
James Dunlop, 1826

This image from Hubble shows light in the visible and ultraviolet light spectrums.

Centaurus A

Only 11 to 12 million light years away, odd-looking Centaurus A is one of the closest active galaxies to planet Earth, and it emits more radio waves than just about anything else in the sky. Spanning more than 60,000 light years, the peculiar elliptical galaxy is also known as NGC 5128, and it resembles a bright blob of stars wrapped in a dark, dusty ribbon.

Centaurus A probably formed from a collision between two otherwise normal galaxies, resulting in a fantastic jumble of star clusters and imposing dust lanes. Near the galaxy's centre, leftover cosmic debris is steadily being consumed by a central black hole containing more than a billion times the mass of the Sun. That process generates intense radio, X-ray, and gamma-ray energy; it also fires jets of extremely energetic particles into space that are more than a million light years long. Astronomers estimate that matter near the base of these jets races outward at about one-third the speed of light. The jets strongly interact with surrounding gas, and may change a galaxy's rate of star formation. These jets made Centaurus A one of the first celestial radio sources identified with a galaxy. Seen in radio waves, Centaurus A is one of the biggest and brightest objects in the sky, nearly 20 times the apparent size of the full moon.

Centaurus A is the fifth-brightest galaxy in the sky, making it a popular observing and imaging target for amateur astronomers. With binoculars or a small telescope, the galaxy's central bulge and warped dust lanes can be seen, most easily by observers in the southern hemisphere and low northern latitudes, who can peer up at this nearby active galaxy.

The long, narrow shape of this galaxy in images gave rise to its unique name.

OBJECT TYPE
Spiral (edge-on) galaxy

WIDTH
37,000 light years

CONSTELLATION
Ursa Major

APPARENT MAGNITUDE
8.4

DISTANCE FROM THE SOLAR SYSTEM
12 million light years

DISCOVERED
Johann Elert Bode, 1774

Cigar Galaxy

This spectacular starbursting galaxy earned its nickname because its spiral shape – which we see edge-on from Earth – looks like a burning cigar. The Cigar Galaxy, or M82, is gravitationally interacting with its galactic neighbour, M81, causing it to form new stars at a notably high rate.

Around this galaxy's centre, young stars are being born 10 times faster than they are inside the entire Milky Way, generating outflows of gas and dust that appear as red puffs above and below the galaxy. Within M82, radiation and energetic particles from newborn stars compress enough gas to make millions more stars. But this rapid rate of star formation is self-limiting: When star formation becomes too vigorous, it consumes or destroys the material needed to make more stars, eventually leaving no fresh ingredients. The starburst then subsides, usually within a few tens of millions of years.

The galaxy's strong winds that blow gas and dust into intergalactic space were studied in greater detail by the Stratospheric Observatory for Infrared Astronomy (SOFIA). Researchers found that the galactic wind flowing from the centre of the Cigar Galaxy is aligned along a magnetic field and transports a very large mass of gas and dust – the equivalent mass of 50 million to 60 million Suns. Observations indicate the powerful winds associated with the starburst phenomenon could be one of the mechanisms responsible for seeding material and injecting a magnetic field into the nearby intergalactic medium. If similar processes took place in the early universe, they would have affected the evolution of the first galaxies.

M82, which is in Ursa Major, has an apparent magnitude of 8.4. Best observed in April, it is visible as a patch of light with binoculars, and larger telescopes can resolve the galaxy's core.

OBJECT TYPE
Spiral Seyfert

WIDTH
1400 light years

CONSTELLATION
Circinus

APPARENT MAGNITUDE
12.1

DISTANCE FROM THE SOLAR SYSTEM
13-14 million light years

DISCOVERED
1975

Brightly glowing filaments can be observed in the Circinus Galaxy.

Circinus Galaxy

Resembling a swirling witch's cauldron of glowing vapours, the Circinus Galaxy is one of the nearest large galaxies to our own – yet it is largely unexplored, as it hides behind the gas and dust in the Milky Way's plane.

At just 13 to 14 million light years away in the Circinus constellation (the Latin name means Compass), Circinus is among the closest active galaxies – meaning that its core powers a large fraction of its brightness, rather than the gleaming light of stars in its disc. As their name suggests, active galaxies contain what astronomers call an 'active galactic nucleus'. At Circinus, stars are rapidly being born in a region encircling the galaxy's core, producing dust and filaments that are also warmed and glows in the infrared. Hence this galaxy's additional classification as a Seyfert galaxy: a type of active galaxy with strong characteristic emission lines.

This core is a supermassive black hole containing millions or billions of times the mass of the Sun that is slowly swallowing a surrounding disc of gas and dust. As that matter falls into the gargantuan black hole at the galaxy's heart, it becomes hot enough to produce intense X-rays and ultraviolet light. Those emissions shine brightly on their own, but they also heat dust further out in the disc, which then begins to glow in the infrared. There is also a smaller black hole closer to the galaxy's edge that belongs to a class called ultraluminous X-ray sources, or ULXs. ULXs consist of black holes actively accreting, or feeding off, material drawn in from a partner star. The ULX was spotted by NASA's Nuclear Spectroscopic Telescopic Array (NuSTAR). Further observations with other telescopes revealed that the galaxy's outerlying smaller black hole is about 100 times the mass of our Sun.

A processed image of the Condor Galaxy based on ESO data.

OBJECT TYPE
Barred spiral galaxy

WIDTH
522,000 – 700,000 light years

CONSTELLATION
Pavo

APPARENT MAGNITUDE
12.7

DISTANCE FROM THE SOLAR SYSTEM
212 million light years

DISCOVERED
John Herschel, June 1835

Condor Galaxy

The spectacular barred spiral galaxy known as the Condor (or NGC 6872) is among the longest stellar systems – if not the longest – ever observed. The galactic giant spans more than 522,000 light years from the tip of one spiral arm to the other, making it more than five times the size of our Milky Way.

It's not normal for galaxies to be as large as the Condor, which has a central bar that's 26,000 light years across - longer than some entire galaxies! Scientists suspect the galaxy's unusual size and appearance stem from its interaction with a much smaller galaxy named IC 4970, which is about one-fifth the mass of NGC 6872. Normally large galaxies, including our own, grow over billions of years by absorbing numerous smaller systems; here, though, the opposite might be happening. Intriguingly, the gravitational interaction of NGC 6872 and IC 4970 may be spawning what could develop into a small, new galaxy near the Condor's northeastern arm.

By analysing the distribution of energy by wavelength, the team uncovered a distinct pattern of stellar age along the galaxy's two prominent spiral arms. The youngest stars appear within the tidal dwarf candidate, which glows brightly in ultraviolet light and is less than 200 million years old. Stellar ages skew progressively older towards the galaxy's centre. The southwestern arm displays the same pattern, likely connected to waves of star formation triggered by the galactic encounter. There is no sign of recent star formation along the central bar, indicating it formed at least a few billion years ago. Its aged stars provide a record of the galaxy's early stellar population.

OBJECT TYPE
Spiral galaxy

WIDTH
200,000 light years

CONSTELLATION
Eridanus

APPARENT MAGNITUDE
10.9

DISTANCE FROM THE SOLAR SYSTEM
60 million light years

DISCOVERED
William Herschel, October 1784

A face-on view of NGC 1232 allows a full perspective of its spiral.

Grand Spiral Galaxy

The Grand Spiral Galaxy, also known as NGC 1232, looks like what you'd draw if someone asked you to illustrate a spiral galaxy – which is, perhaps, why it got its name.

Seen nearly face-on from Earth, the galaxy's beautiful spiralling structure stretches across 200,000 light years and contains a central area with older, redder stars and numerous spiral arms populated by young, blue stars and star-forming regions. Next to NGC 1232 is a distorted companion galaxy that's shaped like the Greek letter theta; it's called NGC 1232A and is rather large for a companion galaxy. In 2013, observations by NASA's Chandra X-ray Observatory revealed a massive cloud of multimillion-degree gas in the Grand Spiral, which is the likely result of a collision with a smaller dwarf galaxy. These collisions result in hot gas-generating shock waves; as the galaxy moves, the hot gas creates a comet-like trail. The mass of this heated gas is estimated to be between 40,000 and a million solar masses, depending on how it is distributed throughout the galaxy. Such impacts are also fertile ground for star formation.

Located south of the celestial equator, the Grand Spiral is in the constellation Eridanus, known since the time of Ptolemy. It was depicted in old times as a winding river, and is the sixth largest modern constellation as it trickles across a long patch of sky in both hemispheres, from the feet of Orion near the equator to Tucana.

Hoag's Object resembles an ouroboros in images.

OBJECT TYPE
Ring galaxy

WIDTH
100,000 light years

CONSTELLATION
Serpens

APPARENT MAGNITUDE
16

DISTANCE FROM THE SOLAR SYSTEM
600 million light years

DISCOVERED
Art Hoag, 1950

Hoag's Object

A nearly perfect ring of hot, blue stars pinwheels around the yellow nucleus of an unusual galaxy known as Hoag's Object.

The entire structure of this distinctive ring galaxy is about 100,000 light years wide, which is slightly larger than our Milky Way Galaxy. The blue ring, dominated by clusters of young, massive stars, contrasts sharply with the bright, yellow nucleus of mostly older stars. What appears to be a 'gap' separating the two stellar populations may actually contain some star clusters that are almost too faint to see.

Though other objects bearing a resemblance to Hoag's have been spotted, they're rare. Astronomers aren't yet entirely sure how these bizarre configurations arise. One possibility invokes an ancient galactic collision; another points to the gravitational effects of a long-vanished central bar, like the structures observed at the centres of some spiral galaxies.

Curiously, an object that bears an uncanny resemblance to Hoag's Object can be seen in the gap at the one o'clock position. The object is probably a background ring galaxy. As for Hoag's Object itself, it lies about 600 million light years away towards the constellation of Serpens (the Snake). And as for the name? It's called Hoag's Object for the straight-forward reason that it was the astronomer Art Hoag who first spied this intriguing galaxy in 1950.

OBJECT TYPE
Irregular dwarf galaxy

WIDTH
14,000 light years (LMC), 7000 light years (SMC)

CONSTELLATION
Dorado and Mensa (LMC), Tucana and Hydrus (SMC)

APPARENT MAGNITUDE
0 (LMC), 2 (SMC)

DISTANCE FROM THE SOLAR SYSTEM
~160,000 light years (LMC), ~200,000 light years (SMC)

DISCOVERED
Known since antiquity

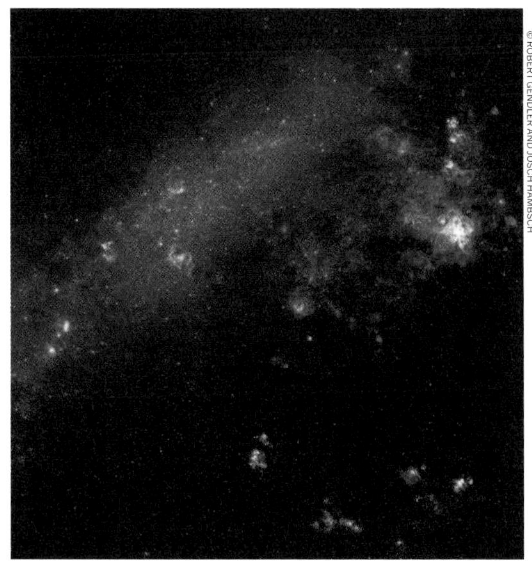

The moniker 'clouds' comes from the cloud-like high-density gas nebulae shown here.

Large and Small Magellanic 'Clouds'

The Magellanic Clouds are the Milky Way's two biggest satellite galaxies. Situated approximately 75,000 light years from one another, the pair is easily visible in the southern hemisphere.

Despite their relatively modest sizes, the galaxies loom large in the sky because they are so close to us. Over 160,000 light years away, the Large Magellanic Cloud (LMC) is one of the most prominent features of the nighttime southern sky. Suspended near the constellations Dorado and Mensa, it looks like a discarded piece of Milky Way fluff. But even though the LMC weighs less than one-tenth the mass of the Milky Way, it is a galaxy in its own right, containing roughly 10 billion solar masses of stars. The LMC is ablaze with star-forming regions. From the Tarantula Nebula, one of the brightest stellar nurseries in our cosmic neighbourhood, to LHA 120-N 11, the small and irregular galaxy is scattered with glowing nebulae, the most noticeable sign of new stars being born here.

The Small Magellanic Cloud (SMC), as its name suggests, is smaller than its dwarf neighbour. It's about half the width of the LMC, and contains about two-thirds of its mass, roughly 7 billion solar masses of stars. Like the LMC, it has a central stellar bar of stars, though it does not have a spiral arm, as the LMC does.

At nearly 200,000 light years away from the Milky Way, the Small Magellanic Cloud – which straddles the constellations Tucana and Hydrus – is one of the most distant objects humans can see with the unaided eye. The SMC, also an irregular galaxy, may be a distorted barred disc, deformed by tidal forces of the Milky Way.

Highlights

Cepheid Variables

1 More than a century ago, pulsing stars in the Small Magellanic Cloud helped astronomers figure out the distances to cosmic objects. In 1912, Henrietta Swan Leavitt noted that 25 stars in the SMC, now called Cepheid variables, would brighten and dim. She measured the pulsating period of each star and determined that the brighter the Cepheid, the longer its period. Now, we know that Cepheids pulse in a way that is uniform with their intrinsic brightness; by timing the periods of faraway Cepheids out to 13 million light years away, astronomers know how bright the stars should be, and therefore how far away they actually are.

Tarantula Nebula

2 The LMC is a hotbed of vigorous star formation. Rich in interstellar gas and dust, the galaxy is home to approximately 60 globular clusters and 700 open star clusters. One of the most notable star-forming regions is the exquisite Tarantula Nebula, where vast clouds of gas are collapsing to form new stars across a width of almost 1000 light years. There, an international team of astronomers recently identified nine monster stars with masses exceeding 100 times that of the Sun, comprising the largest sample of very massive stars identified to date.

The Tarantula Nebula in the LMC measures almost 1000 light years across.

Top Tip

The Magellanic Clouds were the closest known galaxies to our own until 1994, when scientists discovered the Sagittarius Dwarf Elliptical Galaxy (which has since been eclipsed in proximity by the Canis Major Dwarf Galaxy). But unlike those closer stellar conglomerates, the Magellanic Clouds are easy to spot in the southern sky. In fact, they take their name from Ferdinand Magellan, who reportedly used them as navigational aides while he sailed around the globe; but their presence has been noted for much longer by indigenous Australians, the New Zealand Maori, and South Sea Islanders.

Getting There & Away

Though the Magellanic Clouds are our near neighbours, are easily visible from the southern hemisphere, and might be the nearest large galaxy to our own, 160,000 light years (200,000 to the SMC) makes for a pretty mean commute. Sci-fi dreamers will keep hoping for a breakthrough, however.

Supernova 1987A

Three decades ago in the Large Magellanic Cloud, astronomers spotted one of the brightest exploding stars in more than 400 years. The titanic supernova, called Supernova 1987A (SN 1987A), blazed with the power of 100 million suns for several months following its discovery on Feb. 23, 1987. Since then, SN 1987A has continued to fascinate astronomers with its spectacular light show.

SN 1987A detonated when a blue supergiant star exploded and died. For thirty years, powerful space telescopes have been monitoring the explosion's aftermath and watching as multiple, glowing rings formed around the former star. Hubble images have revealed that the dense circle of gas surrounding the supernova is roughly one light year across and is glowing in optical light; the ring was there at least 20,000 years before the star exploded, but a flash of ultraviolet light during the explosion energised the gas in the ring and made it glow for decades.

Astronomers still aren't sure whether the giant star collapsed into a black hole or a neutron star, although neutrinos emitted by the explosion make them certain that some form of dense, compact object remains.

Merger

3 Roughly 160,000 light years from Earth, the Large Magellanic Cloud is performing a long and slow dance around the Milky Way – but that performance won't last forever. Soon, the two will merge. New observations suggest the LMC is heavier than anticipated, and could therefore crash into the Milky Way in two billion years – much sooner than our impending collision with the Andromeda Galaxy – and be obliterated forever; the Small Magellanic Cloud will eventually do the same.

Intergalactic Battle

4 In a cosmic tug-of-war between the Magellanic Clouds, one galaxy's gravity is pulling out a huge amount of gas from its companion. This shredded and fragmented gas, called the Leading Arm, is between 1 and 2 billion years old and is being devoured by the Milky Way. Recently, scientists using the Hubble Space Telescope determined that the Leading Arm originates from the Small Magellanic Cloud and is a casualty of its ongoing battle with the LMC.

Stolen Wing

5 In the southeast region of the SMC is a structure known as 'The Wing' – and at its tip is a dazzling nebula resembling open jaws filled with stars. Recently, astronomers discovered that stars in the Wing are moving away from the body of the dwarf galaxy. Because those stars are all moving at similar speeds and are heading towards the Large Magellanic Cloud, the observations suggest the two galaxies collided, perhaps as recently as several hundred million years ago.

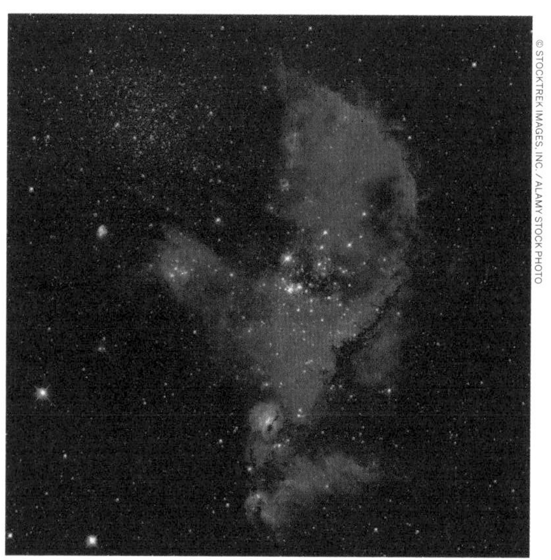

Open cluster NGC 346 in the Small Magellanic Cloud.

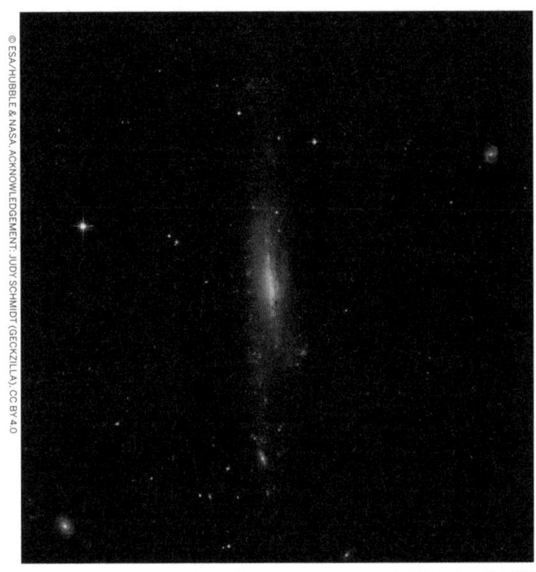

This NASA/ESA Hubble Space Telescope image captures the LSB galaxy UGC 477.

OBJECT TYPE
Spiral LSB (low surface brightness) galaxy

WIDTH
650,000-750,000 light years

CONSTELLATION
Coma Berenices

APPARENT MAGNITUDE
15.8

DISTANCE FROM THE SOLAR SYSTEM
1.19 billion light years

DISCOVERED
David Malin, 1986

Malin 1 Galaxy

Malin 1 is remarkable for what isn't there: brightness. The giant spiral, which is about seven times longer than the Milky Way, exemplifies a class of objects known as giant low surface brightness galaxies or LSBs – it's so faint it wasn't even discovered until 1986!

Because they are very diffuse and dim, giant low surface brightness galaxies are difficult to observe and are still poorly understood. In fact, though the existence of LSB galaxies was posited in 1976 by Mike Disney, it was the discovery of Malin 1 by David Malin in 1986 that confirmed their existence. These massive objects could represent a significant percentage of the galaxies in the universe – especially because scientists could have overlooked such faint dusky structures in galaxy surveys. They tend to be 250 times less bright than what we consider 'standard' galaxies. Newer-generation telescopes and modern detectors, with higher sensitivity to low surface brightnesses, should help answer questions about how these objects formed and evolved.

In 2016, scientists using the Canada-France-Hawaii telescope took a good look at Malin 1 in six different wavelengths. They found that the galaxy's giant disc is a relatively placid place where stars have been slowly forming for billions of years; intriguingly, the observation challenges a long-standing idea suggesting that behemoth galaxies attain their gargantuan sizes by violently swallowing and assimilating smaller galaxies.

OBJECT TYPE
Type-1 Seyfert galaxy

WIDTH
Unknown

CONSTELLATION
Ursa Major

APPARENT MAGNITUDE
13.6

DISTANCE FROM THE SOLAR SYSTEM
~600 million light years

DISCOVERED
1969

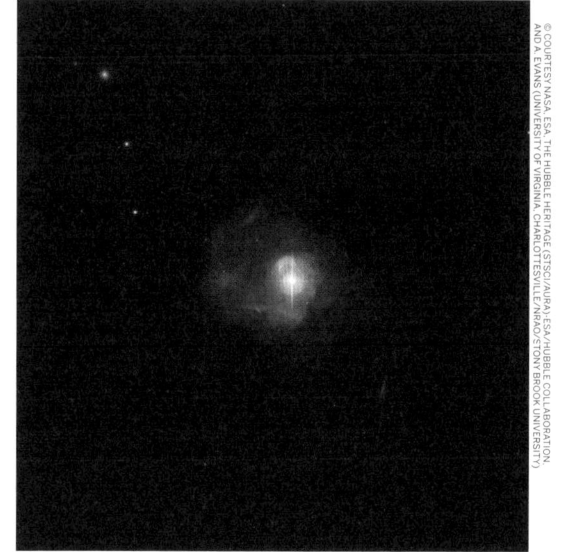

Markarian 231 has long tidal tails and a disturbed shape.

Markarian 231

Markarian 231 (Mrk 231) is the nearest galaxy that hosts a quasar, or an active, churning core that is so bright it appears star-like (quasar is a portmanteau of 'quasi-stellar object'). The galaxy itself has long tidal tails and a disturbed shape, suggesting powerful interactions with a neighbouring galaxy.

Markarian 231 is undergoing an energetic starburst, with a ring of active star formation – estimated to produce more than 100 solar masses of stars per year – circling its centre. In 2015, scientists took a close look at Markarian 231 with the Hubble Space Telescope and seemed to discover that its quasar is powered by two supermassive black holes furiously whirling about one another. The central black hole is estimated to be 150 million times the mass of our Sun, and its companion – which is the remnant core of a galaxy Markarian 231 swallowed – weighs in at 4 million solar masses. The dynamic duo would complete an orbit around one another every 1.2 years, generating tremendous amounts of energy and helping the core of the galaxy to outshine its population of billions of stars. But their dramatic dance wouldn't last forever: The two are predicted to spiral together and merge within a few hundred thousand years.

Sound thrilling? Unfortunately, later analysis published in *The Astrophysical Journal* seemed to indicate that there is only one black hole powering the quasar, rather than a binary pair. It's possible the black hole appears to be 'reflected' on the far side of the quasar.

As a Seyfert galaxy, M77's centre is intensely active and obscured by gas.

OBJECT TYPE
Barred spiral galaxy

WIDTH
100,000-170,000 light years

CONSTELLATION
Cetus

APPARENT MAGNITUDE
9.6

DISTANCE FROM THE SOLAR SYSTEM
47 million light years

DISCOVERED
Pierre Méchain, 1780

M77

Initially misidentified as a nebula by the French astronomer Pierre Méchain in 1780, M77 – or Messier 77 – is one of the largest galaxies in the Messier catalogue. The big, beautiful spiral is located in the constellation Cetus (the Whale), about 45 million light years from Earth.

Messier 77 (M77) is one of the most famous and well-studied galaxies in the sky. It has pockets of star formation bursting to light along its pinwheeling arms, and dark dust lanes stretching across its energetic centre. The galaxy is categorised as a barred spiral, and it is the closest and brightest example of a class of galaxies known as Seyfert galaxies. These are galaxies that are full of hot, highly ionised gas that glows brightly, emitting intense radiation from their quasar like nuclei. In fact, M77 is considered the prototype of the Seyfert type galaxy, as the brightest among them.

That radiation comes from the heart of Messier 77, where a black hole 15 million times as massive as our Sun resides. Material is dragged towards this black hole and circles around it, heating up and glowing strongly and sometimes shining tens of thousands of times brighter than a typical galaxy.

M77 has an apparent magnitude of 9.6 and can be seen using a small telescope. It is most easily observed during December. In November 2018, the galaxy was the site of a 15th magnitude supernova, named SN 2018ivc, the first to be recorded in the galaxy.

OBJECT TYPE
Elliptical galaxy

WIDTH
120,000+ light years

CONSTELLATION
Virgo

APPARENT MAGNITUDE
9.6

DISTANCE FROM THE SOLAR SYSTEM
54 million light years

DISCOVERED
Charles Messier, 1781

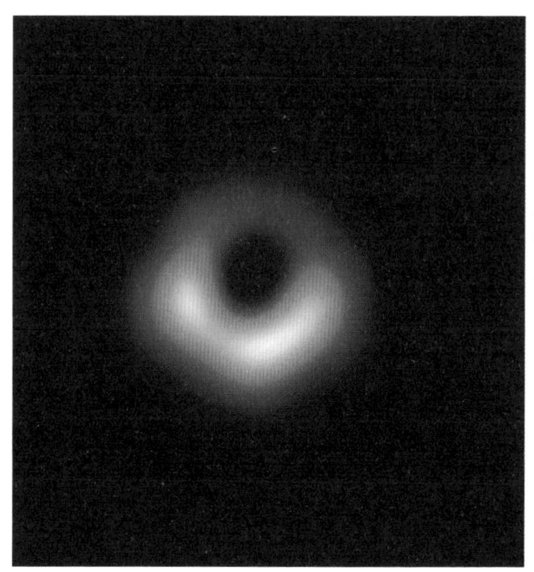

M87's central supermassive black hole, the first ever imaged.

M87

Everything about elliptical galaxy M87 is monstrous: The galaxy is home to several trillion stars and 15,000 ancient star clusters; for comparison, our Milky Way Galaxy contains only a few hundred billion stars and about 150 globular clusters. More impressively, M87 is anchored to a central black hole weighing more than a whopping 3 billion suns. That big, churning bruiser is launching a massive jet of energetic particles into space – a blow-torch spanning 4000 light years.

Spanning 120,000 light years, M87 is the dominant member of the Virgo group of galaxies. It is an elliptical, or egg-shaped galaxy that (unlike spiral galaxies, like our own) lacks the gas and dust to form new stars. Instead, it is populated by randomly swarming older stars, which lend it a distinctive reddish-yellow colour.

Discovered in 1781 by Charles Messier, this galaxy is located 54 million light years away from Earth in the constellation Virgo. It has an apparent magnitude of 9.6 and can be observed using a small telescope, most easily in May. You'd need a pretty special telescope to be able to see the black hole at M87's centre, however. Specifically, the Event Horizon Telescope (EHT), which captured the first-ever image of a black hole in 2019 by focusing on M87. The EHT combines data from a global array of radio telescopes. Having this combined data gives the project a sufficient resolution to peer at a supermassive black hole's event horizon, past which no light can escape – hence the telescope's name. It's this supermassive black hole that powers the 4000-light-years-long plasma jet which dramatically streams out of M87.

M87 Highlights

A Turbulent Core

1 A spiral-shaped disc of hot gas rotates around the core of M87, hiding the central supermassive black hole at the galaxy's heart. Though the black hole weighs at least 3 billion suns, it is concentrated into a space no larger than our solar system. Because it is so large and relatively nearby, M87's black hole is a frequent target for astronomers attempting to understand how these objects power and affect galaxies. X-rays from M87's core reveal a series of outbursts from the galaxy's black hole, seen as loops and bubbles in hot gas. Narrow filaments of X-ray-rich material – likely coming from hot gases trapped by magnetic fields – extend for more than 100,000 light years as well.

Event Horizon Telescope

2 In April 2019, a consortium of astronomers announced that they'd snapped the first actual image of a black hole – and it was the gargantuan, supermassive roiling mass of bottomless space-time parked at M87's core. The team used numerous radio telescopes and observatories, sprinkled across the entire planet, to construct an image of the black hole. Called the Event Horizon Telescope, the project is the first real attempt to photograph these otherwise invisible objects in galactic centres. Next on the group's list of targets: Sagittarius A*, the black hole spinning at the Milky Way's core.

Jet

3 Spiralling out from the centre of M87, like a cosmic searchlight, is one of nature's most amazing phenomena: a jet of superheated gas and subatomic particles travelling at nearly the speed of light, stretching more than 4000 light years into space. When Heber Curtis first observed the jet in 1918, he described it as 'a curious straight ray'. Now we know that it is powered by M87's monstrous central black hole. As gaseous material from the galactic centre falls into the black hole, the energy released produces a stream of subatomic particles, accelerated to velocities approaching light speed.

Globular Clusters

4 Astronomers suspect that galaxies grow to gargantuan sizes by eating their smaller neighbours, and M87 is no exception. It is home to, and is flanked by, an abnormally high number of ancient groups of stars. Called globular clusters, these glittering conglomerates are probably stolen from nearby dwarf galaxies.

Top Tip

M87 has the hottest science destination of recent memory in its massive black hole, the first ever imaged. This black hole is 6.5 billion times the mass of the Sun. Catching its shadow involved eight ground-based radio telescopes around the globe, operating together as if they were one telescope the size of our entire planet. Mysteries remain around black holes, though they are more tantalising in reach than ever with this finding. Why do particles get such a huge energy boost around black holes, forming dramatic jets that surge away from the poles of black holes at nearly the speed of light? When material falls into the black hole, where does the energy go?

Getting There & Away

M87 is located roughly 50 million light years away, in the constellation Virgo. To navigate there: Spot the stars Epsilon Virginis and Denebola, and aim just to the left of the centre point between the two. But steer clear of the central supermassive black hole.

OBJECT TYPE
Barred spiral galaxy

WIDTH
35,000 light years

CONSTELLATION
Horologium

APPARENT MAGNITUDE
11.1

DISTANCE FROM THE SOLAR SYSTEM
38 million light years

DISCOVERED
James Dunlop, 1826

A bright nuclear ring surrounds NGC 1512's centre, itself enclosed by a larger ring.

NGC 1512

NGC 1512 is a magnificent barred spiral galaxy located over 30 million light years away in the constellation Horologium (the Clock). Spanning 35,000 light years, the galaxy has a double ring structure and complex architecture.

Most galaxies don't have any rings, so why does this galaxy have two? The bright nuclear ring surrounds the galaxy centre with recently formed stars, but the bulk of NGC 1512's stars and their accompanying gas and dust orbit the galactic centre in a ring much further out, counter-intuitively called the inner ring. If you look closely, you will see this the inner ring connects ends of a diffuse central bar that runs horizontally across the galaxy.

These ring structures are thought to be caused by NGC 1512's own asymmetries in a drawn-out process called secular evolution. The gravity of these galaxy asymmetries, including the bar of stars, cause gas and dust to fall from the inner ring to the nuclear ring, enhancing this ring's rate of star formation. Some spiral galaxies also have a third ring: an outer ring that circles the galaxy even further out. The inner ring of NGC 1512, which stretches across 2400 light years, comprises bright blue infant stars surrounding the galaxy's core. Slicing through that ring is the galaxy's bar, acting as a pipeline of gas and dust to draw those materials inward from the galaxy's large, outer ring, which is dimmer and more diffuse than the nuclear ring.

Both the bar and the nuclear ring are thought to be at least partially sculpted by an ongoing cosmic scuffle between colossal NGC 1512 and the smaller galaxy next door, a dwarf called NGC 1510. The two galaxies are roughly 45,000 light years apart, but the pair has been slowly merging for about 400 million years. Ultimately, NGC 1512 will swallow its diminutive neighbour.

Amid a backdrop of far-off galaxies, the majestic dusty spiral NGC 3370 looms.

OBJECT TYPE
Spiral galaxy

WIDTH
100,000 light years

CONSTELLATION
Leo

APPARENT MAGNITUDE
12.3

DISTANCE FROM THE SOLAR SYSTEM
98 million light years

DISCOVERED
William Herschel, 1784

NGC 3370

A majestic dusty spiral, NGC 3370 is similar in size and design to the Milky Way. Its intricate spiral arms are spotted with hot areas of new star formation and well-delineated dust lanes – but its nucleus is oddly ill-defined.

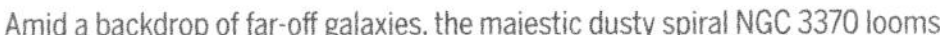

This beautiful spiral galaxy is more than just a pretty face. It's also the site of an ongoing cycle of star formation and death. In November 1994, astronomers caught the light of a star that exploded and died in NGC 3370, a supernova that briefly outshone the tens of billions of other stars in the galaxy. Although these stellar outbursts are common, with one exploding every few seconds somewhere in the universe, this one was special. Designated SN 1994ac, the explosion was one of the nearest and best observed supernova at the time. Plus, it was a Type 1A supernova, which scientists use to measure distances and chart the growth rate of the expanding universe. For two weeks, astronomers closely monitored the explosion's light; when these stars explode and die so close to Earth, it helps astronomers better understand the intricacies of their death spasms, which allows more detailed mapping of the cosmos. The light from SN 1994ae, and NGC 3370 as a whole, arrives at Earth only after travelling for 94 million light years. Sometimes referred to as the Silverado Galaxy, it is located in the Leo constellation (the Lion).

OBJECT TYPE
Spiral galaxy

WIDTH
170,00 light years

CONSTELLATION
Ursa Major

APPARENT MAGNITUDE
7.9

DISTANCE FROM THE SOLAR SYSTEM
20.87 million light years

DISCOVERED
Pierre Méchain, 1781

(GSFC), F. BRESOLIN (UNIVERSITY OF HAWAII), J. TRAUGER (JPL), J. MOULD (NOAO), AND Y.-H. CHU (UNIVERSITY OF ILLINOIS, URBANA); IMAGE PROCESSING: DAVIDE DE MARTIN (ESA/HUBBLE); CFHT IMAGE: CANADA-FRANCE-HAWAII TELESCOPE/J.-C. CUILLANDRE/COELUM; NOAO IMAGE: GEORGE JACOBY, BRUCE BOHANNAN, MARK HANNA/NOAO/AURA/NSF

This infrared and visual-light composite image is made from 51 exposures over 10 years.

Pinwheel Galaxy

One of the most striking galaxies in our sky, M101 – also called the Pinwheel – is a giant spiral disc of stars, dust and gas spanning 170,000 light years, or nearly twice the diameter of our galaxy, the Milky Way.

The Pinwheel's majestic spiral arms are sprinkled with large regions of star-forming nebulae, or areas of intense star formation within giant molecular hydrogen clouds. Brilliant, young clusters of hot, blue, newborn stars trace out the spiral arms. M101 is believed to contain at least one trillion stars, more than twice the number of stars populating our home galaxy. Because our view of the galaxy from Earth is face-on, we can see its full structure in impressive detail.

Pierre Méchain, one of Charles Messier's colleagues, first discovered the Pinwheel Galaxy in 1781. As its high 'M' number indicates, it was one of the last additions to Messier's catalogue. Located 21 million light years away from Earth in the constellation Ursa Major, M101 has an apparent magnitude of 7.9. To find M101's distinctive pinwheel shape, look for the asterism in Ursa Major known as the Big Dipper; then look for the bend in the Big Dipper's handle. M101 is near that crook in the ladle, and can be spotted through a small telescope, most easily in April.

Pinwheel Galaxy Highlights

Asymmetry

1 M101 might look like a pinwheel, but it's a bit of an off-kilter pinwheel. The galaxy's slightly lopsided shape is the result of interactions with nearby galaxies that tug on and compress interstellar hydrogen gas, igniting star formation in the Pinwheel's looping arms. In particular, M101 forms a sextet with five prominent companion galaxies: NGC 5204, NGC 5474, NGC 5477, NGC 5585 and Holmberg IV. Together, these galaxies form the bulk of the M101 group.

Superduper Supernova

2 In 2011, astronomers spotted a supernova going off in M101 – in near real-time. As light from the explosion rippled outwards, astronomers realised it was a Type 1A supernova, or a variety of exploding star that is crucial for understanding how the Universe is expanding. Later, observations of this supernova would help astronomers understand such explosions themselves.

Called supernova 2011fe (or PTF 11kly), it was the nearest, brightest such explosion to Earth since 1987. In fact, the explosion was visible to observers using small, backyard telescopes. Because astronomers caught the star in the act of exploding, it offered them an unprecedented chance to study these exploding stars from their very earliest stages. Astronomers know that Type Ia supernovas originate from remnant stars called white dwarfs, the collapsed stellar corpses of Sun-like stars. When those white dwarfs detonate, they produce a predictable brightness.

Comparing a Type 1A supernova's observed brightness with its predicted brightness lets astronomers calculate how far away the object is. By comparing those luminosities for dozens of Type 1A supernovas, scientists figured out that the Universe is expanding at an accelerating rate; in other words, more distant objects are flying away faster than they should be. The discovery implicated a still-mysterious phenomenon called 'dark energy' and won the Nobel Prize for physics in 2011.

But figuring out exactly why those white dwarf stars explode has been difficult. Astronomers know the dwarfs detonate when too much mass is piled onto them, but exactly what type of star donates that deadly mass has been a mystery. Studying supernova 2011fe thickened the mystery by providing scientists with clues that ruled out an extremely popular scenario, and suggested that the doomed dwarf's companion must have been much less massive than our Sun.

Top Tip

Stars seem to grow up and die at an unusually high rate in M101, where four stellar explosions have been observed in roughly one century. Astronomers estimate that on average, maybe one or two supernovas explode per galaxy per 100 years – so M101 is a bit of an overachiever. Named after the years in which they exploded, M101's supernovas are called SN 1909A, SN 1951H, SN 1970G, and SN 2011fe. This last one offered astronomers a rare treat (more on that to the left).

Getting There & Away

At 21 million light years away, this one might be better to admire from afar.

OBJECT TYPE
Dwarf elliptical galaxy

WIDTH
10,000 light years

CONSTELLATION
Sagittarius

APPARENT MAGNITUDE
4.5

DISTANCE FROM THE SOLAR SYSTEM
70,000 light years

DISCOVERED
Rodrigo Ibata, Gerry Gilmore, Mike Irwin, 1994

The Sagittarius Dwarf Elliptical Galaxy is so dim that it eluded discovery until 1994.

Sagittarius Dwarf Elliptical Galaxy

For almost a decade, the Sagittarius Dwarf Elliptical Galaxy (Sag DEG, not to be confused with the Sagittarius Dwarf) was the closest known galaxy to the Milky Way. The small, dim cluster of stars is a mere 70,000 light years from the solar system, or roughly one-third the distance to the Large Magellanic Cloud.

Diffuse and largely obscured by stars in the Milky Way which block astronomers' views, the Sagittarius Dwarf Elliptical Galaxy (or SagDEG) evaded detection until 1994. Now, astronomers know it is a satellite galaxy orbiting our own, and loops once around the Milky Way every billion years or so. But scientists suspect the small galaxy is in the process of being shredded by our galaxy's gravity – just like the Canis Major Dwarf, which usurped the title of the nearest galaxy to ours in 2003.

The Sagittarius Dwarf Elliptical Galaxy contains several knots of stars known as globular clusters. These are ancient stellar conglomerates that are older than the Milky Way itself. One of these, called M54, is so bright and distinct that astronomers suspect it's the remnant core of the Sagittarius Dwarf, a relic from the galaxy's bigger, brighter days.

This near galaxy is on the other side of the Milky Way from the Sun, about 70,000 light years away from us and 50,000 light years away from the centre of the Milky Way. Some of SagDEG's stars are actually in the outermost regions of the Milky Way!

© STOCKTREK IMAGES, INC./ALAMY STOCK PHOTO

The Sculptor Galaxy, found in its namesake Sculptor constellation.

OBJECT TYPE
Spiral galaxy

WIDTH
90,000 light years

CONSTELLATION
Sculptor

APPARENT MAGNITUDE
8 (one of the brightest galaxies in the sky)

DISTANCE FROM THE SOLAR SYSTEM
11.42 million light years

DISCOVERED
Caroline Herschel, 1783

Sculptor Galaxy

Also known as NGC 253 or the Silver Dollar, the Sculptor Galaxy is part of a galaxy cluster visible to observers in the southern hemisphere. It has spectacular swirling arms and a central bar, and is classified as a starburst galaxy because of the extraordinarily intense star formation in its nucleus.

As a starburst galaxy, newborn stars in NGC 253 warm the surrounding dust clouds, causing a brilliant glow in the galaxy's centre. The closest such starburst galaxy to our own, NGC 253 is also one of the brightest and dustiest galaxies in the sky. It's known as an active galaxy, which means that a significant fraction of its energy output does not come from normal populations of main sequence stars, such as our Sun. Instead radio waves and X-ray sources may be the main sources of energy, perhaps indicating black hole accretion. Closer investigation will surely reveal some dramatic discoveries within the galaxy.

The Sculptor Galaxy was discovered in 1783 by Caroline Herschel, a sister and collaborator of the discoverer of infrared light, Sir William Herschel. The easiest galaxy to spot after Andromeda, it was named after the constellation in which it is found and can be seen by southern hemisphere observers with a pair of good binoculars under the right conditions.

OBJECT TYPE
Lenticular or spiral galaxy

WIDTH
50,000 light years

CONSTELLATION
Virgo

APPARENT MAGNITUDE
8

DISTANCE FROM THE SOLAR SYSTEM
28 million light years

DISCOVERED
Pierre Méchain, 1781

This Hubble mosaic image captures the galaxy's prominent dust lane and halo of stars.

Sombrero Galaxy

The Sombrero Galaxy, aka M104, looks like a hat because we're seeing it nearly edge-on. In reality, the galaxy has an unusually large and extended central bulge of stars that is bisected by dark, prominent dust lanes.

Billions of old stars produce the diffuse glow of the extended central bulge; all together, the galaxy contains the mass of 800 billion suns, making it one of the more massive objects near the Virgo Galaxy Cluster. Some observers believe it to be a lenticular galaxy; others believe it's a spiral galaxy. A faded spiral might appear similar to lenticular galaxies, making the distinction a matter of disagreement among astronomers in some cases. An effect called tidal harassment from the impact of other galaxies' gravitational pull can also move a galaxy from one classification to the other over time.

Many points of light in the Sombrero are actually ancient clumps of stars called globular clusters. They're estimated to be nearly 2000 in number — or 10 times more than the number of globular clusters in our Milky Way. But the ages of the clusters are similar to those closer to home, ranging from 10-13 billion years old.

With a relatively high apparent magnitude of 8, the Sombrero Galaxy cannot be seen with the unaided eye, but can be spotted through small telescopes, most easily during May. M104 is located 28 million light years away in the constellation Virgo, and has an apparent diameter that's one-fifth of the full moon.

Recently formed blue-white giant stars are readily seen in this Hubble image.

OBJECT TYPE
Flocculent spiral galaxy

WIDTH
100,000 light years

CONSTELLATION
Canes Venatici (near Ursa Major)

APPARENT MAGNITUDE
9.3

DISTANCE FROM THE SOLAR SYSTEM
29.3 million light years

DISCOVERED
Pierre Mechain, 1779

Sunflower Galaxy

M63 takes its colloquial name from the sunflower it resembles. To astronomers, it is known as a flocculent spiral galaxy.

Star formation is one of the most important processes in shaping the universe. In addition to birthing new stars, it gives rise to planetary systems and plays a pivotal role in the evolution of galaxies. Yet there is still much that astronomers do not understand about this fundamental process. The driving force behind star formation is particularly important for a type of galaxy called a flocculent spiral. Unlike grand-design spiral galaxies, flocculent spiral galaxies do not have well defined spiral arms. Instead, they appear to have many discontinuous arms, influenced by the galaxy's self-propagating methods of star formation, as theorised in the aptly named SSPSF (stochastic self-propagating star formation) model.

M63 is one such flocculent spiral galaxy. Although it only has two arms, many appear to be winding around its yellow core. The arms shine with the radiation from recently formed blue stars and can be more clearly seen in infrared observations.

The galaxy is located roughly 27 million light years from Earth in the constellation Canes Venatici. Discovered by Pierre Méchain in 1779, it was the first of 24 objects the French astronomer contributed to the Messier catalogue. It has an apparent magnitude of 9.3 and appears as a faint patch of light in small telescopes. The best time to observe M63 is during May. A dominant member of a known galaxy group, M63 has faint, extended features that are likely star streams from tidally disrupted satellite galaxies. The Sunflower Galaxy shines across the electromagnetic spectrum and is thought to have undergone bursts of intense star formation in the past.

OBJECT TYPE
Disrupted barred spiral galaxy

WIDTH
195,000 light years

CONSTELLATION
Draco

APPARENT MAGNITUDE
14.4

DISTANCE FROM THE SOLAR SYSTEM
420 million light years

DISCOVERED
2002

© HUBBLE LEGACY ARCHIVE, ESA, NASA; PROCESSING - BILL SNYDER BILLSNYDERASTROPHOTOGRAPHY.COM

The long extended 'tail' of the galaxy gives rise to its name.

Tadpole Galaxy

A cosmic tadpole hangs in space, a mere 420 million light years away in the northern constellation Draco. Called Arp 188 – or the Tadpole Galaxy – its eye-catching tail is about 280,000 light years long and features massive, bright blue star clusters.

That tail, which is twice as long as the Milky Way's width, is the result of a hit-and-run by an interloping galaxy. Scientists believe that a more compact intruder galaxy crossed in front of Arp 188 and was then slung behind the Tadpole via gravitational attraction. During the close encounter, tidal forces yanked the galaxy's stars, gas, and dust into the spectacular tail. This eye-catching tail is about 280 thousand light-years long and features massive, bright blue star clusters. Two prominent clumps of bright blue stars in the tail, separated by a gap, will likely become dwarf galaxies that orbit in the Tadpole's halo.

The intruder galaxy itself, estimated to lie about 300,000 light years behind the Tadpole, can be seen through the foreground spiral arms of the galaxy it tore through. Like its terrestrial namesake, the Tadpole Galaxy will likely lose its tail as it grows older, the tail's star clusters themselves forming smaller satellites of the large spiral galaxy. Other galaxies with a similar shape have been found, all influenced by the same type of disruption from a passing intruder.

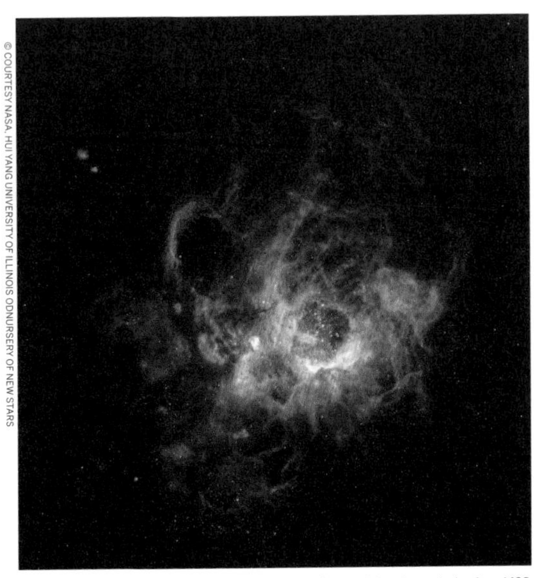

This vast nebula, NGC 604, can be found in the neighbouring spiral galaxy M33.

OBJECT TYPE
Spiral galaxy

WIDTH
60,000 light years

CONSTELLATION
Triangulum

APPARENT MAGNITUDE
5.7

DISTANCE FROM THE SOLAR SYSTEM
2.7 million light years

DISCOVERED
Giovanni Battista Hodierna, ~1654; later by Charles Messier, 1764

Triangulum Galaxy

The Triangulum Galaxy, also called M33, is a beautiful, face-on spiral in the Local Group. Located about 3 million light years away and seen in its eponymous constellation, it is the group's smallest member.

Measuring only about 60,000 light years across, the Triangulum contains between 10 and 40 billion solar masses of stars. It lacks a bright, central bulge and stellar bar, but does contain a huge amount of gas and dust, giving rise to rapid star formation. The orderly nature of Triangulum's spiral suggests to astronomers that the galaxy has been a bit of an introvert, avoiding disruptive interactions with its neighbours, and instead spending eons tending its stars. However, that could change in the future. Only slightly farther away from us than the Andromeda Galaxy, M33 is a suspected gravitational companion to Andromeda, and both galaxies are moving towards our own. M33 could become a third party involved in the impending collision between the Andromeda and Milky Way Galaxies more than 4 billion years from now.

The Local Group of galaxies is dominated by the Milky Way, Andromeda and Triangulum. As the junior member of this trio of spiral galaxies, Triangulum provides the valuable comparisons and contrasts that only a close companion can. Most notably, Triangulum's star formation is 10 times more intense than in the neighbouring Andromeda Galaxy. NGC 604 (shown above), is the largest star-forming region in M33 and one of the largest stellar nurseries in the entire Local Group. Under excellent dark-sky conditions, the Triangulum Galaxy can be seen with the naked eye as a blurry object in the constellation of Triangulum (the Triangle), one of the most distant objects that can be seen with the unaided eye.

OBJECT TYPE
Spiral Hot DOG (Hot dust obscured galaxy)

WIDTH
Unknown

CONSTELLATION
Aquarius

APPARENT MAGNITUDE
Unknown

DISTANCE FROM THE SOLAR SYSTEM
12.4 billion light years

DISCOVERED
WISE survey, 2015

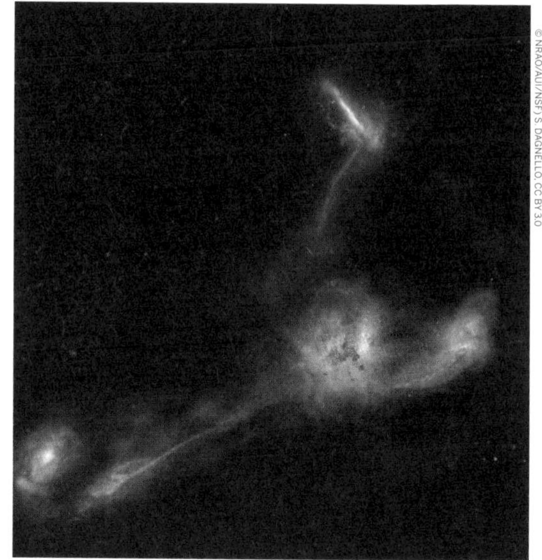

This artist's impression shows luminous galaxy W2246-0526 as it might look.

W2246-0526

Clunkily named WISE J224607.55-052634.9 (or W2246-0526 for short) is the most luminous galaxy ever discovered, radiating 350 trillion times the luminosity of the Sun.

It's also in the process of chomping not one, not two, but three of its galactic neighbours. At 12.4 billion light years from Earth, W2246-0526 is stunningly far away; but if all galaxies were positioned at equal distances from us, this one would shine the brightest.

New observations using an array of telescopes in Chile have revealed distinct trails of dust being yanked into W2246-0526. The trails originate from three smaller, neighbouring galaxies and contain about as much material as the smaller galaxies themselves. At this point, it's unclear whether those galaxies will escape their current fate or will be completely consumed by their luminous neighbour. For now, W2246-0526 uses the siphoned material to power star formation and seed the glowing, ultraluminous cloud surrounding its central supermassive black hole, which is estimated to contain 4 billion times the mass of the Sun.

Most of W2246-0526's record-breaking luminosity comes not only from stars, but also from this collection of hot gas and dust concentrated around the centre of the galaxy. In the intense gravity, matter falls toward the black hole at high speeds, crashing together and heating up to millions of degrees, causing the material to shine with incredible brilliance in quasars. Galaxies that contain these types of luminous, black-hole-fueled structures are known as having an AGN, or active galactic nucleus.

Like any engine on Earth, W2246-0526's enormous energy output requires an equally high fuel input. In this case, that means gas and dust to form stars and to replenish the cloud around the central black hole. The new study shows that the amount of material being accreted by WJ2246-0526 from its neighbours is enough to replenish what is being consumed, thereby sustaining the galaxy's tremendous luminosity.

'It is possible that this feeding frenzy has already been ongoing for some time, and we expect the galactic feast to continue for at least a few hundred million years', says Tanio Diaz-Santos of the Universidad Diego Portales in Santiago, Chile, who studied the galaxy. This kind of galactic cannibalism is not uncommon. Astronomers have previously observed galaxies merging with or accreting matter from their neighbours in the nearby universe. For example, the pair of galaxies collectively known as 'the Mice' are so named because each has a long, thin tail of accreting material stretching away from it.

W2246-0526 is the most distant galaxy ever found to be accreting material from multiple sources. The light from W2246-0526 took 12.4 billion years to reach us, so astronomers are seeing the object as it was when our universe was only a tenth of its present age of 13.8 billion years. At that distance, the streams of material falling into W2246-0526 are particularly faint and difficult to detect.

W2246-0526 falls into a special category of particularly luminous quasars known as hot, dust-obscured galaxies, or Hot DOGs. Astronomers think that most quasars get some of their fuel from external sources. One possibility is that these objects receive a slow trickle of material from the space between galaxies. Another is that they feed in bursts by eating up other galaxies, which appears to be occurring with W2246-0526. It's unclear whether W2246-0526 is representative of other obscured quasars (those with their central engines obscured by thick clouds of dust) or if it is a special case. It certainly has one special feature: the supermassive black hole at the galaxy's centre is 4 billion times the mass of the Sun. This mass is large, but the extreme luminosity of W2246-0526 was thought to require a supermassive black hole with a mass at least three times larger. Solving this apparent contradiction will require more observations.

Ultimately, though, the galaxy's gluttony may only lead to self-destruction. Scientists hypothesise galaxies gathering too much material around them end up vomiting gas and dust back out.

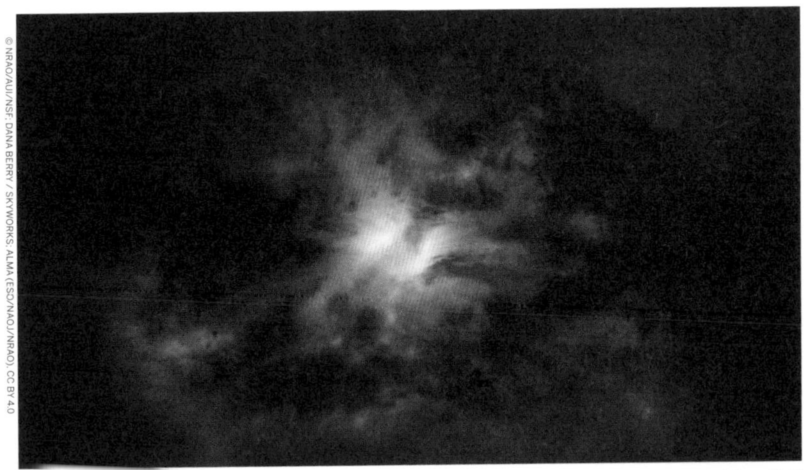

Another artist's impression of W2246-0526, a single galaxy glowing in infrared light.

OBJECT TYPE
Grand design spiral

WIDTH
60,000 light years

CONSTELLATION
Canes Venatici

APPARENT MAGNITUDE
8.4

DISTANCE FROM THE SOLAR SYSTEM
~30 million light years

DISCOVERED
Charles Messier, 1773

The Whirlpool Galaxy and its companion NGC 5195.

Whirlpool Galaxy

The Whirlpool Galaxy, aka M51, was the first spiral galaxy discovered, and it's a beautiful example of the type.

Looking at the majestic spiral galaxy M51 is not unlike staring down a spiral staircase. But instead of railings, the galaxy's two coiled arms are actually lanes of stars and gas laced with dust. Such striking arms are a hallmark of so-called grand-design spiral galaxies, of which this is a classic example. At the spiral galaxy's massive centre, which is about 80 light years across and has a brightness of about 100 million suns, lurks a supermassive black hole.

Some astronomers think that the Whirlpool's arms are particularly prominent because of a close encounter with NGC 5195, the small, yellowish galaxy at the outermost tip of one of the arms. The compact galaxy appears to be tugging on that arm and unleashing tidal forces that trigger new star formation, which helps to define the spiral arms. NGC 5195 is currently passing behind M51 and has been gliding past the Whirlpool for hundreds of millions of years.

M51 is located about 31 million light years from Earth in the constellation Canes Venatici (the Hunting Dogs). It has an apparent magnitude of 8.4 and can be spotted with a small telescope most easily during May. The Whirlpool Galaxy's beautiful face-on view and closeness to Earth allow astronomers to study a classic spiral galaxy's structure and star-forming processes. It's been closely examined via the visible light spectrum; infrared light, which reveals the oldest and coolest stars in the galaxy; and X-rays, which reveal emissions from black holes and neutron stars. Many of these are binary systems in which a black hole or neutron star orbits a main sequence star such as our Sun. Supernovas also heat the surrounding gas to X-ray temperatures.

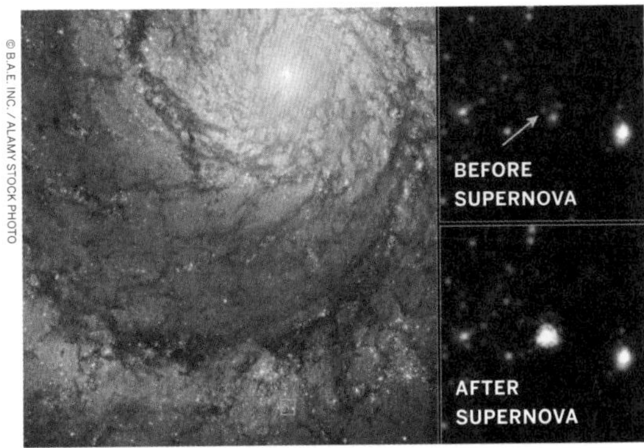

A close-up of SN 2005cs.

This composite image combines X-ray data from Chandra and optical data from Hubble.

Supernova Central

The Whirlpool Galaxy is particularly rich in supernovas. There were observed supernova in 1994, 2005 and 2011, a fast rate for one galaxy! The June 2011 Type II supernova, SN 2011dh, reached at least magnitude 12.7. Supernova 2005cs, another Type II supernova, reached a 14.1 magnitude. It helps that the distinctive visual appeal and easy-to-find location off the Big Dipper makes it a constant target of amateur observers. While SN 2011dh's star wouldn't have been visible before its spectacular end, the shock-waves grabbed the eye of keen observers familiar with the galaxy's usual appearance and able to note the appearance of a new feature. In fact, SN 1994I was discovered by astronomers Jerry Armstrong and Tim Puckett of the Atlanta Astronomy Club. This Type 1c supernova was the result of a massive star exploding, and it was rapidly confirmed and also independently observed by many others – that's the benefit of such a prima donna galaxy experiencing a supernova!

Colliding Galaxies

Interacting galaxies are a true cosmic spectacle, and among the most majestic, compelling vistas the cosmos can offer.

When two galaxies are close enough to one another, their gravitational attraction draws them together. That attraction increases as the galaxies move closer, and they eventually engage in a fateful dance. Ultimately, the two galaxies will become one, mixing their stars and gas together and creating a newer, larger galaxy. When that happens, their central supermassive black holes will collide, unleashing so much energy that it creates gravitational waves, or ripples in the very fabric of the cosmos.

But even before two galaxies merge, their interactions leave a mark. Close passes by one another rip long streamers of stars and gas from the main galactic conglomerates, creating odd and undulating cosmic shapes in their wake. These colliding, compressed gases ignite furious rounds of star formation, birthing areas that glow and radiate immense amounts of heat and light. This might sound like the destiny of faraway galaxies, but the Milky Way and Andromeda are hurtling toward this eventuality and will collide in about 5 billion years; but before that, they'll spiral around one another and permanently alter our night sky while doing so.

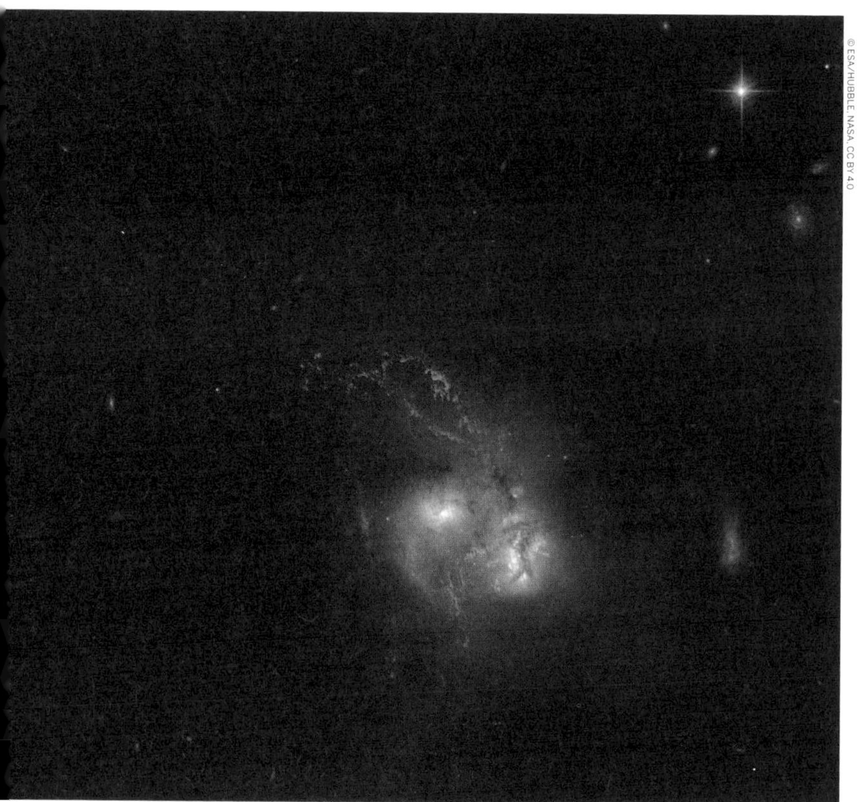

This composite image shows NGC 5256, a pair of galaxies in its final stage of merging.

OBJECT TYPE
Interacting galaxies

WIDTH
61,000 light years

CONSTELLATION
Corvus

DISTANCE FROM THE SOLAR SYSTEM
65 million light years

DISCOVERED
William Herschel, 1765

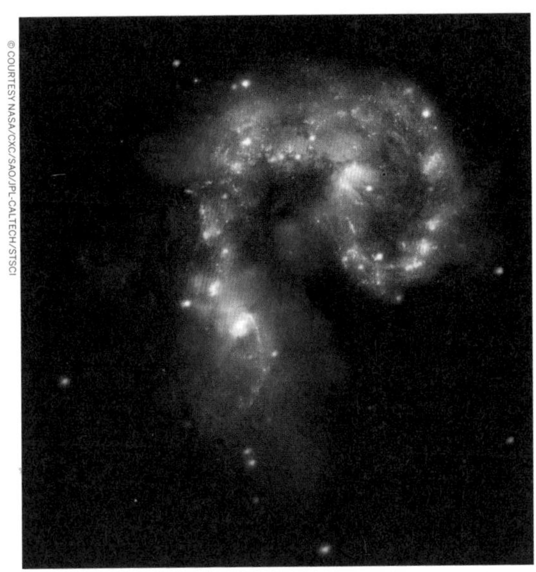

A composite image of the tangled Antennae Galaxies.

Antennae Galaxies

The Antennae Galaxies — also known as NGC 4038 and NGC 4039 — are locked in a deadly embrace.

Once normal, sedate spiral galaxies like the Milky Way, the pair of interacting galaxies that form this mixed-up structure have spent the past few hundred million years sparring with one another. This clash is so violent that stars have been ripped from their host galaxies to form a streaming arc between the two. The rate of star formation is so high that the Antennae Galaxies are said to be in a state of starburst, a period in which all of the gas within the galaxies is being used to form stars. This cannot last forever and neither can the separate galaxies; eventually the nuclei will coalesce, and the galaxies will begin their retirement together as one large elliptical galaxy. The two spiral galaxies started to interact a few hundred million years ago, making the Antennae Galaxies one of the nearest and youngest examples of a pair of colliding galaxies.

Nearly half of the faint objects in most Antennae images are young clusters containing tens of thousands of stars. Astronomers find that only about 10 percent of the newly formed super star clusters in the Antennae Galaxies will survive beyond the first 10 million years. The vast majority of the super star clusters formed during this interaction will disperse, with the individual stars becoming part of the smooth background of the galaxy. It is, however, believed that about a hundred of the most massive clusters will survive to form regular globular clusters, similar to the globular clusters found in our own Milky Way Galaxy.

The collision, which began more than 100 million years ago and is still occurring, has triggered the formation of millions of stars in clouds of dusts and gas in the galaxies. The most massive of these young stars have already sped through their evolution in a few million years and exploded as supernovas. Huge clouds of hot, interstellar gas in the galaxies have been injected with rich deposits of elements from supernova explosions. This enriched gas, which includes elements such as oxygen, iron, magnesium and silicon, will be incorporated into new generations of stars and planets. Also in the starburst, material falls onto black holes and neutron stars that are remnants of the massive stars. Some of these black holes may have masses that are almost one hundred times that of the Sun. Warm dust clouds in the galaxies have been heated by newborn stars, with the brightest clouds lying in the overlap region between the two galaxies. Old stars and star-forming regions remain as well, belonging to clusters containing thousands of stars. In wide-field images of the pair the reason for their name becomes clear – far-flung stars and streamers of gas stretch out into space, creating long tidal tails reminiscent of antennae, the result of tidal forces generated by the collision. These 'tidal tails' were formed during the initial encounter of the galaxies some 200 to 300 million years ago. They give us a preview of what may happen when our Milky Way Galaxy collides with the neighbouring Andromeda Galaxy in several billion years.

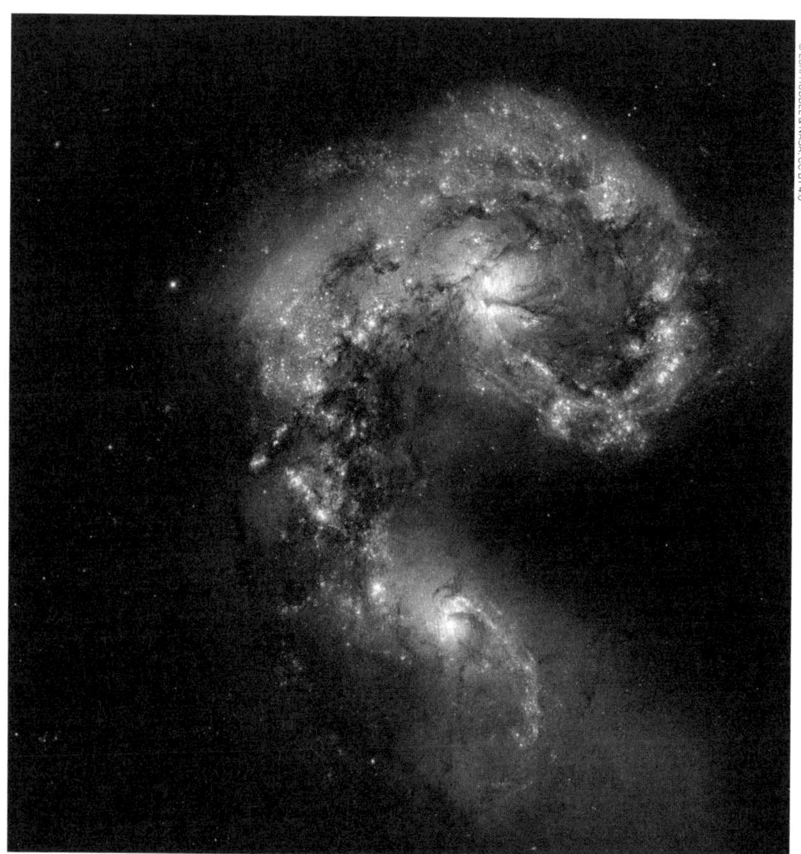

The blue star-forming regions seen in this image indicate that the Antennae Galaxies are starbursting.

Arp 273's distorted shape results from tidal interactions.

OBJECT TYPE
Interacting galaxies

WIDTH
Unknown

CONSTELLATION
Andromeda

DISTANCE FROM THE SOLAR SYSTEM
300 million light years

DISCOVERED
Halton Arp, 1966

Arp 273

This so-called cosmic rose, also known as Arp 273, marks an ongoing interaction between two distant galaxies. Arp 273 lies in the constellation Andromeda and is roughly 300 million light years away from Earth.

The disc of the larger galaxy – UGC 1810 – in this interacting pair is distorted into a floral shape by the gravitational pull of the companion galaxy below it. A swath of blue, jewel-like points across the top is the combined light from clusters of bright and hot young stars, which glow fiercely in ultraviolet light.

The smaller, nearly edge-on companion, called UGC 1813, shows distinct signs of intense star formation at its nucleus, perhaps triggered by the encounter with the companion galaxy. A tenuous bridge of material links the two galaxies, which are separated by roughly 100,000 light years.

A series of uncommon spiral patterns in the large galaxy is a telltale sign of interaction. The large, outer arm appears partially as a ring, a feature seen when interacting galaxies actually pass through one another. This suggests that the smaller companion actually dived deep, but off-centre, through UGC 1810. The inner set of spiral arms is highly warped out of the plane, with one of the arms going behind the bulge and coming back out the other side. How these two spiral patterns connect is still not precisely known. The larger galaxy in the UGC 1810-UGC 1813 pair has a mass that is about five times that of the smaller galaxy. In unequal pairs such as this, the relatively rapid passage of a companion galaxy produces the lopsided or asymmetric structure in the main spiral. In such encounters, the starburst activity typically begins in the minor galaxies earlier than in the major galaxies.

OBJECT TYPE
Interacting galaxies

WIDTH
65,000 light years

CONSTELLATION
Ursa Major

DISTANCE FROM THE SOLAR SYSTEM
450-500 million light years

DISCOVERED
Nicholas Mayall, 1940

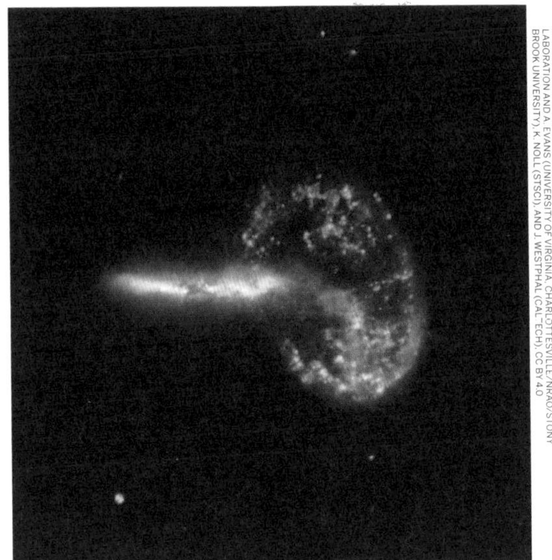

Mayall's Object is composed of a ring-shaped galaxy and a long-tailed companion.

Mayall's Object

Mayall's Object, or Arp 148, is the staggering aftermath of an encounter between two galaxies, resulting in an intertwined ring-shaped galaxy and a long-tailed companion. The galaxies are located in the constellation of Ursa Major, approximately 500 million light years away.

The collision between these two parent galaxies produced a shock-wave effect that first drew matter into the centre and then caused it to propagate outwards into a ring. But Arp 148 is also an image of a collision in action: That ring is being punctured by an elongated companion. The ring's intensely blue light indicates that it's a region of high star-formation. The nucleus of the ring (visible in yellow) also has bands of obscuring dust. Interacting galaxies are fertile ground in the Universe; the shock-waves make them highly active, though the gravitational pull varies from one interacting galaxy pair to another.

No wonder it was included as part of astronomer Halton Arp's *Catalog of Peculiar Galaxies* as number 148. Its other name comes from its discoverer, Nicholas U. Mayall, who studied the deep-sky object from Lick Observatory and made enormous contributions to astronomy's understanding of the age of the Universe and the mass of the Milky Way. Mayall's Object can be found within the Great Bear. This deep-sky object is too faint to be readily observed by non-professional telescopes, however.

The two spiral galaxies NGC 2207 and IC 2163.

OBJECT TYPE
Interacting galaxies

WIDTH
Unknown

CONSTELLATION
Canis Major

DISTANCE FROM THE SOLAR SYSTEM
110 million light years

NGC 2207 & IC 2163

In the direction of the constellation Canis Major, two spiral galaxies pass by each other like majestic ships in the night. The larger and more massive galaxy is catalogued as NGC 2207, and the smaller one is IC 2163. Strong tidal forces from NGC 2207 have distorted the shape of IC 2163, drawing stars and gas into long streamers stretching roughly 100,000 light years.

Computer simulations suggest that IC 2163 is swinging past NGC 2207 in a counterclockwise direction, having made its closest approach 40 million years ago. However, IC 2163 does not have sufficient energy to escape from the gravitational pull of NGC 2207, and is destined to be pulled back and swing past the larger galaxy again in the future. Eventually, billions of years from now, they will merge into a single, more massive galaxy.

NGC 2207 and IC 2163 have hosted three supernova explosions in the past 15 years and have produced one of the most bountiful collections of super bright X-ray lights known. These special objects are known as 'ultraluminous X-ray sources' (ULXs; see next page for more). The scientists involved in studying this system note that there is a strong correlation between the number of X-ray sources in different regions of the galaxies and the rate at which stars are forming in these regions. X-ray sources are concentrated in the spiral arms of the galaxies, where large amounts of stars are known to be forming.

Ultraluminous X-ray Sources (ULXs)

Interacting galaxies tend to be a concentrated spot for ultra-luminous X-ray sources, ULXs, which are found using data from NASA's Chandra X-ray Observatory. Often these are located in the star systems known as X-ray binaries, which consist of a star in a tight orbit around either a neutron star or a 'stellar-mass' black hole. The strong gravity of the neutron star or black hole pulls matter from the companion star. As this matter falls towards the neutron star or black hole, it is heated to millions of degrees and generates powerful X-rays.

ULXs have far brighter X-rays than most 'normal' X-ray binaries. The true nature of ULXs is still debated, but they are likely a peculiar type of X-ray binary. The black holes in some ULXs may be heavier than stellar-mass black holes and could represent a hypothesised, but as yet unconfirmed, intermediate-mass category of black holes.

Scientists now tally a total of 28 ULXs between NGC 2207 and IC 2163. Twelve of these vary over a span of several years, including seven that were not detected before because they were in a 'quiet' phase during earlier observations. Over the course of such quiet periods, when the X-ray source is between outbursts, astronomers are able to take more accurate measurement of the binary companion, which gives information that can hint at the black hole or neutron star's mass. For most ULXs, the main sequence companion star in the binary system is likely young and massive.

Colliding galaxies like this pair are well known to contain intense star formation. Shock waves – like the sonic booms from supersonic aircraft – form during the collision, leading to the collapse of clouds of gas and the formation of star clusters. In fact, researchers estimate that the stars associated with the ULXs are very young and may only be about 10 million years old. In contrast, our Sun is about halfway through its 10-billion-year lifetime. Moreover, analysis shows that stars of various masses are forming in this galaxy pair at a rate equivalent to forming 24 stars the mass of our Sun per year. In comparison, a galaxy like our Milky Way is expected to spawn new stars at a rate equivalent to only about one to three new suns every year.

The icy blue eyes are actually the cores of two merging galaxies, called NGC 2207 and IC 2163.

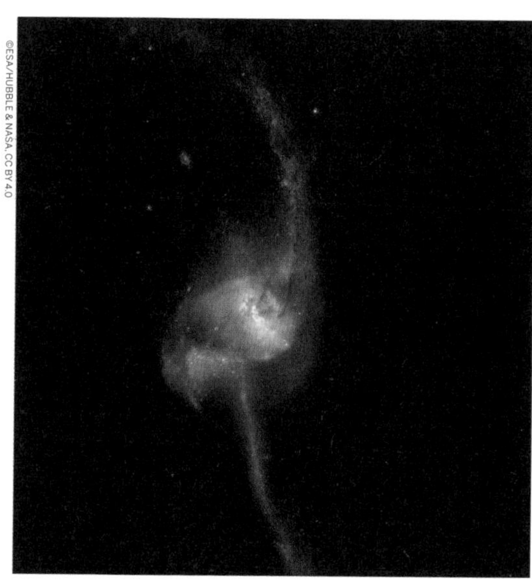

The twisted cosmic knot seen here is NGC 2623, or Arp 243.

OBJECT TYPE
Interacting galaxies

WIDTH
50,000 light years

CONSTELLATION
Cancer

DISTANCE FROM THE SOLAR SYSTEM
250 million light years

DISCOVERED
Édouard Jean-Marie Stephan, 1885

NGC 2623

The confused spiral muddle that is NGC 2623, also known as Arp 243, gained its unusual and distinctive shape as the result of a major collision and subsequent merger between two separate galaxies.

The violent encounter that birthed NGC 2623 caused clouds of gas within the two galaxies to become compressed and stirred up, in turn triggering a sharp spike of star formation that is marked by speckled patches of bright blue; these can be seen clustered both in the centre and along the trails of dust and gas forming NGC 2623's sweeping curves. Known as tidal tails, these curves stretch roughly 50,000 light years from end to end. The prominent lower tail is richly populated with bright star clusters - 100 of them have been found in observations. These star clusters may have formed as part of a loop of stretched material associated with the northern tail, or they may have formed from debris falling back onto the nucleus. In addition to this active star-forming region, both galactic arms harbour very young stars in the early stages of their evolutionary journey.

Unlike some of the collisions caught in action, NGC 2623 is in a very late stage of merging. It is thought that the Milky Way will eventually resemble NGC 2623 when it collides with our neighbouring galaxy, the Andromeda Galaxy, in about five billion years. The collision powers NGC 2623 to issue a huge amount of energy - 400 billion times the Sun's energy in infrared alone! Under that infrared signal is a powerful amount of star creation. As a result, it's considered a ULIRG, or ultraluminous infrared galaxy, huge galaxies that have reached their current size through collisions just like this one.

OBJECT TYPE
Interacting galaxies

WIDTH
Unknown

CONSTELLATION
Vela

DISTANCE FROM THE SOLAR SYSTEM
100 million light years

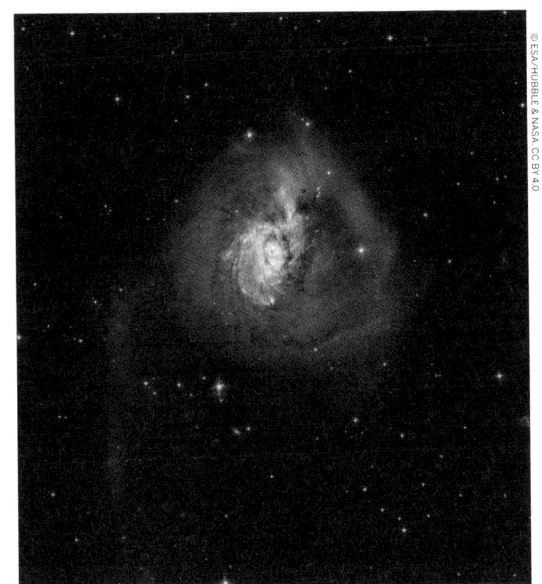

NGC 3256 is the result of a past galactic merger.

NGC 3256

NGC 3256 is an impressive example of a peculiar galaxy that is actually the relic of an ancient collision.

As such, NGC 3256 provides an ideal target to investigate starbursts that have been triggered by galaxy mergers. Two extended luminous tails swirling out from the galaxy betray the collision's occurrence, as well as the galaxy's distorted appearance. These tails are studded with young blue stars, which were born in the frantic but fertile collision of gas and dust.

NGC 3256 belongs to the Hydra-Centaurus supercluster complex and provides a nearby template for studying the properties of young star clusters in tidal tails. As well, the system hides a double nucleus and a tangle of dust lanes in the central region. It has been the subject of much study due to its luminosity, its proximity, and its orientation: astronomers observe its face-on orientation, that shows the disc in all its splendour.

Located about 100 million light years away in the constellation of Vela (the Sails), NGC 3256 is approximately the same size as our Milky Way. The tails are thought to have formed 500 million years ago at the initial encounter between the two galleries. As well as being lit up by over 1000 bright star clusters, the central region of NGC 3256 is also home to criss-crossing threads of dark dust and a large disc of molecular gas spinning around two distinct nuclei – the relics of the two original galaxies. One nucleus is largely obscured, only unveiled in infrared, radio and X-ray wavelengths.

These two initial galaxies were gas-rich and had similar masses, as they seem to be exerting roughly equal influence on one another. Their spiral discs are no longer distinct, and in a few hundred million years' time, their nuclei will also merge and the two galaxies will likely become united as a large elliptical galaxy.

Galaxy Clusters

The galaxy cluster pictured here is SDSS J033+0651.

Galaxy clusters contain thousands of galaxies of all ages, shapes and sizes, which together can total a mass thousands of times greater than that of the Milky Way. These groupings of galaxies are colossal — they are the largest structures in the Universe to be held together by their own gravity. The Virgo Galaxy Cluster alone holds over 1000 galaxies that are packed together.

Clusters are useful in probing mysterious cosmic phenomena like dark energy and dark matter, which can contort space itself. There is so much matter stuffed into a galaxy cluster that its gravity has visible effects on its surroundings. The cluster's gravity warps the very structure of its environment (spacetime), causing light to travel along distorted paths through space. This phenomenon can produce a magnifying effect, allowing us to see faint objects that lie far behind the cluster and are thus otherwise unobservable from Earth. It's a behaviour of light that was predicted in Albert Einstein's Theory of General Relativity. This 1915 theory predicts that objects deform spacetime, causing any light that passes by to be deflected and resulting in a phenomenon known as gravitational lensing. The effect is only noticeable for very massive objects. A few hundred strong gravitational lenses are known, but most are too distant to precisely measure their mass. Hubble observations have provided renewed confirmation of general relativity's predictions in action.

Not only that, but by utilizing Hubble's past observations of six massive galaxy clusters in the Frontier Fields program, astronomers have demonstrated that intracluster light — the diffuse glow between galaxies in a cluster — traces the path of dark matter, illuminating its distribution more accurately than existing methods that observe X-ray light. Intracluster light is the byproduct of interactions between galaxies that disrupt their structures; in the chaos, individual stars are thrown free of their gravitational moorings in their home galaxy to realign themselves with the gravity map of the overall cluster. This is also where the vast majority of dark matter resides. X-ray light indicates where groups of galaxies are colliding, but not the underlying structure of the cluster. This makes it a less precise tracer of dark matter. By monitoring the movements of these collections of galaxy clusters, our knowledge about the larger structure of the Universe can slowly come into clearer view. The Hubble Frontier Fields program was a deep imaging initiative designed to utilise the natural magnifying glass of galaxy clusters' gravity to see the extremely distant galaxies beyond them, and thereby gain insight into the early (distant) universe and the evolution of galaxies since that time. In that study the diffuse intracluster light was an annoyance, partially obscuring the distant galaxies beyond. However, that faint glow around these galaxy clusters could end up shedding significant light on one of astronomy's great mysteries: the nature of dark matter.

CONSTELLATION
Virgo

DISTANCE
2.2 billion light years

GALAXIES
~1000

MASS
Somewhere on the order of a quadrillion Suns

Galaxy cluster Abell 1689 in visible and infrared light.

Abell 1689

Abell 1689 is one of the most massive galaxy clusters known, containing roughly 1000 galaxies and about one quadrillion solar masses of stars. About 2 billion light years from Earth in the constellation Virgo, the cluster is known for acting like a cosmic lens.

Recent studies have shown that Abell 1689 is home to some 10,000 globular clusters of stars – perhaps the largest number of these clusters ever spotted. Material from some of these galaxies is being stripped away, giving the impression that the galaxy is dripping into the surrounding space. As well, astronomers using the Chandra X-ray Observatory spied signs of extremely hot gas – in excess of 100 million degrees – among the galaxy's clusters, which strongly suggests the presence of mergers and collisions.

In fact, Abell 1689 contains so much mass that it magnifies objects behind it, making it possible for astronomers to explore incredibly distant regions of space. The effect is called gravitational lensing. In many images of Abell 1689, its galaxies are interspersed between curved arcs of light; those arcs are images of background galaxies that are being stretched and warped by the cluster's mass. In 2014, scientists used the cluster's magnification power to find 58 remote galaxies; sprinkled behind Abell 1689, they are among the smallest, faintest and most numerous galaxies ever seen in the remote universe.

© NASA/CXC/M. WEISS

This composite image shows the galaxy cluster 1E 0657-56, Bullet Cluster.

CONSTELLATION
Carina

DISTANCE
3.7 billion light years

GALAXIES
40

MASS
Unknown

Bullet Cluster

The Bullet Cluster is named for its distinctive shape, which is the result of a collision between two enormous galaxy clusters.

The speed and shape of the Bullet Cluster, as well as other information, suggest that a smaller cluster passed through the core of a larger cluster about 150 million years ago, producing a gigantic cosmic pileup that could be one of the most energetic events since the Big Bang.

Those two enormous clusters collided at speeds exceeding several million miles an hour, creating such force that it wrenched the clusters' visible matter, such as stars and gas, away from the invisible substance astronomers call 'dark matter.' As the Bullet Cluster's galaxies crossed and merged together, their stars easily continued on their way unscathed. This may seem a bit perplexing, because the bright light of stars makes them appear enormous and crowded together. It would be easy to expect them to smash into one another during their cosmic commute. But the truth is, stars are actually spaced widely apart and pass harmlessly by like ships on an ocean. The gas clouds from the merging galaxies, however, found the going much tougher. As the clouds ran together, the rubbing and bumping of their gas molecules caused friction to develop. The friction slowed the clouds down, while the stars they contained kept right on moving. Before long, the galaxies slipped out of the gas clouds and into clear space.

Comprising some 25 percent of the mass of the universe, dark matter is still fundamentally mysterious; it doesn't interact with normal matter, but astronomers can infer its presence in multiple ways. The fact that the Bullet Cluster's dark matter is offset from its visible matter points to the severity of the collision, as well as the fact that dark matter actually exists.

What Is Dark Matter?

At least 80% of the Universe is thought to be composed of dark matter – now if only we knew what that meant. We are much more certain what dark matter is not than we are what it is. In keeping with the name 'dark', we know that it is not found in the form of stars and planets that we can see or observe in the electromagnetic spectrum. This renders it unobservable except by deduction from what's missing in the Universe's overall mass. Observations show that there is far too little visible matter in the Universe to match projections; 'dark' matter must therefore make up the shortfall. What we can observe is considered 'normal' matter, matter made up of particles called baryons. Baryonic clouds and objects are detectable by their absorption of radiation passing through them. Neither is dark matter antimatter, because we do not see the unique gamma rays that are produced when antimatter collides with matter. Large galaxy-sized black holes are no longer under consideration as black matter candidates, however.

Dark matter in space in a computer-generated fractal background.

El Gordo's subclusters are colliding over seven billion light years from Earth.

CONSTELLATION
Phoenix

DISTANCE
7.2 billion light years

WIDTH
7.72 million light years

GALAXIES
Several hundred

MASS
3 million billion suns (3 quadrillion)

El Gordo

The largest known galaxy cluster in the distant universe, El Gordo certainly lives up to its nickname, which means 'the fat one' in Spanish.

Though equally massive conglomerates - such as the Bullet Cluster - are found in the nearby universe, at over 7 billion light years away, El Gordo is much farther away and much older, having achieved its current size when the universe was roughly half its estimated age of 13.8 billion years.

Like the Bullet Cluster, El Gordo is the result of a collision between two large galaxy clusters. The collision gives a distinct cometary appearance to El Gordo, including two 'tails'. It now stretches 7 million light years across, and in 2014, astronomers calculated that the gigantic cluster contains the mass of roughly 3 quadrillion suns. They did this by measuring how much the cluster's gravity warps images of galaxies in the distant background - because like Abell 1689, El Gordo acts as a foreground lens, stretching and distorting background light.

A fraction of that mass (only an estimated 1%) is locked up in several hundred galaxies that inhabit the cluster, and a larger portion is found in hot gas that fills its volume. The rest is tied up in dark matter, an invisible form of matter that makes up the bulk of the mass of the universe. This ratio of stars to gas is similar to results from other massive clusters. As with the Bullet Cluster, there is evidence that normal matter, mainly composed of hot, X-ray bright gas, has been wrenched apart from the dark matter in El Gordo. The hot gas in each cluster was slowed down by the collision, but the dark matter was not.

CONSTELLATION
Fornax

DISTANCE
65 million light years

WIDTH
Radius 2.62-5.18 mpc
~60 large galaxies,
~60 dwarf galaxies

MASS
0.4-3.32x10^14 suns

© COURTESY NASA, ESA, AND THE HUBBLE HERITAGE TEAM (STSCI/AURA)

NGC 1316, pictured, is one of the galaxies located in the Fornax Cluster.

Fornax Cluster

Located in the southern sky, the Fornax Cluster is the second-richest galaxy cluster known within 100 million light years (our neighbouring Virgo Cluster is the first).

Like most clusters, the Fornax Cluster does not have neatly defined edges, so it is difficult to determine exactly where it begins and ends. However, astronomers estimate that the Fornax Cluster's centre is roughly 65 million light years from Earth, making it one of the closest known such groupings. It contains nearly sixty large galaxies and a similar number of smaller dwarf galaxies.

The centre of the cluster is dominated by the galaxy known as NGC 1399, a large spheroidal galaxy whose light is almost exclusively from old stars and thus appears blue. The most spectacular member of Fornax is the galaxy known as NGC 1365, a giant barred spiral galaxy with dusty spiral arms. The arms contain younger stars that are heating up their dust-enshrouded birth clouds, causing them to glow at longer infrared wavelengths. This galaxy is one of only a few in the Fornax Cluster where prolific star formation can be seen.

Galaxy clusters like this one are common and illustrate the powerful influence of gravity. Even over large distances, the Fornax Cluster's gravity draws the enormous masses of individual galaxies into one region. Optical studies of the cluster have identified a large group of galaxies on the cluster's outskirts that appear to be on a collision course with its core. The motions of these galaxies indicate that they lie along a large, unseen, filamentary structure composed mostly of dark matter that is flowing towards a common centre of gravity. The galactic cluster can be found in Fornax (the Furnace), visible from the southern hemisphere.

Fornax Cluster Highlights

The Fornax Wall

1 The Fornax Cluster is part of a much larger structure called the Fornax Wall. Such superstructures are longer along one axis than the other, and are formed from multiple smaller groups of galaxies. Joining the Fornax Cluster in the Fornax Wall is the Dorado Group, which is also in the southern sky. Containing a loose collection of spirals and ellipticals, the Dorado group – though smaller than a cluster – is one of the richer nearby groupings of galaxies.

A Galactic Cannibal: NGC 1399

2 The Fornax Cluster's centre contains what is known as a cD galaxy – a galactic cannibal – that has grown by swallowing smaller galaxies. Called NGC 1399, this galaxy appears similar to elliptical galaxies, but is bigger and has an extended, faint envelope. Recent observations have revealed a very faint bridge of light between NGC 1399 and its neighbour, a smaller galaxy called NGC 1387. This bridge is bluer than either galaxy, meaning that it contains young stars erupting from the gas NGC 1399 is cannibalising.

A Turbulent History: NGC 1316

3 NGC 1316 stands out from the many galaxies in the Fornax Cluster. It contains a supermassive black hole with as much mass as 150 million suns; as it swallows material from its surroundings, that black hole launches powerful jets into the cosmos, making NGC 1316 one of the brighter objects in the radio sky. As well, the galaxy's structure betrays a turbulent past involving the consumption of multiple smaller galaxies. That adventurous history has carved loops, arcs and rings into its starry outer envelope.

Bathed in Hot Gas

4 A vast cloud of ten-million-degree gas surrounds the Fornax Cluster. In 2003, NASA's Chandra X-ray Observatory spent 143 hours staring at that giant gas cloud – and it revealed that the cloud has a swept-back, comet-like shape that extends for more than half a million light years. Its geometry suggests that the giant, hot gas cloud is moving through a larger, but less dense gassy cloud, creating an intergalactic headwind.

A Greedy Monster: NGC 1365

5 The large spiral galaxy NGC 1365 is a striking example of its type. Spanning some 200,000 light years from end to end, It has a thick, prominent stellar bar passing through its core, with spiral arms emerging from the ends. But there's more to NGC 1365 than meets the eye: It is a classified as a Seyfert galaxy, with a bright, active galactic nucleus powered by a supermassive black hole that is hungrily chomping on stars, gas and dust.

Top Tip

Often, the central galaxy in a cluster is the brightest – but that's not the case for the Fornax Cluster. Its brightest galaxy, NGC 1316, is situated at the edge of the cluster. Also known as Fornax A, it is one of the most powerful sources of radio waves in the sky. Those radio waves can be seen by specialised telescopes that are sensitive to this kind of radiation, which emanates from two enormous lobes extending far into space on either side of the visible galaxy. A supermassive black hole lurking at galaxy's centre power those lobes.

Getting There & Away

With its core just 65 million light years away, the Fornax Cluster is the cosmic equivalent of right next door. Yet you still couldn't visit on vacation (or for any other reason).

CONSTELLATION
All of them

DISTANCE
We're in it

WIDTH
10 million light years

GALAXIES
30+

MASS
Unknown

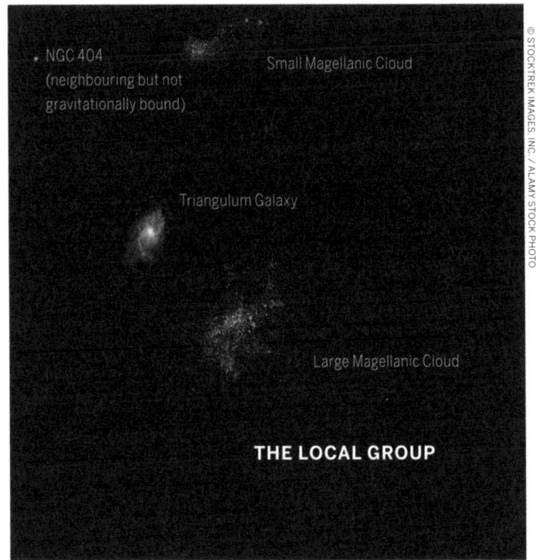

Some of the smaller Local Group members (not to scale).

Local Group

The Milky Way is part of the aptly named 'Local Group' of galaxies. Consisting of more than 30 galaxies that are gravitationally bound to one another, the Local Group stretches roughly 10 million light years across.

Aside from the Milky Way, the most massive galaxy in this group is neighbouring Andromeda, which is bigger both in size and in mass to our own. Also prominent are Leo I, the Triangulum Galaxy (M33) and the Large Magellanic Cloud, a satellite of the Milky Way that's the fourth most massive galaxy in the Local Group. The cluster's centre is somewhere between the Milky Way and Andromeda, both of which have dwarf galaxies associated with them.

The dynamics of the Local Group are changing; one day long off, the Milky Way and Andromeda will collide, forming a giant elliptical galaxy at the cluster's core. In addition, the Local Group and its galaxies are all being yanked towards the Great Attractor, and it's possible that our cluster may one day merge with the next nearest big grouping, called the Virgo Cluster.

Top Tip

The Local Group earned its name in 1936 from the astronomer Edwin Hubble, who described it as 'a typical small group of nebulae which is isolated in the general field'. For a long time, astronomers referred to galaxies as 'nebula', mistakenly believing they were small clusters of stars within the Milky Way. Hubble helped demonstrate that galaxies like Andromeda are not citizens of the Milky Way, but instead are far, far away; he defined the Local Group as comprising 12 galaxies, including Andromeda, the Milky Way, M33, the Magellanic Clouds, M32, NGC 205, NGC 6822, NGC 185, IC 1613, and NGC 147.

Local Group Highlights

Intergalactic Cloud

1 A cloud of neutral hydrogen drifts in space between Andromeda and Triangulum, completely untethered from any galaxy. Called a hyper velocity cloud, the object measures some 20,000 light years across and is 2.3 million light years away. But it is speeding towards Earth at 330 km/s (205 mps). Bearing the distinctly unpoetic name HVC 127-41-330, it's the first such cloud astronomers have spotted in intergalactic space. Because roughly 80 percent of the cloud's mass is made of dark matter, astronomers suspect it could be a very low mass dwarf galaxy that never got around to building stars.

The Milky Way's Minions

2 Dozens of small galaxies swarm the Milky Way. These satellites are gravitationally bound to our home galaxy and are slowly being drawn inward and yanked apart; some of them are already leaving starry entrails in orbit around the Milky Way. Among the satellite galaxies are the Large and Small Magellanic Clouds – two of the largest galaxies in the Local Group – as well as the Sagittarius and Canis Major Dwarfs, which are two of the closest galaxies to our own.

Andromeda's Minions

3 Andromeda is our nearest galactic neighbour, about 2.5 million light years away, and the largest galaxy in the group. Like the Milky Way, Andromeda has numerous, small satellite galaxies gravitationally bound to it. More than a dozen of them, in fact. These hangers-on include M32 and M110, which are the largest, and twelve galaxies named Andromeda I through Andromeda XXII. In another four billion years or so, Andromeda will collide with the Milky Way and the two will merge.

Triangulum

4 The third-largest member of the Local Group, the Triangulum Galaxy is a spiral just half the size of the Milky Way. It's about 3 million light years away, and could be a large, interacting companion of the Andromeda Galaxy. As the logic goes, either Triangulum is on an incredibly long, six-billion-year orbit around Andromeda and has already fallen into it in the past – or it is currently on its very first journey into the much larger galaxy. After studying the motions of individual stars in each galaxy, scientists recently determined that Triangulum is approaching Andromeda for the first time.

Cluster, Supercluster, Superdupercluster

5 The local group of galaxies is part of the larger Virgo Supercluster, which includes 2000 galaxies in the neighbouring Virgo Galaxy Cluster, as well as the M66 group, the M81 group, the M101 group, and a handful of other galaxy clusters. The Virgo Supercluster, in turn, is part of the Laniakea Supercluster, a huge conglomeration of more than 100,000 galaxies that is one of the largest structures in the known Universe.

Farthest Flung

5 Of the galaxies in the local group, the Sagittarius Dwarf Irregular Galaxy is among the farthest from the cluster's core (it's not to be confused with the Sagittarius Dwarf Elliptical Galaxy, which is a Milky Way satellite). Spanning just 1500 light years, this tiny galaxy is more than 3.5 million light years away. It's one of the most metal-poor galaxies observed, containing very few elements heavier than helium. Scientists think the Sagittarius Dwarf Irregular Galaxy is ancient.

Getting There & Away

Travelling to the Local Group is easy: You're already there! On a dark night, the Milky Way's ribbon of stars winds overhead, and from the southern hemisphere, the Magellanic Clouds – two more cluster members – are easily visible. Just don't expect to be able to visit any of our other Local Group members in your lifetime.

CONSTELLATION
Cancer

DISTANCE
5.23 billion light years

WIDTH
8 million light years

GALAXIES
Unknown

MASS
Unknown

Musket Ball Cluster

Only discovered in 2012, the Musket Ball Cluster is the result of a massive galactic pile-up that occurred roughly 700 million years ago.

The newly discovered system has been nicknamed the 'Musket Ball Cluster' because the cluster collision is older and slower than the Bullet Cluster, to which it bears similarities. It also goes by the unpromising label DLSCL J0916.2+2951.

When its two parent galaxy clusters smashed into each other, the collision was so violent it wrenched normal matter (also called baryonic matter), including million-degree gas, from dark matter, the invisible substance that makes up most of the mass of the universe. Most of the time, these two types of matter travel together, with galaxies embedded in haloes of dark matter. It seems the impact may have caused them to separate, though more research is needed into these dissociative galaxy cluster mergers.

This cluster is similar to the Bullet Cluster, another galactic pile-up in which observable, normal matter has been shoved away from dark matter. But at 700 million years post-smashup, the Musket-Ball has had more time to recover, giving scientists a glimpse into a later phase of cluster evolution. Taking into account the uncertainties in the age estimate, the merger that has formed the Musket Ball Cluster is two to five times further along

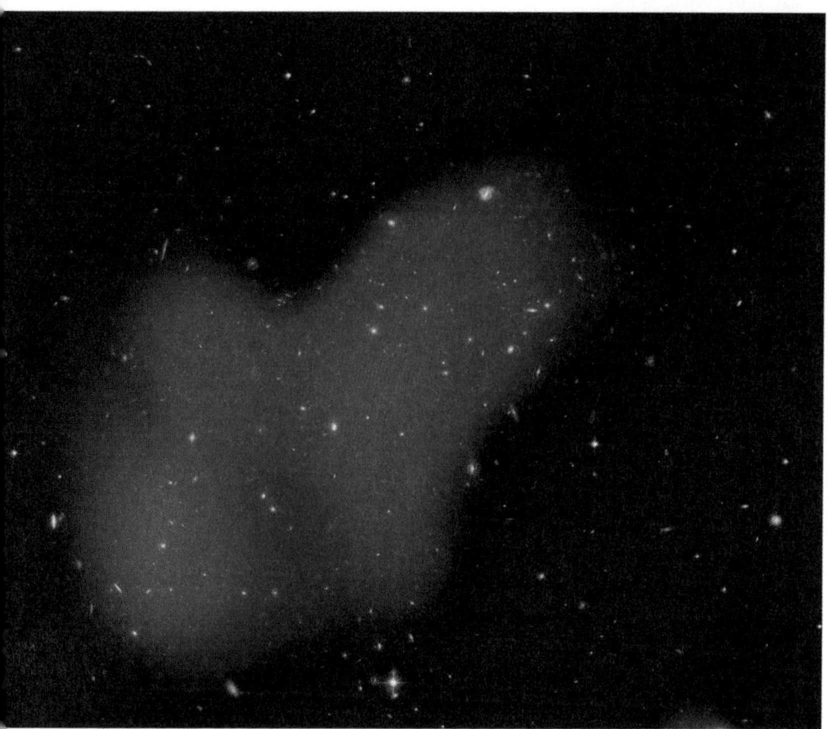

In this composite image, hot gas observed by Chandra is coloured red; galaxies in the optical Hubble data are mostly white/yellow.

than in previously observed systems. Also, the relative speed of the two clusters that collided to form the Musket Ball Cluster was lower than most of the other Bullet Cluster-like objects. In addition to the Bullet Cluster, five other examples of merging clusters with separated normal and dark matter have previously been found. In these six systems, the collision is estimated to have occurred between 170 million and 250 million years earlier.

In composite images of the Musket Ball Cluster, hot gas is coloured red, galaxies are yellow and white, and dark matter is coloured blue.

Cluster Mergers

The special environment of galaxy clusters, including the effects of frequent collisions with other clusters or groups of galaxies and the presence of large amounts of hot, intergalactic gas, is likely to play an important role in the evolution of their member galaxies. However, it is still unclear whether cluster mergers trigger star formation, suppress it, or have little immediate effect. The Musket Ball Cluster holds promise for deciding amongst these alternatives. The Musket Ball Cluster also allows an independent study of whether dark matter can interact with itself. This information is important for narrowing down the type of particle that may be responsible for dark matter. No evidence is reported for self-interaction in the Musket Ball Cluster, consistent with the results for the Bullet Cluster and other similar clusters.

CONSTELLATION
Norma, Triangulum, Australe

DISTANCE
222 million light years

WIDTH
Unknown

GALAXIES
Unknown

MASS
A quadrillion suns

A view towards the Great Attractor.

Norma Cluster

The Norma Cluster is the closest massive galaxy cluster to the Milky Way, and lies about 220 million light years away. It's also called Abell 3627.

The enormous mass concentrated in the Norma Cluster, and the consequent gravitational attraction, mean that this region of space is known to astronomers as the Great Attractor, and it dominates our region of the Universe. Despite being among the heftiest nearby galaxy clusters and relatively close by cosmic standards - just a bit more than 200 million light years away - the Norma Cluster is tough to observe. From Earth, this group of galaxies lies near the plane of the Milky Way and is obscured by a thick smog of cosmic dust. Observing the Great Attractor is difficult at optical wavelengths. The plane of the Milky Way — responsible for the numerous bright stars in this image — both outshines (with stars) and obscures (with dust) many of the objects behind it. There are some tricks for seeing through this — infrared or radio observations, for instance — but the region behind the centre of the Milky Way, where the dust is thickest, remains an almost complete mystery to astronomers.

But over the years, various telescopes have glanced at the cluster, which spans three constellations in the southern sky, on the border of Triangulum Australe (the Southern Triangle) and Norma (the Carpenter's Square). In 2014, the Hubble Space Telescope spied a large, spiral galaxy called ESO 137-001 travelling through the cluster's heart. As the galaxy moves through the superheated gas in the Norma Cluster's centre, it is being violently ripped apart, leaving bright blue galactic entrails spilling into space.

Both our Milky Way Galaxy and its home group, the Local Group, are slowly being hauled towards this mysterious, massive region and its extremely strong gravitational pull.

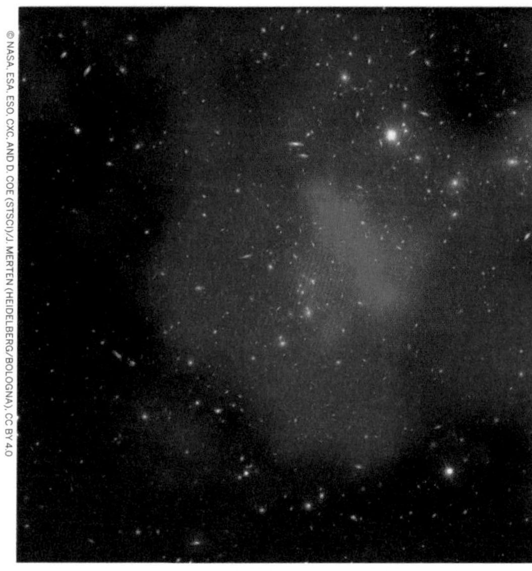

Four separate galaxy clusters have piled-up to form Pandora.

CONSTELLATION
Sculptor

DISTANCE
3.9 billion light years

WIDTH
Unknown

GALAXIES
Unknown

MASS
Unknown

Pandora's Cluster

Pandora's Cluster, or Abel 2744, is a mammoth conglomeration of at least four separate galaxy clusters that merged over 350 million years.

The Pandora Cluster, also called Abell 2744, is nearly 4 billion light years away in the direction of the constellation Sculptor, in the southern sky. Its four smaller galaxy clusters crashed together over a span of 350 million years.

When huge clusters of galaxies crash together, the resulting mess is a treasure trove of information for astronomers. Some of that information, such as the arrangement of the galaxies themselves, is relatively easy to observe. Other parts, not so much.

In Pandora, the galaxies are bright – but they make up less than 5 percent of the cluster's mass. The rest, perhaps 20 percent, is gas so hot it shines only in X-rays; more substantially, invisible dark matter comprises 75 percent of the cluster's mass, but its presence must be inferred by the ways in which it warps and changes the paths of photons zooming by.

As with the Bullet and Musket Ball Clusters, it seems that the complex collision responsible for crafting Pandora's Cluster has separated dark and 'normal' baryonic matter like protons and neutrons (but not electrons or other leptons, subatomic particles that respond to weak, gravitational and electromagnetic forces). As a result, these types of matter now lie apart from one another. The dark matter is not even near the visible galaxies.

Pandora's Cluster Highlights

Dark Matter

1 Dark matter does not emit, absorb, or reflect light, but it makes itself apparent through its gravitational interactions. To pinpoint the location of this elusive substance, scientists exploit a phenomenon known as gravitational lensing. Described by Albert Einstein, gravitational lensing happens when dark matter bends light rays from distant galaxies as they pass through the dark matter's gravitational field. By studying telltale distortions in images of those distant galaxies, scientists can map the regions where a cluster's mass is dominated by dark matter.

Lensed Galaxy

2 In 2014, astronomers spotted an extremely small, faint and ancient galaxy. Some 13 billion light years away, the galactic runt is just 850 light years across and contains the mass of 40 million suns. Normally, such tiny galaxies would be impossible to see at such exquisitely large distances – but astronomers had Pandora's Cluster in their tool box.

Packed with dark matter and acting like a giant, cosmic magnifying glass, Pandora's Cluster deflects light passing through it; as a result, it magnifies, brightens and distorts background objects, allowing astronomers to find many dim, distant structures that would normally be too faint to see.

That tiny primordial galaxy is one of those objects. Pandora's massive gravity produced three magnified images of the galaxy, each appearing 10 times larger and brighter than it normally would. Astronomers believe that such galaxies formed during the very early years of the universe, when stars began to grow and shine, yet didn't amass enough matter to assume a defined shape.

Shot Through the Heart

3 Near the core of Pandora's Cluster is a 'bullet' shape, crafted by the collision of one galaxy cluster with another. Here, the hot gases smashed together and created a characteristic shock wave, normally shaped like a ballistic projectile. However, even though hot gases got hung up and sculpted by this shock wave, dark matter passed through the collision without being affected. In another part of the cluster, galaxies and dark matter can be found, but there's no hot gas; scientists think the gas may have been stripped away during the collision, leaving behind nothing more than a faint trail.

Strange Outskirts

4 Pandora's heart might be bizarre, but even odder features lie in the outer parts of the cluster. One region contains plenty of dark matter, but has no luminous galaxies or hot gas. Instead, a separate ghostly clump of gas has been ejected, which precedes rather than follows the correspondingly associated dark matter. This puzzling arrangement may be telling astronomers something about how dark matter behaves and how the various ingredients of the Universe interact with one another.

Top Tip

Pandora takes its name from the Greek mythological character who, through opening a jar, released evil into the world. 'We nicknamed it Pandora's Cluster because so many different and strange phenomena were unleashed by the collision. Some of these phenomena had never been seen before,' said Renato Dupke of the University of Michigan, a member of the team that pieced together the cluster's violent history in 2011. Among those bizarre treasures is the complicated and uneven distribution of different types of matter, which scientists call 'extremely unusual and fascinating.'

Getting There & Away

Pandora's Cluster is a reach destination by any definition. Even travelling at light-speed, it would take nearly 4 billion years to get there – and the cluster will look very different by the time you arrive.

The 'bay' within the Perseus Cluster.

CONSTELLATION
Perseus

DISTANCE
240.1 million light years

WIDTH
11 million light years

GALAXIES
~190

MASS
660 trillion Suns

Perseus Cluster

Located around 235 million light years from Earth, the Perseus Galaxy Cluster is one of the nearest to us – but it is difficult to see in visible-light wavelengths.

Like all galaxy clusters, most of the Perseus Cluster's observable matter is in the form of superhot gas, warmed to tens of millions of degrees. This gas pervades the cluster and is so hot it only glows in X-rays. Because Perseus is so massive and gassy, it is one of the brightest objects in the X-ray sky.

The cluster contains 190 galaxies, and is dominated by a massive elliptical galaxy called NGC 1275. It also contains roughly 30 dwarf elliptical galaxies, some of which were only spotted within the last decade, and is part of the mammoth Perseus-Pisces Supercluster, which is populated by more than a thousand galaxies.

Observations with the Chanda X-ray Observatory indicate that the Perseus Cluster alone spans roughly 11 million light years. Other recent analysis by a team at Oxford University suggests that dark matter particles in the cluster are both absorbing and emitting X-rays. If the new model turns out to be correct, it could provide a path for scientists to one day identify the true nature of dark matter, that mysterious, hidden substance composing 85% of our known Universe.

True to its name, the Perseus Cluster is located in the northern sky in the direction of the constellation Perseus.

Perseus Cluster Highlights

Superheated Galactic Wave

1 A vast wave of hot gas is rolling through the Perseus Galaxy Cluster. Spanning some 200,000 light years, the wave is about twice the size of the Milky Way Galaxy. Scientists think it formed billions of years ago after a small galaxy cluster containing perhaps 1,000 times the mass of the Milky Way brushed by Perseus, missing its centre by about 650,000 light years. The passing cluster's gravity caused the gas in Perseus to slosh around and form large waves over billions of years.

Dwarf Cluster

2 Four dwarf galaxies live near the heart of the Perseus Cluster. Their smooth, spherical and symmetrical appearance suggests the galaxies have not been tidally disrupted by dense cluster's pull of gravity – yet all around them, larger galaxies are being ripped apart by the tugging and jostling of their companions. Scientists suspect the four dwarfs are undisturbed because they are embedded in a cushion of dark matter that shields them from their otherwise rough-and-tumble neighbourhood.

NGC 1275

3 The Perseus Cluster's dominant galaxy is a giant elliptical conglomerate called NGC 1275. Parked near the cluster's heart, this active galaxy – also known as Perseus A – is chomping its neighbours and is home to a supermassive black hole containing the mass of some 800 million suns. The galaxy is surrounded by numerous filaments of cool gas, each as many as 200,000 light years long. And near NGC 1275's core are some 50 globular clusters of stars.

Mammoth Cold Front

4 A gigantic and resilient 'cold front' is hurtling through the Perseus galaxy cluster. The cosmic weather system spans about two million light years and has been moving through space for more than 5 billion years – longer than the solar system has been around. The long, vertical structure travels at roughly 482,000 km/h (300,000 mph), and is powered by collisions between the Perseus Cluster and smaller groups of galaxies. Scientists suspect that magnetic field lines wrapped around the cold front's mass are preserving its sharpness.

Weird Gassiness

5 Data from the orbiting Chandra X-Ray Observatory have revealed a variety of structures carved into the Perseus Cluster's gas. These include giant, rolling waves; vast bubbles blown by the supermassive black hole in the cluster's central galaxy, called NGC 1275; an enigmatic concave feature known as the 'bay'; and a gigantic, mysterious cold front from the cluster's centre.

Top Tip

In 2002, scientists detected sound waves coming from the supermassive black hole at the Perseus Cluster's heart. The 'note' is the deepest ever detected and translates into a musical B-flat. But a human would have no chance of hearing this cosmic performance. The black hole's pitch is 57 octaves lower than middle-C, or almost a billion times deeper than the limits of human hearing. Scientists detected by the note by reading ripples in the gas filling the Perseus Cluster; these ripples are evidence for sound waves emanating from the central black hole.

Getting There & Away

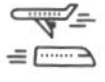

The Perseus Cluster might be the nearest bright galaxy cluster to our own – but even if you're travelling at 25 percent the speed of light, it'll still take you a billion years to get there.

© NASA/CHANDRA SPACE TELESCOPE

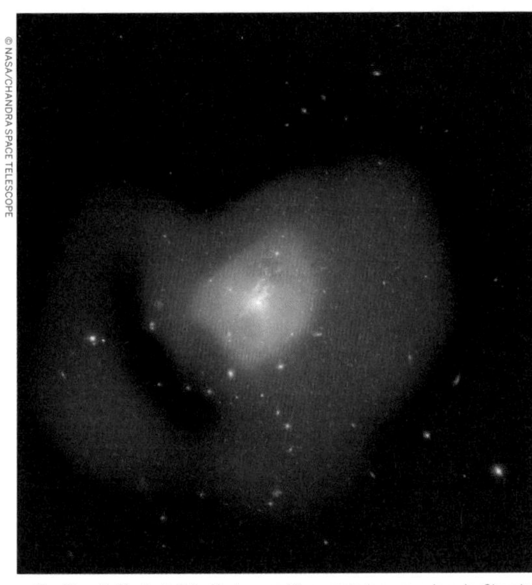

The Phoenix Cluster is full of hot gas and X-ray emissions, seen here by Chandra.

CONSTELLATION
Phoenix

DISTANCE
5.7 billion light years

WIDTH
1.5 million light years

GALAXIES
thousands

MASS
2 quadrillion suns

Phoenix Cluster

Discovered in 2010, the Phoenix Cluster holds several superlatives: It's among the brightest objects in the X-ray sky, it's among the most massive galaxy clusters – containing as much heft as two quadrillion Suns – and the galaxy beating at its heart forms stars more quickly than has ever been seen in a cluster.

Also known as SPT-CLJ2344-4243, the Phoenix Cluster has an immense galaxy with an active galactic nucleus at its very middle, which would be notable on its own. For starters, it's huge. It's also a total bruiser, with a central supermassive black hole that contains as much mass as 20 billion suns, making it one of the bulkiest on record. The event horizon is believed to stretch 39.5 AUs, or as far as Pluto. And the galaxy cluster is undergoing a rapid starburst, converting raw materials into as many as 740 solar masses of stars each year and producing a huge number of X-rays. Compare that to the Milky Way, which is thought to form one star a year.

Those stars are forming in long filaments of cosmic gas and dust that extend for 160,000 to 330,000 light years from the cluster's centre. The tendrils surround large cavities in the hot gas, excavated by powerful jets of energy launched from the environment surrounding the galaxy's black hole. Cavities located farther from the cluster's centre suggest that strong outbursts from the central black hole occurred about 100 million years ago.

CONSTELLATION
Virgo

DISTANCE
55 million light years

WIDTH
10 million light years

GALAXIES
1000-2000

MASS
1.2 quadrillion solar masses

Virgo Cluster member NGC 4388 (above) is under the influence of the Cluster's gravity.

Virgo Cluster

At just 50 million light years away, the Virgo Cluster is the nearest galaxy cluster to our own. It is located in the zodiacal constellation Virgo, spans 15 million light years, and contains as many as 2000 galaxies. Virgo's gravity is so strong that it pulls galaxies and groups of galaxies towards it – including our Local Group.

The Virgo Cluster's constituent members include giant elliptical galaxies, spirals like our own Milky Way, and dwarf galaxies hundreds of times smaller than their bigger brethren. Between those galaxies are hot gas, wayward stars and oodles of dark matter, which lends the cluster the majority of its bulk.

The Virgo Cluster is considered to be a relatively young cluster, with small sub-clusters of galaxies centred around the major galaxies Messier 87, Messier 86 and Messier 49. These sub-clusters have yet to merge and form a denser and smoother galaxy cluster, which is one of the reasons why the Virgo Cluster has an irregular shape – its roughly 2000 galaxies are scattered asymmetrically. It is part of the immense Laniakea Supercluster.

Top Tip

Although the Virgo Cluster spans roughly the same width as the Local Group – between 10 and 15 million light years – it is packed with 50 times more galaxies. As a result, the cluster's immense gravity is pulling external groups of galaxies in its direction, in a process called the Virgo-centric flow. The cluster forms the nucleus of the Virgo Supercluster, which includes the Local Group and a handful of others. Each of the supercluster's groups are moving toward the Virgo cluster, and some may eventually merge and become part of it.

Getting There & Away

The Virgo Cluster is the nearest galaxy cluster to our own, roughly 50 million light years away, but eventually, as it reels in the Local Group, it will be even closer.

Virgo Cluster Highlights

M87

1 The Virgo Cluster's dominant galaxy is M87, which is a monster by all standards. About 53 million light years away, M87 is home to several trillion stars, roughly 15,000 globular star clusters, and has a supermassive black hole weighing as much as 6.5 billion Suns at its centre. (For comparison, our Milky Way Galaxy contains only a few hundred billion stars, about 150 globular clusters, and is anchored to a black hole weighing 4 million Suns). M87's black hole is launching powerful jets of extremely energetic particles into space – jets that stretch at least 4,000 light years from the galaxy's centre. Recently, scientists with the Event Horizon Telescope project made an image of M87's behemoth black hole; it was the first time humans have ever seen the shadow of one of these exotic objects.

M49

2 The cluster's brightest galaxy is an elliptical giant called M49, discovered by Charles Messier in 1771. Not only was it the first object discovered in the Virgo Cluster of galaxies, but it was also the first elliptical galaxy detected outside of the Milky Way's Local Group. M49 also contains a rich collection of globular star clusters — nearly 6,000. It can be observed using a pair of binoculars, and the best time to see it is the month of May.

NGC 4388 – A Hybrid Galaxy

3 Located some 60 million light years away in the Virgo Cluster, galaxy NGC 4388 is undergoing a transformation. The galaxy's outskirts appear smooth and featureless, which is a classic feature of elliptical galaxies. But its centre displays remarkable dust lanes constrained within two symmetric spiral arms, which emerge from the galaxy's glowing core — one of the obvious features of spiral galaxies. Despite the mixed messages, NGC 4388 is classified as a spiral galaxy, and scientists think its unusual features are the result of gravitational interactions with its neighbours.

Galactic Sub-Clumps

4 Three large sub-clusters of galaxies are identifiable in the Virgo Cluster: Virgo A, which has giant galaxy M87 as its nucleus; Virgo B, which is centred on M49 – the brightest galaxy in the entire cluster; and a third clump, grouped around M86. Several smaller groups surround these prominent clumps, and eventually they will all merge and form a smooth cluster. For now, Virgo A's mass dominates the cluster, and it is dragging everything else inward at extremely high rates of speed.

Markarian's Chain

5 North of M87 in the Virgo Cluster, Markarian's Chain is a long stretch of roughly a dozen galaxies that appear to be strung together like beads on a cosmic string. The chain includes bright galaxies M84 and M86, as well as several interacting pairs; one of these is a well-known duo known as Markarian's Eye's. At least seven of the galaxies move together through the cosmos, while the rest just appear to be part of the chain from Earth.

Intergalactic Nebulae

6 As many as 10 percent of the Virgo Cluster's stars lie in intergalactic space – a result of galaxies gravitationally jostling one another and ejecting their stars. Some of those stars have formed the beautiful, wispy structures astronomers call planetary nebulae. These picturesque clouds are actually the result of stellar death spasms, and it turns out, they can be observed from the next cluster over. By tracking the motions of planetary nebulae, astronomers can use them as a fossil record of galaxy interactions over the history of the cluster, and learn more about the numbers, types and motions of stars in various regions.

Known for Centuries

7 The Virgo Cluster has been known of for centuries, although it wasn't always recognised as a massive, external cluster of galaxies. In the late 1700s, French astronomer Charles Messier entered 16 members of the Virgo Cluster in his famous catalogue, classifying them as 'nebulae without stars'. As early as 1784, he noticed an unusual clustering of such nebulae in the constellation Virgo. Today, we know those nebulae are galaxies in their own right, and that they are gravitationally bound to one another.

The Sombrero – An Illusory Virgo Member

8 One of the universe's most stately and photogenic galaxies, the Sombrero Galaxy or M104, lies in the sky at the southern edge of the Virgo Cluster. Although it appears to be a member of the cluster, it's actually in the foreground – although still some 30 million light years from Earth. The galaxy's hallmark is a brilliant white, bulbous core encircled by the thick dust lanes comprising the spiral structure of the galaxy; we call it the Sombrero because of its resemblance to the broad-rimmed, high-topped hat. Just beyond the limit of the unaided eye, M104 is easily seen through small telescopes. It is one of the most massive objects in the Virgo Cluster, containing the equivalent of 800 billion suns.

NGC 4522 in the Virgo Cluster is being impacted by the cluster's winds.

GLOSSARY & INDEX

Glossary

Albedo - The ratio of the light received by a body to the light reflected by that body. Albedo values range from 0 (pitch black) to 1 (perfect reflector).

Aphelion - An orbit's farthest point to the Sun or its star.

Apparent Magnitude - How bright the stars in the sky appear to an observer.

Arcsecond - Abbreviated arcsec. A unit of angular measure used for objects in the sky, in which there are 60 arcseconds in 1 arcminute and therefore 3600 arcseconds in 1 arc degree. One arcsecond is equal to about 725 km (450 mi) on the Sun. Arcdegrees and arcminutes are also measured.

Asterism - A pattern of stars in the sky which appears to be so distinctive that it is easily identifiable and remembered.

Asteroid - Asteroids are small, rocky objects that orbit the Sun. Although asteroids orbit the sun like planets, they are much smaller and less spherical than planets. Historically they may be called minor planets or planetoids.

Asteroid Belt - An area of ancient space rubble orbiting the Sun between Mars and Jupiter within the main asteroid belt. Asteroids range in size from Vesta - the largest at about 530 km (329 mi) in diameter - to bodies that are less than 10 m (33 ft) across. The total mass of all the asteroids combined is less than that of Earth's moon.

Asteroseismology - The study of star oscillations.

Astronomical Unit (AU) - It is approximately the average distance between the Earth and the Sun. The astronomical unit (AU) is defined by the IAU as exactly 149,597,870,700 metres.

Atmosphere - The blanket of gases which surrounds a planet, held near the surface by gravitational attraction.

Baryonic Matter - All material made up of ordinary matter, that is, protons, neutrons and electrons.

Big Bang - The leading theory of the Universe's origins posits a 10-billion degree sea of neutrons, protons, electrons, anti-electrons (positrons), photons, and neutrinos. From an incredibly compact beginning, this went on to inflate explosively. As it cooled it eventually formed stars and galaxies.

Binary Stars - A star system comprised of two stars. In most binary systems, both stars follow an elliptical orbit about their common centre of mass and are gravitationally bound together.

Black Hole - A great amount of matter packed into a very small area, formed when a massive star collapses. The result is a gravitational field so strong that nothing, not even light, can escape. They can merge to create supermassive black holes, believed to be found at the centre of most galaxies.

Centaur - Small celestial bodies orbiting the sun between Jupiter and Neptune. Centaurs are similar in size to asteroids but similar in composition to comets, earning them the name for the mythical Greek half-man, half-horse creature.

Cepheid Variables - Also called Cepheids or standard candles; stars which brighten and dim periodically. This behaviour allows them to be used as cosmic yardsticks out to distances of a few millions of light years.

Comet - Cosmic snowballs of frozen gases, rock and dust that orbit the Sun.

Dark Matter - Non-baryonic matter; a not fully understood component which comprises roughly 27% of the matter in the Universe.

Dayside - The side of a planet that is facing its primary star.

Dwarf Planet - A celestial body that orbits the sun, has enough mass to assume a nearly round shape, has not cleared the neighbourhood around its orbit, and is not a moon.

Eccentricity - One half of the major axis of the elliptical orbit of an object; also the mean distance from the Sun.

Ecliptic Plane - The plane of the earth's orbit in the solar system.

Electromagnetic Spectrum - The range of all types of electromagnetic (EM) radiation. EM radiation includes visible light, radio waves, microwaves, infrared light, ultraviolet light, X-rays, and gamma-rays.

Event Horizon - The 'surface' of a black hole; the boundary where the velocity needed to escape exceeds the speed of light.

Event Horizon Telescope (EHT) - An international collaboration that networked eight ground-based radio telescopes and observatories into a single Earth-size dish which captured an image of a black hole (M87) for the first time in 2019.

Exomoon - A moon orbiting a planet that lies outside our solar system.

Exoplanet - A planet orbiting a star other than our sun, that is, outside our solar system (also sometimes called an extrasolar planet).

Fossae - Ridges, valleys, or depressions formed on a celestial body due to atmospheric forces.

Gas Giant - A large planet of relatively low density consisting predominantly of hydrogen and helium; in our solar system, Jupiter and Saturn are considered gas giants.

Globular Clusters - Islands of several hundred thousand of older stars.

Gravitational Lensing - When the gravity of massive galaxy clusters, which contain dark matter, bends and distorts the light of more distant galaxies located behind the cluster.

Habitable Zone - The range of distances from a star where liquid water might pool on the surface of an orbiting planet.

Heliopause - The outer limits of the Sun's magnetic field and outward flow of the solar wind.

Heliosphere - The solar wind, emanating from the Sun, which creates a bubble that extends far past the orbits of the planets.

Hot Jupiter - Exoplanet gas giants like Jupiter but much hotter, with orbits that take them feverishly close to their stars.

Hot Neptune - A type of giant planet with a mass similar to that of Uranus or Neptune orbiting close to its star, normally within less than 1 AU.

Hubble Constant - How the current rate of expansion in the Universe is usually expressed (in units of kilometers per second per Megaparsec, or just per second). Its exact rate and mechanism is still under debate.

Ice Giant - A category of planet with a massive slushy mantle of ices beneath atmospheric clouds; fundamentally different from the gas giant and terrestrial planets. Uranus and Neptune are considered ice giants in our solar system.

Interacting Galaxy - Two or more galaxies whose gravitational fields are impacted by one another.

Intergalactic Space - The space between clusters of galaxies.

International Astronomical Union (IAU) - An international association of professional astronomers which meets to agree on standards, conventions, and other issues for the astronomical field.

Interstellar Space - The area outside the sun's heliosphere, where the Sun's magnetic field stops affecting its surroundings.

Kuiper Belt - A region of leftovers from the solar system's early history, found beyond Neptune to 50 AU; it's more of a thick disc (like a donut) than a thin belt.

Kuiper Belt Object (KBO) - Dwarf planets and asteroids or particles of frozen ice and rock found within the Kuiper Belt

Light Year - The distance light travels in one Earth year, commonly used to measure distances in the Universe. One light year is about 9 trillion km (6 trillion mi).

Magnetar - Magnetars are a neutron star variant with magnetic fields a thousand times stronger than ordinary neutron stars that measure a million billion Gauss, or about a hundred-trillion refrigerator magnets. For comparison, the Sun's magnetic field is only about 5 Gauss. They erupt without warning, for hours or months, before dimming and disappearing again.

Main Sequence Star - About 90 percent of the stars in the Universe, including the Sun, which fuse hydrogen atoms to form helium atoms in their cores; a classification which spans a wide range of luminosities and colours, from small red dwarfs to hypergiants.

Mass - Mass is the amount of matter in an object. It can also be defined as the property of a body that causes it to have weight in a gravitational field. It is important to understand that the mass of an object is not dependent on gravity. Bodies with greater mass are accelerated less by the same force.

Moon - Also known as natural satellites; a celestial object that orbits a planet or asteroid.

Nebula - A giant cloud of dust and gas in space. They come in many forms: reflection, dark, emission, supernova remnants, or planetary nebula from a dying white dwarf.

Neutron Star - The most dense star structure in the Universe, comprised of a collapsed star whose protons and electrons have been crushed into neutrons.

Nightside - The side of a planet that is facing away from its primary star.

Oort Cloud - A distant region of icy, comet-like bodies that surrounds the solar system. Most comets are thought to originate in the Oort Cloud.

Open Cluster - A group of loosely bound stars which were formed from the same molecular cloud and which are roughly the same age.

Orbit - An orbit is a regular, repeating path that one object in space takes around another one. All objects are elliptical, but they may be more circular or more 'squashed'.

Parsec - Equal to 3.26 light years; defined as the distance at which 1 Astronomical Unit subtends an angle of 1 second of arc (1/3600 of a degree).

Perihelion - An orbit's closest point to the Sun or the object it orbits.

Planet - A round celestial object which orbits the Sun or another star.

Potentially Hazardous Asteroid (PHA) - All asteroids with a minimum orbit intersection distance (MOID) from Earth of 0.05 AU or less and an absolute magnitude (H) of 22.0 or less.

Pulsar - A neutron star that emits beams of radiation.

Quasar - Short for 'quasi-stellar radio source'; the most distant object in the Universe, which is believed to pull and eject its energy from massive black holes in the centre of the galaxies in which the quasar is located.

Runaway Greenhouse Effect - When a planet absorbs more energy from the sun than it can radiate back to space; detected when planetary heat loss drops as surface temperature rises.

Runaway Star - A star moving through space at a faster velocity than the surrounding interstellar matter thus escaping or leaving the area which it was formed.

Satellite - An object in an orbit is called a satellite. It can be natural, like Earth or the moon, or it can be man-made.

Spectroscopy - The study of spectra, which helps astronomers determine what elements are present in a given spectrum of light from a celestial object.

Super Earth - An exoplanet up to 10 times larger Earth, but substantially below the size of ice giants.

Supernova - The explosion of a star which takes place at the end of the star's life-cycle.

Syfert Galaxy - A type of active spiral galaxy with a centre or nucleus which is very bright at visible light wavelengths.

Syzygy - A straight-line alignment of three or more celestial bodies, such as a transit or eclipse.

Terrestrial Planet - A rocky body composed mainly of silicate rock or minerals, such as Mercury or Earth.

Tidal Locking - When one side of a celestial body always faces a star or planet while the other side is in permanent night.

Trans-Neptunian Object (TNO) - Any objects in the solar system that have an orbit beyond Neptune.

Transit - The passage of a planet or object across the disc of the Sun or a star.

Trojans - A small celestial body which stably shares the orbit of a larger body. Jupiter has many trojan asteroids.

Ultraluminous X-ray Source (ULX) - Bright X-ray sources which emit more X-rays than can be produced by accretion onto a stellar-mass black hole, yet are much less bright than accreting supermassive black holes.

Variable Star - A star whose brightness changes, either irregularly or regularly.

Visible Light - The segment of the electromagnetic spectrum that the human eye can view.

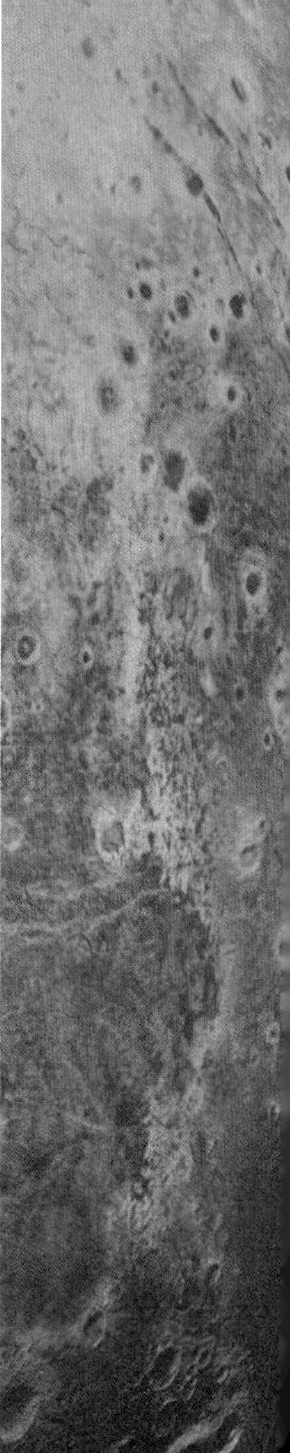

Index

A

P

R

S

T

U

V

W

Acknowledgments

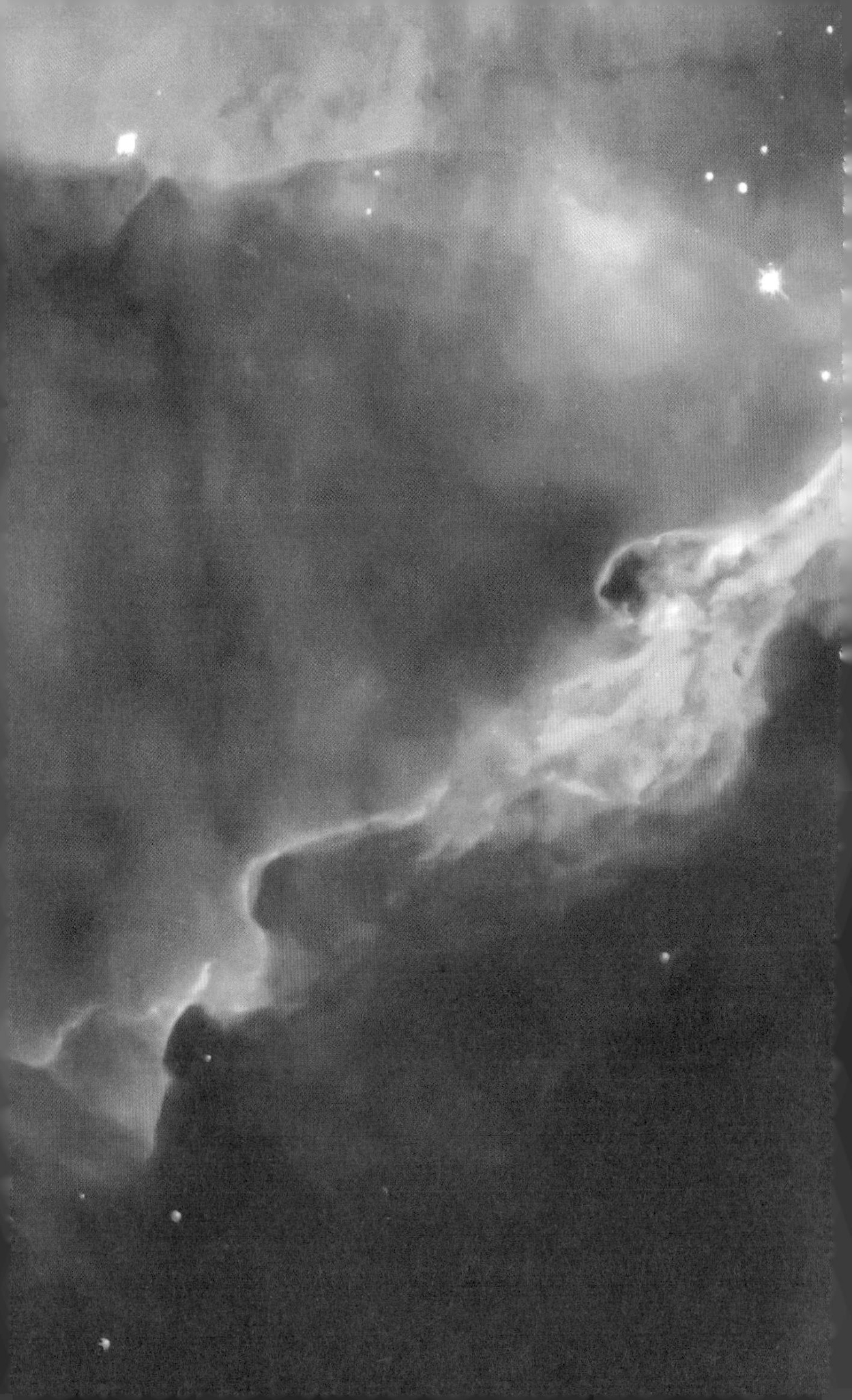

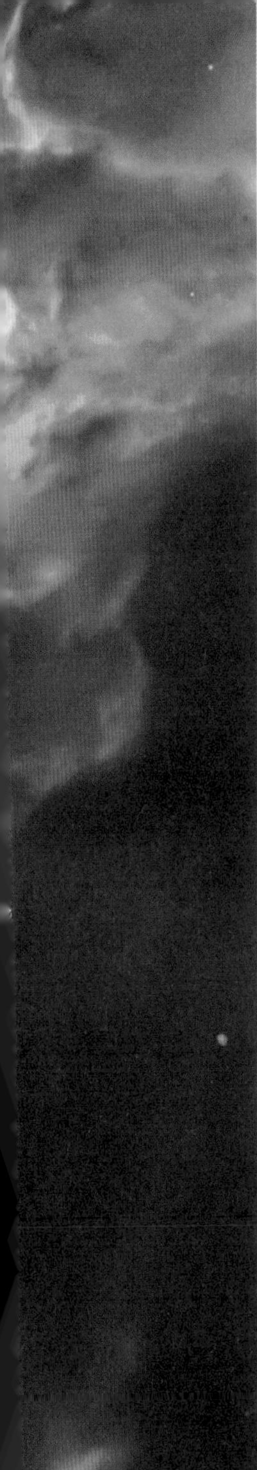

© LUC NOVOVITCH / ALAMY STOCK PHOTO

Author Biographies

Oliver Berry

Oliver Berry is a writer, photographer and filmmaker, specialising in travel, nature and the great outdoors. He has travelled to sixty-nine countries and five continents, and his work has been published by some of the world's leading media organisations, including Lonely Planet, the BBC, Immediate Media, John Brown Media, *The Guardian* and *The Telegraph*.

Dr Mark A. Garlick

Dr. Mark A. Garlick completed his PhD on binary stars in 1993, at the UK's Mullard Space Science Laboratory. After three further years of research at Sussex University he changed careers, and is now a freelance writer, illustrator and computer animator, specializing in astronomy.

Mark Mackenzie

London-based editor and writer Mark Mackenzie also edited the 2019 Lonely Planet anthology *Curiosities and Splendour: An Anthology of Classic Travel Literature*.

Valerie Stimac

Valerie Stimac is an Oakland-based travel writer and astronomy enthusiast. She founded the online Space Tourism Guide and authored Lonely Planet's *Dark Skies: A Practical Guide to Astrotourism*.

The Universe

Published in October 2019
by Lonely Planet Global Limited
CRN 554153
ISBN 9781788686365

Printed in China
10 9 8 7 6 5 4 3 2 1

The Universe was produced by the following:

Managing Director, Publishing
Piers Pickard
Associate Publisher Robin Barton
Senior Editor Nora Rawn
Indexer Nick Mee
Art Direction Daniel Di Paolo
Layout Gwen Cotter
Photo Research Ceri James
Print Production Nigel Longuet

Special Thanks Laura Lindsay, without whom this book wouldn't have happened; Grace Dobell; Laurie Cantillo and Bert Ulrich at NASA; and NASA's chief scientist Dr James Green for his early enthusiasm and insights.

STAY IN TOUCH LONELYPLANET.COM/CONTACT

AUSTRALIA The Malt Store, Level 3, 551 Swanston St, Carlton, Victoria 3053

IRELAND Digital Depot, Roe Lane (Off Thomas Street) The Digital Hub, Dublin 8, D08 TCV4

USA 124 Linden Street, Oakland, CA 94607

UK 240 Blackfriars Road, London SE1 8NW

Front cover photo: Pillars of Creation courtesy of NASA. Spine photo: M106 courtesy of NASA.

twitter.com/lonelyplanet

facebook.com/lonelyplanet

instagram.com/lonelyplanet

youtube.com/lonelyplanet

lonelyplanet.com/newsletter